Constructions of Deviance

Social Power, Context,

and Interaction

PATRICIA A. ADLER

University of Colorado, Boulder

PETER ADLER

University of Denver

Wadsworth Publishing Company
Belmont, California
A Division of Wadsworth, Inc.

Editor: Serina Beauparlant
Editorial Assistants: Marla Nowick, Susan Shook
Production Editor: Jerilyn Emori
Interior and Cover Designer: Andrew Ogus
Print Buyer: Karen Hunt
Permissions Editor: Jeanne Bosschart
Copy Editor: William Waller
Cover Photograph: Malcolm Tarlofsky
Editors Photograph: Martin Natvig, University of Colorado
Compositor: Joan Olson, Wadsworth In-House Composition
Printer: Vail-Ballou Press

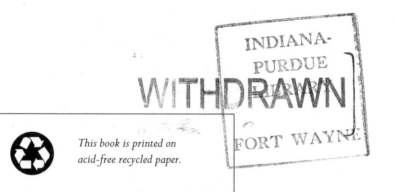

I(T)P

International Thomson Publishing
The trademark ITP is used under license.

Printed in the United States of America

2 3 4 5 6 7 8 9 10—98 97 96 95 94

Library of Congress Cataloging-in-Publication Data

Adler, Patricia A.
    Constructions of deviance : social power, context, and interaction
    Patricia A. Adler, Peter Adler
        p.   cm.
    Includes bibliographical references.
    ISBN 0-534-21342-1
    1. Deviant behavior.   2. Social interaction.   I. Adler, Peter.   II. Title
HM291.A337   1993
302.5'42—dc20   93–11335

To Diane and Dana

Who remind us that the zest of life lies near the margins

Contents

PREFACE

GENERAL INTRODUCTION   1

PART I
Defining Deviance   7

1   *Deviance as Crime, Sin, and Poor Taste*   11
     Alexander B. Smith and Harriet Pollack

2   *Deviance as Fun*   21
     Jeffrey W. Riemer

3   *Defining Deviancy Down*   25
     Daniel Patrick Moynihan

PART II

Studying Deviance   35

4   *Researching Dealers and Smugglers*   39
     Patricia A. Adler

5   *Field Work and Membership with the Hare Krishna*   55
     E. Burke Rochford

PART III

Constructing Deviance   69

Moral Entrepreneurs

6   *Coffee Drinking: An Emerging Social Problem?*   73
     Ronald J. Troyer and Gerald E. Markle

7   *The Social Construction of Drug Scares*   92
     Craig Reinarman

8   *Rhetoric in Claims-Making: Constructing*
     the Missing Children Problem   105
     Joel Best

Differential Social Power

9   *The Police and the Black Male*   122
     Elijah Anderson

10   *Sex Discrimination—Subtle and Covert*   133
     Nijole V. Benokraitis and Joe R. Feagin

11   *Masculinity, Homophobia, and Misogyny*   142
     Michael A. Messner

PART IV

Organizational/Institutional Deviance   149

Creating Deviance

12   *The Social Context of Police Lying*   153
     Jennifer Hunt and Peter K. Manning

13   *State-Organized Crime*   169
     William J. Chambliss

Labeling Deviance

14   *The Moral Career of the Mental Patient*   185
      Erving Goffman

15   *The Making of Blind Men*   204
      Robert A. Scott

16   *Discrediting Victims' Allegations of Sexual Assault:*
      Prosecutorial Accounts of Case Rejections   210
      Lisa Frohmann

P A R T   V

Deviant Identity   229

Identity Development

17   *Divorce and Stigma*   237
      Naomi Gerstel

18   *Anorexia Nervosa and Bulimia:*
      The Development of Deviant Identities   249
      Penelope A. McLorg and Diane E. Taub

Accounts

19   *Convicted Rapists' Vocabulary of Motive:*
      Excuses and Justifications   261
      Diana Scully and Joseph Marolla

20   *The Influence of Situational Ethics*
      on Cheating among College Students   278
      Donald L. McCabe

21   *Denying the Guilty Mind:*
      Accounting for Involvement in a White-Collar Crime   287
      Michael L. Benson

Individual Stigma Management

22   *Return to Sender: Reintegrative Stigma-Management*
      Strategies of Ex-Psychiatric Patients   304
      Nancy J. Herman

23   *Confronting Deadly Disease: The Drama of Identity*
      Construction among Gay Men with AIDS   323
      Kent L. Sandstrom

Collective Stigma Management

24   *Illness Ambiguity and the Search for Meaning:*
     A Case Study of a Self-Help Group for Affective Disorders   338
     David A. Karp

25   *From Stigma to Identity Politics: Political Activism*
     among the Physically Disabled and Former Mental Patients   350
     Renee R. Anspach

PART VI

Organizing Deviants and Deviance   365

Relations among Deviants

Subcultures

26   *Real Punks and Pretenders: The Social*
     Organization of a Counterculture   373
     Kathryn Joan Fox

Gangs

27   *Women in Outlaw Motorcycle Gangs*   389
     Columbus B. Hopper and Johnny Moore

Rings

28   *Con Men: Professional Crews of Card and Dice Hustlers*   401
     Robert Prus and C. R. D. Sharper

Deviant Acts

Individual

29   *Nonmainstream Body Modification: Genital Piercing,*
     Branding, Burning, and Cutting   414
     James Myers

Cooperation

30   *Tearoom Trade: Homosexual Behavior in Public Restrooms*   431
     Laud Humphreys

31   *Turn-Ons for Money: Interactional Strategies of the Table Dancer*   447
     Carol Rambo Ronai and Carolyn Ellis

Conflict

32   *Fraternities and Rape on Campus*   468
     Patricia Yancey Martin and Robert A. Hummer

33   *Transaction Systems and Unlawful Organizational Behavior*   483
     Diane Vaughan

Deviant Careers

Entering Deviance

34 *Joining a Gang*   489
    Martín Sánchez-Jankowski

35 *Marks of Mischief: Becoming and Being Tattooed*   511
    Clinton R. Sanders

Being Deviant

36 *Managing Deviance: Hustling, Homophobia, and*
    the Bodybuilding Subculture   529
    Alan M. Klein

37 *The Intangible Rewards from Crime: The Case*
    of Domestic Marijuana Cultivation   545
    Ralph A. Weisheit

Exiting Deviance

38 *Shifts and Oscillations in Deviant Careers: The Case*
    of Upper-Level Drug Dealers and Smugglers   560
    Patricia A. Adler and Peter Adler

39 *The Professional Ex-: An Alternative*
    for Exiting the Deviant Career   573
    J. David Brown

REFERENCES   587

Preface

In assembling a collection of readings in the sociology of deviance, we have tried to please both students and professors by meeting several goals. First, we have grounded this anthology within the interactionist/social constructionist perspective. Social constructionism—the view that deviance is a product of groups of people forging normative definitions; organizations and their agents applying them; and the actions of individuals, alone or in groups—is particularly well suited to the study of deviance. It goes beyond the absolute legal and moral overtones inherent in addressing crime, thereby wielding particular power in focusing in noncriminal norm violations. This focus differentiates a deviance reader from a criminological one, although considerable overlap remains.

Sociological attitudes toward deviance have evolved considerably over the last century, from simplistic views that regarded deviance as rooted in heredity or pathology to much more complex and multicausal explanations. The social constructionist approach represents the most contemporary and sophisticated perspective on deviance, drawing on elements of social structure and power, cultural and subcultural norms and values, and interpersonal interactions and self-reflections.

We have also chosen to shape this book around the interactionist/social constructionist stance because of its ability to convey the sequential nature of deviance: how it is defined, framed, imputed, enacted, accepted, and managed. Other, less comprehensive theories sometimes offer a more static view of societal, organizational, and individual factors affecting deviance, focusing on fixed cause–effect relationships. In this volume we hope to take students on a journey grounded in everyday life, beginning at the heart of how society operates to create broad social definitions of deviance and following the ways in which these are implemented and experienced. This perspective is designed to offer readers new insights into the many facets of deviance.

Second, we have chosen to convey these ideas through ethnographic studies based on descriptions and analyses of deviance conducted in its natural setting. We have steered away from including articles based on quantitative treatments that reduce behaviors to statistical indexes of prevalence, overviewed correlations, or long tables of numbers. To us, these articles have a harder time coming alive and keeping the reader's attention. Ethnographic studies, on the other hand, more often tend to be the ones that students remember after their undergraduate days are over. We thus offer richly descriptive, qualitative studies of deviant subcultures, deviant behavior, and the management of deviant identities that incorporate vivid images using the subjects' own voices.

Third, our intent here is to convey the many penetrating concepts contained in the sociology of deviance through an application, rather than an abstract explanation, of its principles. Each selection was chosen because it applies or illustrates specific, critical concepts in the deviance literature. In this way, readers can more easily see the operation of these ideas in empirical settings. These studies frame the deviance experience with spirited depictions and stories cast within analytical categories that illustrate sociological concepts and teach the sociological perspective.

Fourth, we aim to address a mixture of societal levels and units of analysis. We begin by looking at how society operates in forging broad collective definitions of deviance. Our focus here is on issues of social structure, social power, and social conflict. We then move to a middle level by addressing organizations and institutions, examining the effects of bureaucratic contexts and constraints on the shaping and application of deviance. Last we look at the individual level, focusing on how people's identities and behavior are affected by their passage through deviance. This broadly encompassing focus of analysis is reflected in the book's subtitle, "Social Power, Context, and Interaction."

Fifth, we have assembled a collection of studies embodying four foci: (1) those close to students' lives (so they can relate to the experience); (2) those far from them (captivating the appeal of the exotic), (3) those showing the presence of deviance at the core of society (in addition to the margins), and (4) those dealing with contemporary social problems. Examples of topics familiar to readers include eating disorders, growing marijuana, drinking coffee, students' excuses for cheating, divorce, homophobia, and fraternity date rape. Exotic studies that will fascinate students focus on genital piercing, homosexual prostitution, stranger rape, motorcycle gangs, and rings of traveling con artists. Ironically, some of our topics fall into both of these first two categories, such as table dancing and becoming tattooed. Studies showing that deviance occurs in the suites as well as the streets include those on white collar crime (Medicare fraud, embezzlement, income tax evasion, and the like) and organizational/institutional deviance (government, bureaucratic, and social welfare deviance). Selections dealing with the emerging social issues of late-twentieth-century America include articles dealing with cultural diversity (racism), social inequality (sexism, rape, homophobia), and urban poverty and crime (gangs).

Although it was easier to find interesting, conceptual, and well-written studies on some topics than others (ethnographic studies of the exotic and powerless were more prevalent than those training their lens on the social elite), our review of the deviance literature reminded us of the enormous quality and breadth of work in this field. Not only is this literature enormously fascinating in its presentation of data, but it is richly theoretical in its analysis. To enliven this anthology, we have mostly tried to move away from older, albeit classic, articles describing aspects of deviance, searching out newer treatments that address the major relevant concepts within the contemporary forms of deviance.

Constructions of Deviance is organized into six parts. In the first section we begin by "Defining Deviance" with some different types and forms of deviance, including the relationship between deviance and crime. We then offer three articles that discuss the nature and relativity of deviance, focusing on the search for thrills, excitement, and rebellion, as distinct from more formal criminological definitions and theories.

Part II, "Studying Deviance," presents another feature not found in previous books. Conducting research on deviant behavior entails taking risks, reflecting on one's self, and making moral and pragmatic decisions that may not be evoked through the study of more conventional topics. Following the ethnographic theme of the articles, we offer two methodological discussions, one of a criminological and one of a deviance setting, that illustrate some of the unique features of this type of challenging research. This should help readers envision the role of researchers in the subsequent readings, as well as to imagine themselves in similar situations.

In Part III, "Constructing Deviance," we look at the process through which definitions of deviance are constructed and the consequences of these definitions. Our two subsections address the *moral entrepreneurs* who campaign for social norms and laws and the *differential social power* possessed by some groups in society. For the first, we build from a simple and accessible base to a more sophisticated discussion of the mechanisms by which norms are constructed at a macro, societal level through the activities of these activists and their minions of "opinion leaders" and "experts." For this section, we have chosen issues that will either resonate with readers' cultural upbringing (drug scares, missing children) or fall into their everyday experience (coffee drinking). We then look at the effects of social inequality, whereby some groups have the power to define their own status, condition, or lifestyle as preferred and that of others as deviant. Readings on how people may be defined as deviant and discriminated against on the basis of their race, gender, or sexual orientation are included.

Deviance is shaped not only by societal contexts but by organizational ones as well, and we examine the twofold effects of this shaping in Part IV, "Organizational/Institutional Deviance." We use two selections on deviance among the police and within the federal government to show how organizational goals and demands lead individuals to violate norms and laws for the sake of the institution rather than themselves. We then look at the way in which three medical and criminal justice organizations function to apply the deviant label to individuals they process, focusing on how routine features of bureaucracies operate to deindividualize and resocialize clients.

In Part V, "Deviant Identity," we follow the themes associated with individuals' development and management of deviant self-conceptions, using articles that apply the core concepts from this literature. We look at the social-psychological effects of the labeling process, incorporating selections that address such themes as primary, secondary, and tertiary deviance; the dynamics of inclusion and exclusion; excuses, justifications, and techniques of neutralization; passing and covering; deviance avowal and deviance disavowal; individual and collective stigma management; and deviant identity politics.

Our last overarching section, "Organizing Deviants and Deviance," addresses the way in which deviant behaviors and people are socially organized. Part VI is divided into three themes: the associations among deviants, the sociological features of the deviant acts, and individuals' passage through deviant careers. We use this section to offer analytically informed research for discussing some of the fascinating studies of deviant subcultures and behavior. This part suggests structural similarities among types of deviance that may be rather diverse empirically, and it helps readers grasp the insights of sociological analysis in an enjoyable way.

ACKNOWLEDGMENTS

Deviance, our first sociological love, has now endured in our imaginations for over twenty years. When we were undergraduates at Washington University in the early 1970s, this topic captured our fancy and enticed us into a career in sociology. Throughout the years, several people have served to solidify our interest in deviant behavior. First, while we were still matriculating at Washington University, Marvin Cummins and David Pittman piqued our curiosity. Later, as their colleagues, we continued the relationship that we had established more than a decade before. During our graduate years, Jack Douglas, whose massive amount of work in this area spurred us to continue thinking about deviance, forced us to stretch our intellectual capacities in both empirical and theoretical ways. Fred Davis, who sadly recently passed away, was a constant source of guidance and inspiration.

Just as these four professors inspired us, several of our students who have worked in the area of deviance have also, reciprocally, motivated us to reconsider the fashion in which we studied this topic. Most especially, Kathy Fox, David Brown, and Martin Kretzmann produced work that has made us proud and taught us new ideas.

We have had the good fortune of being surrounded by many colleagues who took interest in our work. At the University of Denver, Paul Colomy and Bob Granfield discussed with us, during the early inception of this project, some directions for the book. Bob Regoli of the University of Colorado suggested one of the readings we selected. A number of people in the profession have been sounding boards and supporters: Mitch Allen, David Altheide, Joel Best, Norman Denzin, Carolyn Ellis, Andy Fontana, John Irwin, John Johnson, Joe Kotarba, Stanford Lyman, Craig Reinarman, Marsha Rosenbaum, David Snow, Carol Warren, and Jackie Wiseman. Most especially, our dear friend Chuck Gallmeier has been unfaltering in his devotion and love.

Several of our students commented on the articles we have chosen, suggested areas we should cover, and gave us feedback on what students in the 1990s want to read. Lisa Hill, Dina Raichel, Andrea Rosen, Shawn Snow, Jon Umaki, Katie Wagman, and Connie Sze Yu were particularly helpful. The thousands of students in our deviance classes have also been important sources for us.

This book could not have been completed without the stalwart work of Dorene Miller. Her professionalism, dedication, and efficiency in coordinating this project, securing permissions, and making sure that all of the materials were gathered is unsurpassed. For the last six years, she has been the foundation that has permitted us to accomplish so much. Tamara Trueblood, too, helped in preparing the text and providing cheerful, competent, and professional assistance.

During a rushed meeting in a Pittsburgh hotel hallway we met with Serina Beauparlant of Wadsworth about the possibility of editing a collection of readings in the sociology of deviance. Little did we imagine that we would embark on a whirlwind few months that culminated in the completion of this book. Without Serina's enthusiasm, undying conviction that the project was right, and faith in us, we never would have tackled this task. Marla Nowick also helped administratively, editorially, and personally. Jerilyn Emori, production editor, was impeccable in assuring that the book was produced with speed and quality. Numerous reviewers provided comments and criticisms about early drafts. Wherever possible, we have taken their advice: Michael Breci, St. Cloud State University; Clifton D. Bryant, Virginia Polytechnic Institute and State University; Jeffrey Chin, Le Moyne College; Roger Dunham, University of Miami; Stephen Green, North Adams State College; Barbara Heyl, Illinois State University; Ruth Horowitz, University of Delaware; Lloyd Klemke, Oregon State University; Judson Landis, California State University, Sacramento; Jane Malone, University of North Carolina, Greensboro; Suzanne Ortega, University of Nebraska; John Pfuhl, Monmouth College; and Kenrick Thompson, Northern Michigan University.

Finally, our friends Diane and Dana provide images of deviance and respectability that continually make us question the fuzzy lines in society. We respectfully, and with love, dedicate this book to them.

About the Editors

Patricia A. Adler (Ph.D., University of California, San Diego) is Associate Professor of Sociology at the University of Colorado. She has written and taught in the areas of deviance, drugs in society, social theory, and the sociology of children. A second edition of her book *Wheeling and Dealing* (Columbia University Press), a study of upper-level drug traffickers, was released in 1993.

Peter Adler (Ph.D., University of California, San Diego) is Professor of Sociology at the University of Denver, where he served as Chair from 1987 to 1993. His research interests include social psychology, qualitative methods, and the sociology of sport and leisure.

Together the Adlers edit the *Journal of Contemporary Ethnography* and were the founding editors of *Sociological Studies of Child Development*. Their most recent book, *Backboards and Blackboards,* based on a five-year participant-observation study of college athletes, was published by Columbia University Press in 1991. Currently, they are studying the culture of elementary schoolchildren.

About the

Contributors

Elijah Anderson earned his Ph.D. from Northwestern University. He is currently the Charles and William L. Day Professor of the Social Sciences at the University of Pennsylvania. An expert on the sociology of black America, he is the author of *A Place on the Corner* and *Streetwise,* which was honored with the Robert E. Park Award of the American Sociological Association. He is interested in the social psychology of organizations, field methods of social research, social interaction, and social organization.

Renee R. Anspach received her Ph.D. from the University of California, San Diego, and has written extensively on the sociology of disability, health professions and organizations, and medical language. She is the author of *Deciding Who Lives: Fateful Choices in the Intensive-Care Nursery,* (California, 1993). She is currently Associate Professor of Sociology at the University of Michigan.

Nijole V. Benokraitis earned her Ph.D. at the University of Texas at Austin and is currently Professor of Sociology at the University of Baltimore. She is the author of *Marriages and Families: Changes, Choices, and Constraints* and co-editor (with John Macionis) of *Seeing Ourselves: Classic, Contemporary and Cross-Cultural Readings in Sociology.* She has also served as a consultant in the area of sex and race discrimination to women's commissions, business groups, colleges and universities, and federal government programs.

Michael L. Benson is Associate Professor and Head of Sociology at the University of Tennessee at Knoxville. He has a long-standing interest in the causes and consequences of white-collar crime. His most recent project is a national study of local prosecutors and their responses to corporate crime.

Joel Best is Professor of Sociology at Southern Illinois University at Carbondale. His research focuses on deviance and social problems. His books include *Organizing Deviance* (with David F. Luckenbill), *Images of Issues, Threatened Children,* and *The Satanism Scare* (with James T. Richardson and David G. Bromley).

J. David Brown earned his Ph.D. at the University of Denver after working for a number of years as a substance abuse therapist and administrator. He is currently an independent scholar as well as an Adjunct Professor of Sociology at Red Rocks Community College.

William J. Chambliss is Professor of Sociology at George Washington University. He is the author of *Law, Order, and Power* with Robert Seidman, *Making Law* with Marjorie Zatz, and *On the Take: From Petty Criminals to Presidents.* He is the past president of the American Society of Criminology (1989–1990) and current president of the Society for the Study of Social Problems (1992–1993). He has received the American Sociological Association Lifetime Achievement Award and the Academy of Criminal Justice Sciences Outstanding Achievement Award.

Carolyn Ellis received her Ph.D. from SUNY at Stony Brook and currently is Associate Professor of Sociology and Director of the Institute for Interpretive Human Studies at the University of South Florida. She writes about narrative, lived emotional experience, and relationships. Her most recently published book is *Investigating Subjectivity: Research on Lived Experience* (with Michael Flaherty).

Joe R. Feagin (Ph.D., Harvard University, 1966) is Graduate Research Professor in Sociology, University of Florida. His research on racial relations can be found in the best-selling textbook, *Racial and Ethnic Relations* (Prentice-Hall, 1993) and in *Fighting Racism: The Black Middle Class Experience* (Beacon Press, forthcoming). Feagin has served as Scholar-in-Residence at the U.S. Commission on Civil Rights.

Kathryn J. Fox is a doctoral candidate in the Department of Sociology at the University of California, Berkeley. She is currently completing her dissertation, entitled "The Micro-Politics of Expert Knowledge: An Ethnography of Structure and Ideology in the 'West Coast AIDS Project'." Her areas of specialization include medical sociology, "deviant" behavior/criminology, qualitative methods, and the sociology of organizations.

Lisa Frohmann received her Ph.D. in sociology from UCLA and is an Assistant Professor of Criminology at the University of Illinois, Chicago. She coauthored *The Guide to Writing Sociology Papers* with the Sociology Writing Group at UCLA and is currently writing a book about prosecutorial decision making in sexual assault cases.

Naomi Gerstel, Professor of Sociology (University of Massachusetts, Amherst) wrote, with Harriet Gross, *Commuter Marriage, Families and Work,* and has contributed articles to *Social Forces, Gender & Society, Social Problems,* and *Journal of Marriage and the Family.* Funded by the National Science Foundation and Russell Sage, she is studying strategies for surviving homelessness.

Erving Goffman was the Benjamin Franklin Professor of Sociology and Anthropology at the University of Pennsylvania.

Nancy J. Herman earned her Ph.D. from McMaster University and is Associate Professor of Sociology at Central Michigan University. She has published a number of articles on deviance, the sociology of mental illness, ex-psychiatric patients, and symbolic interactionism. With Larry T. Reynolds, she has co-edited *Symbolic Interaction: An Introduction to Social Psychology.*

Columbus B. Hopper is Professor of Sociology and Graduate Director of Sociology at the University of Mississippi. Professor Hopper's writing has been in the areas of sex in prison, the role of pharmacists in hospitals, outlaw motorcycle gangs, and popular culture. He is a past president of the Alabama-Mississippi Sociological Association.

Robert A. Hummer is an Assistant Professor of Sociology at East Carolina University. His research interests include infant and maternal health; race, class, gender, and health; and smoking and health. He has recently published work on the continuing association between race and infant mortality in the United States.

Laud Humphreys was a Professor of Sociology at Pomona College, California.

Jennifer C. Hunt is an Associate Professor of Sociology at Montclair State and a Clinical Instructor of Psychiatry at New York University Medical Center. She has published articles on gender roles, police, qualitative field methods, applied psychoanalysis, and decompression illness among sports divers. A monograph, *Psychoanalytical Aspects of Fieldwork,* was published in 1989 by Sage Publications. She is currently conducting a fieldwork and interviewing study of sports scuba diving with particular attention to the social and psychological dimensions of risk among extended range and recreational divers.

David A. Karp received his Ph.D. from New York University in 1971. He is currently a Professor of Sociology at Boston College. Professor Karp has co-authored several books in the areas of the sociology of everyday life, aging, and urban social psychology. Currently he is working on two research projects, both of which involve in-depth interviewing. One study is concerned with the experience of living with depression and the second involves family dynamics during the year that high school students apply to college.

Alan M. Klein is Professor of Sociology-Anthropology at Northeastern University. He is the author of *Sugarball: The American Game, The Dominican Dream,* and *Little Big Men: Bodybuilding Subculture and Gender Construction.* His new book is on culture, politics, and the role of baseball on the U.S.-Mexican border.

Donald L. McCabe is Associate Professor of Management at the Graduate School of Management, Rutgers University in Newark, New Jersey. He received his Ph.D. in Management from New York University and joined Rutgers in 1988 after a 22-year career in industry. His research interests currently center on business ethics, with particular interest in the moral education of future business leaders.

Penelope A. McLorg is a doctoral student in anthropology at Southern Illinois University at Carbondale. She is specializing in biological anthropology, with interest in women's health, aging, epidemiology, and human adaptability. Recent publications involve infant feeding practices among economically disadvantaged mothers and gender socialization in eating disorders.

Peter K. Manning (Ph.D., 1966, Duke) is Professor of Sociology and Criminal Justice at Michigan State University. He was a Fellow of Balliol and Wolfson Colleges and held a position in the Centre of Socio-Legal Studies, Oxford. His works include *Police Work* (1977), *Narc's Game* (1980), *Symbolic Communication* (1990), and *Organizational Communication* (1992). He is listed in *Who's Who in the World* and *Who's Who in America,* and was given in 1993 the Bruce Smith Research Award of the Academy of Criminal Justice Sciences.

Gerald E. Markle received his Ph.D. from Florida State University. Currently he is Professor of Sociology at Western Michigan University. His most recent book, *Minutes to Midnight: Nuclear Weapons Protest in America* (Sage), won the Distinguished Scholarship Award from the North Central Sociological Association. He is co-editor of the forthcoming *Handbook of Science, Technology and Society* (Sage), and is currently completing a book, *The Gray Zone: Holocaust, Culture and Society.*

Joseph Marolla is currently chair of the Department of Sociology and Anthropology at Virginia Commonwealth University. His primary areas of research and publication have been self esteem and sexual violence.

Patricia Yancey Martin is Daisy Parker Flory Alumni Professor and Professor of Sociology at Florida State University. She specializes in gender and organizations. Martin has a book (edited with Myra Marx Ferree) forthcoming on feminist organizations, is completing a research monograph on rape crisis centers and the politics of rape processing in local communities, and is conducting an ethnographic study of gender in major U.S. corporations.

Michael A. Messner received his Ph.D. in sociology from the University of California, Berkeley. He currently is Associate Professor in the Department of Sociology and the Program for the Study of Women and Men in Society at the University of Southern California. He has co-edited (with Michael S. Kimmel) *Men's Lives* (Macmillan, 1992) and (with Donald F. Sabo) *Sport, Men and the Gender Order: Critical Feminist Perspectives* (Human Kinetics Publishers, 1990). His book, *Power at Play: Sports and the Problem of Masculinity,* was published by Beacon Press in 1992. He is currently conducting research on embodiments of masculinity.

Johnny Moore, a former president of Satan's Dead, a Gulf Coast outlaw motorcycle club, is now employed in the Food Services Division of the University of Mississippi. He has earned a degree from the University of Mississippi and continues research on motorcycle clubs.

Daniel Patrick Moynihan is the senior United States senator from New York. He was elected in 1976, and re-elected in 1982 and 1988. He currently serves as chair of the Senate Committee on Finance. Senator Moynihan received his bachelor's degree (cum laude) from Tufts University in 1948, and his Ph.D. from the Fletcher School of Law and Diplomacy in 1961.

James Myers earned his Ph.D. from the University of California, Berkeley, and is a Professor of Anthropology at the California State University, Chico. He is the coauthor (with Arthur Lehmann) of *Magic, Witchcraft, and Religion: An Anthropological Approach to the Supernatural.*

Harriet Pollack has a Ph.D. in constitutional law from Columbia University. For 25 years she taught Civil Liberties and Criminal Justice courses at John Jay College (CUNY). She has written numerous articles and five books, mostly coauthored with Alexander B. Smith. Her most recent book is *Criminal Justice: An Overview* (West, 1991).

Robert Prus is a sociologist at the University of Waterloo. He (with Styllianoss Irini) coauthored *Hookers, Rounders, and Desk Clerks* (Sheffield Press, 1988, reprinted) and is the author of *Pursuing Customers: An Ethnography of Marketing Activities* and *Making Sales: Influence as Interpersonal Accomplishment* (Sage, 1989). He is presently working on an interactionist analysis of consumer behavior and a study of the pursuit of corporate investors by communities.

Jeffrey W. Riemer is Professor of Sociology at Tennessee Technological University. He received his Ph.D. from the University of New Hampshire, in sociology. He is the author of *Hard Hats: The Work World of Construction Workers* and coauthor of *Framing the Artist: A Social Portrait of Mid-American Artists.* Reimer has also published his work in various sociology journals including: *Social Problems, Work and Occupations, Deviant Behavior, Symbolic Interaction,* and *Urban Life* (now *Contemporary Ethnography*). His areas of interest include occupational sociology, deviant behavior, social problems, and research methodology.

Craig Reinarman earned his Ph.D. in sociology at the University of California, Santa Barbara. He was a post-doctoral Fellow in Public Health at UC Berkeley and is currently Associate Professor of Sociology at UC Santa Cruz. He is author of *American States of Mind* and coauthor of *Cocaine Changes.*

E. Burke Rochford, Jr., is Associate Professor in the Department of Sociology and Anthropology at Middlebury College. He has studied the Hare Krishna movement over the past 17 years. He is author of *Hare Krishna in America.* His present research focuses on the movement's second generation in the United States.

Carol Rambo Ronai received her Ph.D. in sociology from the University of Florida and is currently Assistant Professor of Sociology at Memphis State University. She is exploring racism and child sex abuse using various narrative methodologies like written essays and the "layered account."

Martín Sánchez-Jankowski is Associate Professor of Sociology and Chair of the Chicano/Latino Policy Project of the Institute for the Study of Social Change at the University of California at Berkeley. He has written *City Bound: Urban Life and Political Attitudes Among Chicano Youth* and *Islands in the Street: Gangs and American Urban Society.*

Clinton R. Sanders is Professor of Sociology at the University of Connecticut. He is the author of *Customizing the Body: The Art and Culture of Tattooing* (Temple University Press, 1989) and the editor of *Marginal Conventions: Popular Culture, Mass Media and Social Deviance* (Bowling Green University Popular Press, 1990). Sanders' current research is directed at building a sociologically informed perspective on human relationships with companion animals and the constitution of animal "mind."

Kent Sandstrom earned his Ph.D. from the University of Minnesota and he is currently an Assistant Professor at the University of Northern Iowa. His most recent publication, "Ideology in Action: A Pragmatic Approach to a Contested Concept" (with Gary Alan Fine), appears in *Sociological Theory*.

Robert Scott is Associate Director of the Center for Advanced Study in the Behavioral Sciences and teaches a course on the social psychology of physical disability in the Program on Human Biology at Stanford University.

Diana Scully is Director of Women's Studies and Associate Professor of Sociology at Virginia Commonwealth University in Richmond, Virginia. She received her Ph.D. from the University of Illinois, Chicago. Her books include *Men Who Control Women's Health: The Miseducation of Obstetrician-Gynecologists* and *Understanding Sexual Violence: A Study of Convicted Rapists*.

C. R. D. Sharper is a pseudonym.

Alexander B. Smith practiced law, became a parole officer and then a probation supervisor. At John Jay College (CUNY) he was professor, chair, and dean. He has coauthored 70 articles and 12 books, most of them with Harriet Pollack. Their latest is *Criminal Justice: An Overview* (West, 1991).

Diane E. Taub is Associate Professor of Sociology at Southern Illinois University at Carbondale. Her research primarily involves the sociology of deviance, with a focus on the experiences of women. Recent publications concern eating disorders and athleticism, depictions of Satanism, and women's empowerment. She has also published on teaching and has received several teaching awards.

Ronald J. Troyer is currently Professor of Sociology and Associate Dean of the College of Arts and Sciences at Drake University. He has written *Cigarettes: The Battle Over Smoking,* is the senior editor and author of five chapters of *Social Control in the People's Republic of China* and journal articles on social problems and social movements.

Diane Vaughan is Associate Professor of Sociology, Boston College. She received her Ph.D. from Ohio State University. Her interests are the sociology of organizations, deviance and social control, and transitions. The author of *Controlling Unlawful Organizational Behavior* and *Uncoupling,* she currently is writing a historical ethnography of NASA's *Challenger* launch decision.

Ralph A. Weisheit earned his Ph.D. in sociology from Washington State University and is currently a Professor of Criminal Justice at Illinois State University. His research areas include the drugs-crime connection, female criminality, and rural crime/rural culture. He is the author of several books, including *Domestic Marijuana: A Neglected Industry,* and the editor of several books, including *Drugs, Crime and the Criminal Justice System*. He is currently involved in a study of rural crime and rural policing funded by the National Institute of Justice.

General Introduction

The topic of deviance has held an enduring fascination for students of sociology, gripping their interest for several reasons. Some people plan a career in law or law enforcement and want to expand their base of practical knowledge. Others feel a special affinity for the subject of deviance based on personal experience or inclination. A third group is drawn to deviance merely because it is different, offering the promise of the exciting or the exotic. The sociological study of deviance can fulfill all these goals, taking us deep into the criminal underworld, inward to the familiar, and outward to the fascinating and bizarre. In the following pages we peer into the deviant realm, looking at both deviants and those who define them as such. In so doing, we look at a range of deviant behaviors, discuss why people engage in them, and analyze how they are sociologically organized. We begin in Part I by defining deviance, in particular, its scope.

Reasonable theories and social policies pertaining to deviance must be based on a firm foundation of accurate knowledge. The works in this book are tied together by the belief that researchers must study deviance as it naturally occurs in the real world. This method, called participant observation, or ethnography, advocates that sociologists get as close as possible to the people they are studying in order to understand their worlds. Although deviance is often clandestine and hidden by its very nature, we discuss in Part II some of the techniques sociologists use to penetrate secluded deviant worlds, complete with the perils that may accompany this endeavor.

In Part III we delve further into the origin and definitions of deviant behavior. Many people have traditionally considered defining deviance a simple task, suggesting a widespread agreement on what is deviant and what is not. This view corresponds to the *objectivist* position on deviance, the perspective that something obvious within an act, belief, or condition makes it different from the norm in everybody's eyes. People have backed up this assumption by pointing to seemingly universal taboos such as bans against murder, incest, and lying. Although these acts might clearly seem deviant, other behaviors such as having an abortion, gambling, and consuming illicit drugs may not generate universal agreement so simply. Delving into the former cases, we also find that there are conditions under which these acts would be considered nondeviant. To kill somebody for personal gain, for vengeance, to gain freedom, or through negligence might be deviant (and even criminal), but when these acts are committed by the state (in executions, war, or covert operations), in self-defense, or on one's own property (in states that have "make my day" laws), they are considered not only nondeviant but also heroic. Similarly, anthropologists have found cultures that condone certain forms of incest to quiet or soothe infants (Henry 1964; Weatherford 1986). Often, too, people differentiate between normal lies and acceptable "white" lies, those designed to spare the feelings of someone other than the speaker or differentiated by situational circumstances. These definitional variations suggest that deviance may be lodged in the eye of the beholder rather than in the act itself and that the social context in which an act occurs frames and gives meaning to how it is interpreted. We should thus regard definitions of deviance as relative to their situation and surroundings. This relativistic view of deviance constitutes the *subjectivist* position, and it arises from *interactionist theory,* or the *social constructionist perspective* on deviance.

Assuming shared agreement, the objectivist position on deviance guides researchers to study the structural conditions fostering deviance, leading them to generate causal correlations that might predict and ultimately modify or control deviance through deterrence and social policy. Yet the strength of this simple approach is also its weakness, because general agreement on societal norms and values cannot always be found. Interpretations may be complex and varied, especially in a society as broad and diverse as that of the United States, because people belong to many divergent subcultural groups. Yet even if people shared broad agreement, definitions of deviance would not necessarily be objective. The constructionist perspective, in contrast, focuses attention on the audience that frames and reacts to people's actions. Deviance is not located in the act but in the societal reaction. As Becker (1963:9) argues in advancing *labeling theory,* one of the main components of an interactionist approach to deviance,

> *social groups create deviance by making the rules whose infraction constitutes deviance,* and by applying these rules to particular people and labeling them as outsiders. From this point of view, deviance is *not* a quality of the act the person commits, but rather a consequence of the application by others of rules and sanctions to an "offender."

In labeling theory we see the overlap between the interactionist and the power/conflict approaches to deviance. Deviance is defined through the social meanings collectively applied to people's behavior or conditions, which is rooted in interaction. Those who have the power to make rules and apply them to others control the normative order. The politically, socially, and economically dominant enforce their definitions on the downtrodden and powerless. Deviance is thus a representation of unequal power in society.

With deviance thus defined, one might then ask why do people engage in norm-violating behavior? Explanations for deviant behavior are as divergent as the acts they explain, ranging from acts of delinquency to professional theft, from acts of integrity to a search for kicks, and from acts of desperation to bravado and daring. Some sociologists have advanced structural explanations, suggesting that people break norms because blocked opportunity structures prevent them from getting what they want through legitimate means (Merton 1938) or because some people have greater access to illegitimate avenues of opportunity than others (Cloward and Ohlin 1960). These structural theories are best focused on helping us understand why people from disadvantaged backgrounds, who do not have access to legitimate opportunity structures, engage in deviance. Cultural explanations have been popular with other sociologists, suggesting that people violate norms when the values of their subculture conflict with those of the dominant culture (Miller 1958; Sellin 1938) or when their frustration with the dominant culture leads them to outright rebellion (Cohen 1955). These theories are suited to illustrating the motivations of people from minority or disadvantaged subcultures that are not well aligned with the dominant culture. Although these theories provide insight into some types of deviance, interactional forces inevitably intervene between the causes these theories propose and the way in which deviant behavior takes shape. When people confront the problems, pressures, excitements, and allures of the world, they most often do so in conjunction with their peer group. It is within peer groups that people make decisions about what they will do and how they will do it. Their core feelings about themselves develop and become rooted in such groups. People's actions and reactions are thus guided by the collective interpretations, norms, and values of their peer group. Edwin Sutherland (1934) recognized this point when he proposed his *differential association* theory of deviance, which suggests that deviant behavior is learned from people's intimate friends. The more people associate with others who hold favorable definitions of deviant acts, the more likely they are to develop these same attitudes. The more their friends engage in deviant behavior, the more likely they are to follow suit. Sutherland suggests that people learn a variety of critical elements of deviance from their associates: the norms and values of the deviant subculture, the rationalizations for legitimizing deviant behavior, the techniques necessary to commit the deviant acts, and the status system of the subculture, by which members evaluate themselves and others. As people "drift" into deviant subcultures (Matza 1964), they become enmeshed in a circle of deviant associates; quitting deviance, conversely, requires that they abandon this group of friends. Another element of the interactionist approach to deviance, then, is to understand its roots in human peer interaction.

Peer groups and the social contexts within which they form operate not only in the interactional setting but also within social organizations and institutions. Pressures can arise that shape collective attitudes, propelling people into deviance and providing legitimations that neutralize the perceived consequences of the deviance. In Part IV we examine the political and organizational contexts that surround people and push them into deviance for the sake of the group. We then look at the way in which organizations, especially the bureaucracies that process norm violators, treat their clients and the effects this treatment has on them.

Many people dabble to a greater or lesser degree in forms of deviance. Studies of juvenile delinquency suggest that participation is extremely widespread, nearly universal. How many people can claim to have reached adulthood without having experimented in illicit drinking, drug use, stealing, or vandalism? Yet do all of these people consider themselves deviants? Most do not. Many people "retire" from deviance as they mature, avoiding developing the deviant identity altogether. Others go on to engage in what Becker (1963) has called "secret deviance," conducting their acts of norm violation without ever seriously encountering the deviant label. Yet others, many of them no more experienced in the ways of deviance than the youthful delinquent or the secret deviant, become identified, and identify themselves, as deviants. What causes this difference? One critical difference, interactionist theory suggests, lies in who gets caught. Getting caught sets off a chain reaction of events that leads to profound social and self-conceptual consequences. Tannenbaum (1938) has described how individuals are publicly identified as norm violators and branded with that tag. They may go through official or unofficial social sanctioning in which people identify and treat them as deviant. Returning to labeling theory, Becker (1963:9) has noted that "the deviant is one to whom that label has successfully been applied; deviant behavior is behavior that people so label." Deviance exists at the macro, or societal, level of social norms and definitions through the collective attitudes we assign to certain acts and conditions. But it also comes into being at the micro, or everyday, level when the deviant label is applied to someone. This process and the consequences for people who become labeled are the focus of Part V.

The thrust of labeling theory is thus twofold, focusing on diverse levels and forces. As Schur (1979:160) summarizes its complexity:

> The twin emphases in such an approach are on *definition* and *process* at all the levels that are involved in the production of deviant situations and outcomes. Thus, the perspective is concerned not only with what happens to specific individuals when they are branded with deviantness ("labeling," in the narrow sense) but also with the wider domains and processes of social definitions and collective rule-making that frequently lie behind such concrete applications of negative labels.

We conclude this volume, in Part VI, with a discussion of how deviants and deviance are socially organized. Earlier sections of the book have concentrated on, first, macro and, then, micro levels of addressing deviance. Here, we take a midlevel focus by looking at how deviants organize their social relationships, activities, and lives in conjunction with others. We begin with the study of deviant associations, examining various types of relationships among members of deviant scenes. These range from loose subcultures to more tightly bound rings and, finally, to highly committed gangs. We then consider the organization of deviant acts. Although some forms of deviance can be committed alone, others are more complex, requiring multiple persons performing different roles. Some involve cooperation between the participants, with people exchanging illicit goods or services. Others are characterized by conflict, with some parties to the act taking advantage of others, often against their will. Finally, we look at the contours of deviant careers, beginning with people's entry into the world of deviant behavior and associates, continuing with the way in which they fashion their involvement in deviance, and concluding with their often problematic, occasionally inconclusive, retirements from the compelling world of deviance.

■

PART I

Defining Deviance

In order to study the topic of deviance, we must first clarify what we mean by the term. What behaviors or conditions fall into this category, and what is the relation between deviance and other categories such as the criminal? When we speak of deviance, we refer to violations of social norms. Norms are behavioral codes, or prescriptions, that guide people into actions and self-presentations that conform to social acceptability. Norms need not be agreed on by every member of the group doing the defining, but a clear or vocal majority must agree.

One of the founding sociologists, William Sumner (1906), conceptualized norms into three categories: *folkways, mores,* and *laws.* He defined *folkways* as simple everyday norms based on custom, tradition, or etiquette. Violations of folkway norms do not generate serious outrage but might cause people to think of the violator as odd. Common folkway norms include standards of dress, demeanor, physical closeness to or distance from others, and eating behavior. Someone who came to class dressed in a bathing suit, who never seemed to be paying attention when he was spoken to, who sat or stood too close to people, or who ate with his hands instead of silverware would be violating a folkway norm. We would not arrest him, nor would we impugn his moral character, but we might think that there was something wrong with him.

Mores are norms based on broad societal morals, whose infraction would generate more serious social condemnation. Interracial marriage, illegitimate childbearing, and drug addiction all constitute moral violations. Upholding these norms is seen as critical to the fabric of society, so that their violation threatens the social order. Interracial marriage threatens racial purity and the stratification hierarchy based on race; illegitimate childbearing threatens the institution of marriage and the transference of money, status, and family responsibility from one generation to the next; and drug addiction represents the triumph of hedonism over rationality, threatening the responsible behavior necessary to hold society together and accomplish its necessary tasks. People who violate mores may be considered wicked and potentially harmful to society.

Laws are the strongest norms because they are supported by codified social sanctions. People who violate them are subject to arrest and punishments ranging from fines to imprisonment. Many laws are directed toward behavior that used to be folkways or, especially, moral violations but became encoded into law. Others are regarded as necessary for maintaining social order. Although violating a law will bring the stigma associated with arrest, it will not necessarily brand the violator as deviant.

This discussion returns us to the question of the relationship between *deviance* and *crime*. Are they identical terms, is one a subset of the other, or are they overlapping categories? To answer this question, we must consider one facet of it at a time. First, do some things fall into both categories, crime and deviance? The overlap between these two is extensive, with crimes of violence, crimes of harm, and theft of personal property considered both deviant and illegal. Second, are there types of deviance that are not crimes? Actually, much deviance is noncriminal, such as obesity, stuttering, physical handicaps, racial intermarriage, and unwed pregnancy. Deviance is not a subset of crime, then. Finally, is there crime that is nondeviant? Although much crime is considered deviant and derives from various lesser deviant categories, some criminal violations do not violate norms or bring moral censure. Examples include some white collar crimes commonly regarded as merely aggressive business practices, such as income tax evasion, and forms of civil disobedience, in which people break laws to protest them. Thus, crime is not a subset of deviance. Crime and deviance, then, are overlapping categories with independent dimensions.

People can be labeled deviant as the result of their *behavior, beliefs,* or *condition.* It is the *behavioral* category that is the most familiar, with people coming to be regarded as deviant for their actions. These behaviors may be intentional or inadvertent and include such activities as violating dress or speech conventions, kinky sexual behavior, or murder. People can also be branded deviant for alternative attitudes, or *belief systems.* These beliefs can fall into the religious or political category, with people who hold radical or unusual views of the supernatural (cult members, Satanists, fundamentalists) or hold extreme political attitudes (far leftists or rightists, terrorists) considered deviant. Mental illness also falls into the deviant belief category, as people with deviant attitudes are often considered mentally ill and people with emotional deficiencies are considered deviant. People cast into the deviant realm for their acts or attitudes have an *achieved deviant status:* they have earned the deviant label through something they have done.

People regarded as deviant because of their *condition* often have an *ascribed deviant status:* it is something they acquire at birth. This would include having a deviant socioeconomic status, like being poor; a deviant racial status, such as being a person of color (in a dominantly Caucasian society); a congenital physical handicap; or a height deviance (too tall or too short). Here, there is nothing that such people have done to become deviant and nothing they can do to repair the deviant status. Moreover, there is nothing necessarily inherent in these statuses that make them deviant; they become deviant through the result of a socially defining process that gives unequal weight to powerful and dominant groups in society.

With all of these categories, we can see the differentiation between the deviant and the conventional. Our first reading selection, "Deviance as Crime, Sin, and Poor Taste," by Alexander Smith and Harriet Pollack, is based on some of the variations in deviance discussed above. Smith and Pollack build on Sumner's three categories of norms—folkways, mores, and laws—to typify the patterns of norm violation. Then Jeffrey Riemer writes about "Deviance as Fun." This important dimension of deviance, as both a motivational and definitional category, differentiates it in some important ways from crime. People pursue deviance for the sheer expressiveness of throwing off the shackles of protocol, as well as because their interests are turning in new directions. As such, deviance can also be considered both a harbinger and agent of social change. Finally, Daniel Patrick Moynihan looks at the relativity of deviance in "Defining Deviancy Down." He argues that deviance fulfills an important role in every society. To achieve the maximum benefit, however, a society needs a manageable amount. When the numbers of people declared deviant by current moral standards rises or falls too much, society alters its moral standard to maintain the level of deviance in the optimal range. Moynihan offers three examples of this shifting, and he argues that we are now learning to tolerate a greater level of deviance.

I

Deviance as Crime, Sin,

and Poor Taste

ALEXANDER B. SMITH

AND HARRIET POLLACK

S uperficially, it is very easy to define deviance. A deviant person is one who does something we wouldn't do. In the words of Howard Becker, he is an outsider, one who is outside the consensus of what constitutes proper conduct. The problem is that from someone's point of view we are all outsiders in one respect or another. Discussions of deviance, therefore, really turn on searches for universals, for modes of conduct that all human societies consider unacceptable.

In the classroom, anthropology professors like to upset their students by pointing out that there are no such universally disapproved modes of conduct. Even a killing that we would consider murder is acceptable in some societies: infanticide was common in Sparta, as was deliberate starvation of old people by Eskimos. In actuality, however, assaultive acts against the person or someone else's property, such as murder, assault, rape, and robbery, are considered taboo in almost all human societies, and people who perform such acts are clearly deviant. These acts, however, constitute only a tiny fraction of all the modes of conduct that our own and other societies have from time to time labeled as wrong.

If today we were to ask a middle-class, middle-aged, white American what kinds of acts (outside of assaultive crime) he considered deviant, he might respond as follows:

Being a homosexual; reading dirty books or seeing pornographic movies; going to prostitutes; engaging in sex outside of marriage; having illegitimate children (especially if the children wind up on welfare).

Using drugs—not prescription drugs or over-the-counter items like Alka Seltzer or Geritol or Vitamin E—but heroin, LSD, and pep pills.

Drinking too much; eating enough to make you fat; smoking cigarettes (maybe); smoking marijuana (positively).

Not taking care of your obligations; being lazy or shiftless; losing money at gambling; swearing and using bad language publicly.

From: "Deviance as a Method of Coping," Alexander B. Smith and Harriet Pollack, *Crime and Delinquency,* Vol. 22, No. 1, 1976. Reprinted by permission of Sage Publications, Inc.

If we accept this list as typical, it is as interesting for the conduct it omits as for that which it includes. Many acts which once were or now are attacked as highly immoral are not even mentioned: for example, contraception, abortion, and sexual and racial discrimination. Our Everyman also seems unconcerned about profiteering, sharp dealing, tax evasion, consumer fraud, and other kinds of white-collar crime. To be sure, if questioned specifically about these unmentioned acts, he would disapprove of all of them (except possibly for contraception), but the term "deviant conduct" would not bring them immediately to mind.

The reason for our Everyman's perceiving deviance selectively lies in our description of Everyman: middle-class, middle-aged, and white. From where he stands, some acts affect his world adversely, others have little effect, and some are simply irrelevant. He doesn't care especially about racial or sexual discrimination because he is neither black nor female. He believes in sexual regularity because he is a family man and his world is stabilized by the nuclear families of his friends and neighbors. Furthermore, illegitimacy (as he sees it) is a direct and undeserved burden on taxpayers like him because of its effect on the welfare rolls. On the other hand, contraception doesn't seem wrong to him since his middle-class status probably depends on his success in limiting the size of his family. Even abortion has much to be said for it, since anyone can get into trouble and anyway maybe abortion will keep some of those babies off welfare. He doesn't worry too much about tax evasion because he is not aware of the activities of large-scale tax evaders, such as giant corporations and wealthy individuals whose accountants and tax lawyers have created tax shelters for them; and small-scale tax evasion is probably a fairly common and socially acceptable activity in his milieu. Sharp dealings (such as exploitative landlord-tenant or seller-consumer transactions) are likewise a middle-class way of making a living; and in any case, most middle-class persons are able to cope with dishonest landlords or tradesmen. On the other hand, persons who take or sell drugs are enormously threatening, both because drug use frequently leads to assaultive or dangerous criminal conduct and because drug addicts threaten the stability of the social system by their aberrant attitudes toward work and other social obligations. In fact, if there is one thread that runs through the fabric of Everyman's scheme of desirable social conduct, it is the desire to maintain stability, to preserve the status quo. As a member of the middle class, he has made it, and he recognizes that life is as good for him as it is ever likely to be. He doesn't want to lose what he has. Change is threatening and makes him very uncomfortable.

The laundry list of unacceptable conduct varies with the age and status of the person compiling it. Inner city blacks, for example, might list racial discrimination first and not list gambling at all. Marijuana smoking might be quite acceptable to middle-class university students, but tax evasion, sharp dealing, and profiteering would be high on their list of forbidden conduct. In the Bible Belt of the Deep South, blasphemy, secularism, and atheism are still heinous offenses, yet relatively free use of firearms, moonshining, and blatant racial discrimination are regarded with considerable tolerance.

Obviously, deviance is to some extent in the eye of the beholder—but only to some extent. All classes and status groups reject violent assaultive crime.[1] They differ, however, in respect to other types of unacceptable conduct, some of which in our system are illegal, some of which are immoral, and some of which are merely displays of poor taste. In considering these widely varying perceptions of what constitutes deviant conduct, we must ask not who is right and who is wrong but what kinds of conduct society can tolerate and still exist as a viable society and what kinds it cannot accept. Part of the answer to this basic question must lie in one's perception of a desirable society. For purposes of this discussion we are assuming an ideal closely akin to the traditional Jeffersonian model: an open society predicated on a belief in equality of opportunity and equality before the law, with a reasonable level of material comfort and economic security for all. In such a society, what kinds of behavior are necessarily beyond the pale? In this connection, we propose to discuss three categories of conduct: crime, sin, and actions that are in poor taste.

DEVIANCE: CRIME

Clearly, heading the list are murder, rape, arson, assault, robbery, burglary, and larceny, acts which are totally unacceptable and which can be condoned, if at all, only under very special circumstances.[2] We label these acts *crimes,* meaning that their violation of the public order is so severe that they must be handled punitively and coercively by the police, courts, and prisons. Even those who commit them agree that this type of conduct is wrong. A housebreaker does not want his own house to be burglarized and, except in Robin Hood legends, robbers do not argue that what they do is legitimate. This type of conduct is taboo because if it is tolerated a viable society is not possible. The control of such conduct, indeed, is one of the central problems faced throughout history by philosophers who have attempted to construct model societies. Whatever their point of view and whatever type of Utopia they have created, they all agree at least that this type of act must be forbidden. While Hobbes and Locke, for example, differed radically in their perceptions of the fundamental nature of man and in their prescriptions for social control of human conduct, they agreed that the principal difficulty in human society is the governance of violent assault by one individual against another.

However, assaultive conduct is only one category of crime. So-called "white collar crime," while nonviolent, is basically an attack on legitimate property arrangements in society. Acts such as tax fraud, stock manipulations, commercial bribery, misrepresentation in advertising and salesmanship, short weighting and misgrading of commodities, embezzlement, etc., are all methods of obtaining money or other property illegitimately. Since the function of an economic system is to prescribe how one may properly obtain property, white collar criminals are subversive of accepted economic relationships. As such, like their more violent criminal counterparts, they are a threat to a viable society and it is reasonable that their acts be included in the penal codes.[3]

Although the prescribed penalties for white-collar crimes may sometimes be as severe as those for burglary or larceny, these acts do not carry the stigma or the punishment of violent crimes.

Basically our law is ambivalent. Property crimes are crimes, but they are not really heinous if they are not violent or potentially violent. Far less ambivalence in regard to so-called "economic crimes" is exhibited in the Soviet Union, where some offenses of this type, such as currency manipulation, are punishable by death sentences whereas certain kinds of homicide are treated relatively leniently. This probably reflects the orientation of the legal system toward preservation of the Soviet economic and social order rather than, as in this country, protection of individual rights. From this point of view, the inconsistency of the American system, which punishes personal crimes more severely than property crimes, is understandable. Whatever our ambivalence, however, it is clear that nonviolent crimes of property must be handled punitively, at least to the extent necessary to maintain the legitimacy of both our property arrangements and our system of law. The latent admiration of Americans for Robber Baron types may never disappear from the culture. Nevertheless, if business dealings are to be conducted in an orderly way and if prohibitions on assaultive crimes are to be taken seriously, there must be reasonable enforcement of the law relating to white collar offenses. As the public conscience grows in sensitivity, moreover, the criminal sanction will be extended to dealings which are now considered unsavory but not illegal. The basic push in the developing field of poverty law is to extend the criminal law to cover some actions of landlords against tenants and of merchants against customers that were not considered illegal before. For example, may a landlord be paid rent by his tenants if he has failed to provide the agreed-upon level of services? May a merchant misrepresent the quality of the merchandise he is selling and demand continued performance of a time-payment contract if the goods in question have already deteriorated? These practices are probably permissible at present. The trend, however, is toward making such actions illegal—probably an indication of our feeling that even nonassaultive crimes of property are a threat to the viability of our society.

Our penal law, thus, contains prohibitions against both assaultive crimes against persons and property and nonassaultive crimes against property. Assaultive crimes offend our notions of natural justice; nonassaultive property crimes undermine the economic arrangements that are basic to the stability of society. The penal code contains, however, strictures against a number of modes of conduct which are included because of a relatively parochial cultural determination that they are immoral: drinking, gambling, homosexuality, doing business on Sunday, prostitution, drug addiction, abortion, etc. While at the time these prohibitions were enacted, the particular legislative majority which enacted them doubtless felt they were preventing subversion of the legitimate social system, many societies quite similar to ours do, in fact, tolerate such prohibited conduct quite well or, in any case, handle it nonpunitively. Many of these regulations are, moreover, both inconsistent and incomplete in their regulatory schemes. Prostitutes are punished but not their customers; heroin is forbidden but not amphetamines; it is permissible to bet on a race but not on a football game, etc.

DEVIANCE: SIN

Many of these modes of conduct were originally thought of as sin and were *religiously* prohibited. Our use of secular law to regulate them is a relic of the time when the authority of the state was used to enforce the rules of an established church. That era is past, but we can see our cultural heritage most clearly perhaps in the laws we inherited from the Puritan theocracy in New England. We have (or have had in the recent past) laws against blasphemy, obscenity, contraception, Sabbath breaking, extramarital sexual relations, lewdness, homosexuality, gambling, and drunkenness. We also have inherited a distrust of self-indulgence and hedonism: even a rich man is expected to be constructively, if not gainfully, employed.

This heritage reflects a culture in which religion once was dominant. As our culture has changed, as religion has waned in importance, as our economic system has developed, as scientific discoveries have occurred, and as improved communications and the development of the mass media have reduced both social and cultural isolation, our feelings about what constitutes sin have undergone a marked change. Some behavior once regarded as sinful has become virtually acceptable today—for example, blasphemy; some, like heroin use, is still taboo. About other forms of conduct such as gambling, drinking, homosexuality, and abortion, we have ambivalent feelings. Some of this conduct is still subject to criminal sanction; some is not. If we remove the religious component, the criterion for whether the conduct in question should be forbidden should rest on whether *there is any demonstrable, objectively measurable social harm resulting from it*. To determine this, we must separately consider and evaluate each mode of conduct. In a totally rational world we would expect to find a correlation between the prohibition of conduct and its objective harmfulness. But this is not a rational world and the correlation does not exist.

Of all the modes of conduct in this culturally determined category, drinking is probably the most harmful and also the most widely accepted. Alcohol is involved in at least half of all fatal automobile accidents, a majority of private airline crashes, thousands of industrial accidents, millions of lost man-days annually, etc. We have in this country approximately 9,000,000 alcoholics who are unable to support their families, do their jobs, or function normally in the community. Alcohol use is involved in 55 per cent of the arrests made by police. From a medical point of view, furthermore, even moderate drinking puts a strain on the liver and complicates many maladies.

Yet alcohol consumption is widely accepted today in the United States, where nondrinkers constitute only a small minority of the population. Historically, the temperance movement waxed and waned in strength for over a century before it culminated in the "noble experiment" of Prohibition in

1920. However, within a few years after enactment of the Eighteenth Amendment, it became apparent that Prohibition was a disaster and, since the repeal in 1933, the temperance movement appears to be all but moribund. Thus, drinking has been handled both coercively and noncoercively, and while our current noncoercive approach has fewer adverse effects in the form of enforcement difficulties and police corruption, alcohol abuse still presents a problem—a problem not reflected in public attitudes.

Even more permissive than our attitudes toward drinking are our feelings about cigarette smoking and overeating. The medical evidence against both smoking and obesity is overwhelming but to forbid them by law would be ludicrous, a civil liberty horror. Even attempts to regulate cigarette advertising have met with great resistance. Though smoking and overeating are seriously harmful, medically and sociologically speaking, and though there is considerable consensus that people should not smoke or get fat, Americans who do not smoke and who are not overweight probably constitute a minority.

In contrast to drinking, smoking, and overeating, there is no medical evidence that moderate use of marijuana is harmful and no medical evidence of physiological harm from reasonable heroin consumption. That many heroin or marijuana users exhibit undesirable psychological symptoms is undoubtedly true. It is not clear, however, whether these symptoms are a result of drug use or whether both drug use and behavioral dysfunction result from a prior existing pathological, psychological, or sociological condition. Most of the other adverse sociological effects of drug use, such as crime and prostitution, result from our present coercive handling of the drug problem rather than from drug use per se. Yet few modes of conduct are looked upon with more social disapproval than heroin use and only recently has a similar attitude toward marijuana been softening. Moreover, in certain respects our method of handling drug use has been precisely opposite from our handling of alcohol: alcohol, formerly handled punitively, is now handled nonpunitively; opiates and marijuana, formerly handled nonpunitively, are now handled punitively. Neither punitive handling nor extreme social disapproval has resulted in a decline (or even a stabilization) of the number of marijuana and heroin users in the United States. In 1967 there were in the United States about 100,000 addicts, of whom 50,000 were in New York City; five years later the estimates had precisely tripled: 300,000 addicts in the United States, with 150,000 in New York City.

In contrast to our attitude toward alcohol and drug use, which has fluctuated between acceptance and rejection, our attitude toward deviant sexual conduct has become consistently more permissive. During the eighteenth and nineteenth centuries in this country, man-woman relationships reflected a society that placed high value on premarital chastity and monogamy. Divorce was frowned upon, and premarital dalliance (except possibly for young men who were sowing their "wild oats") was strictly taboo. Prostitution, at least from the middle-class point of view, was considered degrading and abhorrent, and the fallen woman became a stock figure in literature. In the same period, homosexuality was considered so dreadful that there was no public discussion of the subject and, except for some very guarded indirect references, no literary mention of the problem. Today we are permissive in regard to premarital sex, we permit divorce, we have ambivalent attitudes toward prostitution, and we are slowly coming to a grudging acceptance of homosexual conduct. Some of these attitudinal changes have been reflected in changes in either the criminal law or its application; others have not. Nevertheless, few people would dispute the proposition that our attitudes toward sexual conduct have changed substantially even if the conduct in question has not.

To understand this phenomenon one must appreciate that the older rules for sexual conduct were drawn up in a society which had vastly different needs: until the twentieth century the need was for more population rather than less; venereal disease was an uncontrollable plague; and production of goods and services was directly dependent on the family in a way that no longer exists. Twentieth century advances in public health and medical knowledge have changed all this.

Medical knowledge and technology have turned the older rationale for monogamous units upside down. One hundred years ago, a couple might have to produce ten or twelve children to be certain that five or six would survive them; today the parents of two can reasonably expect to raise both to adulthood. Formerly children represented a source of income and social security for one's old age; today children are economic liabilities at least until they reach adulthood, and some thereafter.

In the face of these substantial changes, it is understandable that many of the older rules of sexual conduct are anachronistic. This is not to say that our commitment to monogamous union as the basis of family structure has diminished. Nor does it mean that actual sexual practices (as opposed to the accepted social standard for what those practices should be) have changed very much. What it means is that deviation from these sexual norms is accepted more readily and less fearfully than before. We are not so hysterically defensive about our rules of sexual conduct because we no longer regard deviations from them as subversive of the entire social order. We no longer need a strict sexual code to provide for population maintenance or growth, industrial or agricultural production, or prophylaxis against rampant venereal disease. We adhere to our family structure—and hence our sexual code—not so much to meet societal needs as to fulfill our own, the achievement of personal happiness and an optimal setting in which to raise children. Under these circumstances the desire of some individuals to find personal happiness through premarital sex, homosexuality, prostitution, etc., becomes less terrifying and is, if not acceptable, at least understandable.

Gambling, however, is a mode of conduct which probably has come closest of all to shedding the stigma of immorality inherited from the past. American attitudes toward gambling have always been ambivalent. Even in Puritan times we find mention of gaming and lotteries at the same time the churches were exhorting against such worldly pleasures. Gradually, however, our attitudes have softened, probably because of the general relaxation of the personal standards of behavior and possibly because of the possibilities of relief for the hard-pressed taxpayer through state-sponsored lotteries. In any case, at the present time, not only does Nevada have legalized gambling and New York the OTB (a public corporation to conduct off-track betting) but increasingly the criminal justice system is refusing to use its resources to enforce antigambling laws. The police protest openly at the futility of picking up small-time gamblers who are doing no more than the OTB employees, and such gamblers as are prosecuted are handled by the courts perfunctorily and with minimal penalties. The change in public opinion, the negative attitudes of police and prosecutors toward gambling law enforcement, and general awareness that illegal gambling is a major source of income for organized crime have combined to hasten the repeal of many—perhaps all—gambling statutes. There is virtually no effective interest group in the United States that espouses the retention of gambling laws. Apparently, legislative repeal is retarded only by public apathy and the fear of criticism by zealots.

DEVIANCE: POOR TASTE

In contrast to acts which are crimes or sins are actions which are matters of taste and which, even when disapproved, are rarely regulated by law. Manners and style fall within this category. Pants on women were once an object of scandal; girls' bobbed hair in the 1920's was viewed as dubiously as boy's long hair in the 1960's. In Puritan New England it was a misdemeanor for a man and a woman to kiss in public even if they were married; we think nothing of more overt expressions of affection although we become increasingly offended as the conduct becomes more explicitly sexual. Adults smile benignly at little Boy Scouts and Girl Scouts in their uniforms, but glare at black-jacketed Hell's Angels and similarly dressed members of black and Puerto Rican youth gangs. Frenchmen may kiss each other heartily; American men may not. It is all right to wear a cross or a mezuzah, but a swastika arm-band, a hooded sheet, and a clenched fist salute are perceived with considerable hostility and, under certain circumstances, are forbidden by the authorities.

To the visitor from Mars, all of this can be very confusing. Why, for example, is it all right for an adult to appear in public wearing a skimpy bathing suit but not his underwear? To us, however, it is not confusing at all, although few people when pressed could rationalize all the idiosyncrasies of manners and style that go to make up taste. It is clear that to a great extent these modes of conduct are cultural accidents. Pants are no more ordained by nature for men than skirts are for women, and in some tribal societies men do wear skirts and women pants. There is nothing in the shape of a cross that necessarily suggests Christianity and nothing in the shape of the swastika that necessarily equals fascism. As a method of greeting, handshaking is neither more nor less rational than a kiss on the cheek or a deep curtsy. But while the conduct in question may be irrational, the inferences drawn from it may be highly rational. The wearing of the swastika by American fascists is a reliable indicator of a belief in racial inequality, a totalitarian system of government, etc. A man who appears in public in a woman's dress probably is sexually deviant. What we object to in these modes of conduct, therefore, is that they suggest or anticipate other actions to which we take exception. They are in a sense symbolic conduct, symbolic of some type of overt action to which there is or may be a rational objection. Thus, the objection to the swastika is an objection to fascism; and the more we object to fascism as a mode of conduct, the more we will object to the swastika. Many modes of dress are objectionable because they appear to anticipate undesirable sexual conduct: slacks and bikinis on women, long hair and feminine looking clothing on men. Interpersonal conduct—modes of greeting and communicating with other people—are evaluated by our interpretation of the hidden messages those modes send out. When attempts are made to change matters of manner and style, objection is frequently vigorous simply because such changes are viewed as a precedent to change in more serious forms of nonsymbolic conduct. Opposition fades away when the symbolic conduct loses its symbolism. In Victorian times a woman who showed her ankles freely was considered "fast," aggressively inviting promiscuous conduct. When enough women wore short skirts without the occurrence of the undesirable sexual conduct that had been anticipated, short skirts became acceptable. The first men wearing long hair in the current style were considered to be homosexually inclined. When the majority of adolescent youths and young men adopted the fashion, long hair as a symbol of homosexuality faded.

Thus the problem in regard to matters of taste is to recognize, first of all, that they are cultural accidents and may be intrinsically quite irrational. We must also recognize, however, that such conduct is symbolic conduct and may be the surface manifestation of far more meaningful attitudes and actions. In regulating such matters of taste, then, we must know when the surface conduct is truly symbolic and when it has lost its symbolism. If the symbolism is extant and the conduct to which it refers is truly harmful, it is possible that even symbolic action may need to be regulated socially. . . .

CONCLUSION

To sum up, deviant conduct is ubiquitous in a society such as ours. While deviance lies, to some extent, in the eye of the beholder, certain forms of it that are objectively and measurably harmful to the community or that violate rational institutionalized expectations are always deviant. The roots of deviance lie in sociological and psychological pressures generated within the individual by social forces frequently beyond his control. Since, however, the very notion of a free society is based on the responsibility of each individual for his own conduct, the responsibility for the control of deviant conduct lies with both the individual and the community at large. . . . Physically coercive punishment must be used only as a last resort and for the protection of the community, for it has almost no rehabilitative effect and serves only to keep the offender away from the community. For this reason the criminal process should be reserved almost exclusively for all persons who must be restrained at all costs or who are so seriously disruptive of the peace and good order of the community (e.g., swindlers and embezzlers who commit nonviolent property crimes) that rehabilitative counseling should be carried on in a semicoercive setting such as probation. For all others, either we should attempt education and persuasion by appropriate therapists or, in regard to those whose conduct really harms no one but themselves, *we should let them alone,* recognizing that to some extent we are all deviants.

NOTES

1. An exception might be black revolutionaries such as George Jackson, who, while imprisoned in San Quentin for armed robbery, wrote extensively on the place of blacks in white society. Jackson felt that because "Amerika" was a "society above society" in which blacks were "captive," they were under no obligation to obey the laws. All crime, therefore, was an act of rebellion. Even Jackson concedes, however, that noneconomic crime—e.g., "the rape of a Black woman by a Black man"—is an expression of racial violence turned inward. It is "autodestructive" and hence presumably wrong even if understandable. Tad Szulc, "George Jackson Radicalizes the Brothers in Soledad and San Quentin," *New York Times Magazine,* Aug. 1, 1971, p. 10.

2. We are referring here, of course, to random acts by individuals or small groups such as gangs and are omitting discussion of governmentally organized and sponsored violence such as that practiced during the Hitler period in Germany, the Spanish Inquisition, or any war. Whether this kind of organized violence is ever justifiable depends on one's politics, religion, nationality, and time in history.

3. While many political theorists have attacked the American economic system and consequent property arrangements as illegitimate—as violations of "natural justice"—none has seriously suggested that the types of fraud usually encompassed by the term "white collar crime" are justified as an attempt to remedy economic inequity. The embezzlers and stock manipulators have not yet produced their George Jackson.

2

Deviance as Fun

JEFFREY W. RIEMER

Deviant behavior as a fun activity has been consistently ignored in the work of social scientists. Rarely has deviance been considered a spontaneous, "just for the hell of it" activity, in which the participants engage simply for the pleasure it provides. Rather than incorporating this simple, yet potentially useful, dimension into our understanding of deviance phenomena, sociologists have continued to be preoccupied with elaborating the more sober social, psychological, and sociological explanations for this activity.

Deviance is seldom treated as a frivolous, flippant activity. Yet, this hedonistic "pleasure of the moment" explanation squares well with a common sense interpretation for some of the behavior usually recognized as deviant. In an era of multi-causal explanations we may be inadvertently ignoring an important dimension for better understanding deviant behavior.

Vandalism could be partially explained this way, as could some instances of pre-marital and extra-marital sex, homosexuality (or bisexuality), drug use (or experimentation), drunkenness, shoplifting, auto theft, gambling, fist fights, reckless driving, profanity and prostitution, at least from the client's point of view—to mention a few.

Gustave Ichheiser (1970) has suggested that we are often "blind" to the obvious. We overlook or ignore what we believe everyone already knows. He suggests that what is "taken for granted" is usually neglected or treated as unimportant. This seems to be what has happened with the fun dimension of deviance. When something squares well with common sense knowledge it does not nullify its theoretical importance.

THE LITERATURE

Some social scientists have treated this dimension of deviance in an ancillary way in their writings, while most have totally ignored it. In fact, at least one has been literally forced to recognize this alternative by a respondent who was unwilling to accept a more sophisticated interpretation of her behavior constructed by serious-minded social scientists. Simmons relates the following:

> Several years ago a lesbian interrupted my interview with her and said, "We've kept talking about emotional turmoil and male avoidance. But the truth is I go to bed with her because it's fun." (Simmons, 1969:63)

From: "Deviance as Fun," Jeffrey W. Riemer, *Adolescence,* Vol. 16, No. 61 (Spring 1981). Reprinted by permission of Libra Publishers, Inc.

Garfinkel (1964) has actually created deviant behavior by purposely having his students violate "taken for granted" background expectancies in routine social situations in an attempt to understand the common sense world and how social order emerges. Many of his quasi-experimental researches are quite humorous and, in addition, demonstrate subject reactions to folkway violations as well as experimenter discomfort and pleasure at being "deviant." But Garfinkel fails to interpret any of this research designs as exercises in "fun" or "deviance."

Deviant behavior texts through the years have routinely ignored this dimension with no mention of "fun," "pleasure," "hedonism," or the like, in reference to the causal or contributory factors that influence deviant behavior. Most theorists have continued to delve into the more serious publicly recognized forms of objectionable behavior, often including what is usually recognized as criminal behavior (Thio, 1978; Goode, 1978; Akers, 1977; Sagarin and Montanino, 1977). Certainly, all crime is deviance but not all deviance is crime (Glaser, 1971).

Mundane deviance has been typically neglected. Denzin has argued that sociologists should pay more attention to the "mundane, routine, ephemeral, or normal" forms of deviance that surround all of our everyday life experiences (Denzin, 1970:121). However, his suggestion has fallen on deaf ears or met objection. Most current students of deviant behavior continue to agree with Gibbons and Jones (1971) who argue that "omnibus" definitions of deviance are unfruitful and loaded with triviality. Mundane deviance is seen as inconsequential and not worth consideration.

The major exception is Lofland (1969) who, in passing, recognized the "deviance as fun" alternative.

> At least some kinds of prohibited activities are claimed by some parts of the population to be *in themselves* fun, exciting and adventurous. More than simply deriving pleasant fearfulness from violating the prohibition per se, there can exist claims that the prohibited activity *itself* produces a pleasant level of excitation. (Lofland, 1971:109)

Even so, after eight years we find little incorporation of this argument.

Research studies have also consistently ignored the deviance as fun dimension. Research into deviant behavior regularly employs the more accepted sophisticated conceptual frameworks for describing and explaining the behavior in question. Again, a somber and serious approach is taken.

One exception, Riemer (1978), has looked at deviance in the work place as a fun activity. Four activities that building construction workers routinely engage in while being paid to work are discussed. These include: drinking alcoholic beverages, "girl watching" and other sexually related activities, stealing, and loafing. Each of these activities are engaged in periodically by some workers as a pleasant change of pace from the routine and boredom of work when the opportunities arise. These activities typically arise spontaneously (given the opportunity) with the workers giving a "just for the hell of it" rationale for their participation. This fun explanation is not meant to be the sole reason for this worker behavior but it cannot be ignored as contributory.

Unfortunately, it is the juvenile delinquency theorists who have taken a more assimilating view of deviance as a fun activity. Ferdinand (1966) in developing a typology of delinquency, included the "mischievous-indulgent" and the "disorganized acting-out" types which are both motivated in their actions by some degree of hedonism and spontaneity. Along this same line, Gibbons (1965) developed a series of delinquent categories in which he included the "casual gang delinquent" (who regards himself as a non-delinquent but likes to have fun) and the "auto-thief joyrider."

Similarly, Briar and Piliavin have argued for the influence of situationally induced motives to deviate. They suggest that some delinquency may be

> prompted by short-term situationally induced desires experienced by all boys to obtain valued goods, to portray courage in the presence of, or be loyal to peers, to strike out at someone who is disliked, or simply to "get kicks." (Briar and Piliavin, 1965:36)

Matza (1961) and Cohen (1955) have also built an "adventure," "play," and "fun" dimension into their explanation of delinquent behavior.

To offer a twist on the Matza and Sykes (1961) thesis that much juvenile behavior could be analyzed as an extension of adult behavior, it is argued here that some adult deviance could be analyzed as an extension of juvenile delinquency. Some deviant behavior may arise as a simple pleasure-seeking activity.

IMPLICATIONS

The argument presented here is not meant to dispute any particular theoretical orientation for deviant behavior. Rather, it calls attention to them all for neglecting a seemingly fruitful area of pursuit for helping to explain this behavior. Certainly, the fun dimension should not be construed as a unidimensional explanation for deviance phenomena. It is simply another perspective that should be incorporated into our existing approaches.

A deviance as fun argument suggests that all persons are deviant at least some of the time. Each of us departs from prevailing normative standards on occasion, weighing the possibility of being publicly designated as deviant and sanctioned accordingly, and choosing a seemingly pleasant diversion from our normative routines. Accordingly, this fun dimension implies a "normal" deviance that exists in society (Durkheim, 1964; Goffman, 1963).

For most of us, most of the time, these pleasant diversions remain non-problematic. But occasionally an act that begins innocently evolves into a serious legal infraction, e.g., the party-goer who drinks to excess as "the life of the party" and as a result of the intoxication becomes involved in a serious traffic accident on the journey home. What began as fun turned into tragedy.

What is pleasurable, enjoyable, or fun lies in the eyes of the beholder. The current labeling and political theorists might well address what is considered acceptable fun. "One man's pleasure is another man's pain."

The choice to violate normative standards is a choice we all have. For some, this choice represents an enjoyable and exciting alternative. It is within this context that we may learn more about deviant behavior and social control.

REFERENCES

Akers, Ronald L. 1977. *Deviant Behavior*. Belmont, Calif.: Wadsworth.

Becker, Howard S. 1963. *Outsiders*. New York: Free Press.

Birenbaum, Arnold, and Edward Sagarin. 1976. *Norms and Human Behavior*. New York: Praeger.

Briar, Scott, and Irving Piliavin. 1965. "Delinquency, Situational Inducements, and Commitment to Conformity," *Social Problems,* 13 (Summer): 35–45.

Cohen, Albert K. 1955. *Delinquent Boys: The Culture of the Gang*. Glencoe, Ill.: Free Press.

Davis, Nanette J. 1975. *Sociological Constructions of Deviance*. Dubuque, Iowa: Wm. C. Brown Company.

Denzin, Norman K. 1970. "Rules of Conduct and the Study of Deviant Behavior: Some Notes on the Social Relationship," in *Deviance and Respectability,* Jack D. Douglas (Ed.). New York: Basic Books, pp. 120–159.

Durkheim, Emile. 1964. *The Rules of Sociological Method*. New York: Free Press.

Feldman, Saul D. 1978. *Deciphering Deviance*. Boston: Little, Brown.

Ferdinand, Theodore N. 1966. *Typologies of Delinquency: A Critical Analysis*. New York: Random House.

Finestone, Harold. 1957. "Cats, Kicks and Color," *Social Problems,* 5 (July): 3–13.

Garfinkel, Harold. 1964. "Studies of the Routine Grounds of Everyday Activities," *Social Problems,* 11 (Winter): 225–250.

Gibbons, Don C. 1965. *Changing the Lawbreaker: The Treatment of Delinquents and Criminals*. Englewood Cliffs, N.J.: Prentice-Hall.

Gibbons, Don C., and Joseph F. Jones. 1971. "Some Critical Notes on Current Definitions of Deviance," *Pacific Sociological Review,* 14 (January): 20–37.

Glaser, Daniel. 1971. *Social Deviance*. Chicago: Markham.

Goffman, Erving. 1963. *Stigma*. Englewood Cliffs, N.J.: Prentice-Hall.

Goode, Erich. 1978. *Deviant Behavior*. Englewood Cliffs, N.J.: Prentice-Hall.

Ichheiser, Gustav. 1970. *Appearances and Realities*. San Francisco: Jossey-Bass.

Lofland, John. 1969. *Deviance and Identity*. Englewood Cliffs, N.J.: Prentice-Hall.

Matza, David. 1961. "Subterranean Traditions of Youth," *Annals of the American Academy of Political and Social Science*, 338 (November).

Matza, David, and Gresham M. Sykes. 1961. "Juvenile Delinquency and Subterranean Values," *American Sociological Review*, 26: 712–719.

Riemer, Jeffrey W. 1978. "'Deviance' as Fun—A Case of Building Construction Workers at Work," in *Social Problems—Institutional and Interpersonal Perspectives*, K. Henry (Ed.), Glenview, Ill.: Scott, Foresman, pp. 322–332.

Sagarin, Edward, and Fred Montanino (Eds.). 1977. *Deviants: Voluntary Actors in a Hostile World*. Morristown, N.J.: General Learning Press.

Simmons, J. L. 1978. *Deviants*. Berkeley: Glendessary Press.

Thio, Alex. 1978. *Deviant Behavior*. Boston: Houghton Mifflin.

3

Defining Deviancy Down

DANIEL PATRICK MOYNIHAN

In one of the founding texts of sociology, *The Rules of Sociological Method* (1895), Emile Durkheim set it down that "crime is normal." "It is," he wrote, "completely impossible for any society entirely free of it to exist." By defining what is deviant, we are enabled to know what is not, and hence to live by shared standards. This aperçu appears in the chapter entitled "Rules for the Distinction of the Normal from the Pathological." Durkheim writes:

> From this viewpoint the fundamental facts of criminology appear to us in an entirely new light. . . . [T]he criminal no longer appears as an utterly unsociable creature, a sort of parasitic element, a foreign, inassimilable body introduced into the bosom of society. He plays a normal role in social life. For its part, crime must no longer be conceived of as an evil which cannot be circumscribed closely enough. Far from there being

From: "Defining Deviancy Down," Daniel Patrick Moynihan, *The American Scholar*, Volume 62, No. 1, Winter 1993. Copyright © 1992 Daniel Patrick Moynihan. Reprinted by permission of The American Scholar.

cause for congratulation when it drops too noticeably below the normal level, this apparent progress assuredly coincides with and is linked to some social disturbance.

Durkheim suggests, for example, that "in times of scarcity" crimes of assault drop off. He does not imply that we ought to approve of crime—"[p]ain has likewise nothing desirable about it"—but we need to understand its function. He saw religion, in the sociologist Randall Collins's terms, as "fundamentally a set of ceremonial actions, assembling the group, heightening its emotions, and focusing its members on symbols of their common belonging-ness." In this context "a punishment ceremony creates social solidarity."

The matter was pretty much left at that until seventy years later when, in 1965, Kai T. Erikson published *Wayward Puritans,* a study of "crime rates" in the Massachusetts Bay Colony. The plan behind the book, as Erikson put it, was "to test [Durkheim's] notion that the number of deviant offenders a community can afford to recognize is likely to remain stable over time." The notion proved out very well indeed. Despite occasional crime waves, as when itinerant Quakers refused to take off their hats in the presence of magistrates, the amount of deviance in this corner of seventeenth-century New England fitted nicely with the supply of stocks and whipping posts. Erikson remarks:

> It is one of the arguments of the . . . study that the amount of deviation a community encounters is apt to remain fairly constant over time. To start at the beginning, it is a simple logistic fact that the number of deviancies which come to a community's attention are limited by the kinds of equipment it uses to detect and handle them, and to that extent the rate of deviation found in a community is at least in part a function of the size and complexity of its social control apparatus. A community's capacity for handling deviance, let us say, can be roughly estimated by counting its prison cells and hospital beds, its policemen and psychiatrists, its courts and clinics. Most communities, it would seem, operate with the expectation that a relatively constant number of control agents is necessary to cope with a relatively constant number of offenders. The amount of men, money, and material assigned by society to "do something" about deviant behavior does not vary appreciably over time, and the implicit logic which governs the community's efforts to man a police force or maintain suitable facilities for the mentally ill seems to be that there is a fairly stable quota of trouble which should be anticipated.

In this sense, the agencies of control often seem to define their job as that of keeping deviance within bounds rather than that of obliterating it altogether. Many judges, for example, assume that severe punishments are a greater deterrent to crime than moderate ones, and so it is important to note that many of them are apt to impose harder penalties when crime seems to be on the increase and more lenient ones when it does not, almost as if the power of the bench were being used to keep the crime rate from getting out of hand.

Erikson was taking issue with what he described as a "dominant strain in sociological thinking" that took for granted that a well-structured society "is somehow designed to prevent deviant behavior from occurring." In both authors, Durkheim and Erikson, there is an undertone that suggests that, with deviancy, as with most social goods, there is the continuing problem of demand exceeding supply. Durkheim invites us to

> imagine a society of saints, a perfect cloister of exemplary individuals.
> Crimes, properly so called, will there be unknown; but faults which
> appear venial to the layman will create there the same scandal that the
> ordinary offense does in ordinary consciousness. If, then, this society has
> the power to judge and punish, it will define these acts as criminal and will
> treat them as such.

Recall Durkheim's comment that there need be no cause for congratulations should the amount of crime drop "too noticeably below the normal level." It would not appear that Durkheim anywhere contemplates the possibility of too much crime. Clearly his theory would have required him to deplore such a development, but the possibility seems never to have occurred to him.

Erikson, writing much later in the twentieth century, contemplates both possibilities. "Deviant persons can be said to supply needed services to society." There is no doubt a tendency for the supply of any needed thing to run short. But he is consistent. There can, he believes, be *too much* of a good thing. Hence "the number of deviant offenders a community *can afford* to recognize is likely to remain stable over time." [My emphasis]

Social scientists are said to be on the lookout for poor fellows getting a bum rap. But here is a theory that clearly implies that there are circumstances in which society will choose *not* to notice behavior that would be otherwise controlled, or disapproved, or even punished.

It appears to me that this is in fact what we in the United States have been doing of late. I proffer the thesis that, over the past generation, since the time Erikson wrote, the amount of deviant behavior in American society has increased beyond the levels the community can "afford to recognize" and that, accordingly, we have been re-defining deviancy so as to exempt much conduct previously stigmatized, and also quietly raising the "normal" level in categories where behavior is now abnormal by any earlier standard. This redefining has evoked fierce resistance from defenders of "old" standards, and accounts for much of the present "cultural war" such as proclaimed by many at the 1992 Republican National Convention.

Let me, then, offer three categories of redefinition in these regards: the *altruistic,* the *opportunistic,* and the *normalizing.*

The first category, the *altruistic,* may be illustrated by the deinstitutionalization movement within the mental health profession that appeared in the 1950s. The second category, the opportunistic, is seen in the interest group rewards derived from the acceptance of "alternative" family structures. The third category, the normalizing, is to be observed in the growing acceptance of unprecedented levels of violent crime.

II

It happens that I was present at the beginning of the deinstitutionalization movement. Early in 1955 Averell Harriman, then the new governor of New York, met with his new commissioner of mental hygiene, Dr. Paul Hoch, who described the development, at one of the state mental hospitals, of a tranquilizer derived from rauwolfia. The medication had been clinically tested and appeared to be an effective treatment for many severely psychotic patients, thus increasing the percentage of patients discharged. Dr. Hoch recommended that it be used system wide; Harriman found the money. That same year Congress created a Joint Commission on Mental Health and Illness whose mission was to formulate "comprehensive and realistic recommendations" in this area, which was then a matter of considerable public concern. Year after year, the population of mental institutions grew. Year after year, new facilities had to be built. Never mind the complexities: population growth and such like matters. There was a general unease. Durkheim's constant continued to be exceeded. (In *Spanning the Century: The Life of W. Averell Harriman,* Rudy Abramson writes: "New York's mental hospitals in 1955 were overflowing warehouses, and new patients were being admitted faster than space could be found for them. When he was inaugurated, 94,000 New Yorkers were confined to state hospitals. Admissions were running at more than 2,500 a year and rising, making the Department of Mental Hygiene the fastest-growing, most-expensive, most-hopeless department of state government.")

The discovery of tranquilizers was adventitious. Physicians were seeking cures for disorders that were just beginning to be understood. Even a limited success made it possible to believe that the incidence of this particular range of disorders, which had seemingly required persons to be confined against their will or even awareness, could be greatly reduced. The Congressional Commission submitted its report in 1961; it proposed a nationwide program of deinstitutionalization. . . .

The mental hospitals emptied out. At the time Governor Harriman met with Dr. Hoch in 1955, there were 93,314 adult residents of mental institutions maintained by New York State. As of August 1992, there were 11,363. This occurred across the nation. However, the number of community mental health centers never came near the [federal government's] goal of 2,000 [centers around the country by 1980]. Only some 482 received federal construction funds between 1963 and 1980. The next year, 1981, the program was folded into the Alcohol and Other Drug Abuse block grant and disappeared from view. Even when centers were built, the results were hardly as hoped for. David F. Musto of Yale writes that the planners had bet on improving national mental health "by improving the quality of general community life through expert knowledge, not merely by more effective treatment of the already ill." There was no such knowledge.

However, worse luck, the belief that there *was* such knowledge took hold within sectors of the profession that saw institutionalization as an unacceptable mode of social control. These activists subscribed to a re-defining mode of their own. Mental patients were said to have been "labeled," and were not to be drugged. Musto says of the battles that followed that they were "so intense and dramatic precisely because both sides shared the fantasy of an omnipotent and omniscient mental health technology which could thoroughly reform society; the prize seemed eminently worth fighting for."

But even as the federal government turned to other matters, the mental institutions continued to release inmates. Professor Fred Siegel of Cooper Union observes: "In the great wave of moral deregulation that began in the mid-1960s, the poor and the insane were freed from the fetters of middle-class mores." They might henceforth sleep in doorways as often as they chose. The problem of the homeless appeared, characteristically defined as persons who lacked "affordable housing."

The *altruistic* mode of redefinition is just that. There is no reason to believe that there was any real increase in mental illness at the time deinstitutionalization began. Yet there was such a perception, and this enabled good people to try to do good, however unavailing in the end.

III

Our second, or *opportunistic* mode of re-definition, reveals at most a nominal intent to do good. The true object is to do well, a long-established motivation among mortals. In this pattern, a growth in deviancy makes possible a transfer of resources, including prestige, to those who control the deviant population. This control would be jeopardized if any serious effort were made to reduce the deviancy in question. This leads to assorted strategies for re-defining the behavior in question as not all that deviant, really.

In the years from 1963 to 1965, the Policy Planning Staff of the U.S. Department of Labor picked up the first tremors of what Samuel H. Preston, in the 1984 Presidential Address to the Population Association of America, would call "the earthquake that shuddered through the American family in the past twenty years." *The New York Times* recently provided a succinct accounting of Preston's point:

> Thirty years ago, 1 in every 40 white children was born to an unmarried mother; today it is 1 in 5, according to Federal data. Among blacks 2 of 3 children are born to an unmarried mother; 30 years ago the figure was 1 in 5. . . .

Richard T. Gill writes of "an accumulation of data showing that intact biological parent families offer children very large advantages compared to any other family or non-family structure one can imagine." Correspondingly, the disadvantages associated with single-parent families spill over into other areas of social policy that now attract great public concern. Leroy L. Schwartz, M.D., and Mark W. Stanton argue that the real quest regarding a government-run health system such as that of Canada or Germany is whether it would work "in a country that has social problems that countries like Canada and Germany don't share to the same extent." Health problems reflect ways of living. The way of life associated with "such social pathologies as the breakdown of the family structure" lead to medical pathologies. Schwartz and Stanton conclude: "The United States is paying dearly for its social and behavioral problems," for they have now become medical problems as well. . . .

For a period there was some speculation that, if family structure got bad enough, this mode of deviancy would have less punishing effects on children. In 1991 Deborah A. Dawson, of the National Institutes of Health, examined the thesis that "the psychological effects of divorce and single parenthood on children were strongly influenced by a sense of shame in being 'different' from the norm." If this were so, the effect should have fallen off in the 1980s, when being from a single-parent home became much more common. It did not. "The problems associated with task overload among single parents are more constant in nature," Dawson wrote, adding that since the adverse effects had not diminished, they were "not based on stigmatization but rather on inherent problems in alternative family structures"—*alternative* here meaning other than two-parent families. We should take note of such candor. Writing in the *Journal of Marriage and the Family* in 1989, Sara McLanahan and Karen Booth noted: "Whereas a decade ago the prevailing view was that single motherhood had no harmful effects on children, recent research is less optimistic."

The year 1990 saw more of this lesson. In a paper prepared for the Progressive Policy Institute, Elaine Ciulla Kamarck and William A. Galston wrote that "if the economic effects of family breakdown are clear, the psychological effects are just now coming into focus." They cite Karl Zinsmeister:

> There is a mountain of scientific evidence showing that when families disintegrate children often end up with intellectual, physical, and emotional scars that persist for life. . . .We talk about the drug crisis, the education crisis, and the problems of teen pregnancy and juvenile crime. But all these ills trace back predominantly to one source: broken families.

As for juvenile crime, they cite Douglas Smith and G. Roger Jarjoura: "Neighborhoods with larger percentages of youth (those aged 12 to 20) and areas with higher percentages of single-parent households also have higher rates of violent crime." They add: "The relationship is so strong that controlling for family configuration erases the relationship between race and crime and between low income and crime. This conclusion shows up time and time again in the literature; poverty is far from the sole determinant of crime." But the large point is avoided. In a 1992 essay "The Expert's Story of Marriage," Barbara Dafoe Whitehead examined "the story of marriage as it is conveyed in today's high school and college textbooks." Nothing amiss in this tale.

It goes like this:

The life course is full of exciting options. The lifestyle options available to individuals seeking a fulfilling personal relationship include living a heterosexual, homosexual, or bisexual single lifestyle; living in a commune; having a group marriage; being a single parent; or living together. Marriage is yet another lifestyle choice. However, before choosing marriage, individuals should weigh its costs and benefits against other lifestyle options and should consider what they want to get out of their intimate relationships. Even within marriage, different people want different things. For example, some people marry for companionship, some marry in order to have children, some marry for emotional and financial security. Though marriage can offer a rewarding path to personal growth, it is important to remember that it cannot provide a secure or permanent status. Many people will make the decision between marriage and singlehood many times throughout their life.

Divorce represents part of the normal family life cycle. It should not be viewed as either deviant or tragic, as it has been in the past. Rather, it establishes a process for "uncoupling" and thereby serves as the foundation for individual renewal and "new beginnings."

. . . What is going on here is simply that a large increase in what once was seen as deviancy has provided opportunity to a wide spectrum of interest groups that benefit from re-defining the problem as essentially normal and doing little to reduce it.

IV

Our *normalizing* category most directly corresponds to Erikson's proposition that "the number of deviant offenders a community can afford to recognize is likely to remain stable over time." Here we are dealing with the popular psychological notion of "denial." In 1965, having reached the conclusion that there would be a dramatic increase in single-parent families, I reached the further conclusion that this would in turn lead to a dramatic increase in crime. In an article in *America,* I wrote:

From the wild Irish slums of the 19th century Eastern seaboard to the riot-torn suburbs of Los Angeles, there is one unmistakable lesson in American history: a community that allows a large number of young men to grow up in broken families, dominated by women, never acquiring any stable relationship to male authority, never acquiring any set of rational expectations about the future—that community asks for and gets chaos. Crime, violence, unrest, unrestrained lashing out at the whole social structure—that is not only to be expected; it is very near to inevitable.

The inevitable, as we now know, has come to pass, but here again our response is curiously passive. Crime is a more or less continuous subject of political pronouncement, and from time to time it will be at or near the top of opinion polls as a matter of public concern. But it never gets much further than that. In the words spoken from the bench, Judge Edwin Torres of the New York State Supreme Court, Twelfth Judicial District, described how "the slaughter of the innocent marches unabated: subway riders, bodega owners, cab drivers, babies; in laundromats, at cash machines, on elevators, in hallways." In personal communication, he writes: "This numbness, this near narcoleptic state can diminish the human condition to the level of combat infantrymen, who, in protracted campaigns, can eat their battlefield rations seated on the bodies of the fallen, friend and foe alike. A society that loses its sense of outrage is doomed to extinction." There is no expectation that this will change, nor any efficacious public insistence that it do so. The crime level has been *normalized*.

Consider the St. Valentine's Day Massacre. In 1929 in Chicago during Prohibition, four gangsters killed seven gangsters on February 14. The nation was shocked. The event became legend. It merits not one but two entries in the *World Book Encyclopedia*. I leave it to others to judge, but it would appear that the society in the 1920s was simply not willing to put up with this degree of deviancy. In the end, the Constitution was amended, and Prohibition, which lay behind so much gangster violence, ended.

In recent years, again in the context of illegal traffic in controlled substances, this form of murder has returned. But it has done so at a level that induces denial. James Q. Wilson comments that Los Angeles has the equivalent of a St. Valentine's Day Massacre every weekend. Even the most ghastly re-enactments of such human slaughter produce only moderate responses. On the morning after the close of the Democratic National Convention in New York City in July, there was such an account in the second section of the *New York Times*. It was not a big story; bottom of the page, but with a heading that got your attention. "3 Slain in Bronx Apartment, but a Baby is Saved." A subhead continued: "A mother's last act was to hide her little girl under the bed." The article described a drug execution; the now-routine blindfolds made from duct tape; a man and a woman and a teenager involved. "Each had been shot once in the head." The police had found them a day later. They also found, under a bed, a three-month-old baby, dehydrated but alive. A lieutenant remarked of the mother, "In her last dying act she protected her baby. She probably knew she was going to die, so she stuffed the baby where she knew it would be safe." But the matter was left there. The police would do their best. But the event passed quickly; forgotten by the next day, it will never make *World Book*.

Nor is it likely that any great heed will be paid to an uncanny reenactment of the Prohibition drama a few months later, also in the Bronx. The *Times* story, page B3, reported:

9 MEN POSING AS POLICE
ARE INDICTED IN 3 MURDERS
Drug Dealers Were Kidnapped for Ransom

The *Daily News* story, same day, page 17, made it *four* murders, adding nice details about torture techniques. The gang members posed as federal Drug Enforcement Administration agents, real badges and all. The victims were drug dealers, whose families were uneasy about calling the police. Ransom seems generally to have been set in the $650,000 range. Some paid. Some got it in the back of the head. So it goes.

Yet, violent killings, often random, go on unabated. Peaks continue to attract some notice. But these are peaks above "average" levels that thirty years ago would have been thought epidemic. . . .

A Kai Erikson of the future will surely need to know that the Department of Justice in 1990 found that Americans reported only about 38 percent of all crimes and 48 percent of violent crimes. This, too, can be seen as a means of *normalizing* crime. . . .

V

The hope—if there be such—of this essay has been twofold. It is, first, to suggest that the Durkheim constant, as I put it, is maintained by a dynamic process which adjusts upwards and *downwards*. Liberals have traditionally been alert for upward redefining that does injustice to individuals. Conservatives have been correspondingly sensitive to downward redefining that weakens societal standards. Might it not help if we could all agree that there is a dynamic at work here? It is not revealed truth, nor yet a scientifically derived formula. It is simply a pattern we observe in ourselves. Nor is it rigid. There may once have been an unchanging supply of jail cells which more or less determined the number of prisoners. No longer. We are building new prisons at a prodigious rate. Similarly, the executioner is back. There is something of a competition in Congress to think up new offenses for which the death penalty is deemed the only available deterrent. Possibly also modes of execution, as in "fry the kingpins." Even so, we are getting used to a lot of behavior that is not good for us.

As noted earlier, Durkheim states that there is "nothing desirable" about pain. Surely what he meant was that there is nothing pleasurable. Pain, even so, is an indispensable warning signal. But societies under stress, much like individuals, will turn to pain killers of various kinds that end up concealing real damage. There is surely nothing desirable about *this*. If our analysis wins general acceptance, if, for example, more of us came to share Judge Torres's genuine alarm at "the trivialization of the lunatic crime rate" in his city (and mine), we might surprise ourselves how well we respond to the manifest decline of the American civic order. Might.

■

PART II

Studying Deviance

A ccurate and reliable knowledge about deviance is critically important to many groups of people in society. First, policymakers are very concerned with deviant groups such as the homeless and transient, the chronically mentally ill, high school dropouts, criminal offenders, prostitutes, juvenile delinquents, gang members, runaways, and other members of disadvantaged and disenfranchised populations. These people pose social problems in society that lawmakers and social welfare agencies want to help alleviate. Second, sociologists and other researchers have an interest in deviance based on their goal of understanding human nature, human behavior, and human society. Deviants are a critical group to this enterprise because they reside near the margins of social definition: they help define the boundaries of what is considered acceptable and unacceptable by given groups of people.

Information about deviance in American society can come from one of three primary data sources. Government officials and employees of social service agencies routinely collect information about their clients as they process them. This information includes arrest data that are compiled by the police and published by the FBI (the *Uniform Crime Reports*), census data on various shifting populations (such as the homeless), victim data from helping agencies (such as battered women's shelters), and prosecution data on cases that are tried in the courts. These statistical *precollected data* are then compiled by the various government organizations responsible for collecting them and made available to the public.

Another source of statistical data about deviance is *survey research*. Rather than relying on information the government might collect, sociologists gather their own through large-scale questionnaire surveys. Prominent ones include the National Youth Survey, a self-report questionnaire about delinquent behavior, the annual survey conducted by the National Institute of Drug Abuse on the drug use of high school seniors, and some of the Kinsey surveys about sexual behavior.

A third kind of information, richly descriptive and analytical rather than numerical, comes from sociologists who conduct *participant-observation field research* on deviance. Much like anthropologists who go out to live among native peoples, sociological fieldworkers live among members of deviant groups and become intimately familiar with the nature of their lives. This type of research yields information more deeply based on the subjects' own perspectives, detailing how they see the world, the allure of deviance for them, the problems they encounter, the ways they resolve these, the significant individuals and groups in their lives, and their role among these others. Unlike the other types, participant observation is generally a longitudinal method, entailing years of involvement with subjects. Researchers must gain acceptance by group members, develop meaningful relationships with them, and learn about the deepest core of their thoughts and feelings.

There are many differences among these types of data and among the methods used to gather them. Although the quantitative data, particularly those gathered by survey research, yield information about a broad spectrum of people, they may be fairly shallow and unreliable in nature. It is not likely that people, especially deviants, will fill out a questionnaire and readily disclose information about the covert aspects of their lives. Participant observation, in contrast, cannot reach as many people but obtains more deep and accurate information about research subjects, backed up by the researchers' own observations, to enhance its validity. Survey research is more controlled, as researchers are not influenced by their relationships with research subjects in gathering data about them. Participant observation relies on its insight, attained through the researchers' close relationships with people to get behind false fronts and find out what is really going on. This ability is especially important to studying a topic like deviance, where so much of the behavior is hidden due to its negative social stigma and illicit status. Also critically important is the ability of participant observation to study deviance as it occurs in its natural setting, not via recollections gathered in questionnaire research or from captive populations of prisoners.

The selections in this book are based on participant-observation studies of deviance for two main reasons. First, participation observation is the method of the interactionist perspective, because it offers direct access to the way in which definitions and laws are socially constructed, the way in which people's actions are influenced by their associates, and the way in which people's identities are affected by the deviant labels cast on them. Second, these studies offer a deeper view of people's feelings, experiences, motivations, and social psychological states, giving a richer and more vivid portrayal of deviance than charts of numbers. We begin by discussing how participant-observation studies of deviance are conducted. In the following two selections, "Researching Dealers and Smugglers," by Patricia Adler, and "Field Work and Membership with the Hare Krishna," by Burke Rochford, we get a glimpse of what it is like to carry out research with deviant groups. Both of these natural histories carefully explain the process used by these field researchers, the relationships they formed with setting members, and the feelings the researchers experienced. We see how putting themselves inside a deviant world can profoundly affect researchers as well as cause them serious potential dangers, but we also see that only from this vantage point can researchers fully comprehend the forces at play in deviant worlds. Both articles increase understanding of some of the roles, concerns, and problems and issues in field research.

4

Researching Dealers and Smugglers

PATRICIA A. ADLER

I strongly believe that investigative field research (Douglas 1976), with emphasis on direct personal observation, interaction, and experience, is the only way to acquire accurate knowledge about deviant behavior. Investigative techniques are especially necessary for studying groups such as drug dealers and smugglers because the highly illegal nature of their occupation makes them secretive, deceitful, mistrustful, and paranoid. To insulate themselves from the straight world, they construct multiple false fronts, offer lies and misinformation, and withdraw into their group. In fact, detailed, scientific information about upper-level drug dealers and smugglers is lacking precisely because of the difficulty sociological researchers have had in penetrating into their midst. As a result, the only way I could possibly get close enough to these individuals to discover what they were doing and to understand their world from their perspectives (Blumer 1969) was to take a membership role in the setting. While my different values and goals precluded my becoming converted to complete membership in the subculture, and my fears presented my ever becoming "actively" involved in their trafficking activities, I was able to assume a "peripheral" membership role (Adler and Adler 1987). I became a member of the dealers' and smugglers' social world and participated in their daily activities on that basis. In this chapter, I discuss how I gained access to this group, established research relations with members, and how personally involved I became in their activities.

GETTING IN

When I moved to Southwest County [California] in the summer of 1974, I had no idea that I would soon be swept up in a subculture of vast drug trafficking and unending partying, mixed with occasional cloak-and-dagger subterfuge. I had moved to California with my husband, Peter, to attend graduate school in sociology. We rented a condominium townhouse near the beach and started taking classes in the fall. We had always felt that socializing exclusively with academicians left us nowhere to escape from our work, so we tried to meet people in the nearby community. One of the first friends we made was our closest neighbor, a fellow in his late twenties with a tall, hulking frame and gentle expression. Dave, as he introduced himself, was always dressed rather casually, if not sloppily, in T-shirts and jeans. He spent most of

From: Patricia A. Adler, *Wheeling and Dealing* (New York: Columbia University Press, 1985). © Columbia University Press. Reprinted by permission of the publisher.

his time hanging out or walking on the beach with a variety of friends who visited his house, and taking care of his two young boys, who lived alternately with him and his estranged wife. He also went out of town a lot. We started spending much of our free time over at his house, talking, playing board games late into the night, and smoking marijuana together. We were glad to find someone from whom we could buy marijuana in this new place, since we did not know too many people. He also began treating us to a fairly regular supply of cocaine, which was a thrill because this was a drug we could rarely afford on our student budgets. We noticed right away, however, that there was something unusual about his use and knowledge of drugs: while he always had a plentiful supply and was fairly expert about marijuana and cocaine, when we tried to buy a small bag of marijuana from him he had little idea of the going price. This incongruity piqued our curiosity and raised suspicion. We wondered if he might be dealing in larger quantities. Keeping our suspicions to ourselves, we began observing Dave's activities a little more closely. Most of his friends were in their late twenties and early thirties and, judging by their lifestyles and automobiles, rather wealthy. They came and left his house at all hours, occasionally extending their parties through the night and the next day into the following night. Yet throughout this time we never saw Dave or any of his friends engage in any activity that resembled a legitimate job. In most places this might have evoked community suspicion, but few of the people we encountered in Southwest County seemed to hold traditionally structured jobs. Dave, in fact, had no visible means of financial support. When we asked him what he did for a living, he said something vague about being a real estate speculator, and we let it go at that. We never voiced our suspicions directly since he chose not to broach the subject with us.

We did discuss the subject with our mentor, Jack Douglas, however. He was excited by the prospect that we might be living among a group of big dealers, and urged us to follow our instincts and develop leads into the group. He know that the local area was rife with drug trafficking, since he had begun a life history case study of two drug dealers with another graduate student several years previously. That earlier study was aborted when the graduate student quit school, but Jack still had many hours of taped interviews he had conducted with them, as well as an interview that he had done with an undergraduate student who had known the two dealers independently, to serve as a cross-check on their accounts. He therefore encouraged us to become friendlier with Dave and his friends. We decided that if anything did develop out of our observations of Dave, it might make a nice paper for a field methods class or independent study.

Our interests and background made us well suited to study drug dealing. First, we had already done research in the field of drugs. As undergraduates at Washington University we had participated in a nationally funded project on urban heroin use (see Cummins et al. 1972). Our role in the study involved using fieldwork techniques to investigate the extent of heroin use and distribution in St. Louis. In talking with heroin users, dealers, and rehabilitation personnel, we acquired a base of knowledge about the drug world and the sub-culture of drug trafficking. Second, we had a generally open view toward soft drug use, considering moderate consumption of marijuana and cocaine to be generally nondeviant. This outlook was partially etched by our 1960s-formed attitudes, as we had first been introduced to drug use in an environment of communal friendship, sharing, and counterculture ideology. It also partially reflected the widespread acceptance accorded to marijuana and cocaine use in the surrounding local culture. Third, our age (mid-twenties at the start of the study) and general appearance gave us compatibility with most of the people we were observing.

We thus watched Dave and continued to develop our friendship with him. We also watched his friends and got to know a few of his more regular visitors. We continued to build friendly relations by doing, quite naturally, what Becker (1963), Polsky (1969), and Douglas (1972) had advocated for the early stages of field research: we gave them a chance to know us and form judgments about our trustworthiness by jointly pursuing those interests and activities which we had in common.

Then one day something happened which forced a breakthrough in the research. Dave had two guys visiting him from out of town and, after snorting quite a bit of cocaine, they turned their conversation to a trip they had just made from Mexico, where they piloted a load of marijuana back across the border in a small plane. Dave made a few efforts to shift the conversation to another subject, telling them to "button their lips," but they apparently thought that he was joking. They thought that anybody as close to Dave as we seemed to be undoubtedly knew the nature of his business. They made further allusions to his involvement in the operation and discussed the out-come of the sale. We could feel the wave of tension and awkwardness from Dave when this conversation began, as he looked toward us to see if we understood the implications of what was being said, but then he just shrugged it off as done. Later, after the two guys left, he discussed with us what hap-pened. He admitted to us that he was a member of a smuggling crew and a major marijuana dealer on the side. He said that he knew he could trust us, but that it was his practice to say as little as possible to outsiders about his activities. This inadvertent slip, and Dave's subsequent opening up, were highly significant in forging our entry into Southwest County's drug world. From then on he was open in discussing the nature of his dealing and smug-gling activities with us.

He was, it turned out, a member of a smuggling crew that was importing a ton of marijuana weekly and 40 kilos of cocaine every few months. During that first winter and spring, we observed Dave at work and also got to know the other members of his crew, including Ben, the smuggler himself. Ben was also very tall and broad shouldered, but his long black hair, now flecked with gray, bespoke his earlier membership in the hippie subculture. A large physical stature, we observed, was common to most of the male participants involved in this drug community. The women also had a unifying physical trait: they were extremely attractive and stylishly dressed. This included Dave's ex-wife, Jean, with whom he reconciled during the spring. We therefore became friendly with Jean and through her met a number of women ("dope chicks") who hung around the dealers and smugglers. As we continued to gain the friendship of Dave and Jean's associates we were progressively admitted into their inner circle and apprised of each person's dealing or smuggling role.

Once we realized the scope of Ben's and his associates' activities, we saw the enormous research potential in studying them. This scene was different from any analysis of drug trafficking that we had read in the sociological literature because of the amounts they were dealing and the fact that they were importing it themselves. We decided that, if it was at all possible, we would capitalize on this situation, to "opportunistically" (Riemer 1977) take advantage of our prior expertise and of the knowledge, entree, and rapport we had already developed with several key people in this setting. We therefore discussed the idea of doing a study of the general subculture with Dave and several of his closest friends (now becoming our friends). We assured them of the anonymity, confidentiality, and innocuousness of our work. They were happy to reciprocate our friendship by being of help to our professional careers. In fact, they basked in the subsequent attention we gave their lives.

We began by turning first Dave, then others, into key informants and collecting their life histories in detail. We conducted a series of taped, depth interviews with an unstructured, open-ended format. We questioned them about such topics as their backgrounds, their recruitment into the occupation, the stages of their dealing careers, their relations with others, their motivations, their lifestyle, and their general impressions about the community as a whole.

We continued to do taped interviews with key informants for the next six years until 1980, when we moved away from the area. After that, we occasionally did follow-up interviews when we returned for vacation visits. These later interviews focused on recording the continuing unfolding of events and included detailed probing into specific conceptual areas, such as dealing networks, types of dealers, secrecy, trust, paranoia, reputation, the law, occupational mobility, and occupational stratification. The number of taped interviews we did with each key informant varied, ranging between 10 and 30 hours of discussion.

Our relationship with Dave and the others thus took on an added dimension—the research relationship. As Douglas (1976), Henslin (1972), and Wax (1952) have noted, research relationships involve some form of mutual exchange. In our case, we offered everything that friendship could entail. We did routine favors for them in the course of our everyday lives, offered them insights and advice about their lives from the perspective of our more respectable position, wrote letters on their behalf to the authorities when they got in trouble, testified as character witnesses at their non-drug-related trials, and loaned them money when they were down and out. When Dave was arrested and brought to trial for check-kiting, we helped Jean organize his defense and raise the money to pay his fines. We spelled her in taking care of the children so that she could work on his behalf. When he was eventually sent to the state prison we maintained close ties with her and discussed our mutual efforts to buoy Dave up and secure his release. We also visited him in jail. During Dave's incarceration, however, Jean was courted by an old boy-friend and gave up her reconciliation with Dave. This proved to be another significant turning point in our research because, desperate for money, Jean looked up Dave's old dealing connections and went into the business herself. She did not stay with these marijuana dealers and smugglers for long, but soon moved into the cocaine business. Over the next several years her experiences in the world of cocaine dealing brought us into contact with a different group of people. While these people knew Dave and his associates (this was very common in the Southwest County dealing and smuggling community), they did not deal with them directly. We were thus able to gain access to a much wider and more diverse range of subjects than we would have had she not branched out on her own.

Dave's eventual release from prison three months later brought our involvement in the research to an even deeper level. He was broke and had nowhere to go. When he showed up on our doorstep, we took him in. We offered to let him stay with us until he was back on his feet again and could afford a place of his own. He lived with us for seven months, intimately shar-ing his daily experiences with us. During this time we witnessed, firsthand, his transformation from a scared ex-con who would never break the law again to a hard-working legitimate employee who only dealt to get money for his chil-dren's Christmas presents, to a full-time dealer with no pretensions at legiti-mate work. Both his process of changing attitudes and the community's grad-ual reacceptance of him proved very revealing.

We socialized with Dave, Jean, and other members of Southwest County's dealing and smuggling community on a near-daily basis, especially during the first four years of the research (before we had a child). We worked in their legitimate businesses, vacationed together, attended their weddings, and cared for their children. Throughout their relationship with us, several participants became co-opted to the researcher's perspective[1] and actively sought out instances of behavior which filled holes in the conceptualizations we were developing. Dave, for one, became so intrigued by our conceptual dilemmas that he undertook a "natural experiment" entirely on his own, offering an unlimited supply of drugs to a lower-level dealer to see if he could work up to higher levels of dealing, and what factors would enhance or impinge upon his upward mobility.

In addition to helping us directly through their own experiences, our key informants aided us in widening our circle of contacts. For instance, they let us know when someone in whom we might be interested was planning on dropping by, vouching for our trustworthiness and reliability as friends who could be included in business conversations. Several times we were even awakened in the night by phone calls informing us that someone had dropped by for a visit, should we want to "casually" drop over too. We rubbed the sleep from our eyes, dressed, and walked or drove over, feeling like sleuths out of a television series. We thus were able to snowball, through the active efforts of our key informants,[2] into an expanded study population. This was supplemented by our own efforts to cast a research net and befriend other dealers, moving from contact to contact slowly and carefully through the domino effect.

THE COVERT ROLE

The highly illegal nature of dealing in illicit drugs and dealers' and smugglers' general level of suspicion made the adoption of an overt research role highly sensitive and problematic. In discussing this issue with our key informants, they all agreed that we should be extremely discreet (for both our sakes and theirs). We carefully approached new individuals before we admitted that we were studying them. With many of these people, then, we took a covert posture in the research setting. As nonparticipants in the business activities which bound members together into the group, it was difficult to become fully accepted as peers. We therefore tried to establish some sort of peripheral, social member-ship in the general crowd, where we could be accepted as "wise" (Goffman 1963) individuals and granted a courtesy membership. This seemed an attain-able goal, since we had begun our involvement by forming such relationships with our key informants. By being introduced to others in this wise rather than overt role, we were able to interact with people who would otherwise have shied away from us. Adopting a courtesy membership caused us to bear a cour-tesy stigma,[3] however, and we suffered since we, at times, had to disguise the nature of our research from both lay outsiders and academicians.

In our overt posture we showed interest in dealers' and smugglers' activi-ties, encouraged them to talk about themselves (within limits, so as to avoid acting like narcs), and ran home to write field notes. This role offered us the advantage of gaining access to unapproachable people while avoiding researcher effects, but it prevented us from asking some necessary, probing questions and from tape recording conversations.[4] We therefore sought, at all times, to build toward a conversion to the overt role. We did this by working to develop their trust.

DEVELOPING TRUST

Like achieving entrée, the process of developing trust with member of unor-ganized deviant groups can be slow and difficult. In the absence of a formal structure separating members from outsiders, each individual must form his or her own judgment about whether new persons can be admitted to their confi-dence. No gatekeeper existed to smooth our path to being trusted, although our key informants acted in this role whenever they could by providing intro-ductions and references. In addition, the unorganized nature of this group meant that we met people at different times and were constantly at different levels in our developing relationships with them. We were thus trusted more by some people than by others, in part because of their greater familiarity with us. But as Douglas (1976) has noted, just because someone knew us or even liked us did not automatically guarantee that they would trust us.

We actively tried to cultivate the trust of our respondents by tying them to us with favors. Small things, like offering the use of our phone, were followed with bigger favors, like offering the use of our car, and finally really meaningful favors, like offering the use of our home. Here we often trod a thin line, trying to ensure our personal safety while putting ourselves in enough of a risk position, along with our research subjects, so that they would trust us. While we were able to build a "web of trust" (Douglas 1976) with some members, we found that trust, in large part, was not a simple status to attain in the drug world. Johnson (1975) has pointed out that trust is not a one-time phenomenon, but an ongoing developmental process. From my experiences in this research I would add that it cannot be simply assumed to be a one-way process either, for it can be diminished, withdrawn, reinstated to varying degrees, and re-questioned at any point. Carey (1972) and Douglas (1972) have remarked on this waxing and waning process, but it was especially pronounced for us because our subjects used large amounts of cocaine over an extended period of time. This tended to make them alternately warm and cold to us. We thus lived through a series of ups and downs with the people we were trying to cultivate as research informants.

THE OVERT ROLE

After this initial covert phase, we began to feel that some new people trusted us. We tried to intuitively feel when the time was right to approach them and go overt. We used two means of approaching people to inform them that we were involved in a study of dealing and smuggling: direct and indirect. In some cases our key informants approached their friends or connections and, after vouching for our absolute trustworthiness, convinced these associates to talk to us. In other instances, we approached people directly, asking for their help with our project. We worked our way through a progression with these secondary contacts, first discussing the dealing scene overtly and later moving to taped life history interviews. Some people reacted well to us, but others responded skittishly, making appointments to do taped interviews only to break them as the day drew near, and going through fluctuating stages of being honest with us or putting up fronts about their dealing activities. This varied, for some, with their degree of active involvement in the business. During the times when they had quit dealing, they would tell us about their present and past activities, but when they became actively involved again, they would hide it from us.

This progression of covert to overt roles generated a number of tactical difficulties. The first was the problem of *coming on too fast* and blowing it. Early in the research we had a dealer's old lady (we thought) all set up for the direct approach. We knew many dealers in common and had discussed many things tangential to dealing with her without actually mentioning the subject. When we asked her to do a taped interview of her bohemian lifestyle, she agreed without hesitation. When the interview began, though, and she found out why we were interested in her, she balked, gave us a lot of incoherent jumble, and ended the session as quickly as possible. Even though she lived only three houses away we never saw her again. We tried to move more slowly after that.

A second problem involved simultaneously *juggling our overt and covert roles* with different people. This created the danger of getting our cover blown with people who did not know about our research (Henslin 1972). It was very confusing to separate the people who knew about our study from those who did not, especially in the minds of our informants. They would make occasional veiled references in front of people, especially when loosened by intoxicants, that made us extremely uncomfortable. We also frequently worried that our snooping would someday be mistaken for police tactics. Fortunately, this never happened.

CROSS-CHECKING

The hidden and conflictual nature of the drug dealing world made me feel the need for extreme certainty about the reliability of my data. I therefore based all my conclusions on independent sources and accounts that we carefully verified. First, we tested information against our own common sense and general knowledge of the scene. We adopted a hard-nosed attitude of suspicion, assuming people were up to more than they would originally admit. We kept our attention especially riveted on "reformed" dealers and smugglers who were living better than they could outwardly afford, and were thereby able to penetrate their public fronts.

Second, we checked out information against a variety of reliable sources. Our own observations of the scene formed a primary reliable source, since we were involved with many of the principals on a daily basis and knew exactly what they were doing. Having Dave live with us was a particular advantage because we could contrast his statements to us with what we could clearly see was happening. Even after he moved out, we knew him so well that we could generally tell when he was lying to us or, more commonly, fooling himself with optimistic dreams. We also observed other dealers' and smugglers' evasions and misperceptions about themselves and their activities. These usually occurred when they broke their own rules by selling to people they did not know, or when they comingled other people's money with their own. We also cross-checked our data against independent, alternative accounts. We were lucky, for this purpose, that Jean got reinvolved in the drug world. By interviewing her, we gained additional insight into Dave's past, his early dealing and smuggling activities, and his ongoing involvement from another person's perspective. Jean (and her connections) also talked to us about Dave's associates, thereby helping us to validate or disprove their statements. We even used this pincer effect to verify information about people we had never directly interviewed. This occurred, for instance, with the tapes that Jack Douglas gave us from his earlier study. After doing our first round of taped interviews with Dave, we discovered that he knew the dealers Jack had interviewed. We were excited by the prospect of finding out what had happened to these people and if their earlier stories checked out. We therefore sent Dave to do some investigative work. Through some mutual friends he got back in touch with them and found out what they had been doing for the past several years.

Finally, wherever possible, we checked out accounts against hard facts: newspaper and magazine reports; arrest records; material possessions; and visible evidence. Throughout the research, we used all these cross-checking measures to evaluate the veracity of new information and to prod our respondents to be more accurate (by abandoning both their lies and their self-deceptions).[5]

After about four years of near-daily participant observation, we began to diminish our involvement in the research. This occurred gradually, as first pregnancy and then a child hindered our ability to follow the scene as intensely and spontaneously as we had before. In addition, after having a child, we were less willing to incur as many risks as we had before; we no longer felt free to make decisions based solely on our own welfare. We thus pulled back from what many have referred to as the "difficult hours and dangerous situations" inevitably present in field research on deviants (see Becker 1963; Carey 1972; Douglas 1972). We did, however, actively maintain close ties with research informants (those with whom we had gone overt), seeing them regularly and periodically doing follow-up interviews.

PROBLEMS AND ISSUES

Reflecting on the research process, I have isolated a number of issues which I believe merit additional discussion. These are rooted in experiences which have the potential for greater generic applicability.

The first is the *effect of drugs on the data-gathering process*. Carey (1972) has elaborated on some of the problems he encountered when trying to interview respondents who used amphetamines, while Wax (1952, 1957) has mentioned the difficulty of trying to record field notes while drinking sake. I found that marijuana and cocaine had nearly opposite effects from each other. The latter helped the interview process, while the former hindered it. Our attempts to interview respondents who were stoned on marijuana were unproductive for a number of reasons. The primary obstacle was the effects of the drug. Often, people became confused, sleepy, or involved in eating to varying degrees. This distracted them from our purpose. At times, people even simulated overreactions to marijuana to hide behind the drug's supposed disorienting influence and thereby avoid divulging information. Cocaine, in contrast, proved to be a research aid. The drug's warming and sociable influence opened people up, diminished their inhibitions, and generally increased their enthusiasm for both the interview experience and us.

A second problem I encountered involved *assuming risks while doing research.* As I noted earlier, dangerous situations are often generic to research on deviant behavior. We were most afraid of the people we studied. As Carey (1972), Henslin (1972), and Whyte (1955) have stated, members of deviant groups can become hostile toward a researcher if they think that they are being treated wrongfully. This could have happened at any time from a simple occurrence, such as a misunderstanding, or from something more serious, such as our covert posture being exposed. Because of the inordinate amount of drugs they consumed, drug dealers and smugglers were particularly volatile, capable of becoming malicious toward each other or us with little warning. They were also likely to behave erratically owing to the great risks they faced from the police and other dealers. These factors made them moody, and they vacillated between trusting us and being suspicious of us.

At various times we also had to protect our research tapes. We encountered several threats to our collection of taped interviews from people who had granted us these interviews. This made us anxious, since we had taken great pains to acquire these tapes and felt strongly about maintaining confidences entrusted to us by our informants. When threatened, we became extremely frightened and shifted the tapes between different hiding places. We even ventured forth one rainy night with our tapes packed in a suitcase to meet a person who was uninvolved in the research at a secret rendezvous so that he could guard the tapes for us.

We were fearful, lastly, of the police. We often worried about local police or drug agents discovering the nature of our study and confiscating or subpoenaing our tapes and field notes. Sociologists have no privileged relationship with their subjects that would enable us legally to withhold evidence from the authorities should they subpoena it.[6] For this reason we studiously avoided any publicity about the research, even holding back on publishing articles in scholarly journals until we were nearly ready to move out of the setting. The closest we came to being publicly exposed as drug researchers came when a former sociology graduate student (turned dealer, we had heard from inside sources) was arrested at the scene of a cocaine deal. His lawyer wanted us to testify about the dangers of doing drug-related research, since he was using his research status as his defense. Fortunately, the crisis was averted when his lawyer succeeded in suppressing evidence and had the case dismissed before the trial was to have begun. Had we been exposed, however, our respondents would have acquired guilt by association through their friendship with us.

Our fear of the police went beyond our concern for protecting our research subjects, however. We risked the danger of arrest ourselves through our own violations of the law. Many sociologists (Becker 1963; Carey 1972; Polsky 1969; Whyte 1955) have remarked that field researchers studying deviance must inevitably break the law in order to acquire valid participant observation data. This occurs in its most innocuous form from having "guilty knowledge": information about crimes that are committed. Being aware of major dealing and smuggling operations made us an accessory to their commission, since we failed to notify the police. We broke the law, secondly, through our "guilty observations," by being present at the scene of a crime and witnessing its occurrence (see also Carey 1972). We knew it was possible to get caught in a bust involving others, yet buying and selling was so pervasive that to leave every time it occurred would have been unnatural and highly suspicious. Sometimes drug transactions even occurred in our home, especially when Dave was living there, but we finally had to put a stop to that because we could not handle the anxiety. Lastly, we broke the law through our "guilty actions," by taking part in illegal behavior ourselves. Although we never dealt drugs (we were too scared to be seriously tempted), we consumed drugs and possessed them in small quantities. Quite frankly, it would have been impossible for a nonuser to have gained access to this group to gather the data presented here. This was the minimum involvement necessary to obtain even the courtesy membership we achieved. Some kind of illegal action was also found to be a necessary or helpful component of the research by Becker (1973), Carey (1972), Johnson (1975), Polsky (1969), and Whyte (1955).

Another methodological issue arose from the *cultural clash between our research subjects and ourselves*. While other sociologists have alluded to these kinds of differences (Humphreys 1970, Whyte 1955), few have discussed how the research relationships affected them. Relationships with research subjects are unique because they involve a bond of intimacy between persons who might not ordinarily associate together, or who might otherwise be no more than casual friends. When fieldworkers undertake a major project, they commit themselves to maintaining a long-term relationship with the people they study. However, as researchers try to get depth involvement, they are apt to come across fundamental differences in character, values, and attitudes between their subjects and themselves. In our case, we were most strongly confronted by differences in present versus future orientations, a desire for risk versus security, and feelings of spontaneity versus self-discipline. These differences often caused us great frustration. We repeatedly saw dealers act irrationally, setting themselves up for failure. We wrestled with our desire to point out their patterns of foolhardy behavior and offer advice, feeling competing pulls between our detached, observer role which advised us not to influence the natural setting, and our involved, participant role which called for us to offer friendly help whenever possible.[7]

Each time these differences struck us anew, we gained deeper insights into our core, existential selves. We suspended our own taken-for-granted feelings and were able to reflect on our culturally formed attitudes, character, and life choices from the perspective of the other. When comparing how we might act in situations faced by our respondents, we realized where our deepest priorities lay. These revelations had the effect of changing our self-conceptions: whereas we, at one time, had thought of ourselves as what Rosenbaum (1981) has called "the hippest of non-addicts" (in this case nondealers), we were suddenly faced with being the straightest members of the crowd. Not only did we not deal, but we had a stable, long-lasting marriage and family life, and needed the security of a reliable monthly paycheck. Self-insights thus emerged as one of the unexpected outcomes of field research with members of a different cultural group.

The final issue I will discuss involved the various *ethical problems* which arose during this research. Many fieldworkers have encountered ethical dilemmas or pangs of guilt during the course of their research experiences (Carey 1972; Douglas 1976; Humphreys 1970; Johnson 1975; Klockars 1977, 1979; Rochford 1985). The researchers' role in the field makes this necessary because they can never fully align themselves with their subjects while maintaining their identity and personal commitment to the scientific community. Ethical dilemmas, then, are directly related to the amount of deception researchers use in gathering the data, and the degree to which they have accepted such acts as necessary and therefore neutralized them.

Throughout the research, we suffered from the burden of intimacies and confidences. Guarding secrets which had been told to us during taped interviews was not always easy or pleasant. Dealers occasionally revealed things about themselves or others that we had to pretend not to know when interacting with their close associates. This sometimes meant that we had to lie or build elaborate stories to cover for some people. Their fronts therefore became our fronts, and we had to weave our own web of deception to guard their performances. This became especially disturbing during the writing of the research report, as I was torn by conflicts between using details to enrich the data and glossing over description to guard confidences.[8]

Using the covert research role generated feelings of guilt, despite the fact that our key informants deemed it necessary, and thereby condoned it. Their own covert experiences were far more deeply entrenched than ours, being a part of their daily existence with non–drug world members. Despite the universal presence of covert behavior throughout the setting, we still felt a sense of betrayal every time we ran home to write research notes on observations we had made under the guise of innocent participants.

We also felt guilty about our efforts to manipulate people. While these were neither massive nor grave manipulations, they involved courting people to procure information about them. Our aggressively friendly postures were based on hidden ulterior motives: we did favors for people with the clear expectation that they could only pay us back with research assistance. Manipulation bothered us in two ways: immediately after it was done, and over the long run. At first, we felt awkward, phony, almost ashamed of ourselves, although we believed our rationalization that the end justified the means. Over the long run, though, our feelings were different. When friendship became intermingled with research goals, we feared that people would later look back on our actions and feel we were exploiting their friendship merely for the sake of our research project.

The last problem we encountered involved our feelings of whoring for data. At times, we felt that we were being exploited by others, that we were putting more into the relationship than they, that they were taking us for granted or using us. We felt that some people used a double standard in their relationship with us: they were allowed to lie to us, borrow money and not repay it, and take advantage of us, but we were at all times expected to behave honorably. This was undoubtedly an outgrowth of our initial research strategy where we did favors for people and expected little in return. But at times this led to our feeling bad. It made us feel like we were selling ourselves, our sincerity, and usually our true friendship, and not getting treated right in return.

CONCLUSIONS

The aggressive research strategy I employed was vital to this study. I could not just walk up to strangers and start hanging out with them as Liebow (1967) did, or be sponsored to a member of this group by a social service or reform organization as Whyte (1955) was, and expect to be accepted, let alone welcomed. Perhaps such a strategy might have worked with a group that had nothing to hide, but I doubt it. Our modern, pluralistic society is so filled with diverse subcultures whose interests compete or conflict with each other that each subculture has a set of knowledge which is reserved exclusively for insiders. In order to serve and prosper, they do not ordinarily show this side to just anyone. To obtain the kind of depth insight and information I needed, I had to become like the members in certain ways. They dealt only with people they knew and trusted, so I had to become known and trusted before I could reveal my true self and my research interests. Confronted with secrecy, danger, hidden alliances, misrepresentations, and unpredictable changes of intent, I had to use a delicate combination of overt and covert roles. Throughout, my deliberate cultivation of the norm of reciprocal exchange enabled me to trade my friendship for their knowledge, rather than waiting for the highly unlikely event that information would be delivered into my lap. I thus actively built a web of research contacts, used them to obtain highly sensitive data, and carefully checked them out to ensure validity.

Throughout this endeavor I profited greatly from the efforts of my husband, Peter, who served as an equal partner in this team field research project. It would have been impossible for me to examine this social world as an unattached female and not fall prey to sex role stereotyping which excluded women from business dealings. As a couple, our different genders allowed us to relate in different ways to both men and women (see Warren and Rasmussen 1977). We also protected each other when we entered the homes of dangerous characters, buoyed each others' initiative and courage, and kept the conversation going when one of us faltered. Conceptually, we helped each other keep a detached and analytical eye on the setting, provided multiperspectival insights, and corroborated, clarified, or (most revealingly) contradicted each other's observations and conclusions.

Finally, I feel strongly that to ensure accuracy, research on deviant groups must be conducted in the settings where it naturally occurs. As Polsky (1969:115–16) has forcefully asserted:

> This means—there is no getting away from it—the study of career crimi-nals *au natural,* in the field, the study of such criminals as they normally go about their work and play, the study of "uncaught" criminals and the study of others who in the past have been caught but are not caught at the time you study them. . . . Obviously we can no longer afford the conve-nient fiction that in studying criminals in their natural habitat, we would discover nothing really important that could not be discovered from criminals behind bars.

By studying criminals in their natural habitat I was able to see them in the full variability and complexity of their surrounding subculture, rather than within the artificial environment of a prison. I was thus able to learn about otherwise inaccessible dimensions of their lives, observing and analyzing first-hand the nature of their social organization, social stratification, lifestyle, and motivation.

NOTES

1. Gold (1958) discouraged this methodological strategy, cautioning against overly close friendship or intimacy with informants, lest they lose their ability to act as informants by becoming too much observers. Whyte (1955), in contrast, recommended the use of informants as research aides, not for helping in conceptualizing the data but for their assistance in locating data which supports, contradicts, or fills in the researcher's analy-sis of the setting.

2. See also Biernacki and Waldorf 1981; Douglas 1976; Henslin 1972; Hoffman 1980; McCall 1980; and West 1980 for discussions of "snow-balling" through key informants.

3. See Kirby and Corzine 1981; Birenbaum 1970; and Henslin 1972 for more detailed discussion of the nature, problems, and strategies for deal-ing with courtesy stigmas.

4. We never considered secret tapings because, aside from the ethical prob-lems involved, it always struck us as too dangerous.

5. See Douglas (1976) for a more detailed account of these procedures.

6. A recent court decision, where a federal judge ruled that a sociologist did not have to turn over his field notes to a grand jury investigating a suspi-cious fire at a restaurant where he worked, indicates that this situation may be changing (Fried 1984).

7. See Henslin 1972 and Douglas 1972, 1976 for further discussions of this dilemma and various solutions to it.

8. In some cases I resolved this by altering my descriptions of people and their actions as well as their names so that other members of the dealing and smuggling community would not recognize them. In doing this, however, I had to keep a primary concern for maintaining the sociological integrity of my data so that the generic conclusions I drew from them would be accurate. In places, then, where my attempts to conceal peoples' identities from people who know them have been inadequate, I hope that I caused them no embarrassment. See also Polsky 1969; Rainwater and Pittman 1967; and Humphreys 1970 for discussions of this problem.

REFERENCES

Adler, Patricia A., and Peter Adler. 1987. *Membership Roles in Field Research.* Beverly Hills, Calif.: Sage.

Becker, Howard. 1963. *Outsiders.* New York: Free Press.

Biernacki, Patrick, and Dan Waldorf. 1981. "Snowball sampling." *Sociological Methods and Research* 10:141–63.

Birenbaum, Arnold. 1970. "On managing a courtesy stigma." *Journal of Health and Social Behavior* 11:196–206.

Blumer, Herbert. 1969. *Symbolic Interactionism.* Englewood Cliffs, N.J.: Prentice-Hall.

Carey, James T. 1972. "Problems of access and risk in observing drug scenes." In Jack D. Douglas, ed., *Research on Deviance,* pp. 71–92. New York: Random House.

Cummins, Marvin, et al. 1972. *Report of the Student Task Force on Heroin Use in Metropolitan Saint Louis.* Saint Louis: Washington University Social Science Institute.

Douglas, Jack D. 1972. "Observing deviance." In Jack D. Douglas, ed., *Research on Deviance, pp. 3–34. New York: Random House.*

———. 1976. *Investigative Social Research.* Beverly Hills, Calif.: Sage.

Fried, Joseph P. 1984. "Judge protects waiter's notes on fire inquiry." *New York Times,* April 8:47.

Goffman, Erving. 1963. *Stigma.* Englewood Cliffs, N.J.: Prentice-Hall.

Gold, Raymond. 1958. "Roles in sociological field observations." *Social Forces* 36:217–23.

Henslin, James M. 1972. "Studying deviance in four settings: research experiences with cabbies, suicides, drug users and abortionees." In Jack D. Douglas, ed., *Research on Deviance,* pp. 35–70. New York: Random House.

Hoffman, Joan E. 1980. "Problems of access in the study of social elites and boards of directors." In William B. Shaffir, Robert A. Stebbins, and Allan Turowetz, eds., *Fieldwork Experience,* pp. 45–56. New York: St. Martin's.

Humphreys, Laud. 1970. *Tearoom Trade.* Chicago: Aldine.

Johnson, John M. 1975. *Doing Field Research.* New York: Free Press.

Kirby, Richard, and Jay Corzine. 1981. "The contagion of stigma." *Qualitative Sociology* 4:3–20.

Klockars, Carl B. 1977. "Field ethics for the life history." In Robert Weppner, ed., *Street Ethnography,* pp. 201–26. Beverly Hills, Calif.: Sage.

———. 1979. "Dirty hands and deviant subjects." In Carl B. Klockars and Finnbarr W. O'Connor, eds., *Deviance and Decency,* pp. 261–82. Beverly Hills, Calif.: Sage.

Liebow, Elliott. 1967. *Tally's Corner.* Boston: Little, Brown.

McCall, Michal. 1980. "Who and where are the artists?" In William B. Shaffir, Robert A. Stebbins, and Allan Turowetz, eds., *Fieldwork Experience,* pp. 145–58. New York: St. Martin's.

Polsky, Ned. 1969. *Hustlers, Beats, and Others.* New York: Doubleday.

Rainwater, Lee R., and David J. Pittman. 1967. "Ethical problems in studying a politically sensitive and deviant community." Social Problems 14:357–66.

Riemer, Jeffrey W. 1977. "Varieties of opportunistic research." *Urban Life* 5:467–77.

Rochford, E. Burke, Jr. 1985. *Hare Krishna in America.* New Brunswick, N.J.: Rutgers University Press.

Rosenbaum, Marsha. 1981. *Women on Heroin.* New Brunswick, N.J.: Rutgers University Press.

Warren, Carol A. B., and Paul K. Rasmussen. 1977. "Sex and gender in field research." *Urban Life* 6:349–69.

Wax, Rosalie. 1952. "Reciprocity as a field technique." *Human Organization* 11:34–37.

———. 1957. "Twelve years later: an analysis of a field experience." *American Journal of Sociology* 63:133–42.

West, W. Gordon. 1980. "Access to adolescent deviants and deviance." In William B. Shaffir, Robert A. Stebbins, and Allan Turowetz, eds., *Fieldwork Experience,* pp. 31–44. New York: St. Martin's.

Whyte, William F. 1955. *Street Corner Society.* Chicago: University of Chicago Press.

5

Field Work and Membership
with the Hare Krishna

E. BURKE ROCHFORD

Since the 1960s, there has been an increasing interest by field researchers in the personal dimensions of field work (Emerson 1981, 1983). Attention has centered on how various personal characteristics and motives and feelings of the field worker influence his or her field relations and thereby the reliability of the data collected and the validity of the subsequent analyses. While the scholarly discussion has generally centered on the sources of distortion and bias that result from such influences (i.e., the problem of reactivity) more recently, some field workers have recast these personal and subjective experiences as sources of insight and understanding. Indeed, many of these field workers insist that access to members' worlds can only be gained by actively participating in the everyday lives of those under study (Bodemann 1978; Douglas 1976; Jules-Rosette 1975; Wolff 1964).

In this chapter, I describe the natural history of my field work and involvement with the Hare Krishna movement. This reflexive account of my field work with ISKCON [International Society for Krishna Consciousness] speaks to the unique problems faced by field workers when they attempt to research highly charged ideological settings, such as are often found in social movements (see Thorne 1979). Because of strong pressures to participate in the activities of the group, it often becomes difficult to work out a role that is acceptable both to the researcher and to those under study. My discussion highlights the negotiated quality of field work roles and questions the principal prescriptions that are generally offered to define appropriate field relations, both those stressing role neutrality and those stressing active participation. In addition, I also discuss the personal consequences for the researcher of participation in a stigmatized group.

FROM OBSERVER TO PARTICIPANT:
EARLY FIELD RELATIONS WITH ISKCON

My initial research involvement with ISKCON began in the fall of 1975. At the time I was looking to become involved in a field research project as part of my graduate studies. Remembering two long-time friends who had become

Krishna devotees several years earlier, I contacted them about the possibilities of conducting research on the Hare Krishna movement in Los Angeles. Without hesitation, they invited me to visit them at the nearby ISKCON community. My first entry into the community was thus informal, and although community authorities became aware of my presence and purposes, they never confronted me about my research. In large part, the reason for the lack of official interest in my research reflects the fact that during this early period I was seen by many, if not most, ISKCON members as a *potential convert* rather than as a researcher.

My efforts to conduct my research were very difficult and personally trying during the first year of my involvement with ISKCON. Instead of accepting my repeated assertions that I was a researcher, the devotees invariably refused to accept my explanation, seeing me instead as a "spirit-soul" who had been sent by Krishna. If I saw myself as being there to conduct research, that was my problem, and was of little or no concern to them. But being new to field work and somewhat frightened by the prospect of researching people who were attempting to convert me, I felt it to be imperative personally that I establish myself within the setting as a researcher:

> One of the most critical aspects of this research on the Hare Krishna devotees has been my inability to establish a "researcher's role." Since this was my first research experience, I did not have a clear conception of what a researcher's role actually was. Nevertheless, I had a vague idea of how people in my research setting were "supposed" to respond to me. If I said that I was a researcher, I assumed that, after a period of getting to know the setting, I ultimately would be accepted in that role. The need to establish such a notion in the minds of those under study seemed particularly important. I thought the role of researcher might serve as a protective shield against forseen attempts at conversion on the part of movement members. I assumed that if the Hare Krishna devotees knew that my interest in the movement was purely for purposes of research, they would discontinue their initial attempts at conversion and accept me as a researcher (Rochford 1976).

I had hoped that my research role would become a "retreat" (Thorne 1979:87) allowing me to escape the risks involved in participating in the lives of the Krishna devotees. At this early point in my research, I wanted to maintain a role that was more like being a strict observer than a quasi-participant.

Early on in my research I was invited by the devotees to take a more active part in the life of the community, in particular, in the religious practices of Krishna Consciousness. For many weeks I steadfastly refused to participate in this way, however, fearing that such participation would only increase the strain and pressures I was feeling. Over and over the devotees asked if I was chanting the Hare Krishna mantra. While on some occasions I suggested that I was, to avoid a confrontation, I nevertheless refused to participate with the

devotees as they chanted in the temple. Chanting was appropriate for someone who was becoming a devotee, not for a researcher:

> In the morning, when all the devotees were chanting their rounds in the
> temple, I refrained from doing so. It was very uncomfortable to be among
> all the devotees while they chanted. I observed for a while, but then real-
> ized that I was being observed more than I was observing. I could see that
> all the devotees were noticing that I wasn't chanting. I felt that I was being
> avoided (looked down upon) because of my refusal to take part in the
> chanting. I started wishing that I was somewhere else—anywhere. I had
> visions of the study coming apart at the seams. Several devotees stopped
> and suggested that I try chanting the mantras. I continued to refuse
> (Rochford 1976).

As the pressures continued to mount, on one such occasion, one of my main
informants entered the temple. Within a few minutes he also noticed that I
was not chanting. He walked over and asked if I had *japa* beads. Responding
that I did not, he offered me an extra string of beads.

> At this point I felt I had better take the beads and try chanting. I suc-
> cumbed to the pressure. I knew that by chanting, the devotees would be
> pleased and the pressures I was feeling would subside. Soon after I began
> chanting, things began to happen. First, a devotee came over to me and put
> a flower garland around my neck (flowers that had previously been on the
> altar and thus were very special to the devotees). He welcomed me to
> Krishna Consciousness. A few minutes later, another devotee stopped and
> showed me the correct way to hold the beads while chanting. Immediately
> after that still another devotee came over and asked if I wanted to water the
> spiritual plant *Tulasi Devi*. We went over to the plant and he showed me
> the proper way to "offer" the water to *Tulasi* (Rochford 1976).

But while I had given in to the devotees' pressures, I nevertheless continued to
interpret my actions in terms of my researcher's role. The following statement
from my field notes, however, points to my growing sense of confusion:

> I should make the point that I was giving in to the pressures I felt, but not
> for the reasons the devotees imagined. I wasn't concerned with demon-
> strating my devotion to Krishna. No doubt many of the devotees observ-
> ing my actions thought that I was beginning to "surrender to Krishna."
> Actually, I felt that I was surrendering more to the thought that if I didn't
> show my interest (as a potential convert) my relationship with the devo-
> tees would suffer, and my research would also. I sense that I am getting
> into a bind. I've got to think this out some (Rochford 1976).

After this initial period of frustration and anxiety about my role within the
Los Angeles ISKCON community, I began more easily to accept the devotees'
recognition of me as a potential recruit. After the chanting episode, I began to

participate more freely in the ceremonial aspects of Krishna Consciousness, but again, the decision felt more like a research choice than like a personal decision reflecting my changing relationship to God (Krishna) or the movement. Rather than fighting the devotees' interpretation and treatment of me as a potential convert, I finally was able to see these as data allowing me access to the processes of recruitment and conversion to Krishna Consciousness. Indeed, because of the community's definition of me as a potential convert, I would have been hard pressed to have studied systematically any other aspect of the social life of the devotees. So, while recognition as a potential convert sustained my presence in the devotee community, it also cut me off from exploring other issues of sociological interest.

While being treated as a potential convert allowed me to gain an acceptable role within the setting, such a role has only a limited life for most persons who come into an ISKCON community. Because I was seen as a somewhat special recruit, I was allowed a longer period of time to convert. The devotee in charge of the *bhakta* program referred to me as one of his longer-term conversions. He spent considerable time explaining Krishna Consciousness to me because of my graduate status at the university. In short, he thought I was a "prize catch" (my words) and worth the extra effort. As he told me one day: "You're not like the young kids off the street who come to the community because they are in need of shelter. We expect them to convert quickly. We're willing to wait a while for you."

FROM PARTICIPANT TO MEMBER

After my initial year researching the movement, my relationship to ISKCON and Krishna Consciousness began to undergo a change. Once I had stopped actively researching the movement, I felt more at ease taking part in the religious practices of Krishna Consciousness. While it would be easy to say that I continued my involvement as a way of maintaining my research opportunities, I think it more accurate to say that I wanted to explore my spiritual self as I never had before. The research experience had not convinced me to join ISKCON, but it did awaken me as a spiritual being. For the next several years, I attended the evening *arati* ceremony in the temple once a week and continued to maintain my friendships with various devotees. It was during this time that ISKCON members began to see me as a "fringe" Krishna devotee (a person who appreciates Krishna Consciousness but who nevertheless refuses to make the commitment necessary to move into the community and become a disciple). Given the nature of this role within the community, I was not subject to the devotees' proselytizing efforts thereafter. Indeed, for the most part, I was left alone and allowed to come and go from the community without a great deal of interaction with the devotees.

Although my personal relationship with Krishna Consciousness changed during these initial years, changes also took place within the movement itself that facilitated my taking on a membership role. Prior to 1976 there had been little differentiation in the kinds of members in ISKCON; one was either an insider (a devotee) or an outsider (a *karmie*). Persons residing in the Los Angeles ISKCON community were either on the path to becoming a disciple of Srila Prabhupada's, or they had already made the commitment to the spiritual master and were working toward their self-realization in Krishna Consciousness. Other kinds of members were not recognized. (This situation, of course, provided the source of the role strain I was feeling during the early period of my research.) The exclusivist structure of the movement made it nearly impossible for less committed persons to find an acceptable role within the community. But as the movement began to undergo structural and organizational changes, less committed persons were increasingly extended formal membership status within the devotee community, even if this status was not recognized as being in good standing. . . . Given the ability of such a role within the ISKCON community, I ultimately came to be defined as a fringe devotee.

By contrast, in less demanding and inclusivist movement organizations, such as the Nichiren Shoshu Buddhist movement, a researcher can more easily take on a role more closely aligned to being a core member. This consideration weighed heavily in Snow's decision to research Nichiren Shoshu rather than Hare Krishna:

> My involvement with Nichiren Shoshu had been participatory and much more consistent and intense than my involvement in Hare Krishna, which had been observational and occasional. [My] continued participation in Nichiren Shoshu wouldn't require me to drop out, shave my head, and assume an austere, communal lifestyle as would have continued participation in Hare Krishna. An additional and equally important consideration was my marriage, which made the latter course a highly untenable proposition. . . . Again, gaining membership in Nichiren Shoshu is not contingent on any such initial sacrifices. All that is required is that one pay a $5.00 shipping fee for the "Gohonzon," and be willing to risk, perhaps, a negative reaction by nonmovement significant others (1976:10, 16). . . .

PERSONAL DILEMMAS CREATED BY MEMBERSHIP IN A DEVIANT LIFESTYLE

Researchers studying deviant lifestyles often come to be seen by their friends, acquaintances, and associates as having been somehow contaminated by their involvements with such denigrated populations. Investigators studying the lives of sexual deviants, such as the gay community, for example, often have their own sexuality called into question because of their research interest in these alternate lifestyles (Kirby and Corzine 1981). Researchers have sometimes chosen not to study such deviant groups precisely because of the resulting personal stigma, with the corresponding threat it poses to their professional lives. My involvement with Hare Krishna is perhaps different from other investigations of deviant lifestyles, because my membership role within that setting was consciously decided upon in the course of my research, instead of being imputed to me by others based on pure association.

Membership in any deviant group is strongly dependent upon the relationships that an individual forms with other group members. However, while forming affective ties within the group is critical to the membership and conversion process (Lofland and Stark 1965; Snow and Phillips 1980) this process of assimilation is furthered, and reinforced, by the reactions of people outside the group or movement (Turner and Killian 1972; Snow 1976). As with other people who have chosen to become Krishna devotees, I was subjected to many of the suspicions that are often associated with taking part in a deviant lifestyle. Friends, family, acquaintances, and fellow students all questioned and evaluated the extent of my involvement with ISKCON. One long-time friend, whose brothers are former ISKCON members, told of his concern about my involvement with the movement and the means by which he calculated the degree to which I was being pulled into the Krishna lifestyle:

> Friend: I saw the ways that you were becoming real defensive about the movement. The slightest criticisms, that before you just let pass by, now became a cause. You wouldn't allow me to criticize. . . . I remember having a couple of conversations where I would jokingly criticize ISKCON, but I really meant it. I just wanted to see what your reactions would be so that I could judge the level of your involvement. I was gauging what your relationship was by how you would react. If you reacted very defensively, which you did, then I knew that you were real involved. But even more, I could see the ways that you were becoming more interested in the philosophy; you talked a lot more about it. Again that was a gauge that allowed me to see how involved you were. I could see the way that you were getting involved actually was skewing the way that you were interpreting what was going on. I just felt that it was increasingly difficult for you to be, not objective, but just critical. I mean ISKCON was certainly worth a little bit of criticism, but you wouldn't make it. You were trying to couch things in a way that wouldn't seem critical. So I just looked for markers because I could see that you were really becoming involved.
>
> EBR: What were some of the other markers?
>
> Friend: One was dress; you used to wear a lot of Krishna shirts. And your beads, are you still wearing your [Krishna] beads?
>
> EBR: No.
>
> Friend: Other things were that you were always saying "Hare Bol" [a greeting used by the devotees] at the beginning and end of a conversation, just like the devotees. You said things like "Ah

Krishna," or you would be singing various Krishna songs, things like that (Los Angeles 1981).

An exchange with my wife during the course of my field work points further to the ways in which I was becoming involved in the movement, and her fears about it. Late in the evening my wife (JB) and I sat down to a steak dinner. Two things came out of this.

EBR: I want to give up eating beef. I am just going to eat poultry and fish.

JB: Why?

EBR: Well, it's better for you if you eat less red meat. I have always heard it is bad for you, lots of cholesterol.

JB: [not satisfied pushes further] Is that the only reason?

EBR: [I stumble around in my mind looking for another reason, but then just say] I don't want to eat cows.

JB: That's what I thought!

JB: [After our talking a bit more, she finally says] Do what you like, but don't expect me to stop eating meat.

A few minutes later after a long silence,

JB: One thing I want to make clear.

EBR: What's that?

JB: None of our kids are ever going into the *gurukula*. I want you to understand how I feel about that.

EBR: I have no intention of ever putting a kid of mine in the *gurukula*.

JB: The way you are getting involved with these people, I don't know what you'll feel like in the future, so I just want you to know how I feel about this now (Los Angeles 1981).

Involvement in the movement created a variety of other personal problems. On numerous occasions during the course of my research, I was party to conversations about cults in general and Hare Krishna in particular. I often chose to avoid taking part in these conversations because to have done so would have brought my involvement with Hare Krishna into public light. My personal involvement made it difficult for me to take part in discussions about cults because of the burden of defense that I felt I would be forced to argue:

Towards the end of the seminar (group psychotherapy) a discussion emerged about cults. Members of the class began talking about the factors that led young people to take part in cults. The thrust of the discussion centered around the type of personality that would be attracted to the structured existence of cult life. There was considerable discussion as to whether the members of cults were brainwashed, or whether they had been psychologically "unstable" to begin with. Many members of the class seemed to think that cults like Hare Krishna, which was mentioned specifically, were entrapping young people who were "suffering psychologically." During the discussion, I remained silent, but uncomfortable, feeling that I should intervene and set straight some of the notions being discussed, which I felt were incorrect. I continued to hold back, however. Finally, the woman sitting next to me, who was an acquaintance of mine, turned and said "I really would like to hear what Burke has to say on this issue, since he is studying Hare Krishna." Having been exposed, I then went on to talk about the changes in the kinds of people who were entering movements like Hare Krishna, and how it would be incorrect to assume that all, or most, of these people were in any way pathological. Persons joining the cults did so for many reasons. At this point, one member of the class remarked "So why didn't you say something earlier?" (Los Angeles 1979).

But perhaps the most dramatic and emotional event that took place during my research occurred when I faced the Human Subjects Committee at the university in order to gain clearance to distribute my questionnaire. Having been awarded a small grant, I was required to get a "human subjects clearance" prior to administering the questionnaire. After several unsatisfactory written communications about specific procedures to be employed in distributing and having the forms returned, I was asked to appear before the committee. Initially, they addressed specific questions regarding the questionnaire to me. However, the committee's questions quickly turned to the Krishnas themselves and to my involvement with them. They asked questions like: Weren't these people involved in subjecting their members to something akin to brainwashing? How did I ever get involved in this movement to begin with? What did I think of their activities at the airport? Finally, after several minutes of being questioned in this manner, the chair of the committee announced that he had a question to ask on behalf of one member of the committee who wished to remain anonymous. At first I objected, saying that he/she should deliver the question in person. The chair refused and proceeded to say: "One member of the committee feels that this research is not for the purposes that you have stated. Specifically, he or she feels that you are doing the research for the Krishna movement, rather than for academic reasons. Further, he or she believes that you are involved with the Krishnas and hence would like you to specifically address the question of your relationship with them." I did not immediately recognize that the committee's work did not have anything to do with the specifics of my relationship with the movement. I went ahead explaining my religious views and describing the ways they had been influenced by my association with Krishna Consciousness. I assured the committee that, while I had gained an appreciation of the philosophy and lifestyle of Krishna Consciousness, I did not consider myself either a devotee or a "front" for ISKCON.

Several days later, after I had again been asked to address myself to still further questions regarding the questionnaire, I met personally with the chair of the committee. I told him in no uncertain terms that I felt that *my* rights had been compromised by the committee's reaction to my research. He then told me that two members of the committee (who were not themselves researchers) were afraid that I was about to involve the university in a possible Jonestown situation.[1]

LEAVING THE FIELD:
MEMBERSHIP RECONSIDERED

Although my membership status within ISKCON posed a number of personal and research-related problems during the course of my investigation, these problems continued, and became more complex and emotional, after I left the field and began analyzing my data. For several months after leaving the field, I struggled to find a way to write about ISKCON. While many students beginning the data analysis and writing phase of their dissertation feel overwhelmed for a time, I felt an additional sense of depression, which grew from my longstanding relationship with ISKCON. Being a group member and doing sociology are very different enterprises, and I felt that in many ways I was betraying my relationship with the movement and its members by now turning the devotees into objects of analysis. Furthermore, sociology often promotes a cynical way of seeing and analyzing the world, and I wanted to be respectful of ISKCON. As I began analyzing my materials, the analysis that began to take shape seemed critical of the movement, too critical, I thought. To write about the movement's controversial public solicitation efforts, for example, seemed important for sociological purposes, but wasn't I betraying the devotees by pursuing such a line of analysis? For a number of months I struggled with the issues to be analyzed as well as the proper tone to use in the analysis.

After I began analyzing my data and writing the dissertation, I avoided attending functions at the temple and more or less dropped out of sight. While I lived only blocks away from the ISKCON community, I would not allow myself even to walk past the community for fear that I would come upon a devotee on the street. At first I was afraid that the devotees would ask about the dissertation, what topics I was writing about and so on. Because I increasingly came to see the dissertation taking a more critical line of analysis, I wanted to avoid going into any details. But after several months these feelings gave way to more intense concerns. I felt that the devotees would in all probability feel betrayed by me. After all, many of them had given me open access to their lives primarily because they had seen me as a member and had understood me to be sincere in my feelings toward ISKCON and Krishna Consciousness. Also, by now I was unsure and confused about exactly what my true feelings toward ISKCON and Krishna Consciousness were. If I was a member with any conviction at all, shouldn't I be attending *arati* and other functions at the temple? Had I been lying to myself all this time? Had I really been sincere about my interest in Krishna Consciousness, or had I simply tricked myself into believing that I was for the sake of the research?

During the year that I was writing my dissertation, I only visited the ISKCON community in Los Angeles twice. On each occasion I was very uncomfortable; I consciously avoided devotees whom I thought would be inquisitive about my work or whom I thought might be particularly hurt by my apparent lack of sincerity.

> My wife and I walked to the temple this afternoon. We wanted to show our new daughter to one of the Krishna women who was a long-time friend. As we came into the community, I spotted the headmaster of the *gurukula* and his son on the other side of the street. He waved and with a smile said "Hare Krishna." I replied the same but continued walking. The young boy then ran over and asked when I was going to come and visit the *gurukula*. I didn't know what to say, but then said: "Well I will be moving soon, so I'm not sure if I will be able to." With that I excused myself, saying I wanted to attend the *arati*. I walked away feeling very guilty, wishing I hadn't come at all. The headmaster and his son were friends and yet I couldn't be friendly. While my wife went looking for the Krishna woman, I went to the temple where I would be less conspicuous, since this was Sunday and there would a large number of visitors present.

CONCLUSION

Field work, unlike other research strategies employed by investigators of social life, generally involves some level of participation in the everyday life of the people being studied. While the degree of participation is often a matter of choice or preference for an individual field worker, there are some settings which require active participation as a condition of the ability to do research. My investigation of ISKCON highlights the negotiated quality of field work roles. Instead of being predetermined, my relations with ISKCON members evolved over time as I became more deeply involved in Krishna Consciousness and as changes took place within ISKCON itself that accommodated the participation of less-committed members. My relationship with ISKCON was further reinforced as I became subjected to the suspicions and judgments of persons outside the Krishna movement.

But my field research on the Hare Krishna movement raises an additional issue regarding field work and the need and efforts made to gain understanding of a new social world. In the past several years a number of field workers have advocated more active, fully participatory roles for conducting field research. Some more ethnomethodologically oriented field workers have even proposed becoming a member of the studied group as a strategy for gaining insight into the members' worlds (Mehan and Wood 1975; Jules-Rosette 1975).[2] In this model of field work, emphasis is placed on the investigator participating as the members participate. While field workers may or may not actually become formal members of the groups they study, they nevertheless need to participate and to experience the members' everyday lives by "immersion in the natural situation" (Douglas 1976). This approach to field research is thought to yield the subjectively meaningful world of members, rather than an objective analytic account of their world. While I believe active involvement does facilitate access to the members' world, there are nevertheless limitations to this kind of field work.

First, many, if not most, settings that field workers study are characterized by local politics. To take on a membership role necessarily involves making choices about what sort of member the researcher wants to be. Field workers who embrace membership as a research strategy are forced to decide from what position they will gain an understanding of the events in a particular setting. For example, towards the end of my field work with ISKCON, I became a charter member of an organization that was composed of ex-ISKCON members who were challenging ISKCON policies under the new leadership. To become involved in this group was to publicly acknowledge and show my feelings toward ISKCON and toward the policies of the leadership. While I sought to hide my involvement in this splinter group from ISKCON members, some of the devotees did become aware of my membership in the counterorganization. As a result, my relations with some ISKCON members became strained. In the end, this became one factor that pressured me to leave the field when I did.

Second, while field workers may seek to become members in good faith, it seems unlikely that they can ever become more than marginal members. The interests and concerns of field workers ultimately are not the same as those of the people who are committed to the group. As Bittner (1973:121) states, the field worker "is the only person in the setting who is solely and specifically interested in what things are for 'them'." While I learned the religious practices of Krishna Consciousness and took part in the life of the ISKCON community, I nevertheless lacked the commitment that characterized other persons in the setting. Ultimately, I had come to study *their* world, even though I had become a member of it as a means to better understand it.

NOTES

1. My troubles with the university did not stop there. Because ISKCON was in the middle of serious succession problems in 1980, which caused great factionalism, splintering, and the defection of many long-time devotees and the purging of others by the leadership . . . , I felt it important that my dissertation not be made publicly available at that time. My advisors and I petitioned the university to hold my dissertation out of circulation for a one-year period because of the likelihood that it would be used in the ongoing political struggle taking place within ISKCON. The university ruled against my petition and I was forced to wait for a full year before finally filing for my degree.

2. Harold Garfinkel has taken this approach to an extreme by encouraging his students to become trained in one of the professions as a way of gaining access to the everyday details that constitute peoples' work. In place of studies about work, which typify sociological accounts of the professions, Garfinkel is pushing instead for studies of work. These kinds of studies can only come from persons who have fully committed themselves to the work as practitioners (Garfinkel 1981).

REFERENCES

Bittner, E. 1973. "Objectivity and Realism in Sociology." *Phenomenological Sociology: Issues and Applications.* George Psathas (ed.) New York: Wiley, pp. 109–125.

Bodemann, Y. M. 1978. "A Problem of Sociological Praxis: The Case for Interventive Observation in Field Work." *Theory and Society* 5:387–420.

Douglas, J. D. 1976. *Investigative Social Research: Individual and Team Field Research.* Beverly Hills: Sage.

Emerson, Robert M. 1981. "Observational Field Work." *Annual Review of Sociology* 7:351–378.

————. 1983. *Contemporary Field Research.* Boston: Little, Brown.

Garfinkel, H. 1981. Lectures in ethnomethodological methods. Unpublished. University of California, Los Angeles.

Jules-Rosette, B. 1975. *Vision and Realities: Aspects of Ritual and Conversion in an African Church.* Ithaca: Cornell University Press.

Kirby, R., and Corzine, J. 1981. "The Contagion of Stigma: Fieldwork Among Deviants." *Qualitative Sociology* 4 (1):3–20.

Lofland, John, and Stark, Rodney. 1965. "Becoming a World-Saver: A Theory of Conversion to a Deviant Perspective." *American Sociological Review* 30:862–874.

Mehan, H., and Wood, H. 1975. *The Reality of Ethnomethodology.* New York: Wiley.

Rochford, E. Burke, Jr. 1976. "World View Resocialization: Commitment Building Processes and the Hare Krishna Movement." Unpublished M.A. thesis. Department of Sociology, University of California, Los Angeles.

Snow, David A. 1976. "The Nichiren Shoshu Buddhist Movement in America: A Sociological Examination of its Value Orientation, Recruitment Efforts, and Spread." Ann Arbor, Michigan: University Microfilms.

Snow, David A., and Phillips, Cynthia. 1980. "The Lofland-Stark Conversion Model: A Critical Reassessment. *Social Problems* 27:430–447.

Thorne, Barrie. 1979. "Political Activist as Participant Observer: Conflicts of Commitment in a Study of the Draft Resistance Movement of the 1960s." *Symbolic Interaction* 2(1):73–88.

Turner, Ralph, and Killian, Lewis. 1972. *Collective Behavior*. Englewood Cliffs, N.J.: Prentice-Hall.

Wolff, K. 1964. "Surrender and Community Study: The Study of Loma." *Reflections of Community Studies*. A. J. Vidich, J. Bensman, and M. R. Stein (eds.). New York: Wiley, pp. 233–263.

■

PART III

Constructing Deviance

A s we noted in the General Introduction, the social constructionist perspective suggests that deviance should be regarded as lodged in a process of definition, rather than in some objective feature of an object, person, or act. It therefore guides us to look at the process by which a society constructs definitions of deviance and applies them to specific groups of people associated with these objects or acts.

MORAL ENTREPRENEURS

The process of constructing and applying definitions of deviance can be understood as a *moral enterprise*. That is, it involves the constructions of moral meanings and the association of them with specific acts or conditions. The way people make deviance is similar to the way they manufacture anything else, but because deviance is an abstract concept rather than a tangible product, this process involves individuals drawing on the power and resources of organizations, institutions, agencies, symbols, ideas, communication, and audiences. Becker (1963) has suggested that we call the people involved in these activities *moral entrepreneurs*. The deviance-making enterprise has two facets: rule-creating (without which there would be no deviant behavior) and rule-enforcing (applying these rules to specific groups of people). We thus have two kinds of moral entrepreneurs: rule creators and rule enforcers.

Rule-creating can be done by individuals acting either alone or in groups. Prominent individuals who have been influential in campaigning for definitions of deviance include Nancy Reagan, with her "Just Say No" antidrug campaign; former Surgeon General C. Everett Koop with his campaign to marginalize cigarette smoking; and Tipper Gore, with her campaign with the Parents' Media Resource Center (PMRC) to label and stigmatize heavy-metal and rap music. More commonly, however, individuals band together to utilize their collective energy and resources to change social definitions and create norms and rules. These groups of moral entrepreneurs represent interest groups that can be galvanized and activated into *pressure groups*. Rule creators ensure that our society is supplied with a constant supply of deviance and deviants by defining the behavior of others as immoral. They do this because they perceive threats in and feel fearful, distrustful, and suspicious of the behavior of these others. In so doing, they seek to transform private troubles into public issues and their private morality into the normative order.

Moral entrepreneurs manufacture public morality through a multistage process. Their first goal is to generate broad *awareness* of a problem; they must create a sense that certain conditions are problematic and pose a present or future potential danger to society. Because no rules exist to deal with the threatening condition, they construct the impression that these are necessary. In so doing, they draw on the testimonials of various "experts" in the field, such as scholars, doctors, eyewitnesses, ex-participants, and others with specific knowledge of the situation. These testimonials are disseminated to society via the media as "facts."

Secondly, rule creators must bring about a *moral conversion,* convincing others of their views. With the problem outlined, they have to convert neutral parties and previous opponents into supporting partisans. Their successful conversion of others further legitimates their own beliefs. To effect a moral conversion, rule creators gain support from the endorsements of opinion leaders. These spokespersons need not have expert knowledge on any particular subject, they merely need to be liked and respected. Moral entrepreneurial groups thus often turn to athletes, actors, religious leaders, and media personalities for endorsements.

Once the public viewpoint has been swayed and a majority (or vocal and powerful enough minority) of people have adopted a social definition, it may remain at the level of a norm or become elevated to the status of law through a legislative effort. In some cases both situations occur. For example, although the antismoking campaign has been successful in banning cigarette use in airplanes and various public places, it most effectively relies on normative informal sanctions.

Once norms or rules have been enacted, rule enforcers ensure that they are applied. In our society, this process often tends to be selective. Various individuals or groups have greater or lesser power to resist the enforcement of rules against them due to their socioeconomic, racial, religious, gender, political, or other status. Whole battles may begin anew over groups' strength to resist the enforcement of norms and laws, with this arena becoming once again a moral entrepreneurial combat zone.

Three examples of moral entrepreneurial crusades are presented to illustrate this process. In "Coffee Drinking: An Emerging Social Problem?" Ronald Troyer and Gerald Markle take a historical view of social attitudes toward the widespread use of coffee. Detailing the research, both pro and con, on the medical effects of caffeine, they show the role of both lay participants and sympathetic professionals in laying the groundwork for a potential construction of coffee's deviant status. Craig Reinarman then takes up moral attitudes toward drugs in "The Social Construction of Drug Scares." He briefly offers a salient history of drug scares, the major players engineering them, and the social contexts that enhanced their growth. He then outlines seven factors common to drug scares that propel them toward realization. These enable him to dissect the essential processes at work in the rule-creation and enforcement phases of drug scares, despite the contradictory cultural values of temperance and hedonistic consumptionism. Finally, Joel Best offers a sophisticated illustration of the role of language in the work of moral entrepreneurs in his "Rhetoric in Claims-Making: Constructing the Missing Children Problem." He further details a stage model whereby private attitudes become transformed into public ones and describes the vehicles through which this transformation occurs.

DIFFERENTIAL SOCIAL POWER

Specific behavioral acts are not the only things that can be constructed as deviant; this definition can also be applied to a social status or lifestyle. When entire groups of people become relegated to a deviant status through their social condition (especially if it is ascribed through birth rather than voluntarily achieved), we see the force of inequality and differential social power in operation. In the last three selections of this section, Elijah Anderson's "The Police and the Black Male," Nijole Benokraitis and Joe Feagin's "Sex Discrimination—Subtle and Covert," and Michael Messner's "Masculinity,

Homophobia, and Misogyny," we examine three statuses that are often deviant in American society: race, gender, and sexual preference. People may find themselves discriminated against or blocked from the mainstream of society by virtue of this basic feature of their existence, unrelated to an individual situation or act. This application of the deviant label powerfully shows the role of power in the deviance-defining enterprise, as those positioned closer to the center of society, holding the greatest social, economic, political, and moral resources, can turn the force of the deviant stigma onto others less fortunately placed. In so doing, they use the definition of deviance to reinforce their favored position. This politicization of deviance and the power associated with its use serve to remind us that deviance is not a category inhabited only by those on the marginal outskirts of society: the exotics, erotics, and neurotics. Instead, any group can be pushed into this category by the exercise of another group's greater power.

Anderson gives us a glimpse into the perspective of the black man in the ghetto as he is handled by the police, the agents of social control. Benokraitis and Feagin discuss some of the inequities women face in the world of work as they are relegated to the role of second-class citizens. The routine acceptance of exploitation and harassment attests to their lower, deviant status. Finally, Messner investigates the fundamental bastion of male culture, sport, and shows how this basis for male bonding and the development of masculine identity is built not only on homophobia, the revulsion of homosexuality, but also on misogyny, the treatment of women as base sex objects and the disrespect for anything feminine.

6

Coffee Drinking: An Emerging Social Problem?

RONALD J. TROYER AND

GERALD E. MARKLE

offee has a special place in the lives of most people in the United States. As a social custom, coffee confers adult status, since children are not allowed to drink it. Its daily use stimulates social interaction. The coffee pot is present at most work places, business meetings, and social gatherings. The coffee break is a time for relaxed conversation during the workday. Coffee drinking is widespread in the United States: in 1976 some 101 million people 20 years or older—80 percent of the U.S. adult population—drank an average of 3.2 cups of coffee a day (U.S. Department of Health, Education and Welfare, 1980).

Despite its ubiquity and popularity, coffee drinking has come under attack. The morning newspaper, over which most people drink their first cup of coffee, has carried stories relating coffee and caffeine—the active drug in coffee, tea, and some soft drinks—to cancer, birth defects, and heart disease. Some groups have begun describing the United States as a nation of "caffeine addicts" and the Food and Drug Administration has warned pregnant women to limit their consumption of coffee and other substances that contain caffeine. Is the once socially accepted "cup of coffee" undergoing redefinition? Is coffee drinking becoming a social problem?

Most people do not yet define coffee as a social problem. However, the consumption of caffeine has become the object of controversy. Some groups are pressing government agencies to regulate caffeine in products and urging the public to reduce or eliminate their caffeine consumption. Coffee and soft drink industry groups are waging a counter campaign, even trying to persuade people to consume more of their products.

These activities are examples of the initial stages of the social problem process, during which some groups attempt to transform some putative condition into a public issue while others seek to prevent such public recognition. As Blumer (1971) and Spector and Kitsuse (1977:143) have noted, during this emergent phase groups attempt to gain publicity and arouse controversy.

From: "Coffee Drinking: An Emerging Social Problem?" Ronald J. Troyer and Gerald E. Markle, *Social Problems,* Vol. 31, No. 4, 1984. © 1984 Society for the Study of Social Problems. Reprinted by permission of University of California Press Journals and Ronald J. Troyer.

Although coffee drinking has not progressed beyond the early stages of the social problem process, attempts to redefine coffee drinking offer sociologists an opportunity to examine the dynamics involved. In this paper we review the historical claims against coffee and caffeine, look at the medical and psychological research and how it has been popularized, and discuss the partisan groups in the dispute. We conclude by noting parallels between cigarette smoking and coffee drinking.

HISTORICAL CLAIMS

No one knows just when coffee was first drunk. Indigenous to Ethiopia, Arabs used it primarily as a medicine until the 14th century (Ukers, 1935a:12). Apparently first cultivated in Yemen, the development of coffee bean roasting and grinding techniques during the 15th century led to its common consumption as a beverage throughout the Mideast (Jacob, 1935:27). Travelers and explorers introduced coffee to Europe during the 17th century (Ukers, 1935a). It came under attack almost immediately from several quarters. In Germany, doctors argued that women who drank coffee could not bear children (Uribe, 1954:19). In England and France, government officials linked coffee drinking with political radicalism and idle leisure (Jacob, 1935:144; Ukers, 1935a:68). In 1675, Charles II unsuccessfully attempted to close coffee houses, alleging they helped spread political discontent. In England and Germany, brewers opposed coffee because it detracted from beer sales (Jacob, 1935:93). In spite of such opposition, coffee consumption increased in Europe throughout the 18th and 19th centuries.[1]

Apparently the early colonists brought coffee to North America. Its use in the United States was much less controversial than in Europe, at least prior to the 20th century. During the second half of the 19th century the *New York Times* published the claim that coffee caused rheumatism (1878:4) and a German physician's assertion that coffee was responsible for the "nervousness and peevishness of our time" (1879b:4). However, the *New York Times* (1879a:4) stated in an editorial that "coffee is an absolute necessity to us."[2] Social reformers actively promoted coffee and tea as substitutes for alcohol, arguing that coffee houses would help solve the "saloon problem" (Howerth, 1895:589). The *New York Times* (1880:4) noted in an editorial that coffee houses in England were effectively countering the problem of the "working class drinking intemperately" and concluded: "we trust that our philanthropic citizens will come forward liberally to support this wise undertaking." During the last two decades of the 19th century and the early years of the 20th century, coffee houses were set up by temperance and church groups in New York, Boston, Philadelphia (Howerth, 1895), and a number of cities in California (Fox, 1904).

U.S. physicians began warning against coffee at the beginning of the 20th century. Dr. T. D. Crothers, a hospital superintendent, editor of the *Journal of Inebrity,* and professor of nervous and mental diseases at the New York School

of Clinical Medicine, attributed many nervous disorders to caffeine in an article entitled "Coffee and tea drunkeness" (1902). The *New York Times* (1912)
quoted a physician who said "the tea and coffee habit is decidedly pathologic"
and "tea and coffee are baneful drugs, and their sale and use ought to be prohibited by law."[3] In 1916 a chemistry professor, writing in the *Ladies Home
Journal,* warned nursing mothers to avoid coffee because it would interfere
with milk production (Hawk, 1916).

From 1920 to 1960, warnings against coffee in the United States were
infrequent, although not entirely absent. During the 1920s, for example, the
New York Times published two stories reporting scientists' claims that coffee
was responsible for tension and neurosis (1928, 1929a) and three items in
which scientists proclaimed its innocence (1924, 1929b, 1929c). Furthermore,
articles indexed by *Reader's Guide to Periodical Literature*[4] under coffee concluded that the drink was safe for adults, but provided too much stimulation for
children (Johnson, 1925; Osborne, 1924).

Claims against coffee were virtually nonexistent in the United States during the 1930s. Articles extolled coffee's virtues as an aid to digestion (*New
York Times,* 1931) and vitality (*New York Times,* 1938). The same themes
dominated the 1940s, although *Science Digest* (1940:43) discussed coffee
drinkers' "headache" and Wassersug (1946:26) noted that coffee can cause
ulcers "in a few susceptible persons."

A similar pattern prevailed in the 1950s. Although the *New York Times*
(1954) noted that an article published in the *New England Journal of Medicine*
had classified coffee drinking as a drug habit similar to alcohol, opiate, and
barbiturate addiction, the newspaper also carried claims that coffee was an
appropriate stimulant for athletes (*New York Times,* 1953) and that coffee
breaks were important for worker productivity (*New York Times,* 1956). The
Science Newsletter (1955:216) reported a physician's comment that drinking 30
cups of coffee a day was probably harmless; *Science Digest* (1956:49) said that a
Wisconsin physician's research found that coffee contained niacin and B-vita-
mins; and *Americas* (1956:39) published a scientist's comment that drinking
two cups of coffee a day contributed to longevity.

The pattern began to change in the 1960s. While most stories about coffee
continued to focus on its preparation, price, and consumption, reports linking
coffee to heart disease were published (*Newsweek,* 1964:69, 71; *New York
Times,* 1963; *Science Digest,* 1963:55). This new-found link set the stage for
increased controversy during the 1970s.

MEDICAL RESEARCH ON CAFFEINE

Caffeine occurs naturally in plants widely distributed throughout the world.
Coffee, made from the seeds of *Coffea arabica* and related species, contains
approximately 100–150 mg caffeine per 5 oz cup, depending on the method
of preparation,[5] the strength of brew, and the specific variety of bean (Dews,
1982). Coffee accounts for 75 percent of all caffeine consumed in the United

States (Graham, 1978:98). Tea, brewed from the leaves of *Thea sinesis,* contains approximately one-half (50-75 mg) the caffeine of coffee. Cola soft drinks, made from the nuts of the tree *Cola acuminata,* contain 30–50 mg caffeine per 12 oz cup.[6] Cocoa, derived from the seeds of *Theobroma cocao,* has only about 10 mg caffeine per 6 oz cup. Chocolate, also derived from *Theobroma cocao,* is the only other normal dietary source of caffeine;[7] milk chocolate and baking chocolate contain 6 mg and 35 mg caffeine per ounce, respectively.

Caffeine was first discovered in 1820 by Friedlieb Runge in Germany (Anft, 1955). Its formula (1, 3, 7 trimethylxanthine) was established in 1882 by Emil Fisher, also in Germany. Fisher's laboratory method became the basis for the industrial production of caffeine and was part of the work for which he was awarded the Nobel prize in 1902 (Gillespie, 1972:2).

According to Goodman and Gilman (1975:368), caffeine is a psychoactive drug which acts as a "powerful CNS [central nervous system] stimulant."

> Its main action is to produce a more rapid and clearer flow of thought, and to allay drowsiness and fatigue. After taking caffeine one is capable of a greater sustained intellectual effort and a more perfect association of ideas. There is also a keener appreciation of sensory stimuli and reaction time is appreciably diminished.

Caffeine also increases the output of urine and gastric secretions, causes a slight increase in the basal metabolism rate, and increases the capacity for muscular work (Goodman and Gilman, 1975:370).

Because of its ubiquity and because it is a stimulant, caffeine has long been the subject of medical curiosity. However, the medical dispute over caffeine in the 1970s and 1980s had its roots not in pharmacology, but in genetics. In the 1920s and 1930s geneticists discovered that irradiation and various chemicals could cause mutations, the permanent genetic alteration of living cells and their progeny. By the 1940s geneticists were testing large numbers of chemicals for their mutagenicity in micro-organisms. Fries and Kihlman (1948) were the first to report evidence suggesting that caffeine caused mutations in living cells, a finding corroborated by Demerc *et al.* (1951).

When the structure of DNA—the molecule which contains the genetic code—was discovered in 1953, researchers had further reason to suspect caffeine as a mutagen. The chemical structure of caffeine is similar to one of the nucleic acids in the DNA molecule; when such similar substances are accidentally incorporated into DNA, as occasionally happens, mutations occur (Koch, 1956:181). In Germany, Ostertag *et al.*(1965:273) demonstrated that "caffeine is effective in breaking chromosomes of human cells in culture," but concluded that "calculations place the amount of damage (breakage) in the neighborhood of the natural mutation rate." However Kuhlmann *et al.* (1968:2375) warned that caffeine was "one of the most dangerous mutagens in man," a claim most researchers view as highly exaggerated (Adler, 1970; Epstein, 1970).

Thus, on both theoretical and experimental grounds, caffeine became a suspected mutagen. Moreover, because substances which cause mutations often cause cancer and birth defects, caffeine became a suspect in these two areas of research as well. By the 1970s heart disease was added to the list of ills linked to caffeine.

Cancer

In 1971 researchers from the Harvard School of Public Health reported a relationship between coffee drinking and bladder cancer (Cole, 1971). Their method was epidemiological, in which dietary practices and drug consumption were examined to see if they correlated with various disease outcomes. Four years later, these same researchers withdrew their allegations against coffee, reporting that they had found no association between coffee and cancer of the lower urinary tract (Simon *et al.*, 1975). This retraction was corroborated by Morrison *et al.* (1982), who claimed that the original study had not controlled for cigarette smoking. However, Marrett *et al.* (1983:113), working at Yale, wrote that "after adjusting for age and smoking a significant elevation in risk [of bladder cancer] was found for males (odds ratio = 1.5) but not for females (odds ratios = 1.0)." They concluded that "one-quarter of bladder tumors . . . might be attributed to drinking more than one cup of coffee per day" (1983:126).

At Harvard, MacMahon *et al.* (1981) reported a strong association between caffeine consumption and cancer of the pancreas, the fourth most fatal form of cancer. Two cups of coffee a day, they claimed, increased the risk by a factor of 1.8, while three or more cups a day increased the risk by 2.7. They estimated "the proportion of pancreatic cancer that is potentially attributable to coffee consumption to be slightly more than 50 percent" (1981:632), an estimate which translates to more than 10,000 deaths per year.[8]

Birth Defects

Beginning in the early 1960s several studies claimed that coffee induces birth defects in laboratory animals; yet others have denied this link or reported ambiguous findings. A 1978 study by the Food and Drug Administration, the largest that had ever been done on animals, force-fed caffeine to pregnant rats. Those fed the equivalent of 12 to 24 cups of coffee a day gave birth to offspring missing toes or parts of toes; skeletal growth was retarded in rats whose mothers received the equivalent of two cups a day (*Science,* 1980).

Citing this and other studies, but with no direct evidence of harm to humans, the Food and Drug Administration in 1980 warned pregnant women to restrict their intake of coffee as a precaution against birth defects. The warning stimulated new studies, virtually all of which have found coffee, in moderation, to be innocent. For example, a study of 12,205 pregnant women concluded that there was "no association between coffee consumption and adverse outcomes of pregnancy" (Linn *et al.,* 1982).

Heart Disease

Most research has been devoted to the relationship between coffee and heart disease. Between 1970 and 1980, *Index Medicus* listed 19 articles on the subject, compared with only 12 on the coffee and cancer issue. The first serious claim was made by the Boston Collaborative Drug Surveillance Program (1972). They concluded that drinking five cups of coffee a day doubled the risk of heart attack, but cautioned: "Coffee drinking might merely be a feature of the striving, competitive personality which some workers believe to be characteristics of those prone to coronary artery disease" (1972:1280).

Other researchers agreed with, and amplified, this caution. In the mid-1970s one study after another absolved coffee and pointed to tobacco as the culprit. An editorial in the *New England Journal of Medicine* concluded: "There is nothing to suggest, for the present, that we must give up either coffee or alcohol in moderation to avoid heart attack" (Kannell, 1977:444). Even among patients with heart problems, coffee drinking might be allowable. As another editorial stated: "If a caffeinated beverage does not provoke arrhythmia in a given patient, common sense dictates that its use not be forbidden" (Graboys and Lown, 1983:836).

PSYCHOLOGICAL RESEARCH

Psychologists have increasingly focused attention on the effects of caffeine. *Psychological Abstracts* lists 10 studies[9] between 1970 and 1972 which examine the influence of caffeine on human and animal behavior; between 1979 and 1981, 61 studies are listed which evaluate its effect on motor performance (swimming, rate of bar pressing, vasomotor coordination, sleep) and mental performance (attention span, arithmetic, solving crossword puzzles). In the late 1970s, research began to focus on the relationship between caffeine consumption and personality traits. Consumption among college students was associated with lower grades, higher levels of anxiety, and depression (Gilliland and Andress, 1981). Other studies linked caffeine to occupational stress (Ayers *et al.*, 1976), anxiety about grade point average (Hire, 1978), and muscle tension (White *et al.*, 1980). Two studies attempted to establish a correlation between caffeine use and introversion and extroversion (Gilliland, 1980; Smith *et al.*, 1981). Finally, some psychologists associated caffeine consumption with the use of psychotropic drugs (Greden *et al.*, 1981) and alcoholism (Ayers *et al.*, 1976).

Psychiatrists have linked caffeine consumption to mental disorders. Winstead (1976:1450) reported that acute psychiatric ward patients who displayed high levels of anxiety and had been diagnosed as psychotics were also more often "high users" of coffee. He concluded that coffee is a "poor choice of beverage for the schizophrenic or psychotic patient." Greden *et al.* (1978) reported that both moderate and high caffeine consumers among a sample of hospitalized psychiatric patients had poorer physical health and were more anxious and depressed.

Psychiatrists have also begun to study caffeine consumption as a problem for the general population (Foxx and Rubinoff, 1979; Gilliland and Andress, 1981; Greden, 1974). They use the diagnostic label *caffeinism* to characterize behavioral patterns attributed to caffeine consumption. The American Psychiatric Association's *Diagnostic and Statistical Manual* describes the symptoms:

> . . . restlessness, nervousness, excitement, insomnia, flushed face, diuresis, and gastrointestinal complaints. These symptoms appear in some individuals following ingestion of as little as 250 mg per day [two cups of coffee], whereas others require much larger doses. At levels of more than 1 g/day there may be muscle twitching, periods of inexhaustibility, psychomotor agitation, rambling flow of thought and speech, and cardiac arrhythmia (American Psychiatric Association, 1980:160).

This new label, particularly as it applies vague symptoms such as "restlessness" to a potential majority of the adult population of the United States, may be portentous for the future redefinition of coffee drinking.

POPULAR VERSIONS OF THE SCIENTIFIC LITERATURE

During the late 1970s and early 1980s, scientists' findings were debated and evaluated in the popular media. Some publications concluded that coffee presented no danger to humans. The *Reader's Digest* quoted a Swedish scientist as saying that "caffeine can be acquitted" (Ponte, 1983:76). *USA Today* (1980:4) observed that, "for the average person, there isn't much to worry about." *Science Digest* concluded that "normal" caffeine consumption "does not appear to involve an appreciable risk" (Timson, 1978:48). *Science News* titled an article on the danger of birth defects: "Coffee linked birth problems premature" (Herber, 1982:68). Jacobs (1982) concluded in a newspaper story that there was "virtually no evidence" that drinking caffeine beverages in moderation was related to pregnancy complications. And a physician, writing on the controversy in *America,* concluded his essay: "Now, how about a coffee break?" (Sullivan, 1981).

Some publications have taken a more cautious approach. An editorial in the *New York Times* (1981) acknowledged that research linking coffee drinking to pancreatic cancer could not be dismissed and said additional studies were needed before one should "give up a pleasant habit." *Consumer Reports* (1982) and SCIQUEST (1981) reviewed the birth defects and cancer studies and called for more research. The major U.S. news magazines also reviewed the research and called for more studies (*Time,* 1981; *U.S. News and World Report,* 1981).

Several publications concluded that caffeine was dangerous. In *Vogue,* Bennett (1981:438) argued that "cutting down on this 'picker-upper' can add to your health and well-being." In *Moneysworth,* Hodges (1976:17) concluded "It does seem a bit silly, after all, to risk heart disease and cancer for the sake of hot and legal speed in the morning." In syndicated newspaper articles, Rovner (1980) and Mollen (1982) suggested readers limit their consumption of caffeinated beverages, as have writers in *Prevention* (Fredericks, 1982; Mazer, 1982).

Several popular periodicals have begun to use terminology formerly reserved for the "hard drugs." "Are you a caffeine addict?" asked Zander (1982) in the *Saturday Evening Post*. *Current Health* (1981) characterized the coffee break as "Drug break time." *Harpers Bazaar* published a story titled "Coffee break addiction" (Benzaia, 1978). And *Prevention* carried articles entitled "Six ways to kick the caffeine habit" (Mazer, 1982) and "Coffee head" (1979). Many of these articles use terms such as "java junkies," "tea tipplers," "cola cravers," and "caffeine withdrawal syndrome."

PARTISANS

The most effective anti-caffeine interest group in the United States has been the Center for Science in the Public Interest (CSPI). Established in Washington, DC, in 1971 to "promote people-oriented science" by publicizing research findings and petitioning and suing federal agencies (Gale Research Company, 1978:871), CSPI has focused on food and nutrition issues. The organization has a staff of 15 with 28,000 contributing members.[10] The executive director since 1971 has been Michael Jacobson, who holds a Ph.D. in molecular biology from the Massachusetts Institute of Technology. Before helping found CSPI, Jacobson was a food safety consultant for Ralph Nader. He has appeared on radio and television, been quoted extensively in the press, and written for consumer publications.

CSPI has petitioned federal agencies to ban certain food advertising and food additives and requested greater detail in food product labeling. It also appeals directly to the public through its monthly magazine, *Nutrition Action,* which contains articles on the dangers of the U.S. diet and offers recipes for healthy meals. In April of 1975, 1976, and 1977, CSPI also sponsored "National Food Day" in the United States and released a list of the "ten terribles"[11] (foods) to the media (Cerra, 1975:50).

CSPI and Jacobson have attracted the attention of the press. The *New York Times Index* mentioned CSPI or Jacobson by name 59 times from 1971 to 1981. Many of these references were favorable; some quoted CSPI publications almost word for word (Ferretti, 1979) or gave CSPI's address as a source for reliable nutrition information (*New York Times,* 1977).

CSPI launched their campaign against caffeine in 1976. Citing animal studies which linked birth defects to caffeine consumption, CSPI wrote to the Food and Drug Administration charging that caffeine consumption had harmful consequences for the human fetus. In 1978, CSPI filed a petition with the Food and Drug Administration requesting warning labels on coffee. This prompted the department's study of caffeine, which in turn led to the warning that pregnant women should minimize their coffee consumption (Cohn,

1980, *Science,* 1980). CSPI established a special caffeine and birth defects project in 1980 and continued warning the public about the consumption of coffee, tea, and soft drinks containing caffeine, as well as distributing written materials. For example, Jacobson (1981:39) wrote in *Family Health* that he had "identified several babies whose birth defects may be linked to their mother's coffee consumption." He also warned readers that other dangers may be associated with caffeine consumption.

Initially, pro-caffeine forces were not as active as the anti-caffeine partisans. This changed in the early 1980s. The International Life Sciences Institute (ILSI), the National Soft Drink Association (NSDA), and the National Coffee Association of the U.S.A. (NCA) became involved in the debate. The NSDA has been the least active, though it has distributed a pamphlet and a research paper written by a staff member which have attempted to counter the argument that caffeine is harmful. NSDA members have supported the ILSI, a foundation established and supported by corporations which produce food, drugs, and chemicals. This foundation has taken a more active role sponsoring Annual International Caffeine Workshops, conferences where the latest scientific research is discussed (Gale Research Company, 1983:966). The fourth such workshop was held in Athens, Greece, in the fall of 1982. Three U.S. scientists who attended held a press conference afterwards announcing that there was no evidence that caffeine was harmful to humans (Treichel, 1982:311). ILSI also publishes and distributes to the public a newsletter (ILSI *News*), scientific papers, and articles. In 1983 it was preparing a book on "all aspects of the psychology, pharmacology, and toxicology of caffeine."[12]

The NCA appears to have been most active in countering anti-caffeine claims. This industry-supported group seeks "to promote sound business relations and mutual understanding among members of the trade and increase coffee consumption" (Gale Research Company, 1983:518). Besides responding to media inquiries, NCA has convened expert panels for press conferences, sponsored research on the health effects of coffee and caffeine, and prepared papers for the public discussing the scientific issues involved.[13] Its newsletter, *Coffee Update,* reaches an audience of 2500.[14] Launched in January 1982, it is sent unsolicited to "scientists, health professionals, food and nutrition specialists, and consumer, science and food journalists" (National Coffee Association, 1983). The newsletter has reprinted comments critical of the link between caffeine and health and publicized industry-sponsored as well as impartial research that reported no relationship between the two.

COFFEE AND CIGARETTES

The history of the controversy over coffee in the United States is similar to that of cigarettes (Markle and Troyer, 1979; Nuehring and Markle, 1974; Troyer and Markle, 1983). Both substances have been linked to moral degeneration and medical problems. The U.S. government has issued warnings against both. Industry has responded by sponsoring medical research and publicizing counter claims. Psychologists and psychiatrists have associated consumption of both substances with negatively valued behavior and personality traits.

The strong moral overtones in arguments against both substances are similar. Cigarette smoking and coffee drinking are defined, in the narrow sense, as medical problems. Attacks on them, however, invoke a strong environmental or holistic health ideology. Defenses likewise invoke moral suasion; anti-cigarette and anti-coffee partisans are depicted as attacking one of life's innocent pleasures. Even medical researchers fall back on moral suasion in defending coffee and cigarettes. As Vaisrub (1978:1471) commented in the *Archives of Internal Medicine* "coffee competes successfully with the older pleasures of the table, the bed, and the wine cellar . . . it is a way of life." An editorial in the *British Journal of Medicine* (1976:1031) asked, "What twist is it in man's devious make-up that makes him round on the seemingly more wholesome and pleasurable aspects of his environment and suspects them of being causes of his misfortunes?" An editorial in the *New England Journal of Medicine* observed that, "It is encouraging to note that not everything one enjoys predisposes one to cardiovascular disease" (Kannell, 1977:444). Six years later another editorial in the same journal concluded that: "The physicians should take care not to diminish life's pleasures when there is no sound basis to do so, lest he or she be deemed a modern-day Savonarola" (Graboys and Lown, 1983:836).[15]

At the same time, there are several key differences between cigarette smoking and coffee drinking. By the early 1980s, there was scientific consensus that smoking was more harmful to health than caffeine. Physicians and researchers have not been leaders in the anti-coffee and anti-caffeine campaign. Indeed, a number of studies have challenged the accuracy of allegations leveled against caffeine (Linn *et al.*, 1982; Morrison *et al.*, 1982; Simon *et al.*, 1975). Furthermore, medical advice, as reflected in medical journal editorials (Graboys and Lown 1983; Kannell, 1977), has not advocated limiting caffeine consumption.

Another crucial difference is the effects of cigarette and coffee consumption on non-users. The anti-smoking movement argues that non-smokers suffer from tobacco consumed by others. While coffee consumption generally does not effect others, caffeine consumption by pregnant women exposes the fetus to a psychoactive drug. It seems likely that anti-coffee forces will focus on the "innocent victim."

There are both similarities and differences in the patterns and rates of consumption for coffee and cigarettes. After the federal government forced tobacco companies to print a health warning on their products in 1965, cigarette consumption dropped, especially among men. In 1955, 56.9 percent of all U.S. males over 18 years of age smoked cigarettes (U.S. Department of Health, Education, and Welfare, 1970). By 1975, only 39.3 percent did so (U.S. Department of Health, Education, and Welfare, 1977). Coffee consumption also declined dramatically in the United States. In 1980, consumption was 68.9 percent of what it was in 1957, measured in pounds per capita (Maxwell, 1981). The International Coffee Organization (1982) reported that 56.3 percent of the U.S. population drank coffee in 1982, compared with 74.7 percent in 1962. In 1962 coffee drinkers in the United States consumed 4.17 cups a day; by 1982 they drank only 3.38 cups a day. Though consumption was down in all age groups, the young reduced their coffee intake most dramatically; between 1957 and 1980 consumption declined 75.2 percent for the 15 to 19 age group (Maxwell, 1981); in 1982 those between 20 and 39 consumed two cups less per day than people in the same age group had 20 years previously.

The decline in coffee consumption does not appear to be related to health concerns. Although the proportion of the U.S. population drinking decaffeinated coffee has increased from 4 percent in 1962 to 15.7 percent in 1982, the consumption of tea and caffeinated soft drinks has increased dramatically over the past 20 years (International Coffee Organization, 1982). The coffee industry attributes coffee's decline to poor advertising. Whereas soft drinks have been associated with youth, fun, and sexuality, coffee advertising has emphasized maturity and marriage. The industry has been trying to change the image of coffee so "it will appeal more to young people and appear better-suited to modern lifestyles" (International Coffee Organization, 1982:2). As Reddicliffe (1983) observed, coffee ads have begun to feature relatively young people in situations emphasizing feelings, relationships, and active lifestyles.

CONCLUSION

To some, coffee drinking poses what Brecher (1972:205) has called a paradox:

> Thus we come to the coffee paradox—the question of how a drug so fraught with potential hazard can be consumed in the United States at the rate of more than a hundred billion doses a year without arousing the kind of hostility, legal repression, and social condemnation aroused by the illicit drugs.

Apparently "the morning cup of coffee is so much a part of American and European dietary habit that one seldom looks upon its consumption as a drug habit" (Goodman and Gillman, 1975:376).

The acceptability of coffee might be changing. During the 1970s and 1980s, some experts alleged that caffeine caused cancer, heart disease, and birth defects. Citing these studies, a consumer group—the Center for Science in the Public Interest—spurred the Food and Drug Administration to warn pregnant women against caffeine, and various professional and popular periodicals began using negative terms to characterize caffeine consumers. Vested interest groups and independent researchers have challenged the original claims, blunting the anti-caffeine movement and casting doubt on the definition of coffee as a social problem.

Our study of the history of the controversy over coffee sheds some light on the social problems process. The anti-coffee groups were unable to document claims or to gain allies or support from other groups, scientists, experts, or the medical community. This suggests that support from the acknowledged experts is crucial when making a medical claim. Finally, the importance of opposition from vested interest groups cannot be underestimated. These groups' immediate response and launching of a counter claims-making drive may negate the impact of initial claims.

The complexity and ambiguity surrounding the controversy over coffee is not surprising. Rather, it might be instructive for social scientists. Many health-related activities—some of which have been indicted, others of which have been exonerated—have a history of such confusion. Only by studying how claims succeed or fail to create problems can we come to understand and appreciate the social problems process.

NOTES

1. The early history of tea is similar. In England, tea was first sold publicly in 1657 and quickly gained devotees. Although the nephew of a secretary in the royal government publicly characterized consumption as "a base Indian practice" (Ukers, 1935b:47), tea became popular among royalty. Thomas Garway, the keeper of a coffee house known as a center for great mercantile transactions, extolled its virtues (Ukers, 1935b:38). During the 18th century, however, tea became the subject of fierce controversy. In 1745, the *Female Spectator* denounced it as "the bane of Housewifery" (Ukers, 1935b:47). A leading economist described its effects on the economy as evil (1935b:47). John Wesley, the founder of Methodism, urged his followers to abstain from drinking tea and give the money saved to charity (1935b:48). Jonas Hanway, a wealthy London merchant, labeled tea as "pernicious to health, obstructing industry, and impoverishing the nation" (1935b:48). On the other side, such writers as Samuel Johnson, Alexander Pope, and Samuel Coleridge defended the beverage. In spite of the controversy, tea consumption increased; by 1765 90 percent of families in England drank tea (Coffey, 1966).

2. "Necessity" here appears to mean a pervasive social custom since the author notes that coffee is used by the poor and well-to-do alike.

3. The same article quoted other physicians who said they did "not believe tea and coffee [are] harmful . . ." The editor of the *Tea and Coffee Trade Journal* wrote to the *New York Times,* identifying medical authorities who had pronounced coffee safe and arguing that "coffee is good for most people" (Ukers, 1913:10).

4. We searched the *New York Times* and the *Reader's Guide to Periodical Literature* because they are the best record of media attention in the United States since the 1890s. Schwartz and Leitko (1977) have described newspapers as "thermometers" for the rise of social problems. That is exactly what we are trying to measure here. Clearly, these sources have limitations. The *New York Times* probably reflects regional concerns and not all periodicals are indexed by the *Reader's Guide to Periodical Literature.*

5. Automatic drip coffee has more caffeine (110–150 mg) than does percolated (64–124 mg) or instant (40–108 mg).

6. Caffeine is added to some non-cola drinks such as Mountain Dew and Mellow Yellow.

7. Caffeine is added to various prescription and non-prescription drugs. Stimulants, such as Nodoz, contain 200 mg caffeine per tablet; headache remedies, such as Excedrin, contain 65 mg caffeine per tablet.

8. Goldstein (1982) denies entirely the link between coffee and cancer of the pancreas.

9. The subject headings indexing these studies have varied over the years. From 1970 to 1972, reports of psychological research on the effects of both caffeine and coffee were listed under "drug effects." Since 1973, these studies have been indexed under "Caffeine" and "coffee" subject headings with many reports listed under both titles. The source of caffeine in most of these studies was either tablets or coffee.

10. Claire Feinson, CSPI staff member, Washington, DC, May 4, 1983; telephone interview.

11. The "ten terribles" of 1975 were grapes, Coca-Cola, breakfast squares, bacon, Wonderbread, sugar, Gerber baby food, Frute Brute, prime grade beef, and Pringles. Reasons for identifying these products ranged from a political consumer boycott of grapes, to additives in bacon, to high sugar or fat content in the other seven products.

12. Mary Rita Prah, Administrator of ILSI, Washington, DC, April 7, 1983: personal communication.

13. Among the unpublished papers are "Caffeine and birth defects: Scientific update;" "Methylxanthine consumption and fibrocystic breast disease: A scientific update;" "Coffee consumption and pancreatic cancer: A scientific update;" and "Decaffeinated coffee and methylene chloride: A scientific update." Available from National Coffee Association of U.S.A., 120 Wall Street, New York, NY 10005.

14. David Kuhnert, NCA staff member, New York, August 1, 1983: telephone interview.

15. Girolama Savonarola was a 15th century Florentine priest who led campaigns against various pleasures he blamed for moral decline. His efforts led to new laws against gambling and sodomy.

REFERENCES

Adler, I.D. 1970. "The problem of caffeine mutagenicity." Pg. 383–403 in F. Vogel and G. Rohrborn (eds.), Chemical Mutagenesis in Mammals and Man. New York: Springer-Verlag.

American Psychiatric Association. 1980. Diagnostic and Statistical Manual of Mental Disorders. Third edition. Washington, DC: American Psychiatric Association.

American Psychological Association. 1970–1982. Psychological Abstracts, volumes 44–68. Arlington, VA: American Psychological Association.

Americas. 1956. "How to live longer." Americas (Washington, DC) 8 (September):39.

Anft, Berthold. 1955. "Friedlieb Ferdinand Runge: A forgotten chemist of the nineteenth century." Journal of Chemical Education 32:566–574.

Ayers, Joyce, Carol F. Ruff, and Donald I. Templer. 1976. "Alcoholism, cigarette smoking, coffee drinking, and extroversion." Journal of Studies on Alcohol 37(7):983–985.

Bennett, William. 1981. "Coffee, tea, cola and you: Reasons for caffeine caution." Vogue 171 (October):438.

Benzaia, Diana. 1978. "Coffee break addiction." Harper's Bazaar 112 (November):62.

Blumer, Herbert. 1971. "social problems as collective behavior." Social Problems 18 (Winter):298–306.

Boston Collaborative Drug Surveillance Program. 1972. "Coffee drinking and acute myocardial infarction." Lancet 2(7790):1278–1279.

Brecher, Edward M. 1972. Licit and Illicit Drugs. Boston: Little, Brown.

British Journal of Medicine. 1976. "Caffeine, coffee and cancer." Editorial. British Medical Journal 1(6017):1031–1032.

Cerra, Frances. 1975. "Food day: The focus of a cause for everybody." New York Times, January 26:50.

Coffey, Timothy G. 1966. "Beer street: gin lane: Some views of 18th-century drinking." Quarterly Journal of Studies on Alcohol 27(4):669–692.

Cohn, Victor. 1980. "Caffeine, birth defect link feared." Des Moines Register, September 4:sec A, p. 16.

Cole, Phillip. 1971. "Coffee drinking and cancer of the lower urinary tract." Lancet 2(7732):1335–1337.

Consumer Reports. 1982. "Caffeine in pregnancy: A risk or false alarm?" Consumer Reports 47 (May):228.

Crothers, T. D. 1902. "Coffee and tea drunkeness." Current Literature 32 (June):740–741. First published in T. D. Crothers, Morphinism and Narcomaniacs From Other Drugs. Philadelphia: W. B. Saunders, 1902.

Current Health. 1981. "Drug break time?" Current Health 8 (November):12–13.

Demerc, M., G. Bertani, and J. Flint. 1951. "A survey of chemicals for mutagenic action on E. Loli." American Naturalist 85(821):119–136.

Dews, P. B. 1982. "Caffeine." Annual Review of Nutrition 2:223–241.

Epstein, Samuel. 1970. "The failure of caffeine to induce mutation effects or to synergize the effects of known mutagens in mice. Pp. 404–419 in

F. Vogel and G. Rohrborn (eds.), Chemical Mutagenesis in Mammals and Man. New York: Springer-Verlag.

Ferretti, Fred. 1979. "Eyeing what goes into food." New York Times, July 26:sec 3, p. 10.

Fredericks, Carlton. 1982. "Mental symptoms from excessive coffee intake: Hotline to health." Prevention 34 (March):26.

Fox, Ernest. 1904. "The coffee-club movement in California." Arena 32 (November):519–521.

Foxx, R. M., and Andrea Rubinoff. 1979. "Behavioral treatment of caffeinism: Reducing excessive coffee drinking." Journal of Applied Behavior Analysis 12(12):335–344.

Fries, Nils, and Bengt Kihlman. 1948. "Fungal mutations obtained with methylxanthime." Nature 162(4119):573–574.

Gale Research Company. 1978. Encyclopedia of Associations, Volume 1. 12th edition. Detroit: Gale Research.

————. 1983. Encyclopedia of Associations, Volume 1. 17th edition. Detroit: Gale Research.

Gillespie, Charles C. 1972. Dictionary of Scientific Biography, Volume 5. New York: Charles Scribner and Sons.

Gilliland, Kirby. 1980. "The interactive effect of introversion-extroversion with caffeine induced arousal on verbal performance." Journal of Research in Personality 14 (December):482–492.

Gilliland, Kirby, and Dana Andress. 1981. "Ad lib caffeine consumption, symptoms of caffeinism, and academic performance." American Journal of Psychiatry 138(4):512–514.

Goldstein, H. R. 1982. "No association between coffee and cancer of the pancreas." New England Journal of Medicine 306(16):997.

Goodman, Louis, S., and Alfred Gilman. 1975. The Pharmacological Basis of Therapeutics. 5th edition. New York: MacMillan.

Graboys, Thomas B., and Bernard Lown. 1983. "Coffee, arrhythmias, and common sense." New England Journal of Medicine 308(13):835–836.

Graham, D. M. 1978. "Caffeine: Its identity, dietary sources, intake, and biological effects." Nutrition Reviews 36(4):97–102.

Greden, John F. 1974. "Anxiety or caffeinism: A diagnostic dilemma." American Journal of Psychiatry 131(10):1089–1092.

Greden, John F., Patricia Fontaine, Martin Lubetsky, and Kenneth Chamberlin. 1978. "Anxiety and depression associated with caffeinism among psychiatric inpatients." American Journal of Psychiatry 135(8):963–966.

Greden, John F., Andrew Procter, and Bruce S. Victer. 1981. "Caffeinism associated with greater use of other psychotropic agents." Comprehensive Psychiatry 22 (November-December):565–571.

Hawk, Philip B. 1916. "What we eat: And what happens to it." Ladies Home Journal 33 (December):37, 74.

Herber, W. 1982. "Coffee-linked birth problems premature." Science News 121 (January):68.

Hire, Jerome N. 1978. "Anxiety and caffeine." Psychological Reports 42(3):833–834.

Hodges, Parker. 1976. "Drinking coffee: Grounds for concern." Moneysworth 6 (October):16–17.

Howerth, I. W. 1895. "The coffee-house as a rival of the saloon." American Magazine of Civics 6 (June):589–602.

International Coffee Organization. 1982. United States of America Coffee Drinking Study, Winter, 1982. London: International Coffee Organization.

Jacob, Heinrich. 1935. Coffee. Translated by Eden Paul and Cedar Paul. New York: Viking.

Jacobs, Paul. 1982. "Study: Birth problems not tied to coffee drinking." Des Moines Tribune, January 25:18.

Jacobson, Michael F. 1981. "The caffeine catch." Family Health 13 (April):20–21, 39.

Johnson, Helen Louise. 1925. "Scientific research in coffee brewing." Journal of Home Economics 17 (April):199–205.

Kannell, William B. 1977. "Coffee, cocktails, and coronary candidates." New England Journal of Medicine 297(8):443–444.

Koch, Arthur L. 1956. "The metabolism of methylpurines by E. Coli. I Tracer studies." Journal of Biological Chemistry 219:181–188.

Kuhlmann, W., H. G. Fromme, E. M. Hegge, and W. Ostertag. 1968. "The mutagenic action of caffeine in higher organisms." Cancer Research 28(3):2375–2389.

Linn, Shai, Stephen C. Schoenbaum, Richard Monroe, Bernard Rosner, Philip Stubblefield, and Kenneth Ryan. 1982. "No association between coffee drinking and adverse outcomes of pregnancy." New England Journal of Medicine 306(3):141–145.

MacMahon, Brian, Stella Yen, Dimitrios Trichopolous, Kenneth Warren, and George Nardi. 1981. "Coffee and cancer of the pancreas." New England Journal of Medicine 304(11):630–633.

Markle, Gerald E., and Ronald J. Troyer. 1979. "Smoke gets in your eyes: Cigarette smoking as deviant behavior." Social Problems 26(June):611–625.

Marrett, Loraine D., Stephen D. Walter, and J. W. Meigs. 1983. "Coffee drinking and bladder cancer in Connecticut." American Journal of Epidemiology 117(2):113–126.

Maxwell, J. C. 1981. "Coffee drinking drops." Advertising Age 52:74.

Mazer, Eileen. 1982. "Six ways to kick the caffeine habit." Prevention 34 (August):83–88.

Mollen, Art. 1982. "Health risks are seen in drinking too much coffee." Des Moines Register, November 23:21.

Morrison, A. S., J. E. Buring, and W. G. Verhock. 1982. "Coffee drinking and cancer of the lower urinary tract." Journal of the National Cancer Institute 68:91–94.

National Coffee Association. 1983. "Editor's note." Coffee Update 2 (September):2.

New York Times. 1878. Untitled. September 9:4.

———. 1879a. Untitled. April 14:4.

———. 1879b. "Moral effects of tea and coffee." August 16:4.

———. 1880. "Coffee against alcohol." March 7:4.

———. 1912. "Is the drinking of tea and coffee harmful to health?" September 15: sec. 5, p. 11.

———. 1924. "Prove coffee is not intoxicating." August 26:24.

———. 1928. "Tea and tension." August 9:18.

———. 1929a. "Savant lays ills to giving up beer." July 25:6.

———. 1929b. "Topics of the times." August 24:12.

———. 1929c. "Coffee." September 1:5.

———. 1931. "Finds coffee, if fresh, is an aid to digestion." May 5:29.

———. 1938. "Tea and coffee praised as aids to vitality; Prof. Eddy also says they will induce sleep." June 11:17.

———. 1953. "Tea sparks intellect, cigar the ego, according to a report on stimulants." January 17:17.

———. 1954. "Notes on science: Coffee drinking as drug habit." August 29: sec. 4, p. 9.

———. 1956. "Coffee break justified." September 6:27.

———. 1963. "Coffeepot tempest." July 12:30.

———. 1977. "How parents can get involved." November 16: sec. C, p. 5.

———. 1981. "Coffee and cancer." March 14:22.

Newsweek. 1964. "Break with coffee?" Newsweek, February 17:68, 71.

Nuehring, Elaine, and Gerald E. Markle. 1974. "Nicotine and norms: The re-emergence of a deviant behavior." Social Problems 21 (April):513–526.

Osborne, Oliver T. 1924. "Children should not drink coffee or tea." Good Housekeeping 79 (October):284–286.

Ostertag, W., E. Dusberg, and H. Sturman. 1965. "The mutagenic activity of caffeine in man." Mutation Research 2(3):293–296.

Ponte, Lowell. 1983. "All about caffeine." Readers Digest 122 (January):72–76.

Prevention. 1979. "Coffee head." Prevention 31 (March):42.

Reddicliffe, Steven. 1983. "Eets thuh reechest kand, don't yoo see." Des Moines Sunday Register, April 10: TV section, p. 3.

Rovner, Sandy. 1980. "Some wives' tales of pregnancy true." Des Moines Register, November 23: sec. E, p. 4.

Schwartz, T. P., and Thomas Leitko. 1977. "The rise of social problems: Newspapers as 'thermometers'." Pp. 427–436 in Armand L. Mauss and Julie Camile Wolfe (eds.), This Land of Promise, Philadelphia: J. B. Lippincott.

Science. 1980. "FDA caffeine decision too early, some say." Science 209 (September 26):1500.

Science Digest. 1940. "Coffee drinkers' #1 headache." Science Digest 7 (May):43.

———. 1956. "Coffee has vitamins." Science Digest 40 (July):49.

———. 1963. "Coffee and your heart." Science Digest 54 (October):55.

Science News Letter. 1955. "30 cups of coffee daily are probably harmless." Science News Letter 67 (April 2):216.

SCIQUEST. 1981. "Coffee and cancer." SCIQUEST (Columbus, Ohio) 54 (May/June):4–5.

Simon, David, Stella Yen, and Philip Cole. 1975. "Coffee drinking and cancer of the lower urinary tract." Journal of the National Cancer Institute 54(3):587–591.

Smith, Barry D., Craig B. Rypma, and Raymond J. Wilson. 1981. "Dishabituation and spontaneous recovery of electrodermal orienting response: Effect of extraversion, impulsivity, sociability, and caffeine." Journal of Research in Personality 15 (June):233–240.

Spector, Malcolm, and John I. Kitsuse. 1977. Constructing Social Problems. Menlo Park, CA: Cummings.

Sullivan, Philip R. 1981. "Coffee, cancer, reporting: The anatomy of a recent medical report." America 145 (September):93–94.

Time. 1981. "Coffee nerves." Time, March 23:73.

Timson, John. 1978. "That caffeindishly addictive drug earns a B+ on the bill of health." Science Digest 48 (November):44–48.

Treichel, J. A. 1982. "Good news for caffeine consumers." Science News 122 (November):311.

Troyer, Ronald J., and Gerald E. Markle. 1983. Cigarettes: The Battle over Smoking. New Brunswick, NJ: Rutgers University Press.

Ukers, William H. 1913. Letter to the Editor. New York Times, April 18:10.

———. 1935a. All about Coffee. New York: The Tea and Coffee Trade Journal Company.

———. 1935b. All about Tea, Volume 1. New York: The Tea and Coffee Trade Journal Company.

Uribe, C. Andress. 1954. Brown, Gold: The Amazing Story of Coffee. New York: Random House.

USA Today. 1980. "Morning cup of coffee and heartburn." USA Today 109 (August):4.

U.S. Department of Health, Education and Welfare. 1970. Changes in smoking habits between 1955 and 1966. Washington, DC: U.S. Government Printing Office.

———. 1977. The Smoking Digest: Progress Report on a Nation Kicking the Habit. Washington, DC: U.S. Government Printing Office.

———. 1980. "Use habits of cigarettes, coffee, aspirin, and sleeping pills, United States, 1976." Vital Health and Statistics, series 10, 131:80–1559. Washington, DC.

U.S. News and World Report. 1981. "Coffee and cancer: A controversial link." U.S. News and World Report, March 23:8.

Vaisrub, Samule. 1978. "Coffee—Grounds for reassurance." Archives of Internal Medicine 138 (October):1471–1478.

Wassersug, Joseph D. 1946. "Coffee can cause ulcers." Science Digest 20 (December):25–28.

White, Brent C., Cynthia A. Lincoln, Nell W. Pearce, Robin Reeb, and Chris Vaida. 1980. "Anxiety and muscle tension as a consequence of caffeine withdrawal." Science 209 (September):1547–1548.

Winstead, Daniel K. 1976. "Coffee consumption among psychiatric inpatients." American Journal of Psychiatry 133(12):1447–1450.

Zaner, Richard H. 1982. "Are you a caffeine addict?" Saturday Evening Post 254 (May/June):50–53

7

The Social Construction

of Drug Scares

CRAIG REINARMAN

Drug "wars," anti-drug crusades, and other periods of marked public concern about drugs are never merely reactions to the various troubles people can have with drugs. These drug scares are recurring cultural and political phenomena *in their own right* and must, therefore, be understood sociologically on their own terms. It is important to understand why people ingest drugs and why some of them develop problems that have something to do with having ingested them. But the premise of this chapter is that it is equally important to understand patterns of acute societal concern about drug use and drug problems. This seems especially so for U.S. society, which has had *recurring* anti-drug crusades and a *history* of repressive anti-drug laws.

Many well-intentioned drug policy reform efforts in the U.S. have come face to face with staid and stubborn sentiments against consciousness-altering substances. The repeated failures of such reform efforts cannot be explained solely in terms of ill-informed or manipulative leaders. Something deeper is involved, something woven into the very fabric of American culture, something which explains why claims that some drug is the cause of much of what is wrong with the world are *believed* so often by so many. The origins and nature of the *appeal* of anti-drug claims must be confronted if we are ever to understand how "drug problems" are constructed in the U.S. such that more enlightened and effective drug policies have been so difficult to achieve.

In this chapter I take a step in this direction. First, I summarize briefly some of the major periods of anti-drug sentiment in the U.S. Second, I draw from them the basic ingredients of which drug scares and drug laws are made. Third, I offer a beginning interpretation of these scares and laws based on those broad features of American culture that make *self-control* continuously problematic.

DRUG SCARES AND DRUG LAWS

What I have called drug scares (Reinarman and Levine, 1989a) have been a recurring feature of U.S. society for 200 years. They are relatively autonomous from whatever drug-related problems exist or are said to exist.[1] I call them "scares" because, like Red Scares, they are a form of moral panic

From: Reprinted by permission of Craig Reinarman.

ideologically constructed so as to construe one or another chemical bogey-man, à la "communists," as the core cause of a wide array of pre-existing pub-lic problems.

The first and most significant drug scare was over drink. Temperance movement leaders constructed this scare beginning in the late 18th and early 19th century. It reached its formal end with the passage of Prohibition in 1919.[2] As Gusfield showed in his classic book *Symbolic Crusade* (1963), there was far more to the battle against booze than long-standing drinking prob-lems. Temperance crusaders tended to be native born, middle-class, non-urban Protestants who felt threatened by the working-class, Catholic immi-grants who were filling up America's cities during industrialization.[3] The latter were what Gusfield termed "unrepentant deviants" in that they continued their long-standing drinking practices despite middle-class W.A.S.P. norms against them. The battle over booze was the terrain on which was fought a cornucopia of cultural conflicts, particularly over whose morality would be the dominant morality in America.

In the course of this century-long struggle, the often wild claims of Temperance leaders appealed to millions of middle-class people seeking explanations for the pressing social and economic problems of industrializing America. Many corporate supporters of Prohibition threw their financial and ideological weight behind the Anti-Saloon League and other Temperance and Prohibitionist groups because they felt that traditional working-class drinking practices interfered with the new rhythms of the factory, and thus with productivity and profits (Rumbarger, 1989). To the Temperance cru-saders' fear of the bar room as a breeding ground of all sorts of tragic immorality, Prohibitionists added the idea of the saloon as an alien, subver-sive place where unionists organized and where leftists and anarchists found recruits (Levine, 1984).

This convergence of claims and interests rendered alcohol a scapegoat for most of the nation's poverty, crime, moral degeneracy, "broken" families, ille-gitimacy, unemployment, and personal and business failure—problems whose sources lay in broader economic and political forces. This scare climaxed in the first two decades of this century, a tumultuous period rife with class, racial, cultural, and political conflict brought on by the wrenching changes of indus-trialization, immigration, and urbanization (Levine, 1984; Levine and Reinar-man, 1991).

America's first real drug law was San Francisco's anti-opium den ordinance of 1875. The context of the campaign for this law shared many features with the context of the Temperance movement. Opiates had long been widely and legally available without a prescription in hundreds of medicines (Brecher, 1972; Musto, 1973; Courtwright, 1982; cf. Baumohl, 1992), so neither opiate use nor addiction was really the issue. This campaign focused almost exclu-sively on what was called the "Mongolian vice" of opium *smoking* by Chinese immigrants (and white "fellow travelers") in dens (Baumohl, 1992). Chinese immigrants came to California as "coolie" labor to build the railroad and dig

the gold mines. A small minority of them brought along the practice of smoking opium—a practice originally brought to China by British and American
traders in the 19th century. When the railroad was completed and the gold
dried up, a decade-long depression ensued. In a tight labor market, Chinese
immigrants were a target. The white Workingman's Party fomented racial
hatred of the low-wage "coolies" with whom they now had to compete for
work. The first law against opium smoking was only one of many laws enacted to harass and control Chinese workers (Morgan, 1978).

By calling attention to this broader political-economic context I do not
wish to slight the specifics of the local political-economic context. In addition
to the Workingman's Party, downtown businessmen formed merchant associations and urban families formed improvement associations, both of which
fought for more than two decades to reduce the impact of San Francisco's vice
districts on the order and health of the central business district and on family
neighborhoods (Baumohl, 1992).

In this sense, the anti-opium den ordinance was not the clear and direct
result of a sudden drug scare alone. The law was passed against a specific form
of drug use engaged in by a disreputable group that had come to be seen as
threatening in lean economic times. But it passed easily because this new
threat was understood against the broader historical backdrop of long-standing
local concerns about various vices as threats to public health, public morals,
and public order. Moreover, the focus of attention were dens where it was
suspected that whites came into intimate contact with "filthy, idolatrous"
Chinese (see Baumohl, 1992). Some local law enforcement leaders, for example, complained that Chinese men were using this vice to seduce white
women into sexual slavery (Morgan, 1978). Whatever the hazards of opium
smoking, its initial criminalization in San Francisco had to do with both a
general context of recession, class conflict, and racism, and with specific local
interests in the control of vice and the prevention of miscegenation.

A nationwide scare focusing on opiates and cocaine began in the early
20th century. These drugs had been widely used for years, but were first criminalized when the addict population began to shift from predominantly white,
middle-class, middle-aged women to young, working-class males, African-
Americans in particular. This scare led to the Harrison Narcotics Act of 1914,
the first federal anti-drug law (see Duster, 1970).

Many different moral entrepreneurs guided its passage over a six-year campaign: State Department diplomats seeking a drug treaty as a means of expanding trade with China, trade which they felt was crucial for pulling the economy out of recession; the medical and pharmaceutical professions whose
interests were threatened by self-medication with unregulated proprietary ton-
ics, many of which contained cocaine or opiates; reformers seeking to control
what they saw as the deviance of immigrants and Southern Blacks who were
migrating off the farms; and a pliant press which routinely linked drug use
with prostitutes, criminals, transient workers (e.g., the Wobblies), and African-
Americans (Musto, 1973). In order to gain the support of Southern Congress-

men for a new federal law that might infringe on "states' rights," State Department officials and other crusaders repeatedly spread unsubstantiated suspicions, repeated in the press, that, e.g., cocaine induced African-American men to rape white women (Musto, 1973:6-10, 67). In short, there was more to this drug scare, too, than mere drug problems.

In the Great Depression, Harry Anslinger of the Federal Narcotics Bureau pushed Congress for a federal law against marijuana. He claimed it was a "killer weed" and he spread stories to the press suggesting that it induced violence—especially among Mexican-Americans. Although there was no evidence that marijuana was widely used, much less that it had any untoward effects, his crusade resulted in its criminalization in 1937—and not incidentally a turnaround in his Bureau's fiscal fortunes (Dickson, 1968). In this case, a new drug law was put in place by a militant moral-bureaucratic entrepreneur who played on racial fears and manipulated a press willing to repeat even his most absurd claims in a context of class conflict during the Depression (Becker, 1963). While there was not a marked scare at the time, Anslinger's claims were never contested in Congress because they played upon racial fears and widely held Victorian values against taking drugs solely for pleasure.

In the drug scare of the 1960s, political and moral leaders somehow reconceptualized this same "killer weed" as the "drop out drug" that was leading America's youth to rebellion and ruin (Himmelstein, 1983). Bio-medical scientists also published uncontrolled, retrospective studies of very small numbers of cases suggesting that, in addition to poisoning the minds and morals of youth, LSD produced broken chromosomes and thus genetic damage (Cohen et al., 1967). These studies were soon shown to be seriously misleading if not meaningless (Tjio et al., 1969), but not before the press, politicians, the medical profession, and the National Institute of Mental Health used them to promote a scare (Weil, 1972:44–46).

I suggest that the reason even supposedly hard-headed scientists were drawn into such propaganda was that dominant groups felt the country was at war—and not merely with Vietnam. In this scare, there was not so much a "dangerous class" or threatening racial group as multi-faceted political and cultural conflict, particularly between generations, which gave rise to the perception that middle-class youth who rejected conventional values were a dangerous threat.[4] This scare resulted in the Comprehensive Drug Abuse Control Act of 1970, which criminalized more forms of drug use and subjected users to harsher penalties.

Most recently we have seen the crack scare, which began in earnest *not* when the prevalence of cocaine use quadrupled in the late 1970s, nor even when thousands of users began to smoke it in the more potent and dangerous form of freebase. Indeed, when this scare was launched, crack was unknown outside of a few neighborhoods in a handful of major cities (Reinarman and Levine, 1989a) and the prevalence of illicit drug use had been dropping for several years (National Institute on Drug Use, 1990). Rather, this most recent scare began in 1986 when freebase cocaine was renamed crack (or "rock") and sold in pre-cooked, inexpensive units on ghetto streetcorners (Reinarman and Levine, 1989b). Once politicians and the media linked this new form of cocaine use to the inner-city, minority poor, a new drug scare was underway and the solution became more prison cells rather than more treatment slots.

The same sorts of wild claims and Draconian policy proposals of Temperance and Prohibition leaders re-surfaced in the crack scare. Politicians have so outdone each other in getting "tough on drugs" that each year since crack came on the scene in 1986 they have passed more repressive laws providing billions more for law enforcement, longer sentences, and more drug offenses punishable by death. One result is that the U.S. now has more people in prison than any industrialized nation in the world—about half of them for drug offenses, the majority of whom are racial minorities.

In each of these periods more repressive drug laws were passed on the grounds that they would reduce drug use and drug problems. I have found no evidence that any scare actually accomplished those ends, but they did greatly expand the quantity and quality of social control, particularly over subordinate groups perceived as dangerous or threatening. Reading across these historical episodes one can abstract a recipe for drug scares and repressive drug laws that contains the following *seven ingredients:*

1. A Kernel of Truth   Humans have ingested fermented beverages at least since human civilization moved from hunting and gathering to primitive agriculture thousands of years ago (Levine, forthcoming). The pharmacopia has expanded exponentially since then. So, in virtually all cultures and historical epochs, there has been sufficient ingestion of consciousness-altering chemicals to provide some basis for some people to claim that it is a problem.

2. Media Magnification   In each of the episodes I have summarized and many others, the mass media has engaged in what I call the *routinization of caricature*—rhetorically re-crafting worst cases into typical cases and the episodic into the epidemic. The media dramatize drug problems, as they do other problems, in the course of their routine news-generating and sales-promoting procedures (see Brecher, 1972:321–34; Reinarman and Duskin, 1992; and Molotch and Lester, 1974).

3. Politico-Moral Entrepreneurs   I have added the prefix "politico" to Becker's (1963) seminal concept of moral entrepreneur in order to emphasize the fact that the most prominent and powerful moral entrepreneurs in drug scares are often political elites. Otherwise, I employ the term just as he intended: to denote the *enterprise,* the work, of those who create (or enforce) a rule against what they see as a social evil.[5]

In the history of drug problems in the U.S., these entrepreneurs call attention to drug using behavior and define it as a threat about which "something must be done." They also serve as the media's primary source of sound bites on the dangers of this or that drug. In all the scares I have noted, these entrepreneurs had interests of their own (often financial) which had little to do with drugs. Political elites typically find drugs a functional demon in that (like "outside agitators") drugs allow them to deflect attention from other, more systemic sources of public problems for which they would otherwise have to take some responsibility. Unlike almost every other political issue, however, to be "tough on drugs" in American political culture allows a leader to take a firm stand without risking votes or campaign contributions.

4. Professional Interest Groups In each drug scare and during the passage of each drug law, various professional interests contended over what Gusfield (1981:10–15) calls the "ownership" of drug problems—"the ability to create and influence the public definition of a problem" (1981:10), and thus to define what should be done about it. These groups have included industrialists, churches, the American Medical Association, the American Pharmaceutical Association, various law enforcement agencies, scientists, and most recently the treatment industry and groups of those former addicts converted to disease ideology.[6] These groups claim for themselves, by virtue of their specialized forms of knowledge, the legitimacy and authority to name what is wrong and to prescribe the solution, usually garnering resources as a result.

5. Historical Context of Conflict This trinity of the media, moral entrepreneurs, and professional interests typically interact in such a way as to inflate the extant "kernel of truth" about drug use. But this interaction does not by itself give rise to drug scares or drug laws without underlying conflicts which make drugs into functional villains. Although Temperance crusaders persuaded millions to pledge abstinence, they campaigned for years without achieving alcohol control laws. However, in the tumultuous period leading up to Prohibition, there were revolutions in Russia and Mexico, World War I, massive immigration and impoverishment, and socialist, anarchist, and labor movements, to say nothing of increases in routine problems such as crime. I submit that all this conflict made for a level of cultural anxiety that provided fertile ideological soil for Prohibition. In each of the other scares, similar conflicts—economic, political, cultural, class, racial, or a combination—provided a context in which claims makers could viably construe certain classes of drug users as a threat.

6. Linking a Form of Drug Use to a "Dangerous Class" Drug scares are never about drugs per se, because drugs are inanimate objects without social consequence until they are ingested by humans. Rather, drug scares are about the use of a drug by particular groups of people who are, typically, *already* perceived by powerful groups as some kind of threat (see Duster, 1970; Himmelstein, 1978). It was not so much alcohol problems *per se* that most animated the drive for Prohibition but the behavior and morality of what dominant groups saw as the "dangerous class" of urban, immigrant, Catholic, working-class drinkers (Gusfield, 1963; Rumbarger, 1989). It was *Chinese* opium smoking dens, not the more widespread use of other opiates, that prompted California's first drug law in the 1870s. It was only when smokable cocaine found its way to the African-American and Latino underclass that it made headlines and prompted calls for a drug war. In each case, politico-moral entrepreneurs were able to construct a "drug problem" by linking a substance to a group of users perceived by the powerful as disreputable, dangerous, or otherwise threatening.

7. Scapegoating a Drug for a Wide Array of Public Problems   The final ingredient is scapegoating, i.e., blaming a drug or its alleged effects on a group of its users for a variety of pre-existing social ills that are typically only indirectly associated with it. Scapegoating may be the most crucial element because it gives great explanatory power and thus broader resonance to claims about the horrors of drugs (particularly in the conflictual historical contexts in which drug scares tend to occur).

Scapegoating was abundant in each of the cases noted above. To listen to Temperance crusaders, for example, one might have believed that without alcohol use, America would be a land of infinite economic progress with no poverty, crime, mental illness, or even sex outside marriage. To listen to leaders of organized medicine and the government in the 1960s, one might have surmised that without marijuana and LSD there would have been neither conflict between youth and their parents nor opposition to the Vietnam War. And to believe politicians and the media in the past 6 years is to believe that without the scourge of crack the inner cities and the so-called underclass would, if not disappear, at least be far less scarred by poverty, violence, and crime. There is no historical evidence supporting any of this.

In short, drugs are richly functional scapegoats. They provide elites with fig leaves to place over unsightly social ills that are endemic to the social system over which they preside. And they provide the public with a restricted aperture of attribution in which only a chemical bogeyman or the lone deviants who ingest it are seen as the cause of a cornucopia of complex problems.

TOWARD A CULTURALLY-SPECIFIC
THEORY OF DRUG SCARES

Various forms of drug use have been and are widespread in almost all societies comparable to ours. A few of them have experienced limited drug scares, usually around alcohol decades ago. However, drug scares have been *far* less common in other societies, and never as virulent as they have been in the U.S. (Brecher, 1972; Levine, 1992; MacAndrew and Edgerton, 1969). There has never been a time or place in human history without drunkenness, for example, but in *most* times and places drunkenness has not been nearly as problematic as it has been in the U.S. since the late 18th century (Levine, forthcoming). Moreover, in comparable industrial democracies, drug laws are generally less repressive. Why then do claims about the horrors of this or that consciousness-altering chemical have such unusual power in American culture?

Drug scares and other periods of acute public concern about drug use are not just discrete, unrelated episodes. There is a historical pattern in the U.S. that cannot be understood in terms of the moral values and perceptions of individual anti-drug crusaders alone. I have suggested that these crusaders have benefitted in various ways from their crusades. For example, making claims about how a drug is damaging society can help elites increase the social control of groups perceived as threatening (Duster, 1970), establish one class's moral code as dominant (Gusfield, 1963), bolster a bureaucracy's sagging fiscal fortunes (Dickson, 1968), or mobilize voter support (Reinarman and Levine, 1989a,b). However, the recurring character of pharmaco-phobia in U.S. history suggests that there is something about our *culture* which makes citizens more vulnerable to anti-drug crusaders' attempts to demonize drugs. Thus, an answer to the question of America's unusual vulnerability to drug scares must address why the scapegoating of consciousness-altering substances regularly *resonates* with or appeals to substantial portions of the population.

There are three basic parts to my answer. The first is that claims about the evils of drugs are especially viable in American culture in part because they provide a welcome *vocabulary of attribution* (cf. Mills, 1940). Armed with "DRUGS" as a generic scapegoat, citizens gain the cognitive satisfaction of having a folk devil on which to blame a range of bizarre behaviors or other conditions they find troubling but difficult to explain in other terms. This much may be true of a number of other societies, but I hypothesize that this is particularly so in the U.S. because in our political culture individualistic explanations for problems are so much more common than social explanations.

Second, claims about the evils of drugs provide an especially serviceable vocabulary of attribution in the U.S. in part because our society developed from a *temperance culture* (Levine, 1992). American society was forged in the fires of ascetic Protestantism and industrial capitalism, both of which demand *self-control*. U.S. society has long been characterized as the land of the individual "self-made man." In such a land, self-control has had extraordinary importance. For the middle-class Protestants who settled, defined, and still dominate the U.S., self-control was both central to religious world views and a characterological necessity for economic survival and success in the capitalist market (Weber, 1930 [1985]). With Levine (1992), I hypothesize that in a culture in which self-control is inordinately important, drug-induced altered states of consciousness are esp~ally likely to be experienced as "loss of control," and thus to be inordinately feared.[7]

Drunkenness and other forms of drug use have, of course, been present everywhere in the industrialized world. But temperance cultures tend to arise only when industrial capitalism unfolds upon a cultural terrain deeply imbued with the Protestant ethic.[8] This means that only the U.S., England, Canada, and parts of Scandanavia have Temperance cultures, the U.S. being the most extreme case.

It may be objected that the influence of such a Temperance culture was strongest in the 19th and early 20th century and that its grip on the American *zeitgeist* has been loosened by the forces of modernity and now, many say, postmodernity. The third part of my answer, however, is that on the foundation of a Temperance culture, advanced capitalism has built a *postmodern, mass consumption culture* that exacerbates the problem of self-control in new ways.

Early in the 20th century, Henry Ford pioneered the idea that by raising wages he could simultaneously quell worker protests and increase market demand for mass-produced goods. This mass consumption strategy became central to modern American society and one of the reasons for our economic success (Marcuse, 1964; Aronowitz, 1973; Ewen, 1976; Bell, 1978). Our economy is now so fundamentally predicated upon mass consumption that theorists as diverse as Daniel Bell and Herbert Marcuse have observed that we live in a mass consumption culture. Bell (1978), for example, notes that while the Protestant work ethic and deferred gratification may still hold sway in the workplace, Madison Avenue, the media, and malls have inculcated a new indulgence ethic in the leisure sphere in which pleasure-seeking and immediate gratification reign.

Thus, our economy and society have come to depend upon the constant cultivation of new "needs," the production of new desires. Not only the hardware of social life such as food, clothing, and shelter but also the software of the self—excitement, entertainment, even eroticism—have become mass consumption commodities. This means that our society offers an increasing number of incentives for indulgence—more ways to lose self-control—and a decreasing number of countervailing reasons for retaining it.

In short, drug scares continue to occur in American society in part because people must constantly manage the contradiction between a Temperance culture that insists on self-control and a mass consumption culture which renders self-control continuously problematic. In addition to helping explain the recurrence of drug scares, I think this contradiction helps account for why in the last dozen years millions of Americans have joined 12-Step groups, more than 100 of which have nothing whatsoever to do with ingesting a drug (Reinarman, forthcoming). "Addiction," or the generalized loss of self-control, has become the meta-metaphor for a staggering array of human troubles. And, of course, we also seem to have a staggering array of politicians and other moral entrepreneurs who take advantage of such cultural contradictions to blame new chemical bogeymen for our society's ills.

NOTES

1. In this regard, for example, Robin Room wisely observes "that we are living at a historic moment when the rate of (alcohol) dependence as a cognitive and existential experience is rising, although the rate of alcohol consumption and of heavy drinking is falling." He draws from this a more general hypothesis about "long waves" of drinking and societal reactions

to them: "[I]n periods of increased questioning of drinking and heavy drinking, the trends in the two forms of dependence, psychological and physical, will tend to run in opposite directions. Conversely, in periods of a "wettening" of sentiments, with the curve of alcohol consumption beginning to rise, we may expect the rate of physical dependence . . . to rise while the rate of dependence as a cognitive experience falls" (1991:154).

2. I say "formal end" because Temperance ideology is not merely alive and well in the War on Drugs but is being applied to all manner of human troubles in the burgeoning 12-Step Movement (Reinarman, forthcoming).

3. From Jim Baumohl I have learned that while the Temperance movement attracted most of its supporters from these groups, it also found supporters among many others (e.g., labor, the Irish, Catholics, former drunkards, women), each of which had its own reading of and folded its own agenda into the movement.

4. This historical sketch of drug scares is obviously not exhaustive. Readers interested in other scares should see, e.g., Brecher's encyclopedic work *Licit and Illicit Drugs* (1972), especially the chapter on glue sniffing, which illustrates how the media actually created a new drug problem by writing hysterical stories about it. There was also a PCP scare in the 1970s in which law enforcement officials claimed that the growing use of this horse tranquilizer was a severe threat because it made users so violent and gave them such super-human strength that stun guns were necessary. This, too, turned out to be unfounded and the "angel dust" scare was short-lived (see Feldman et al., 1979). The best analysis of how new drugs themselves can lead to panic reactions among users is Becker (1967).

5. Becker wisely warns against the "one-sided view" that sees such crusaders as merely imposing their morality on others. Moral entrepreneurs, he notes, do operate "with an absolute ethic," are "fervent and righteous," and will use "any means" necessary to "do away with" what they see as "totally evil." However, they also "typically believe that their mission is a holy one," that if people do what they want it "will be good for them." Thus, as in the case of abolitionists, the crusades of moral entrepreneurs often "have strong humanitarian overtones" (1963:147-8). This is no less true for those whose moral enterprise promotes drug scares. My analysis, however, concerns the character and consequences of their efforts, not their motives.

6. As Gusfield notes, such ownership sometimes shifts over time, e.g., with alcohol problems, from religion to criminal law to medical science. With other drug problems, the shift in ownership has been away from medical science toward criminal law. The most insightful treatment of the medicalization of alcohol/drug problems is Peele (1989).

7. See Baumohl's (1990) important and erudite analysis of how the human will was valorized in the therapeutic temperance thought of 19th-century inebriate homes.

8. The third central feature of Temperance cultures identified by Levine (1992), which I will not dwell on, is predominance of spirits drinking, i.e., more concentrated alcohol than wine or beer and thus greater likelihood of drunkenness.

REFERENCES

Aronowitz, Stanley, *False Promises: The Shaping of American Working Class Consciousness* (New York: McGraw-Hill, 1973).

Baumohl, Jim, "Inebriate Institutions in North America, 1840–1920," *British Journal of Addiction* 85:1187–1204 (1990).

Baumohl, Jim, "The 'Dope Fiend's Paradise' Revisited: Notes from Research in Progress on Drug Law Enforcement in San Francisco, 1875–1915," *Drinking and Drug Practices Surveyor* 24:3–12 (1992).

Becker, Howard S., *Outsiders: Studies in the Sociology of Deviance* (Glencoe, IL: Free Press, 1963).

Becker, Howard S., "History, Culture, and Subjective Experience: An Exploration of the Social Bases of Drug-Induced Experiences," *Journal of Health and Social Behavior* 8: 162–176 (1967).

Bell, Daniel, *The Cultural Contradictions of Capitalism* (New York: Basic Books, 1978).

Brecher, Edward M., *Licit and Illicit Drugs* (Boston: Little Brown, 1972).

Cohen, M. M., K. Hirshorn, and W. A. Frosch, "In Vivo and in Vitro Chromosomal Damage Induced by LSD-25," *New England Journal of Medicine* 227:1043 (1967).

Courtwright, David, *Dark Paradise: Opiate Addiction in America Before 1940* (Cambridge, MA: Harvard University Press, 1982).

Dickson, Donald, "Bureaucracy and Morality," *Social Problems* 16:143–156 (1968).

Duster, Troy, *The Legislation of Morality: Law, Drugs, and Moral Judgement* (New York: Free Press, 1970).

Ewen, Stuart, *Captains of Consciousness: Advertising and the Social Roots of Consumer Culture* (New York: McGraw-Hill, 1976).

Feldman, Harvey W., Michael H. Agar, and George M. Beschner, *Angel Dust* (Lexington, MA: Lexington Books, 1979).

Gusfield, Joseph R., *Symbolic Crusade: Status Politics and the American Temperance Movement* (Urbana: University of Illinois Press, 1963).

Gusfield, Joseph R., *The Culture of Public Problems: Drinking-Driving and the Symbolic Order* (Chicago: University of Chicago Press, 1981).

Himmelstein, Jerome, "Drug Politics Theory," *Journal of Drug Issues* 8 (1978).

Himmelstein, Jerome, *The Strange Career of Marihuana* (Westport, CT: Greenwood Press, 1983).

Levine, Harry Gene, "The Alcohol Problem in America: From Temperance to Alcoholism," *British Journal of Addiction* 84:109–119 (1984).

Levine, Harry Gene, "Temperance Cultures: Concern About Alcohol Problems in Nordic and English-Speaking Cultures," in G. Edwards et al., Eds., *The Nature of Alcohol and Drug Related Problems* (New York: Oxford University Press, 1992).

Levine, Harry Gene, *Drunkenness and Civilization* (New York: Basic Books, forthcoming).

Levine, Harry Gene, and Craig Reinarman, "From Prohibition to Regulation: Lessons from Alcohol Policy for Drug Policy," *Milbank Quarterly* 69:461–494 (1991).

MacAndrew, Craig, and Robert Edgerton, *Drunken Comportment* (Chicago: Aldine, 1969).

Marcuse, Herbert, *One-Dimensional Man: Studies in the Ideology of Advanced Industrial Society* (Boston: Beacon Press, 1964).

Mills, C. Wright, "Situated Actions and Vocabularies of Motive," *American Sociological Review* 5:904–913 (1940).

Molotch, Harvey, and Marilyn Lester, "News as Purposive Behavior: On the Strategic Uses of Routine Events, Accidents, and Scandals," *American Sociological Review* 39:101–112 (1974).

Morgan, Patricia, "The Legislation of Drug Law: Economic Crisis and Social Control," *Journal of Drug Issues* 8:53–62 (1978).

Musto, David, *The American Disease: Origins of Narcotic Control* (New Haven, CT: Yale University Press, 1973).

National Institute on Drug Abuse, *National Household Survey on Drug Abuse: Main Findings 1990* (Washington, DC: U.S. Department of Health and Human Services, 1990).

Peele, Stanton, *The Diseasing of America: Addiction Treatment Out of Control* (Lexington, MA: Lexington Books, 1989).

Reinarman, Craig, "The 12-Step Movement and Advanced Capitalist Culture: Notes on the Politics of Self-Control in Postmodernity", in B. Epstein, R. Flacks, and M. Darnovsky, Eds., *Contemporary Social Movements and Cultural Politics* (New York: Oxford University Press, forthcoming).

Reinarman, Craig, and Ceres Duskin, "Dominant Ideology and Drugs in the Media," *International Journal on Drug Policy* 3:6–15 (1992).

Reinarman, Craig, and Harry Gene Levine, "Crack in Context: Politics and Media in the Making of a Drug Scare," *Contemporary Drug Problems* 16:535–577 (1989a).

Reinarman, Craig, and Harry Gene Levine, "The Crack Attack: Politics and Media in America's Latest Drug Scare," pp. 115-137 in Joel Best, Ed., *Images of Issues: Typifying Contemporary Social Problems* (New York: Aldine de Gruyter, 1989b).

Room, Robin G. W., "Cultural Changes in Drinking and Trends in Alcohol Problems Indicators: Recent U.S. Experience," pp. 149–162 in Walter B. Clark and Michael E. Hilton, Eds. *Alcohol in America: Drinking Practices and Problems* (Albany: State University of New York Press, 1991).

Rumbarger, John J., *Profits, Power, and Prohibition: Alcohol Reform and the Industrializing of America. 1800–1930* (Albany: State University of New York Press, 1989).

Tijo, J. H., W. N. Pahnke, and A. A. Kurland, "LSD and Chromosomes: A Controlled Experiment," *Journal of the American Medical Association* 210:849 (1969).

Weber, Max *The Protestant Ethic and the Spirit of Capitalism* (London: Unwin, 1985 [1930]).

Weil, Andrew, *The Natural Mind* (Boston: Houghton Mifflin, 1972).

8

Rhetoric in Claims-Making: Constructing the Missing Children Problem

JOEL BEST

RHETORIC AND CLAIMS-MAKING

Claims-makers inevitably hope to persuade. Typically, they want to convince others that X is a problem, that Y offers a solution to that problem, or that a policy of Z should be adopted to bring that solution to bear. While the success of claims-making may well depend, in part, on the constellation of interests and resources held by various constituencies in the process, the way claims are articulated also affects whether they persuade and move the audiences to which they are addressed. Claims-making, then, is a rhetorical activity. Rhetoric—the study of persuasion—can be used to analyze claims. . . .

MISSING CHILDREN: A CASE STUDY

By the mid-1980s, the missing children problem had achieved extraordinary visibility. Americans saw photographs of missing children on milk cartons and grocery bags, billboards and televised public service messages. Toy stores and fast-food restaurants distributed abduction-prevention tips for both parents and children. Parents could have their children fingerprinted or videotaped to make identification easier; some dentists even proposed attaching identification disks to children's teeth. Commercial child identification kits were available, and at least one catalog offered a transmitter which could be attached to a child's clothing:

> Guardian Angel surrounds your child with an invisible circle of protection—in or around the house, in shopping areas, amusement parks, on picnics or outings. The bright yellow finish warns would-be abductors. . . . (Sharper Image, 1986:33).

In short, ordinary citizens may have encountered explicit reminders of missing children more often than for any other social problem.

From: "Rhetoric in Claims-Making: Constructing the Missing Children Problem," Joel Best, *Social Problems,* Vol. 34, No. 2, 1987. © 1987 The Society for the Study of Social Problems. Reprinted by permission of University of California Press Journals and Joel Best.

This problem achieved prominence quickly; the term "missing children" seems to have been coined in 1981.[1] The term encompassed three familiar phenomena: runaways (children—most often adolescents—who chose to leave home and usually returned within a few days); child-snatchings (non-custodial parents who illegally took their own children without the custodial parent's permission); and abductions by strangers (who might keep, sell, ransom, molest, or kill the child).[2]

During 1979–81, several cases in which children presumably were abducted by strangers received national publicity. These included the 1979 revelations that John Wayne Gacy had murdered 33 youths, the series of 28 murders of Atlanta schoolchildren during 1979–81, the 1981 murder of Adam Walsh (age 6, who disappeared from a shopping mall and whose severed head was later recovered), the 1979 disappearance of Etan Patz (age 6, who left for school one morning and was not seen again), and the 1980 return of Steven Stayner (age 14, who had lived with his kidnapper for seven years).[3]

Assisting in some of these cases were existing child-search organizations, such as Child Find, which had been established primarily to help locate children taken during custody disputes. By 1981, child-snatching had received a good deal of attention, but the response of the press, law enforcement, and the general public was ambiguous. Magazine articles and television programs sometimes portrayed the kidnapping parent favorably (e.g., Anonymous, 1979; Jares, 1981). Federal legislation—the 1980 Parental Kidnapping Prevention Act—had little effect, because the Justice Department refused to pursue cases without evidence that the child was in danger of abuse or neglect (Spangler, 1982). Moreover, public opinion remained split; only 64 percent of the respondents to a 1981 CBS/*New York Times* poll fully agreed that federal kidnapping laws should apply to child-snatching.

Organizations like Child Find clearly found it advantageous to link their cause to widespread sympathy for parents whose children were abducted by strangers.[4] The label "missing children" made this link possible. The U.S. Senate held its first hearing on missing children in October, 1981.[5] Magazine and newspaper articles began to appear. On October 10, 1983, NBC broadcast "Adam," a televised drama about Adam Walsh which ended with a roll call of 55 missing children.[6] The National Center for Missing and Exploited Children—funded by a federal grant—opened in 1984. By the year's end, pictures of missing children were being printed on milk cartons, shopping bags, and other paper products. The concept of missing children had become common currency.

Because the missing children problem emerged quickly, and because the construction of the problem involved extensive claims-making in the press, in testimony before Congressional committees, and in billboards, pamphlets, posters, and the like, claims about the problem are accessible for analysis. This paper will classify these claims . . . as a framework. While these claims refer specifically to the missing children problem, their rhetorical structure parallels claims-making for many social problems.

GROUNDS

In any argument, statements about grounds provide the basic facts which serve as the foundation for the discussion which follows. Obviously, facts are themselves socially-constructed knowledge. Claims-makers and their audiences may agree to accept grounds statements without question, or one or both parties may have reservations about the statements' truth, their relevance, the methods used to establish them, and so on.[7] Although the specific facts at issue depend upon the particular claims being made, some types of grounds statements recur in many claims-making campaigns.[8] Three types of grounds statements consistently appeared in claims-making about missing children: definitions, examples, and numeric estimates.

Definitions

Perhaps the most fundamental form of claims-making is to define a problem—to give it a name. Identifying the topic under discussion limits what can be said; a definition makes some issues relevant, while relegating others out of bounds (Gusfield, 1981; Lake, 1986). Definitions can both establish a topic's domain and offer an orientation toward that topic.

Domain Statements A definition identifies a phenomenon, setting its boundaries or domain. Someone who understands a definition can examine phenomena and determine which do, and which do not, fall within the defined category's domain.

Domain statements are particularly important when claims-makers hope to call attention to a previously unacknowledged social problem. Following Pfohl (1977), we might speak of discovery movements—claims-makers who announce that they have discovered a new problem. They may argue that the problem is in fact new (e.g., modern technology is destroying the ozone layer) or they may describe the problem as something which presumably has existed for some time but has only now been recognized for what it is (e.g., child abuse). In either case, domain statements have the power of novelty; they attract interest because they claim to identify a new phenomenon.

Claims-making about missing children involved the creation of a new domain. In this case, there were few efforts to offer precise, technical definitions; claims-makers usually did not specify the age at which one stops being a child, or the length of time a child must be gone to be considered missing. The term was intended to be broad, inclusive, to encompass several misadventures which might befall children. Thus, in the initial U.S. Senate hearings on missing children, Kristin Cole Brown of Child Find argued:

> It is absolutely critical that we establish a policy which guarantees that the various criteria used to determine whether or not a child is to be considered a missing child be subject to the most generous interpretation. We must not begin by discriminating against *kinds* of missing children (U.S. Senate, 1981:75—emphasis in the original).

Most claims-makers preferred an inclusive definition of missing children. Some child-search organizations distributed photos of individuals who were in their twenties (cf., Fritz, 1986). Others argued that children who returned home after even a brief disappearance should be considered missing children. As John Walsh put the case: "If it was your daughter, Mr. McCloskey, and you were waiting for her and she didn't come home for 4 hours and after that time she came home with bloody underpants and she had been raped, was she a missing child? Damn well she was" (U.S. House, 1985:18).

Orientation Statements Typically, claims-makers' definitions give an orientation to the problem. That is, in addition to specifying a problem's domain, there is some assessment of the sort of problem it is. The literature on medicalization offers a standard example; during the twentieth century, medical authorities have frequently argued that specific problems should be recognized as diseases (Conrad and Schneider, 1980). Even when a domain is not in question, orientation can be at issue. We can contrast discovery movements with reorientation movements—claims-makers who cast familiar problems in the fresh light of a new perspective. Thus, by arguing that rape should be viewed as a violent crime, not a sex crime, feminist reformers offered a different perspective on the topic (Rose, 1977). Definitions, then, guide the way we interpret the problem, suggesting appropriate avenues for response.

Claims-making about missing children emphasized that "missing" meant endangered, that even runaways and child-snatchings involved terrible risks:

> . . . once they are on the street they are fair game for child molestation, prostitution, and other exploitation. To label them "runaways" and disregard their safety is to suggest our own lack of compassion and real understanding for this difficult problem (Sen. Hawkins in U.S. Senate, 1981).

> Parental child abuse is a fact. Parents hurt, and kill, their children every day. . . . we must assume that a parent who breaks civil laws [by child-snatching] will break the laws of responsible parents (Kristin C. Brown in U.S. Senate, 1981:76).

This theme of potential violence and exploitation received support by coupling definitions to horrific examples of missing children.

Examples

Although definition might seem to be the logical first step in claims-making, it frequently follows an introductory example. Newspaper and magazine articles which called attention to the missing children problem routinely began with accounts of one or more atrocity tales.[9] Similarly, the initial U.S. Senate (1981) hearings on the problem began with the parents of Etan Patz, Adam Walsh, and Yusuf Bell (age 9, one of the murdered Atlanta schoolchildren) describing their children's disappearance.[10]

Opening with an emotionally-riveting "grabber" is a standard journalistic technique.[11] By focusing on events in the lives of specific individuals, these stories make it easier to identify with the people affected by the problem. Selecting horrific examples gives a sense of the problem's frightening, harmful dimensions.

In addition to these obvious effects, atrocity tales perform another, less visible function. The atrocity—usually selected for its extreme nature—becomes the referent for discussions of the problem in general. Claims-makers routinely used stranger abductions—which they acknowledged to be the least common type of missing child—as referents. Between 1981 and 1985, Congressional hearings featured testimony by parents of eleven missing children; eight of these children had been abducted by strangers. John Walsh became especially prominent, testifying before several Congressional committees as well as many state legislatures. Similarly, Etan Patz's mother, Julie, was the only parent of a missing child on the U.S. Attorney General's Advisory Board on Missing Children. As visible reminders of the horrors of stranger abduction, these parents implicitly helped define the problem. More overtly, sociologist Michael Agopian, Director of the Child Stealing Research Center, testified: "We are all aware of the Adam Walsh case, but please recognize that there are tens of thousands of additional Adams that are not so prominently reported by the media" (U.S. House, 1984:102). Similarly, John Walsh warned, "This country is littered with mutilated, decapitated, raped, strangled children" (U.S. Senate, 1983:33). Atrocity tales do not merely attract attention; they also shape the perception of the problem.

Estimating the Problem's Extent

Once examples establish a problem's human dimensions, claims-makers often try to assess its magnitude. The bigger the problem, the more attention it can be said to merit, so most claims-makers emphasize a problem's size. For missing children claims-makers, these estimates took three forms.

Incidence Estimates Perhaps the most straightforward way to establish a social problem's dimensions is to estimate the number of cases, incidents, or people affected. Claims-makers argue that a widespread problem demands attention. Of course, social phenomena are often difficult to measure; the familiar problems of working with official statistics therefore come into play (Kitsuse and Cicourel, 1963).

Claims-making about the missing children problem relied heavily on estimates of the number of children affected. These were necessarily inexact. Missing children need not come to official attention. Moreover, the various official agencies which might learn about missing children had no standard set of criteria for defining cases (some police departments, for instance, required that 24 or 72 hours pass before a child could be deemed missing). And, even if local agencies kept records, there was no national clearinghouse to compile statistics.

One commonly cited figure was 1.8 million cases per year—a total which included runaways (which claims-makers acknowledged constituted the vast majority of all cases), abductions by non-custodial parents (frequently estimated at 100,000 annually), and abductions by strangers. The latter category received most of the attention, although estimates generally declined over time:

> . . . the most conservative estimate is that 50,000 young people disappear each year, because of "stranger kidnappings." That is the most conservative estimate you will get anywhere (Rep. Simon in U.S. House, 1981b:10).
>
> . . . 20,000 to 50,000 are snatched by strangers—most never to be seen again (Thornton, 1983:63).
>
> Somewhere between 4,000 and 20,000 children each year are abducted by non-family members. . . . (Cerra, 1986:14).

Frequently, these estimates were buttressed by claims about the number of missing children found murdered: "There are about 4,000–8,000 of these children each year who are found dead and probably a majority have experienced some type of sexual exploitation" (Rep. Simon in U.S. House, 1981b:11).

Although the initial claims-making about the missing children problem did distinguish among runaways, custodial kidnappings, and stranger kidnappings, secondary claims-makers often lost sight of these distinctions. Thus, a pamphlet, "To Save a Child," distributed by a Chicago television station, began:

> Nearly 2,000,000 children in this country disappear from their homes each year. Many end up raped, forced into prostitution and pornography. Many are never heard from again.

By including runaways to estimate the number of missing children and then focusing attention on atrocity tales about stranger abduction, claims-makers led many people to infer that the most serious cases were commonplace. Studies of public opinion found that most people assumed stranger abductions were common, accounting for a large share of missing children (Field Institute, 1987; Fritz, 1986; Miller, 1985).[12]

During 1985, the larger estimates of the number of missing children came under attack. Led by the *Denver Post* (which earned a Pulitzer Prize for its coverage of the issue), the press used FBI and other criminal justice statistics to challenge the estimates' accuracy (Abrahms, 1985; Griego and Kilzer, 1985; Karlen, 1985).[13] These articles emphasized that stranger kidnappings accounted for a very small proportion of missing children, concluding that the actual number of children kidnapped by strangers was 100–200 per year.[14] These counterclaims were newsworthy, in part, because the originally acknowledged distinction among types of missing children had become lost in later claims-making.

Growth Estimates A second claim which may be made about a social problem's dimensions is that things are getting worse, that the problem is growing and, unless action is taken, there will be further deterioration. "Everyone close to the missing-child problem agrees that it is a large one—and growing" (Turbak, 1982:61). Claims-makers often describe problems as epidemic—a metaphor which suggests that more people will be affected as the problem spreads (cf., U.S. Senate, 1983:74).

Range Claims The epidemic metaphor also suggests that people may be indiscriminately affected, that the problem extends throughout the social structure. Often, this claim is explicit:

> [Missing] children come from small towns in rural America, and from our largest cities. They are from all races and ethnic backgrounds. They grow up in upper class neighborhoods, in the suburbs, and in the inner cities (Sen. Howard Metzenbaum in U.S. House, 1985:9).

Claims that a social problem's range extends throughout society serve an important rhetorical function. By arguing that anyone might be affected by a problem, a claims-maker can make everyone in the audience feel that they have a vested interest in the problem's solution. Thus, recommendations that all parents maintain files with current pictures and fingerprints which could be used to identify their children cast every child as a potential victim.[15] Like other grounds statements, these claims presented missing children as a problem which could not be ignored. . . .

CONCLUSIONS

Like other forms of argument, claims-making presents conclusions—typically calls for action to alleviate or eradicate the social problem. Claims-makers may have an agenda with several goals. In the case of the missing children problem, claims-makers hoped to affect the general public and, in particular, parents, as well as official policy.

Awareness

Initially, missing children claims-makers sought to bring the problem to public attention. Their goal was not merely greater public awareness; they hoped to enlist the public in searching for children. Originally these efforts reached relatively few people. Reporters who covered the story were urged to accompany their articles with photos of missing children, while Child Find and other child-search organizations distributed directories filled with pictures of missing children. With the 1983 broadcast of "Adam," the problem became well-established; commercial and government agencies began printing billions of images on milk cartons, utility bills, and so on.[16] These efforts had some success; the advertising campaigns were credited with locating some children, and polls showed considerable public awareness of the missing children problem (Field Institute, 1987; Fritz, 1986; Miller, 1985).

Prevention

In addition to locating children who were already missing, claims-makers emphasized the importance of prevention. They urged parents to assemble files of recent photographs, fingerprints, and other material which could be used by investigators if a child disappeared. Pamphlets gave lists of safety tips—ways to protect children from abduction—and parents could choose among many new books and videotapes designed to teach children to protect themselves.[17]

Atrocity tales about stranger abductions shaped the construction of these preventive measures. Typically, both the safety tips and the materials intended for children focused on warnings about strangers. Prevention campaigns had such titles as "Too Smart for Strangers" and "Strangers and Dangers," while one standard tip warned that having a child's name on clothing might let a stranger approach and call the child by name. However useful such advice might be in preventing stranger abductions, it had limited relevance for child-snatchings and runaways.

Social Control Policies

Dissatisfied with official efforts to locate missing children, claims-makers demanded new social control policies. The 1982 federal Missing Children's Act insured that parents could list a missing child with the National Crime Information Center's computers; if a local police department refused to enter the child's name, federal agents were required to do so. Claims-makers also campaigned for state laws to require police to begin searching immediately upon receiving a report of a missing child, and to list all reported children with NCIC. Other recommended policies included: requiring schools to notify parents whenever children were absent, and to transfer school records and birth certificates whenever children changed school districts; modifying FBI practices so that all reported stranger abductions would lead to federal investigations; and giving police greater authority to apprehend and hold runaways.[18]

Two themes united these recommendations. First, believing that the thousands of local law enforcement agencies made it harder to coordinate searches, claims-makers sought to centralize police power. John Walsh described the states as "50 little feudal kingdoms" (U.S. Senate, 1982:46), and the Dee Scofield Awareness Program argued:

> We maintain that every missing child deserves the protection of specially trained investigators who are authorized to transcend every local, county and state boundary. . . . In our opinion, every missing child under 18 should have the benefit of FBI jurisdiction, whether or not voluntary flight was involved (U.S. Senate, 1981:196).

Second, claims-makers argued that children and adolescents should be subject to greater social control. Reinarman (1985) notes that Mothers Against Drunk Driving—another 1980s movement in which parents of child victims played a leading role—did not advocate restricted access to alcohol—except for youth. Both MADD and the missing children movement shared this theme of greater control over the young with a broader "anti-youth movement" which advocated expanding school and juvenile justice authority, while restricting young people's access to birth control and abortion (Carpenter, 1985).

Other Objectives

In addition to advancing ways to prevent abductions and recover missing children, claims-makers promoted several other causes. Because everyone agreed that statistics on the number of missing children were little more than guesses, claims-makers sought a federal study to count missing children.[19] Although their critics charged that most estimates were exaggerated, the claims-makers hinted that an accurate survey might reveal that the estimates were too low (cf., U.S. Senate, 1985:30).

Other objectives wandered further afield. Senator McConnell noted, "when you are talking about the exploitation of children, the missing children issue has been a good peg to focus attention on . . . sexual assaults against children by nonstrangers" (U.S. Senate, 1985:2). In fact, claims-makers used missing children as a peg for attacking a wide range of evils. The U.S. Attorney General's Advisory Board on Missing Children's (1986:22–30) recommendations included: prosecuting the "adult offender who abuses children, leads them into prostitution, victimizes them by pedophilic conduct or pornography, or pushes them into street crime;" extending the statute of limitations for "child sexual abuse crimes;" "careful screening of people who work with children;" and studying the relationship between popular culture and child exploitation. In these cases, "missing children" apparently served as a rubric for addressing a range of putative threats to children.

DISCUSSION
Rhetoric in Social Problems Construction

Most constructionist research pays little attention to rhetoric, focusing instead on substantive matters—the recognition of problematic conditions, the interests of claims-makers and those they address, and so on. In comparison, the construction and presentation of the claims themselves seem relatively unimportant.

Yet this analysis of the missing children problem suggests that rhetoric can play a central role in claims-making about social problems. In particular, atrocity tales gave shape to the missing children problem; case histories of stranger abduction, coupled with estimates that there were two million missing children per year, convinced many people that the problem could not be ignored. People responded empathically to the horrors experienced by parents and children. The claims-makers' warrants—the emphasis on the priceless, blameless nature of children, the association of missing children with other evils (especially exploitative sexual deviance), and the assertion that existing policies and resources could not cope with the problem—drew much of their power from the examples of atrocities. The campaign's conclusions about the need for increased public awareness, prevention, and social control also followed from the perception that atrocities were in some sense representative. Kidnapping, mutilation, and murder had few defenders; presented this way, the missing children problem was uncontroversial.

Nor is the missing children problem atypical. Claims-makers routinely use examples or case histories to shape perceptions of social problems: where the Federal Bureau of Narcotics once personified marijuana smokers as homicidal drug fiends, advocates of decriminalization describe users as well-adjusted, otherwise law-abiding citizens; where pro-choice advocates insist that abortions must be available for victims of rape and incest, pro-life crusaders portray women who callously abort for their own convenience; and where welfare advocates focus on the deserving poor, their opponents speak of welfare Cadillacs. Such images become a convenient shorthand for describing and typifying complex social conditions. By characterizing a problem in terms of an individual's experiences, the claims-maker helps the audience imagine how they might respond under the same circumstances. The portrait may invite sympathy and understanding, or it may encourage the audience to feel that they would never succumb to the same pressures or temptations; in either case, the problem becomes less abstract, the claims easier to comprehend. . . .

Rhetoric is central, not peripheral, to claims-making. Claims-makers intend to persuade, and they try to make their claims as persuasive as possible. Claims-making inevitably involves selecting from available arguments, placing the arguments chosen in some sequence, and giving some arguments particular emphasis. These are rhetorical decisions. Moreover, as claims-makers assess the response to their claims, or as they address new audiences, claims may be revised and reconstructed in hopes of making them more effective. In such cases, even the most ingenuous claims-maker must become conscious of doing rhetorical work. . . .

Context and Claims-Making

Claims do not emerge from a social and historical vacuum. The missing children problem appeared during the 1980s—a period featuring considerable discussion about the victimization of children. The same newspapers and magazines which printed articles about missing children also explored child abuse, child prostitution, child pornography, Halloween sadism, incest, child molestation, satanic cults, and harmful rock lyrics. Even the drinking-driving issue was defined most effectively by Mothers Against Drunk Drivers. Certainly child-saving has a long history; concern for protecting children did not begin in the 1980s. But, where earlier movements warned that children might be harmed by inappropriate socialization, inadequate nutrition, or hazards in the urban environment, recent claims-makers focused on threats deviants posed to children.

Why do claims emerge at particular moments? Constructionist case studies usually finesse this question,[20] or they refer to social strain (Best and Horiuchi, 1985; Troyer and Markle, 1983), or "features of the period" (Reinarman, 1985). And why do similar claims sometimes emerge at roughly the same time? American history reveals other periods of intense claims-making by related social movements, such as the Progressive Era attempts to solve urban problems, and the various 1960s and 1970s movements advocating equal rights for blacks, women, and others.

These considerations affect the rhetoric of claims-making. Just as people's decisions to make claims emerge from a larger social context, so do their rhetorical choices. Claims-makers articulate their claims in ways which they find (and believe their audiences will find) persuasive. The larger cultural context—the weight assigned to various sorts of evidence, the relative importance given to different values, current standards for appropriate social policies, and the degree of consensus about these various judgments—affects rhetorical work. Would-be claims-makers may rely on their own sense of what ought to be said, or they may learn from watching what happens to other claims-makers. These links between rhetoric and cultural context deserve systematic attention.[21]

Sociologists of social problems and social movements cannot afford to ignore the rhetoric used in making claims. Rhetoric reflects both the nature of the interaction between particular claims-makers and their audience, and the larger cultural context within which claims-making occurs. In turn, rhetorical choices affect the success or failure of specific claims. The message—as well as the medium of claims-making—merits further study.

NOTES

1. There were several Congressional hearings on parental kidnapping during 1979–81. Although those testifying sometimes spoke of "missing children," they clearly meant child-snatching. Neither runaways nor children abducted by strangers figured prominently in these deliberations (U.S. House, 1980, 1918a; U.S. Senate, 1979).

2. Occasionally, claims-makers spoke of other categories of missing children—e.g., infants who had been given up for adoption and were now sought by a biological parent (U.S. Senate, 1981:189).

3. Folklorist Jan Brunvand (1981, 1984, 1986) reports that an urban legend about a child being abducted from an amusement park or department store also began circulating around 1978. On urban legends and the constructionist perspective, see Best and Horiuchi (1985).

4. Like many child-search organizations, Child Find was established by the parent of a missing child—Gloria Yerkovich, whose daughter was snatched by her ex-husband. Kristin Cole Brown, Child Find's Information Director, handled press interviews and Congressional testimony. The Dee Scofield Awareness Program was an early effort begun after a stranger abduction. After 1981, John Walsh, who founded the Adam Walsh Child Resource Center in his son's memory, became the most visible claims-maker among the parents.

5. A non-controversial issue—one Congressional aide called it "the apple pie of the '80's"—missing children received widespread, bipartisan support. Leading Congressional claims-makers included Republican Senators Paula Hawkins and Mitchell McConnell, and Democratic Representative (after 1984, Senator) Paul Simon. In 1984, both McConnell and Simon campaigned on their record of work for missing children.

6. Twelve of these children (all abducted by non-custodial parents) were recovered in the weeks following the program's initial broadcast (U.S. House, 1984:52–53). Later broadcasts had similar results. Several media figures became claims-makers themselves. Television journalist Kenneth Wooden, who founded the National Coalition for Children's Justice, was especially prominent.

7. The discussion which follows treats the empirical truth of claims as problematic. As will be seen, some claims made about the missing children problem came under attack, and there were counter-claims. Examining the rhetoric of claims-making does not require establishing the validity of any particular claim. For one discussion of constructionist problems with the issue of empirical truth, see Woolgar and Pawluch (1985). The use of grounds statements in claims-making resembles the process of belief amplification in recruiting members for social movements (Snow et al., 1986), or scientists' cognitive claims (Aronson, 1984).

8. Obviously, criteria for legitimate grounds vary over time and space. Where claims grounded in divine revelation once carried considerable rhetorical power, today's policy-makers more often view official statistics or scientific results as the means of establishing facts (cf., Gusfield, 1981).

9. During 1981–86, the *Reader's Guide to Periodical Literature* listed fourteen articles longer than one page about the missing children problem generally (rather than specific cases). Eight began with one or more horrific examples. Journalists relied on this device when the issue was fresh; only one of the five articles published in 1986 opened with an example.

10. In contrast, Kristin C. Brown of Child Find testified before the same hearing: "Our Executive Director is Gloria Yerkovich. Her daughter, Joanna, has been missing since December 20, 1974" (U.S. Senate, 1981:73). Brown did not explain that this was a child-snatching. This sort of reticence serves to keep attention focused on atrocity tales which should enlist everyone's sympathy. Even accounts of stranger kidnappings

sometimes gloss over details which might strike some as discrediting. Thus, a fundraising mailing from the National Child Safety Council contains a letter from a mother describing her son's disappearance: "My husband was at work and I was in the kitchen." This neglects to mention that the couple was divorced and living in different states (Kurczewski and Lewis, 1986).

11. Johnson (1986) explores the importance of "horror stories" in shaping public perceptions of child abuse. On the use of violent episodes to typify wife abuse, see Loseke (1986).

12. This confusion extended to private sector campaigns to distribute pictures of missing children. Jay Howell, Executive Director of the National Center for Missing and Exploited Children, noted that some organizations limited their displays to children taken by strangers (U.S. House, 1985:31).

13. In Congressional testimony, legislators and representatives of the missing children movement estimated that stranger abductions were common. In contrast, the law enforcement officials who testified at the same hearings either did not speak to this issue, or suggested that the estimates were inflated (cf., U.S. House, 1984:118–19; U.S. Senate, 1984:89–96).

14. In response, missing children claims-makers countered with new statistics of their own, arguing that their critics undercounted abductions by strangers (National Center, 1986) and even suggesting that the total number of missing children might be "well over the 1.8 million estimated by various national groups" (U.S. Senate, 1985:30). For a summary of the conflicting claims and counter-claims, see Spitzer (1986).

15. Occasional claims did identify particular segments of the population as being most at risk—e.g., rural children (Smith, 1986); children of single mothers (U.S. House, 1984:108); or upper and middle-class children (Jordan, 1985). Claims that a problem has an extensive range also keep attention focused on particular aspects of the problem. Pelton (1981) suggests that claims that child abuse occurred throughout the social structure promoted a "myth of classlessness" in spite of evidence that child abuse was concentrated among the poor. While this myth circumvented potential critics who might have refused to support further social programs for the poor, it also implied that the causes of child abuse lay in individual pathology, rather than the social structure. As a consequence, programs aimed at combating child abuse focused on individual treatment, rather than on the social conditions within which abuse occurred (Nelson, 1984).

16. As the search for missing children became more public, problems emerged. Child-search organizations began to compete for contributions and media coverage (Spitzer, 1986). Some organizations also came under attack for deceptive practices (cf., Blumenthal, 1984; Kilzer, 1985; McNair, 1985).

17. By 1985, both awareness and prevention efforts had become the focus of counter-claims. Arguing that the danger of stranger abductions had been exaggerated, critics charged that the missing children campaign was frightening children. Pediatricians and child psychologists played a prominent role in these warnings about psychological damage (Kantrowitz and Leslie, 1986; Vobejada, 1985).

18. For a critical interpretation of one state's missing children legislation, see Elliott and Pendleton (1986).

19. Results of this study are not expected until 1988.

20. "The central problem for a theory of social problems is to account for the *emergence,* nature, and maintenance of claims-making and responding activities" (Spector and Kitsuse, 1977:76—emphasis added). However, Spector and Kitsuse go on to warn against approaching claims-making's emergence from some angles. Studying "participants to find individual and social characteristics that predict participation in [claims-making] activities deflects attention away from the organization of claims-making itself" (1977:82). Similarly, values should be seen as a resource for claims-makers, not a cause of their behavior. Spector and Kitsuse (1977:96) conclude: "Premature consideration of the causes of any phenomenon turns attention to the supposed antecedents of the activity, rather than to the activity itself." Many constructionist studies seem to posit that claims-making emerges from a constant pool of potentially articulatable claims, just as resource mobilization theorists assume the ubiquity of grievances among potential social movement participants.

21. These topics receive some attention in the rhetorical studies of Ernest Bormann (1972), (1973). His fantasy theme analysis argues that rhetoric often presents "composite dramas" which draw on the larger culture. For a related sociological analysis, see Klapp (1964).

REFERENCES

Abrahms, Sally. 1985. "The villain in child snatching is usually . . . a parent." Wall Street Journal (March 19):28.

Anonymous. 1979. "I kidnapped my own son." Good Housekeeping 189 (October):26–36.

Aronson, Naomi. 1984. "Science as a claims-making activity." Pp. 1–30 in Joseph W. Schneider and John I. Kitsuse (eds.), Studies in the Sociology of Social Problems. Norwood, NJ: Ablex.

Best, Joel, and Gerald T. Horiuchi. 1985. "The razor blade in the apple: the social construction of urban legends." Social Problems 32:488–99.

Blumenthal, Ralph. 1984. "Registry of missing children faces state inquiry." New York Times (May 21):B2.

Bormann, Ernest. 1972. "Fantasy and rhetorical vision: the rhetorical criticism of social reality." Quarterly Journal of Speech 58:396–407.

———. 1973. "The Eagleton affair: a fantasy theme analysis." Quarterly Journal of Speech 59:143–59.

Brunvand, Jan Harold. 1981. The Vanishing Hitchhiker. New York: Norton.

———. 1984. The Choking Doberman. New York: Norton.

———. 1986. The Mexican Pet. New York: Norton.

Carpenter, T. G. 1985. "The new anti-youth movement." Nation 240 (January 19):39–41.

Cerra, Frances. 1986. "Missing children." Ms. 14 (January):14–16.

Conrad, Peter, and Joseph W. Schneider. 1980. Deviance and Medicalization. St. Louis: Mosby.

Elliott, Susan Newhart, and Diana L. Pendleton. 1986. "S. 321: the Missing Children Act—legislation by hysteria." University of Dayton Law Review 11:671–708.

Field Institute. 1987. "Threats to children." California Opinion Index 2 (March):1–4.

Fritz, Noah, J. 1986. "The impact of television on the 'missing children' issue." Paper presented before the Pacific Sociological Association.

Griego, Diana, and Louis Kilzer. 1985. "Truth about missing kids: exaggerated statistics stir national paranoia." Denver Post (May 12):1A, 12A.

Gusfield, Joseph R. 1981. The Culture of Public Problems: Drinking-Driving and the Symbolic Order. Chicago: University of Chicago Press.

Jares, Sue Ellen. 1981. "A victim and a childnapper describe the agonizing problem of the 'stolen' kids of divorce." People 15 (February 9):40, 42.

Johnson, John M. 1986. "Formal properties of child abuse horror stories." Unpublished manuscript.

Jordan, Mary. 1985. "The search for missing children." Washington Post (May 25):A1, A10.

Kantrowitz, Barbara, and Connie Leslie. 1986. "Teaching fear." Newsweek 107 (March 10):62–63.

Karlen, Neal. 1985. "How many missing kids?" Newsweek 106 (October 7):30, 35.

Kilzer, Louis. 1985. "Public often not told the facts in missing-children cases." Denver Post (September 22):14A.

Kitsuse, John I., and Aaron Cicourel. 1963. "A note on the use of official statistics." Social Problems 11:131–39.

Klapp, Orrin E. 1964. Symbolic Leaders. Chicago: Aldine.

Kurczewski, Norbert A., and Glenn R. Lewis. 1986. Someone Took Him. Kent, WA: Profiles Mini-books.

Lake, Randall A. 1986. "The metaethical framework of anti-abvortion rhetoric." Signs 11:478–99.

Loseke, Donileen R. 1986. "Lived realities and the construction of social problems: the case of wife abuse." Paper presented before the Society for the Study of Social Problems.

McNair, Jean. 1985. "Va. probing advertising tactics of publication on missing children." Washington Post (August 20):B3.

Miller, Carl. 1985. "Child-abduction worries excessive, area poll finds." Denver Post (May 13):1A.

National Center for Missing and Exploited Children. 1986. "An evaluation of the crime of kidnapping as it is committed against children by nonfamily members." Unpublished report.

Nelson, Barbara J. 1984. Making an Issue of Child Abuse. Chicago: University of Chicago Press.

Pelton, Leroy H. 1981. "Child abuse and neglect: the myth of classlessness." Pp. 23–38 in Leroy H. Pelton (ed.) The Social Context of Child Abuse. New York: Human Sciences Press.

Pfohl, Stephen J. 1977. "The 'discovery' of child abuse." Social Problems 24:310–23.

Reinarman, Craig. 1985. "Social movements and social problems." Paper presented before the Society for the Study of Social Problems.

Rose, Vicki McNickle. 1977. "Rape as a social problem: a byproduct of the feminist movement." Social Problems 25:75–89.

Schneider, Joseph W. 1985. "Social problems theory: the constructionist view." Annual Review of Sociology 11:209–29.

Sharper Image. 1986. Catalog (April).

Smith, Preston. 1986. "Rural abductions." Successful Farming 84 (February):10.

Snow, David A., E. Burke Rochford, Jr., Steven K. Worden, and Robert D. Benford. 1986. "Frame alignment processes, micromobilization, and movement participation." American Sociological Review 51:464–81.

Spangler, Susan E. 1982. "Snatching legislative power: the Justice Department's refusal to enforce the Parental Kidnapping Prevention Act." Journal of Criminal Law and Criminology 73:1176–203.

Spector, Malcolm, and John I. Kitsuse. 1977. Constructing Social Problems. Menlo Park, CA: Cummings.

Spitzer, Neil. 1986. "The children's crusade." Atlantic 257 (June):18–22.

Thornton, Jeannye. 1983. "The tragedy of America's missing children." U.S. News and World Report 95 (October 24):63–64.

Troyer, Ronald J., and Gerald E. Markle. 1983. Cigarettes. New Brunswick, NJ: Rutgers University Press.

Turbak, Gary. 1982. "Missing: 100,000 children a year." Reader's Digest 121 (July):60–64.

U.S. Attorney General's Advisory Board on Missing Children. 1986. America's Missing and Exploited Children. Washington, D.C.: Office of Juvenile Justice and Delinquency Prevention.

U.S. House of Representatives. 1980. Parental Kidnaping. Hearings held by the Subcommittee on Crime, Committee on the Judiciary, June 24.

———. 1981a. Implementation of the Parental Kidnaping Prevention Act of 1980. Hearings held by the Subcommittee on Crime, Committee on the Judiciary. September 24.

———. 1981b. Missing Children's Act. Hearings held by the Subcommittee on Civil and Constitutional Rights, Committee on the Judiciary. November 18, 30.

———. 1984. Title IV, Missing Children's Assistance Act. Hearings held by the Subcommittee on Human Resources, Committee on Education and Labor. April 9.

———. 1985. Photograph and Biography of Missing Child. Hearings held by the Subcommittee on Postal Personnel and Modernization, Committee on Post Office and Civil Service. June 25.

U.S. Senate. 1979. Parental Kidnapping, 1979. Hearings held by the Subcommittee on Child and Human Development, Committee on Labor and Human Resources. April 17.

———. 1981. Missing Children. Hearings held by the Subcommittee on Investigations and General Oversight, Committee on Labor and Human Resources. October 6.

———. 1982. Exploited and Missing Children. Hearings held by the Subcommittee on Juvenile Justice, Committee on the Judiciary. April 1.

———. 1983. Child Kidnaping. Hearings held by the Subcommittee on Juvenile Justice, Committee on the Judiciary. February 2.

———. 1984. Missing Children's Assistance Act. Hearings held by the Subcommittee on Juvenile Justice, Committee on the Judiciary. February 7, 21; March 8, 13, 21.

———. 1985. Missing and Exploited Children. Hearings held by the Subcommittee on Juvenile Justice, Committee on the Judiciary. August 21.

Vobedja, Barbara. 1985. "Abduction publicity could scare children." Washington Post (November 25):A1, A6.

Woolgar, Steve, and Dorothy Pawluch. 1985. "Ontological gerrymandering: the anatomy of social problems explanations." Social Problems 32:214–27.

9

The Police and the Black Male

ELIJAH ANDERSON

The police, in the Village-Northton [neighborhood] as elsewhere, represent society's formal, legitimate means of social control.[1] Their role includes protecting law-abiding citizens from those who are not law-abiding, by preventing crime and by apprehending likely criminals. Precisely how the police fulfill the public's expectations is strongly related to how they view the neighborhood and the people who live there. On the streets, color-coding often works to confuse race, age, class, gender, incivility, and criminality, and it expresses itself most concretely in the person of the anonymous black male. In doing their job, the police often become willing parties to this general color-coding of the public environment, and related distinctions, particularly those of skin color and gender, come to convey definite meanings. Although such coding may make the work of the police more manageable, it may also fit well with their own presuppositions regarding race and class relations, thus shaping officers' perceptions of crime "in the city." Moreover, the anonymous black male is usually an ambiguous figure who arouses the utmost caution and is generally considered dangerous until he proves he is not. . . .

There are some who charge— . . . perhaps with good reason—that the police are primarily agents of the middle class who are working to make the area more hospitable to middle-class people at the expense of the lower classes. It is obvious that the police assume whites in the community are at least middle class and are trustworthy on the streets. Hence the police may be seen primarily as protecting "law-abiding" middle-class whites against anonymous "criminal" black males.

To be white is to be seen by the police—at least superficially—as an ally, eligible for consideration and for much more deferential treatment than that accorded blacks in general. This attitude may be grounded in the backgrounds of the police themselves.[2] Many have grown up in Eastern City's "ethnic" neighborhoods. They may serve what they perceive as their own class and neighborhood interests, which often translates as keeping blacks "in their place"—away from neighborhoods that are socially defined as "white." In trying to do their job, the police appear to engage in an informal policy of monitoring young black men as a means of controlling crime, and often they seem to go beyond the bounds of duty. The following field note shows what pressures and racism young black men in the Village may endure at the hands of the police:

From: Elijah Anderson, *Streetwise* (Chicago: The University of Chicago Press, 1990). Reprinted by permission of the publisher and the author.

At 8:30 on a Thursday evening in June I saw a police car stopped on a side street near the Village. Beside the car stood a policeman with a young black man. I pulled up behind the police car and waited to see what would happen. When the policeman released the young man, I got out of my car and asked the youth for an interview.

"So what did he say to you when they stopped you? What was the problem?" I asked. "I was just coming around the corner, and he stopped me, asked me what was my name, and all that. And what I had in my bag. And where I was coming from. Where I lived, you know, all the basic stuff, I guess. Then he searched me down and, you know, asked me who were the supposedly tough guys around here? That's about it. I couldn't tell him who they are. How do I know? Other gang members could, but I'm not from a gang, you know. But he tried to put me in a gang bag, though." "How old are you?" I asked. "I'm seventeen, I'll be eighteen next month." "Did he give any reason for stopping you?' "No, he didn't. He just wanted my address, where I lived, where I was coming from, that kind of thing. I don't have no police record or nothin'. I guess he stopped me on principle, 'cause I'm black." "How does that make you feel?" I asked. "Well, it doesn't bother me too much, you know, as long as I know that I hadn't done nothin', but I guess it just happens around here. They just stop young black guys and ask 'em questions, you know. What can you do?"

On the streets late at night, the average young black man is suspicious of others he encounters, and he is particularly wary of the police. If he is dressed in the uniform of the "gangster," such as a black leather jacket, sneakers, and a "gangster cap," if he is carrying a radio or a suspicious bag (which may be confiscated), or if he is moving too fast or too slow, the police may stop him. As part of the routine, they search him and make him sit in the police car while they run a check to see whether there is a "detainer" on him. If there is nothing, he is allowed to go on his way. After this ordeal the youth is often left afraid, sometimes shaking, and uncertain about the area he had previously taken for granted. He is upset in part because he is painfully aware of how close he has come to being in "big trouble." He knows of other youths who have gotten into a "world of trouble" simply by being on the streets at the wrong time or when the police were pursuing a criminal. In these circumstances, particularly at night, it is relatively easy for one black man to be mistaken for another. Over the years, while walking through the neighborhood I have on occasion been stopped and questioned by police chasing a mugger, but after explaining myself I was released.

Many youths, however, have reason to fear such mistaken identity or harassment, since they might be jailed, if only for a short time, and would have to post bail money and pay legal fees to extricate themselves from the mess (Anderson 1986). When law-abiding blacks are ensnared by the criminal justice system, the scenario may proceed as follows. A young man is arbitrarily stopped by the police and questioned. If he cannot effectively negotiate with the officer(s), he may be accused of a crime and arrested. To resolve this situation he needs financial resources, which for him are in short supply. If he does not have money for any attorney, which often happens, he is left to a public defender who may be more interested in going along with the court system than in fighting for a poor black person. Without legal support, he may well wind up "doing time" even if he is innocent of the charges brought against him. The next time he is stopped for questioning he will have a record, which will make detention all the more likely.

Because the young black man is aware of many cases when an "innocent" black person was wrongly accused and detained, he develops an "attitude" toward the police. The street word for police is "the man," signifying a certain machismo, power, and authority. He becomes concerned when he notices "the man" in the community or when the police focus on him because he is outside his own neighborhood. The youth knows, or soon finds out, that he exists in a legally precarious state. Hence he is motivated to avoid the police, and his public life becomes severely circumscribed.

To obtain fair treatment when confronted by the police, the young man may wage a campaign for social regard so intense that at times it borders on obsequiousness. As one streetwise black youth said: "If you show a cop that you nice and not a smartass, they be nice to you. They talk to you like the man you are. You gonna get ignorant like a little kid, they gonna get ignorant with you." Young black males often are particularly deferential toward the police even when they are completely within their rights and have done nothing wrong. Most often this is not out of blind acceptance or respect for the "law," but because they know the police can cause them hardship. When confronted or arrested, they adopt a particular style of behavior to get on the policeman's good side. Some simply "go limp" or politely ask, "What seems to be the trouble, officer?" This pose requires a deference that is in sharp contrast with the youth's more usual image, but many seem to take it in stride or not even to realize it. Because they are concerned primarily with staying out of trouble, and because they perceive the police as arbitrary in their use of power, many defer in an equally arbitrary way. Because of these pressures, however, black youths tend to be especially mindful of the police and, when they are around, to watch their own behavior in public. Many have come to expect harassment and are inured to it; they simply tolerate it as part of living in the Village-Northton.

After a certain age, say twenty-four, a black man may no longer be stopped so often, but he continues to be the object of policy scrutiny. As one twenty-seven-year-old black college graduate speculated:

> I think they see me with my little bag with papers in it. They see me with penny loafers on. I have a tie on, some days. They don't stop me so much now. See, it depends on the circumstances. If something goes down, and they hear that the guy had on a big black coat, I may be the one. But when I was younger, they could just stop me, carte blanche, any old time.

Name taken, searched, and this went on endlessly. From the time I was about twelve until I was sixteen or seventeen, endlessly, endlessly. And I come from a lower-middle-class black neighborhood, OK, that borders a white neighborhood. One neighborhood is all black, and one is all white. OK, just because we were so close to that neighborhood, we were stopped endlessly. And it happened even more when we went up into a suburban community. When we would ride up and out to the suburbs, we were stopped every time we did it.

If it happened today, now that I'm older, I would really be upset. In the old days when I was younger, I didn't know any better. You just expected it, you knew it was gonna happen. Cops would come up, "What you doing, where you coming from?" Say things to you. They might even call you nigger.

Such scrutiny and harassment by local police makes black youths see them as a problem to get beyond, to deal with, and their attempts affect their overall behavior. To avoid encounters with the man, some streetwise young men camouflage themselves, giving up the urban uniform and emblems that identify them as "legitimate" objects of police attention. They may adopt a more conventional presentation of self, wearing chinos, sweat suits, and generally more conservative dress. Some youths have been known to "ditch" a favorite jacket if they see others wearing one like it, because wearing it increases their chances of being mistaken for someone else who may have committed a crime.

But such strategies do not always work over the long run and must be constantly modified. For instance, because so many young ghetto blacks have begun to wear Fila and Adidas sweat suits as status symbols, such dress has become incorporated into the public image generally associated with young black males. These athletic suits, particularly the more expensive and colorful ones, along with high-priced sneakers, have become the leisure dress of successful drug dealers, and other youths will often mimic their wardrobe to "go for bad" in the quest for local esteem. Hence what was once a "square" mark of distinction approximating the conventions of the wider culture has been adopted by a neighborhood group devalued by that same culture. As we saw earlier, the young black male enjoys a certain power over fashion: whatever the collective peer group embraces can become "hip" in a manner the wider society may not desire (see Goffman 1963). These same styles then attract the attention of the agents of social control.

THE IDENTIFICATION CARD

Law-abiding black people, particularly those of the middle class, set out to approximate middle-class whites in styles of self-presentation in public, including dress and bearing. Such middle-class emblems, often viewed as "square," are not usually embraced by young working-class blacks. Instead, their connections with and claims on the institutions of the wider society seem to be symbolized by the identification card. The common identification card associates its holder with a firm, a corporation, a school, a union, or some other institution of substance and influence. Such a card, particularly from a prominent establishment, puts the police and others on notice that the youth is "somebody," thus creating an important distinction between a black man who can claim a connection with the wider society and one who is summarily judged as "deviant." Although blacks who are established in the middle class might take such cards for granted, many lower-class blacks, who continue to find it necessary to campaign for civil rights denied them because of skin color, believe that carrying an identification card brings them better treatment than is meted out to their less fortunate brothers and sisters. For them this link to the wider society, though often tenuous, is psychically and socially important. The young college graduate continues:

> I know [how] I used to feel when I was enrolled in college last year, when I had an ID card. I used to hear stories about the blacks getting stopped over by the dental school, people having trouble sometimes. I would see that all the time. Young black male being stopped by the police. Young black male in handcuffs. But I knew that because I had that ID card that I would not be mistaken for just somebody snatching a pocketbook, or just somebody being where maybe I wasn't expected be. See, even though I was intimidated by the campus police—I mean, the first time I walked into the security office to get my ID they all gave me the double take to see if I was somebody they were looking for. See, after I got the card, I was like, well, they can think that now, but I have this [ID card]. Like, see, late at night when I be walking around, and the cops be checking me out, giving me the looks, you know. I mean, I know guys, students, who were getting stopped all the time, sometimes by the same officer, even though they had the ID. And even they would say, "Hey, I got the ID, so why was I stopped?"

The cardholder may believe he can no longer be treated summarily by the police, that he is no longer likely to be taken as a "no count," to be prejudicially confused with that class of blacks "who are always causing trouble on the trolley." Furthermore, there is a firm belief that if the police stop a person who has a card, they cannot "do away with him without somebody coming to his defense." This concern should not be underestimated. Young black men trade stories about mistreatment at the hands of the police; a common one involves policemen who transport youths into rival gang territories and release them, telling them to get home the best way they can. From the youth's perspective, the card signifies a certain status in circumstances where little recognition was formerly available.

"DOWNTOWN" POLICE AND LOCAL POLICE

In attempting to manage the police—and by implication to manage themselves—some black youths have developed a working connection of the police in certain public areas of the Village-Northton. Those who spend a good amount of their time on these corners, and thus observing the police, have come to distinguish between the "downtown" police and the "regular" local police.

The local police are the ones who spend time in the area; normally they drive around in patrol cars, often one officer to a car. These officers usually make a kind of working peace with the young men on the streets; for example, they know the names of some of them and may even befriend a young boy. Thus they offer an image of the police department different from that displayed by the "downtown" police. The downtown police are distant, impersonal, and often actively looking for "trouble." They are known to swoop down arbitrarily on gatherings of black youths standing on a street corner; they might punch them around, call them names, and administer other kinds of abuse, apparently for sport. A young Northton man gave the following narrative about his experiences with the police.

> And I happen to live in a violent part. There's a real difference between the violence level in the Village and the violence level in Northton. In the nighttime it's more dangerous over there.
>
> It's so bad now, they got downtown cops over there now. They doin' a good job bringin' the highway patrol over there. Regular cops don't like that. You can tell that. They even try to emphasize to us the certain category. Highway patrol come up, he leave, they say somethin' about it. "We can do our job over here." We call [downtown police] Nazis. They about six feet eight, seven feet. We walkin', they jump out. "You run, and we'll blow your nigger brains out." I hate bein' called a nigger. I want to say somethin' but get myself in trouble.
>
> When a cop do somethin', nothing happen to 'em. They come from downtown. From what I heard some of 'em don't even wear their real badge numbers. So you have to put up with that. Just keep your mouth shut when they stop you, that's all. Forget about questions, get against the wall, just obey 'em. "Put all that out right there"—might get rough with you now. They snatch you by the shirt, throw you against the wall, pat you hard, and grab you by the arms, and say, "Get outta here." They call you nigger this and little black this, and things like that. I take that. Some of the fellas get mad. It's a whole different world.
>
> Yeah, they lookin' for trouble. They gotta look for trouble when you got five, eight police cars together and they laughin' and talkin', start teasin' people. One night we were at a bar, we read in the paper that the downtown cops comin' to straighten things out. Same night, three police cars, downtown cops with their boots on, they pull the sticks out, beatin' around the corner, chase into bars. My friend Todd, one of 'em grabbed him and knocked the shit out of him. He punched 'im, a little short white guy. They start a riot. Cops started that shit. Everybody start seein' how wrong the cops was—they start throwin' bricks and bottles, cussin' 'em

out. They lock my boy up; they had to let him go. He was just standin'
on the corner, they snatch him like that.

One time one of 'em took a gun and began hittin' people. My boy had
a little hickie from that. He didn't know who the cop was, because there
was no such thing as a badge number. They have phony badge numbers.
You can tell they're tougher, the way they dress, plus they're bigger. They
have boots, trooper pants, blond hair, blue eyes, even black [eyes]. And
they seven feet tall, and six foot six inches and six foot eight inches. Big!
They are the rough cops. You don't get smart with them or they beat the
shit out of you *in front of everybody,* they don't care.

We call 'em Nazis, Even the blacks among them. They ride along with
'em. They stand there and watch a white cop beat your brains out. What
takes me out is the next day you don't see 'em. Never see 'em again, go
down there, come back, and they ride right back downtown, come back,
do their little dirty work, go back downtown, and put their real badges
on. You see 'em with a forty-five or fifty-five number: "Ain't no such
number here, I'm sorry, son." Plus, they got unmarked cars. No sense
takin' 'em to court. But when that happened at that bar, another black cop
from the sixteenth [local] district, ridin' a real car, came back and said,
"Why don't y'all go on over to the sixteenth district and file a complaint?"
Them musclin' cops was wrong. Beatin' people." So about ten people
went over there; sixteenth district knew nothin' about it. They come in
unmarked cars, they must have been downtown cops. Some of 'em do it.
Some of 'em are off duty, on their way home. District commander told us
they do that. They have a patrol over there, but them cops from downtown have control of them cops. Have bigger ranks and bigger guns. They
carry .357s and regular cops carry little .38s. Downtown cops are all
around. They carry magnums.

Two cars the other night. We sittin' on the steps playing cards.
Somebody called the cops. We turn around and see four regular police
cars and two highway police cars. We drinkin' beer and playin' cards.
Police get out and say you're gamblin'. We say we got nothin' but cards
here, we got no money. They said all right, got back in their cars, and
drove away. Downtown cops dressed up like troopers. That's intimida-
tion. Damn!

You call a cop, they don't come. My boy got shot, we had to take him
to the hospital ourselves. A cop said, "You know who did it?" We said
no. He said, "Well, I hope he dies if y'all don't say nothin'." What he say
that for? My boy said, "I hope your mother die," he told the cop right to
his face. And I was grabbin' another cop, and he made a complaint about
that. There were a lot of witnesses. Even the nurse behind the counter
said the cop had no business saying nothin' like that. He said it loud, "I
hope he dies." Nothin' like that should be comin' from a cop.

Such behavior by formal agents of social control may reduce the crime rate, but it raises questions about social justice and civil rights. Many of the old-time liberal white residents of the Village view the police with some ambivalence. They want their streets and homes defended, but many are convinced that the police manhandle "kids" and mete out an arbitrary form of "justice." These feelings make many of them reluctant to call the police when they are needed, and they may even be less than completely cooperative after a crime has been committed. They know that far too often the police simply "go out and pick up some poor black kid." Yet they do cooperate, if ambivalently, with these agents of social control.

In an effort to gain some balance in the emerging picture of the police in the Village-Northton, I interviewed local officers. The following edited conversation with Officer George Dickens (white) helps place in context the fears and concerns of local residents, including black males:

> I'm sympathetic with the people who live in this neighborhood [the Village-Northton], who I feel are victims of drugs. There are a tremendous number of decent, hardworking people who are just trying to live their life in peace and quiet, not cause any problems for their neighbors, not cause any problems for themselves. They just go about their own business and don't bother anyone. The drug situation as it exists in Northton today causes them untold problems. And some of the young kids are involved in one way or another with this drug culture. As a result, they're gonna come into conflict even with the police they respect and have some rapport with.
>
> We just went out last week on Thursday and locked up ten young men on Cherry Street, because over period of about a week, we had under-cover police officers making drug buys from those young men. This was very well documented and detailed. They were videotaped selling the drugs. And as a result, right now, if you walk down Cherry Street, it's pretty much a ghost town; there's nobody out. [Before, Cherry Street was notorious for drug traffic.] Not only were people buying drugs there, but it was a very active street. There's been some shock value as a result of all those arrests at one time.
>
> Now, there's two reactions to that. The [television] reporters went out and interviewed some people who said, "Aw, the police overreacted, they locked up innocent people. It was terrible, it was harassment." One of the neighbors from Cherry Street called me on Thursday, and she was out-raged. Because she said, "Officer, it's not fair. We've been working with the district for well over a year trying to solve some of the problems on Cherry Street." But most of the neighbors were thrilled that the police came and locked all those kids up. So you're getting two conflicting reactions here. One from the people that live there that just wanta be left alone, alright? Who are really being harassed by the drug trade and

everything that's involved in it. And then you have a reaction from the people that are in one way or another either indirectly connected or directed connected, where they say, "You know, if a young man is selling drugs, to him that's a job." And if he gets arrested, he's out of a job. The family's lost their income. So they're not gonna pretty much want anybody to come in there to make arrests. So you've got contradicting elements of the community there. My philosophy is that we're going to try to make Northton livable. If that means we have to arrest some of the residents of Northton, that's what we have to do.

You talk to Tyrone Pitts, you know the group that they formed was formed because of a reaction to complaints against one of the officers of how the teenagers were being harassed. And it turned out that basically what he [the officer] was doing was harassing drug dealers. When Northton against Drugs actually formed and seemed to jell, they developed a close working relationship with the police here. For that reason, they felt the officer was doing his job.

I've been here eighteen months. I've seen this neighborhood go from . . . Let me say, this is the only place I've ever worked where I've seen a rapport between the police department and the general community like the one we have right now. I've never seen it any place else before coming here. And I'm not gonna claim credit because this happened while I happened to be here. I think a lot of different factors were involved. I think the community was ready to work with the police because of the terrible situation in reference to crack. My favorite expression when talking about crack is "crack changed everything." Crack changed the rules of how the police and the community have to interact with each other. Crack changed the rules about how the criminal justice system is gonna work, whether it works well or poorly. Crack is causing the prisons to be overcrowded. Crack is gonna cause the people that do drug rehabilitation to be overworked. It's gonna cause a wide variety of things. And I think the reason the rapport between the police and the community in Northton developed at the time it did is very simply that drugs to a certain extent made many areas in this city unlivable.

In effect the officer is saying that the residents, regardless of former attitudes, are now inclined to be more sympathetic with the police and to work with them. And at the same time, the police are more inclined to work with the residents. Thus, not only are the police and the black residents of Northton working together, but different groups in the Village and Northton are working with each other against drugs. In effect, law-abiding citizens are coming together, regardless of race, ethnicity, and class. He continues:

Both of us [police and the community] are willing to say, "Look, let's try to help each other." The nice thing about what was started here is that it's spreading to the rest of the city. If we don't work together, this problem is gonna devour us. It's gonna eat us alive. It's a state of emergency, more or less.

In the past there was significant negative feeling among young black men about the "downtown" cops coming into the community and harassing them. In large part these feelings continue to run strong, though many young men appear to "know the score" and to be resigned to their situation, accommodating and attempting to live with it. But as the general community feels under attack, some residents are willing to forgo certain legal and civil rights and undergo personal inconvenience in hopes of obtaining a sense of law and order. The officer continues:

> Today we don't have too many complaints about police harassment in the community. Historically there were these complaints, and in almost any minority neighborhood in Eastern City where I ever worked there was more or less a feeling of that [harassment]. It wasn't just Northton; it was a feeling that the police were the enemy. I can honestly say that for the first time in my career I don't feel that people look at me like I'm the enemy. And it feels nice; it feels real good not to be the enemy, ha-ha. I think we [the police] realize that a lot of problems here [in the Village-Northton] are related to drugs. I think the neighborhood realizes that too. And it's a matter of "Who are we gonna be angry with? Are we gonna be angry with the police because we feel like they're this army of occupation, or are we gonna argue with these people who are selling drugs to our kids and shooting up our neighborhoods and generally causing havoc in the area? Who deserves the anger more?" And I think, to a large extent, people of the Village-Northton decided it was the drug dealers and not the police.
>
> I would say there are probably isolated incidents where the police would stop a male in an area where there is a lot of drugs, and this guy may be perfectly innocent, not guilty of doing anything at all. And yet he's stopped by the police because he's specifically in that area, on that street corner where we know drugs are going hog wild. So there may be isolated incidents of that. At the same time, I'd say I know for a fact that our complaints against police in this division, the whole division, were down about 45 percent. If there are complaints, if there are instances of abuse by the police, I would expect that our complaints would be going up. But they're not; they're dropping.

Such is the dilemma many Villagers face when they must report a crime or deal in some direct way with the police. Stories about police prejudice against blacks are often traded at Village get-togethers. Cynicism about the effectiveness of the police mixed with community suspicion of their behavior toward blacks keeps middle-class Villagers from embracing the notion that they must rely heavily on the formal means of social control to maintain even the minimum freedom of movement they enjoy on the streets.

Many residents of the Village, especially those who see themselves as the "old guard" or "old-timers," who were around during the good old days when antiwar and antiracist protest was a major concern, sigh and turn their heads when they see the criminal justice system operating in the ways described here. They express hope that "things will work out," that tensions will ease, that crime will decrease and police behavior will improve. Yet as incivility and crime become increasing problems in the neighborhood, whites become less tolerant of anonymous blacks and more inclined to embrace the police as their heroes.

Such criminal and social justice issues, crystallized on the streets, strain relations between the newcomers and many of the old guard, but in the present context of drug-related crime and violence in the Village-Northton, many of the old-timers are adopting a "law and order" approach to crime and public safety, laying blame more directly on those they see as responsible for such crimes, though they retain some ambivalence. Newcomers can share such feelings with an increasing number of old-time "liberal" residents. As one middle-aged white woman who has lived in the Village for fifteen years said:

> When I call the police, they respond. I've got no complaints. They are fine for me. I know they sometimes mistreat black males. But let's face it, most of the crime is committed by them, and so they can simply tolerate more scrutiny. But that's them.

Gentrifiers and the local old-timers who join them, and some traditional residents continue to fear, care more for their own safety and well-being than for the rights of young blacks accused of wrong-doing. Yet reliance on the police, even by an increasing number of former liberals, may be traced to a general feeling of oppression at the hands of street criminals, whom many believe are most often black. As these feelings intensify and as more yuppies and students inhabit the area and press the local government for services, especially police protection, the police may be required to "ride herd" more stringently on the youthful black population. Thus young black males are often singled out as the "bad" element in an otherwise healthy diversity, and the tensions between the lower-class black ghetto and the middle and upper-class white community increase rather than diminish.

NOTES

1. See Rubinstein (1973); Wilson (1978); Fogelson (1977); Reiss (1971); Bittner (1967); Banton (1964).

2. For an illuminating typology of police work that draws a distinction between "fraternal" and "professional" codes of behavior, see Wilson (1968).

REFERENCES

Anderson, Elijah. 1986. Of old heads and young boys: Notes on the urban black experience. Unpublished paper commissioned by the National Research Council, Committee on the Status of Black Americans.

Banton, Michael. 1964. *The policeman and the community*. New York: Basic Books.

Bittner, Egon. 1967. The police on Skid Row. *American Sociological Review* 32(October): 699–715.

Fogelson, Robert. 1977. *Big city police*. Cambridge: Harvard University Press.

Goffman, Erving. 1963. *Behavior in public places*. New York: Free Press.

Reiss, Albert J. 1971. *The police and the public*. New Haven: Yale University Press.

Rubinstein, Jonathan. 1973. *City police*. New York: Farrar, Straus and Giroux.

Wilson, James Q. 1968. The police and the delinquent in two cities. In *Controlling delinquents,* ed. Stanton Wheeler. New York: John Wiley.

10

Sex Discrimination—Subtle and Covert

NIJOLE V. BENOKRAITIS

AND JOE R. FEAGIN

HOW SUBTLE SEX DISCRIMINATION WORKS

Because most of us are still almost exclusively concerned with documenting and identifying the more visible and widespread types of overt sex discrimination, we are inattentive to other forms of inequality. . . . Subtle sex discrimination is considerably more harmful than most of us realize. Subtle sex discrimination has the following characteristics: (1) It can be intentional or unintentional, (2) it is visible but often goes unnoticed (because it has been built into norms, values, and ideologies), (3) it is communicated both verbally and behaviorally, (4) it is usually informal rather than formal, and (5) it is most visible on individual (rather than organizational) levels. . . .

Condescending Chivalry

Condescending chivalry refers to superficially courteous behavior that is protective and paternalistic but treats women as subordinates. This behavior ranges from simple, generally accepted rules of etiquette regarding sex (for example, opening doors for women) to more deeply entrenched beliefs that women are generally helpless and require protection and close supervision.

Chivalrous behavior implies respect and affection. That is, many men assume that referring to women as "little girl," "young lady," "little lady," and "kiddo" is a compliment—especially if the woman is over thirty. Some women may be flattered by such terms of endearment. Yet, comparable references to men ("little boy," "little man") are considered insulting, demeaning, or disrespectful because they challenge men's adulthood and authority. Thus, it is acceptable to refer to women, but not men, as children.

Even when women are clearly in positions of authority, their power may be undercut through "gentlemanly" condescension. For example, one woman dean (who is responsible for, among other things, collecting, reviewing, and coordinating course schedules every semester) complained that some chairmen refuse to take her seriously. When chairs are late in submitting schedules and she calls them into her office, some emphasize her gender and ignore her administrative power: "They do things like put their arm around me, smile, and say, 'You're getting prettier everyday' or 'You shouldn't worry your pretty little head about these things.' "

Chivalrous, paternalistic and "protective" behavior also limits women's employment opportunities. . . . A number of women we talked to said they were automatically excluded from some jobs because men still assume that women won't want to travel, will be unwilling to set up child-care arrangements, and "don't want to be in the public eye." Or, when women already have jobs, they will be excluded from important meetings or not considered for promotions because they should be "protected." Consider the experience of a thirty-three-year-old, unmarried store manager provided by one of our respondents:

> [Mary's] male counterparts in the company frequently were invited to out-of-town business meetings and social functions from which she was excluded. These occasions were a source for information on business trends and store promotions and were a rich source of potentially important business contacts. When [Mary] asked why she was not invited to these meetings and social gatherings, the response was that her employer thought it was "too dangerous for her to be driving out of town at night by herself. . . ."

In most cases, it is still assumed that women need, want, or should want protection "for their own good." During a recent lunch with colleagues, for example, one of the authors was discussing prospective faculty who could fill a dean's position that was about to be vacated. The comments, from both male and female faculty, were instructive:

> Mary Ann is a very good administrator, but she plans to get married next year. I don't think she'll have time to be both a wife and a dean.

> Well, Susan has the respect of both faculty and administration but hasn't she been talking about having children?

Tracy's been a great faculty leader and she's done an outstanding job on committees, but she's got kids. What if they get sick when important decisions have to be made in the dean's office?

Sara has been one of the best chairs in the college, a good researcher and can handle faculty. [A pause.] On the other hand, now that her kids are grown, she probably wants some peace and quiet and wouldn't want to take on the headaches of a dean's office. . . .

In effect, every prospective female candidate was disqualified from serious consideration because it is generally assumed that women should stay in presumably "safe" positions where their femininity, motherhood, and ability to fulfill wifely duties will remain intact.

Whether well-intentioned or malicious, chivalrous behavior is dysfunctional because it reinforces sex inequality in several ways. First, treating women as nonadults stunts their personal and professional growth. "There are problems harder to put a finger on: . . . suggestions initiated by a woman are listened to, but always a bit more reluctantly than those initiated by a man. People, sure, will listen, but we are not urged to suggest. Women, very simply, are not actively encouraged to develop."[1]

Second, chivalry justifies keeping women in low-paying jobs. Some have argued, for example, that because some women (for example, nurses and cleaning women) are encouraged to work long hours or late at night, state protective laws do not represent progressive reform but have been designed to reduce competition from female workers and to save the premium overtime and better jobs for men. Finally, chivalrous behavior can limit women's opportunities. Men's belief that women should be protected may result, for example, in men's reluctance to criticize women:

> A male boss will haul a guy aside and just kick ass if the subordinate performs badly in front of a client. But I heard about a woman here who gets nervous and tends to giggle in front of customers. She's unaware of it and her boss hasn't told her. But behind her back he downgrades her for not being smooth with customers.[2]

Thus, not receiving the type of constructive criticism that is exchanged much more freely and comfortably between men can lead to treating women like outsiders rather than colleagues.

Benevolent Exploitation

Women are often exploited. Much of the exploitation is carried off so gracefully, however, it often goes unnoticed.

Dumping One of the most common forms of exploitation is dumping—getting someone else (i.e., a woman) to do a job you don't want to do and then taking credit for the results:

Whenever my supervisor gets a boring, tedious job he doesn't want to do, he assigns it to me. He praises my work and promises it will pay off in his next evaluation. Then, he writes the cover letter and takes full credit for the project. . . . I've never been given credit for any of the projects—and some were praised very highly by our executives. But, I suppose it's paid off because my boss has never given me negative evaluations. [Female engineer in aerospace industry]

Another form of dumping—much more elusive—is to segregate top workers by sex and depend on the women to get the work done while the men merely critique the work and implement the results in highly visible and prestigious ways. An aide in a highly placed political office said that one of the reasons her boss was extremely successful politically was because he recognized that his female aides were better, harder working, more committed, and more responsible than the male aides. Thus, he surrounds himself with such women, gives them fancy titles, and gets 60 to 70 hours of work out of them at much lower salaries than those of men. When the projects are finished, he gives a lunch for all his aides and praises the women's work. Even as the dessert is served, new projects for the women are announced. The men, however, publicize the projects and get widespread recognition.

Showcasing   "Showcasing" refers to placing women in visible and seemingly powerful positions in which their talents, abilities, and intelligence can be pulled out, whenever necessary, for the public's consumption and the institution's credibility.

One form of showcasing is to make sure that the institution's token women are present (though not participating) in the institution's meetings with "the outside." Thus, in higher education, a woman faculty member is often expected to serve on national committees (recruitment, articulation with high schools, community colleges, and colleges), grant proposals (to show the involvement of women), search committees (just in case affirmative action officers are lurking around), and a variety of external "women's-type" activities such as panels, commissions, and advisory boards. There is no compensation for these additional duties. Moreover, the women are not rewarded in later personnel reviews because this is "women's work" and because "women's work" has low status.

If an occasional committee is an important one, the women chosen are typically nonfeminists who won't "embarrass" the institution/agency/organization by taking women's issues seriously. Instead, they are Queen Bees,[3] naive neophytes, women who are either not powerful or are insensitive to sex discrimination.

Another form of showcasing is giving women directorships in dead-end jobs which are considered a "natural" for women:

There's probably less discrimination in personnel offices because the job needs a person with traditionally female skills—being nice to people, having verbal abilities, and not being a threat to anyone because a director of personnel is a dead-end job. (Director of personnel in higher education) . . .

Technologically Based Abuses   Americans place a high value on progress, product improvements, and technological advances. "New," "improved," or "better than ever" detergents, toothpaste, and shampoos appear on the market annually, and many people go into debt purchasing such "necessities" as home computers, microwave ovens, electronic games, and VCRs. The profits generated by such "discoveries" are not translated into higher salaries for the many women who work in "high-tech" industries. In the case of new technologies, for example, employers often convince women that their newly developed skills are inadequate and should not be rewarded:

> While office technology creates opportunities for higher pay for some of us, for many others it is used as an excuse for keeping salary levels down. An employer may ignore the new skills you have learned in order to operate your machine and argue it's the machine itself that does all the work so that you are worthless.[4]

Employers/supervisors may discourage women from pursuing personal or professional development programs that might make them more dissatisfied with or question their current subordination:

> Under the negotiated rules, secretaries were entitled to take whatever courses they wanted at a state university, tuition reimbursed. We had . . . no application forms . . . although our immediate supervisors and the office supervisor knew and approved the tuition provision (after we educated them). The Queen Bee three places up on the hierarchy professed ignorance and had to be convinced anew when one of the secretaries wanted to take a history course . . . only in the last three years has anyone gone through the hassle and taken the courses. (Ex-secretary in higher education)

The implication here is that some groups of workers (especially office workers) are presumptuous in assuming that their professional development is significant enough to warrant the institution's attention or expenditures. Perhaps more importantly, college courses might lead to office workers wondering why they're performing high-tech jobs at low-tech salaries. . . .

Finally, "progress" has been a higher priority than the job hazards resulting from new technologies and automation. In most cases, the people using the new technologies are office workers—almost all of them women. The most commonly used new office equipment is the video-display terminal (VDT). There is evidence that long-term exposure to VDTs may be dangerous. Operators of VDTs experience eyestrain, neck and back pain, headaches, and blurred vision; the radiation and chemical fumes emitted by the terminals are believed to cause stress, cataracts, miscarriages, and birth defects.[5]

Yet, management has done little to improve work conditions even when many of the improvements are not costly. As one respondent put it, "Why save labor when it's cheap?"

Nudity in Advertising   One of the most widespread forms of exploitation is to use female nudity to sell everything from toothpaste to tractors. Such advertising may not be seen as exploitation because women are expected to be "decorative." The implicit message to men and women is that the primary role of women is to provide pleasure, sex, or sexual promise:

> A sexual relationship is . . . implied between the male product user and his female companion, such that the advertisement promises, in effect, that the product will increase his appeal to her. Not only will it give him a closer shave, it will also provide a sexually available woman.
>
> Often the advertisements imply that the product's main purpose is to improve the user's appeal to men, as the panty-hose advertisement which claims "gentlemen prefer Hanes." The underlying advertising message for a product advertised in this manner is that the ultimate benefit of product usage is to give men pleasure.[6]

The consistent and continuous message that advertisements send—to both men and women—is that women's roles in society are limited to two—that of housewife or sex partner. Other roles are not taken or presented seriously. Thus, women may dominate advertising space, but they are not dominant. . . .

HOW COVERT SEX DISCRIMINATION WORKS

Covert sex discrimination refers to unequal and harmful treatment of women that is hidden, clandestine, and maliciously motivated. Unlike overt and subtle sex discrimination, covert sex discrimination is very difficult to document and prove because records are not kept or are inaccessible, the victim may not even be aware of being a "target," or the victim may be ignorant of how to secure, track, and record evidence of covert discrimination. . . .

Tokenism

Despite its widespread usage since the 1970s, the term "tokenism" is rarely defined. For our purposes, tokenism refers to the unwritten and usually unspoken policy or practice of hiring, promoting, or otherwise including a miniscule number of individuals from underrepresented groups—women, minorities, the handicapped, the elderly. Through tokenism, organizations maintain the semblance of equality because no group is totally excluded. Placing a few tokens in strategically visible places precludes the necessity of practicing "real" equality—that is, hiring and promoting individuals regardless of their sex. . . .

How Tokenism Works   There are three types of commonly practiced tokenism that limit women's equal participation in the labor force. A popular form is based on *numerical exclusion,* which uses quotas to maintain a predominantly male work force:

> As soon as they come into my office, a lot of recruiters tell me exactly how many women they plan to hire and in which departments. They say things like, "This year we need two women in accounting, one in marketing, and one in data processing." Some [of the recruiters] have fairly detailed data showing exactly how many women they should be hiring for their company.
>
> [What if the most qualified candidates are all women?]
>
> Most recruiters automatically assume that women are *not* the most qualified—they got high grades because they slept around, they're not serious about long-term job commitments, they don't understand the business world and so on. . . . They interview the [women] students we schedule, but rarely hire more than the one or two they're told to hire. (College job placement director)

Because male quotas are high—95 to 99 percent—it is not difficult to fill the low percentage of slots allocated to token women. . . .

Sabotage

Through sabotage, employers and employees purposely and consciously undermine or undercut a woman's position. Although sabotage can be contrived and carried out by individuals, it usually involves covert agreements between two or more persons. Because sabotage is difficult to prove, it is also easy to deny. In almost all cases, it comes down to "my word against yours" because saboteurs do not leave a "paper trail."

Sabotage strategies vary by degree of sophistication, which depends on whether the woman is in a traditionally female job, a traditionally male job, or a job in which boundaries are, in principle, nonexistent because they are, in practice, not job related.

Traditionally Female Jobs In traditionally female jobs (domestic, service, clerical), male sabotage is normative, because men at a comparable job level have higher status (owing to higher wages) or because men have supervisory positions. In terms of the latter, for example, there is a substantial literature documenting male supervisors' sexual harassment of women subordinates because, among other things, men expect women to service all their (real or imagined) needs at all levels. Thus, office workers are the most common targets of sabotage if they don't "put out."

In comparable job levels, men can use sabotage because their job functions are less vulnerable to inspection and represent higher control than those of women:

> I was hassled by the bartender and the male kitchen staff. When you're a waitress, you have to keep in the good books of the guys backing you up.

> If the bartender takes a dislike to you, he can slow down on your orders to the point where you get no tips at all. The kitchen staff can sabotage you in other ways. The food can be cold, it can arrive late, and orders can be all mixed up.[7]

In traditionally female jobs, male sabotage is blatant, unmasked, raw, and unsophisticated. It is used openly to control and take advantage of women's inferior job status.

Traditionally Male Jobs   In traditionally male-dominated jobs, sabotage strategies are more sophisticated. In contrast to the "good ole boys" mentality, which literally and proudly espouses a "women-are-good-for-only-one-thing" rhetoric, traditionally male job occupants react to women negatively because women are seen as potentially threatening the "old gang" cohesion, camaraderie, and esprit de corps. . . .

In an effort to preserve long-accepted strongholds over men's jobs, men use a variety of sabotage techniques to discourage women's participation and success in traditionally male jobs:

> My co-workers would watch me talking to customers. When I went in to get the paperwork, they'd ridicule me to the customers. "She hasn't been here that long," "Women don't know much about cars." Then, they'd go over the same questions with the customers and get the sale. (Automobile salesperson)

> Every time there's a promotion, I put my name in. I always get rejected even though I have seniority, have put in the same number of years on the street as the guys, and have the same firing range results as the men. When there's a temporary opening, a sergeant from another precinct is pulled into the temporary spot even when I request the assignment. . . . I think my supervisor is trying to mess up my work record purposely—I'm the last one to find out about special events and new cases, and I have been late for important meetings because I was told about them five or ten minutes before they start. (Female police sargeant)

> Ever since I became a meter reader, the guys have always teased me that I'd be attacked by dogs, raped, kidnapped, or not return. . . . That's scary, but I tried to ignore it. . . . What gets me is that sometimes I get to the customer's house and none of the keys I picked up fit. I have to go back to the company to get the right keys. I don't know who's doing it, but someone doesn't want me in this job. (Meter reader for a gas and electric company)

In contrast to women in female-dominated jobs, women in male-dominated jobs find that they are "set up" to fail but are not told, openly, that this is due to their gender.

Sex-Neutral Jobs   The most sophisticated sabotage strategies occur in professional, technical, and administrative (and sometimes sales) jobs where sex is totally irrelevant to job performance. Because these occupations do not require physical strength but require professional or academic credentials (Ph.D., J.D., M.A., M.S.) and longer and more specialized training, there is presumably a greater objective reliance on sex-neutral qualifications. . . . One would expect, then, that sex-neutral jobs would be the least discriminatory. Such expectations have not proven to be true.

Quite to the contrary, sex-neutral jobs are often the most discriminatory because they are the most threatening to males dominating the higher echelons of the economy. The sabotage techniques are so subtle and covert that women see the sabotage long after it is too late to do anything about the discrimination:

> One mid-level manager [at a nationally known company] said she had gotten excellent ratings from her supervisors throughout her first year of employment. In the meantime, the company psychologist had called her in about once a month and inquired "how things were going." She was pleased by the company's interest in its employees. At the end of the year, one of her male peers (whose evaluations were known to be very mediocre) got the promotion and she didn't. When she pursued the reasons for her non-promotion, she was finally told, by one of the company's vice presidents, that "anyone who has to see the company psychologist once a month is clearly not management material." She had no way of proving she had been sabotaged.

In other examples, a female insurance agent is directed by the manager to nonelitist client accounts (in contrast to her male counterparts) and then not promoted because her clients take out only "policies for the poor"; and an urban renewal administrative assistant who is more qualified than her supervisor (and is frank about wanting his job) finds the information in her folders scrambled over a period of months and is told that her "administrative chaos" will lead to a demotion.

NOTES

1. Ethel Strainchamps. Ed., *Rooms with No View: A Woman's Guide to the Man's World of the Media* (New York: Harper & Row, 1974), p. 146.

2. Susan Fraker. "Why Top Jobs Elude Female Executives," *Fortune,* Apr. 16, 1984, p.46.

3. "Queen Bees" refer to women who are convinced that they have been successful solely because of their efforts and abilities rather than recognizing that their success could not have become a reality without the sacrifices, pioneering efforts, and achievements of their female predecessors.

Because of their adamant "I'm-terrific-because-I-pulled-myself-up-by-my-bootstraps" beliefs, Queen Bees typically either ignore or resist helping women become upwardly mobile. Thus, Queen Bees openly support men who reject sex equality and provide men (and other Queen Bees) with public rationalizations for keeping women in subordinate positions (in other words, as female drones).

4. Ellen Cassedy and Karen Nussbaum, *9 to 5: The Working Woman's Guide to Office Survival* (New York: Penguin Books, 1983), pp. 93–94.

5. Ibid., pp. 77–78.

6. Alice E. Courtney and Thomas W. Whipple, *Sex Stereotyping in Advertising* (Lexington, MA: Lexington Books, 1983), pp. 103–104.

7. Constance Backhouse and Leah Cohen, *Sexual Harassment on the Job* (Englewood Cliffs, NJ: Prentice-Hall, 1981), p. 9.

11

Masculinity, Homophobia,

and Misogyny

MICHAEL A. MESSNER

The second element of organized sport that undermines intimate connection with other people is homophobia, the "irrational fear or intolerance of homosexuality."[1] The extent of homophobia in the sports world is staggering. Boys learn early that to be gay, to be suspected of being gay, or even to be unable to prove one's heterosexual status is not acceptable. Though athletes are cultural symbols of masculine heterosexual virility, however, it is not true that there are no gay men and boys in sport. There is growing evidence that many (mostly closeted) gay males are competing in organized sport at all levels. Former professional football player David Kopay wrote in his book, "Recently, I've come to the conclusion that a lot of my extra drive came from the same forces that brought black athletes out of the ghettos to the forefront of professional sports. They were out to prove that they were not inferior because of their race. I was out to prove that I was in no way less a man because I was homosexual."[2]

Indeed, gay men, having been raised *as men* in this society, tend to have most of the same ambivalence and insecurities that all men have, compounded by the knowledge that if their secret becomes known, they will be considered less than a man.[3] Furthermore, since talk of "queer bashing" is common in male locker rooms, a closeted gay athlete knows that to reveal the truth about his sexuality can be outright dangerous. Mike T. lived a closeted existence for years while competing as a world-class track-and-field athlete. At a very early age, Mike discovered that athletic participation was a way to construct a public masculine identity and to hide the fact that he was gay:

> When I was a kid, I was tall for my age and I was very thin, but very strong. And I was usually faster than most people. But I discovered rather early that I liked gymnastics and dance. I was very interested in, and studied, ballet. And something became obvious to me right away—that male ballet dancers were effeminate, that they were what most people would describe as faggots. It suddenly occurred to me that this was dangerous territory for me—I'm from a small town, and I was totally closeted and very concerned about being male. This was the 1950's, a terrible time to live, and everything was stacked against me. I realized that I had to do something to protect my image of myself, that I was male—because at that time, homosexuals were thought of primarily as men who really preferred to be women—that was the stereotype. And so I threw myself into athletics. I played football, gymnastics, track and field. . . . I was a jock, and that's how I was viewed, and I was comfortable with that.

Years later, Mike T. realized that were it not for a homophobic world, he might have become a dancer: "I wanted to be viewed as a male, otherwise I would be a dancer today. I wanted the male, macho image of an athlete. So I was protected with a very hard shell. And I was *clearly* aware of what I was doing. I was just as aggressive and hostile on the football field as anyone else."

Mike's story illustrates how homophobia often leads gay males to consciously create identities that conform to narrow definitions of masculinity. But homophobia also serves to limit the self-expression and relationships of nongay males. Psychologist Joseph Pleck has written that "our society uses the male heterosexual-homosexual dichotomy as a central symbol for all the rankings of masculinity, for the division on any grounds between males who are 'real men' and have power and males who are not. Any kind of powerlessness or refusal to compete becomes imbued with the imagery of homosexuality."[4]

Indeed, boys learn early that if it is difficult to define masculinity in terms of what it *is*, it is at least clear what it is *not*. A boy is not considered masculine if he is feminine. In sport, to be told by coaches, fathers, or peers that one throws "like a girl" or plays like a "sissy" or a "woman" is among the most devastating insults a boy can receive, and such words can have a powerful impact upon his actions, relationships, and self-image. Sociologist Gary Alan Fine spent three years studying Little League baseball teams and noted clear and persistent patterns of homophobic banter and sexual talk about females among eleven- and twelve-year-old boys.[5] To give but one example, Fine described an occasion when Dan, a "high-status twelve year old," announced to his teammates that he was going to pay his girlfriend, Annie, ten cents for each sexual "base" she allowed him to reach. The boys speculated as to whether Dan would reach "third base" or "second base." In the end, a disappointed Dan reported to his peers that he had gotten only "half-way to second." But Dan added that he *would* have gone further, but for an interrupting phone call from Gordy, a "low-status" boy in the group, who was then ridiculed as "a gay one" for his transgression. As Fine concludes,

> Dan's reputation is preserved at the expense of Gordy's. Although Dan did
> not meet his objectives, his reputation increased through the talk . . .
> whereas Gordy, a younger boy, was roundly condemned. Significantly,
> Annie's reputation was also affected, as players started calling her a "prostie,"
> and Dan stopped considering her his girlfriend. Aside from Dan and
> Annie, no one knew the truth about what happened between the two
> of them; yet, it is not private actions but public talk that is ultimately sig-
> nificant.[6]

Fine's observations demonstrate how homophobia and the sexual objectification of females together act as a glue that solidifies the male peer group as separate from females, while at the same time establishing and clarifying hierarchical relations within the male peer group. In short, homophobia polices the boundaries of narrow cultural definitions of masculinity and keeps boys—especially those in all-male environments such as organized sport—from getting too close. Through sport, a young boy learns that it is risky—psychologically as well as physically—to become too emotionally open with his peers: he might be labeled a "sissy," a "fag," or even be beaten up or ostracized from the group. He also finds that he had better not become too close to girls; he must, of course, establish his masculine status by making (and laughing at) heterosexist jokes, but he "must never let girls replace boys as the focus of [his] attention."[7]

For the men I interviewed, early experiences in play and sport held the promise of greater attachment with fathers, older males, and peers. For most of them, this promise was partially realized. But ultimately, the specific kinds of relationships that they developed were distorted by the hierarchical structure of sport and by the homophobic and sexist banter within the peer group.

A particular kind of masculinity was being forged among these boys and young men, a masculinity based upon status-seeking through successful athletic competition and through aggressive verbal sparring which is both homophobic and sexist. But it is important not to over-generalize the motivations and experiences of all men. As we have seen with Mike T., a young gay male might bring some very different psychological and interpersonal motivations to his athletic experiences. Furthermore, since boys from varied backgrounds are interacting with substantially different familial and educational contexts, we can expect these contexts will lead them to make choices and define situations in different ways. Boys from middle- and lower-class families and communities, for instance, relate to organized sport in quite distinct ways. . . .

In 1973, conservative writer George Gilder, later to become a central theorist of the antifeminist family policies of the Reagan administration, was among the first to sound the alarm that the contemporary explosion of female athletic participation might threaten the very fabric of civilization. "Sports," Gilder wrote, "are possibly the single most important male rite in modern society." The woman athlete "reduces the game from a religious male rite to a mere physical exercise, with some treacherous danger of psychic effect." Athletic performance, for males, embodies "an ideal of beauty and truth," while women's participation represents a "disgusting perversion" of this truth.[8] In 1986, over a decade later, a similar view was expressed by John Carroll in a respected academic journal. Carroll lauded the masculine "virtue and grace" of sport, and defended it against its critics, especially feminists. He concluded that in order to preserve sport's "naturally conserving and creating" tendencies, especially in the realms of "the moral and the religious, . . . women should once again be prohibited from sport: they are the true defenders of the humanist values that emanate from the household, the values of tenderness, nurture and compassion, and this most important role must not be confused by the military and political values inherent in sport. Likewise, sport should not be muzzled by humanist values: it is the living arena for the great virtue of manliness."[9]

The key to Gilder's and Carroll's chest-beating about the importance of maintaining sport as a "male rite" is their neo-Victorian belief that male-female biological differences predispose men to aggressively dominate public life, while females are naturally suited to serve as the nurturant guardians of home and hearth. As Gilder put it, "The tendency to bond with other males in intensely purposeful and dangerous activity is said to come from the collective demands of pursuing large animals. The female body, on the other hand, more closely resembles the body of nonhunting primates. A woman throws, for example, very like a male chimpanzee."[10] This perspective ignores a wealth of historical, anthropological, and biological data that suggest that the equation of males with domination of public life and females with the care of the domestic sphere is a cultural and historical construction.[11] In fact, Gilder's and Carroll's belief that sport, *a socially constructed institution,* is needed to sustain male-female difference contradicts their assumption that these differences are "natural." As R. W. Connell has argued, social practices that exaggerate male-female difference (such as dress, adornment, and sport) "are part of a continuing effort to sustain a social definition of gender, an effort that is necessary precisely *because the biological logic . . . cannot sustain the gender categories.*"[12]

Indeed, throughout this book I have argued against the view that sees sport as a natural realm within which some essence of masculinity unfolds. Rather, sport is a social institution that, in its dominant forms, was created by and for men. It should not be surprising, then, that my research with male athletes reveals an affinity between the institution of sport and men's developing identities. As the young males in this study became committed to athletic careers, the gendered values of the institution of sport made it extremely unlikely that they would construct anything but the kinds of personalities and relationships that were consistent with the dominant values and power relations of the larger gender order. The competitive hierarchy of athletic careers encouraged the development of masculine identities based on very narrow definitions of public success. Homophobia and misogyny were the key bonding agents among male athletes, serving to construct a masculine personality that disparaged anything considered "feminine" in women, in other men, or in oneself. The fact that winning was premised on physical power, strength, discipline, and willingness to take, ignore, or deaden pain inclined men to experience their own bodies as machines, as instruments of power and domination—and to see other people's bodies as objects of their power and domination.

In short, my research findings are largely consistent with previous feminist analyses of sport. Whether men continue to pursue athletic careers (as many lower-class males do), or whether they shift away from sport toward education and nonathletic careers (as many middle-class males do), they are likely to continue to feel most comfortable, at least into middle adulthood, constructing identities and relationships primarily through their public achievements. As adults, these men will in all likelihood continue to need women (and may even use their power to keep women) in their "feminine" roles as nurturers, emotion-workers, and mothers, even when these women also have jobs or careers of their own. The result is not only that different "masculine" and "feminine" personality structures are perpetuated, but also that institutional inequities (men's control of public life, women's double workday, etc.) persist.

What my research adds to existing feminist analyses of sport is the recognition that sport does not simply and unambiguously reproduce men's existing power and privilege. Though sport clearly helps to produce culturally dominant conceptions of masculinity, my interviews reveal several strains within the sport/masculinity relationship. Most obviously, men's experiences of athletic careers are not entirely positive, nor are they the same for all men. These facts strongly suggest that sport is not a smoothly functioning, seamless institution, nor is masculinity a monolithic category. There are three factors that undermine sport's ability to construct a single dominant conception of masculinity: (1) the "costs" of athletic masculinity to men; (2) men's different experiences with athletic careers, according to social class, race, and sexual orientation; and (3) current challenges to the equation of sport and heterosexual masculinity, as posed by the rise of women's athletics. . . .

NOTES

1. G. K. Lehne, "Homophobia among Men: Supporting and Defining the Male Role," in M. S. Kimmel and M. A. Messner, eds., *Men's Lives* (New York: Macmillan, 1989), 416–29.

2. See D. Kopay and P. D. Young, *The Dave Kopay Story* (New York: Arbor House, 1977), p. 12; and B. Pronger, "Gay Jocks: A

Phenomenology of Gay Men in Athletics," in M. A. Messner and D. F. Sabo, *Sport, Men and the Gender Order* (Champaign, Ill.: Human Kinetics Press, 1990), pp. 141–52.

3. Brian Pronger has written that "because gay men grow up in a predominantly heterosexual world, they have learned the standard language of masculinity." However, gay men also know that they do not conform to the heterosexual standards that are the cornerstone of hegemonic masculinity. They thus tend to develop a sense of "irony" about their own "masculinity" in public arenas such as organized sport (Pronger, "Gay Jocks," p. 145).

4. J. H. Pleck, "Men's Power with Women, Other Men, and in Society," in E. H. Pleck and J. H. Pleck, eds., *The American Man* (Englewood Cliffs, N.J.: Prentice-Hall, 1980), pp. 417–33.

5. See especially chapter 5 of Gary Alan Fine, *With the Boys: Little League Baseball and Preadolescent Culture* (Chicago: University of Chicago Press, 1987). Though Fine utilizes a symbolic interactionist perspective, his observations concerning homophobia and misogyny among young boys are consistent with Chodorow's observation that the devaluation of women and the "feminine" in boys and men is part of the establishment of male identity and the construction of men's power and privilege over women. For a "relational" conception of the role of sexuality in children's culture, see B. Thorne and Z. Luria, "Sexuality and Gender in Children's Daily Worlds," *Social Problems* 33 (1986): 176–90.

6. Fine, *With the Boys,* p. 109.

7. *Ibid.,* p. 107.

8. G. Gilder, *Sexual Suicide* (New York: Bantam Books, 1973), pp. 216, 218.

9. J. Carroll, "Sport: Virtue and Grace," *Theory Culture and Society 3* (1986), 91–98. Jennifer Hargreaves delivers a brilliant feminist rebuttal to Carroll's masculinist defense of sport in the same issue of the journal. See Je. Hargreaves, "Where's the Virtue? Where's the Grace? A Discussion of the Social Production of Gender through Sport," pp. 109–21.

10. Gilder, *Sexual Suicide,* p. 221.

11. For a critical overview of the biological research on male-female difference, see A. Fausto-Sterling, *Myths of Gender: Biological Theories about Men and Women* (New York: Basic Books, 1985). For an overview of the historical basis of male domination, see R. Lee and R. Daly, "Man's Domination and Woman's Oppression: The Question of Origins," in M. Kaufman, ed., *Beyond Patriarchy: Essays by Men on Pleasure, Power, and Change* (Toronto: Oxford University Press, 1987), pp. 30–44.

12. R. W. Connell, *Gender and Power* (Stanford: Stanford University Press, 1987), p. 81 (emphasis in original text).

■

PART IV

Organizational/ Institutional Deviance

We have just looked at how some categories of people and behavior become defined as deviant. Yet interactionist theory, as we noted in the General Introduction, suggests that a deviant label floating around abstractly in society is not meaningful unless it gets attached to people. Groups in society not only work to create definitions of deviance but also create situations where deviance occurs and is labeled. In this section we will examine the role of organizations and institutions in creating a context in which deviance is evoked and shaped.

CREATING DEVIANCE

Society is not made up solely of individuals taking care of their own business and needs. It is filled with many organizations and institutions, such as hospitals, courts, prisons, social welfare agencies, corporations, and even the government. People who work within these organizations are guided, constrained, or pressured to act in ways that help the organization accomplish its goals. Organizational goals may range from performing complex operations to simply staying financially or politically afloat. In our intricate society, organizations sometimes have difficulty achieving these mandates through legitimate means. Whether out of convenience or through necessity, organizations often create a climate in which deviance becomes accepted as the best or only way to accomplish the tasks they deem desirable or vital. In these contexts, people's deviant behavior is not committed to further their own ends, but those of the organization's.

Jennifer Hunt and Peter Manning illustrate the organizational constraints fostering deviance among law enforcement officers in "The Social Context of Police Lying." They outline how the police are taught to lie as part of their job. This instruction begins in the academy, as rookies are socialized by older officers and detectives, and expands once they graduate. They learn that the "by the book" techniques they learned in the academy are not practically effective and that, on the street, they have to lie in order to accomplish good police work. This lying begins with their having to account for their behavior when they were in places where they shouldn't have been or when they weren't in places where they should have been. It builds to lying in court about the extent of their probable cause for action in situations of search, seizure, and arrest. Hunt and Manning carefully document the police norms that excuse and justify lying, so that this act is redefined as the accomplishment of institutional goals rather than as irresponsible and illegal individual manipulation.

William Chambliss then shows the pervasiveness of the institutional mandate to deviance at even greater heights in examining the activities of the most elite group in society: the federal government. Although government agents and agencies have the power to make and enforce laws, Chambliss shows in "State-Organized Crime" that they do not always seek to further their goals by legitimate means. Partly due to their power and position, they hold themselves separate from ordinary citizens and above the law. As their position frames their interests, they justify the necessity of violating the norms and laws they expect others to follow. This includes the smuggling of drugs (Vietnam) and arms (Contragate); murder, assassinations, and coups d'état (both domestically and internationally); and other activities such as spying on citizens, including opposing political parties (Watergate). We see, thus, that deviance is found not only on the margins of society but also within its mainstream and among its elites.

LABELING DEVIANCE

When not pursuing their own deviant activities, organization members often apply the deviant label to others. Many societal institutions, especially those in the medical and criminal justice arenas, are charged with the mandate of processing potential deviants. In so doing, they establish specific rules and procedures for handling people and assigning them to various categories of deviance. As in many bureaucracies, these rules and procedures tend to become organized, stabilized, and systematized, often functioning informally as organizational ends in themselves. This practice aids organizations in processing large numbers of people in a smooth and efficient manner, thus establishing and maintaining order while protecting themselves from outside criticism.

Organizations at work want to handle tasks in routine ways with a minimum of variation. People assigned to intake processing will assess individuals and assign them to bureaucratic categories based on *typifications* they hold about models of deviants and deviant behavior. Differences among individuals must then be minimized or neutralized to facilitate moving people through institutional mechanisms. Most institutions, therefore, incorporate some way of "stripping down" individuals' self-conceptions and replacing them with ones useful to the organization. These procedures vary in severity and intensity according to the bureaucratic organization involved, culminating in what Erving Goffman (1961) has called *"total institutions."* These organizations, such as prisons, military boot camps, and monasteries, demand the total submission of clients, isolate clients from outside populations, and operate for the benefit of a small, elite group rather than for the good of the large group of clients. They typically deindividualize people through a haircut, clothing change, seizure of possessions, and loss of rights, subjecting them to individual and collective humiliations and degradations. Although all of these experiences foster the smooth functioning of deviance-processing organizations, the result is a deep rooting of the deviant label.

Goffman's "The Moral Career of the Mental Patient" offers us some insight into a classic total institution: a state mental hospital. He considers the needs of this organization, how they result in its handling of people, and the consequences for individuals so treated. He examines the effects of people's betrayals by others as they are involuntarily committed and the mortifying *"degradation ceremonies"* they undergo once admitted. Restricted in movement and freedom and punished for breaking rules designed to ease the accomplishment of staff goals, clients are told that these treatments are for their "needs." These experiences frame and shape people's changing conceptions of self.

In "The Making of Blind Men," Robert Scott outlines the way in which social welfare agencies serving the blind label people deviant. Although they have services and benefits they can offer, before they will make any of these available they require potential clients to change their self-perceptions. Clients must show themselves "rehabilitated" rather than "uncooperative" and must submit to having their attitudes "created, shaped, and molded" by the agency. This process fundamentally entails accepting a role of total dependence and organizing their lives around the agency, thus fulfilling the institution's organizational needs for survival.

The judicial labeling process is the focus of Lisa Frohmann's "Discrediting Victims' Allegations of Sexual Assault: Prosecutorial Accounts and Case Rejections." Frohmann discusses the organizational dynamics of the prosecutor's office, focusing on the ways in which rape cases and rape victims are "typified," with those displaying discrepant (that is, individualized) features discredited. Complainants are made to feel deviant and devalued if their cases do not mold to the stereotypical norm, and their cases are dropped from further legal pursuit.

Individuals operating, then, within institutional frameworks are likely to be affected by the organizational dynamics shaping these frameworks, so that they are either spurred to deviance or labeled as deviant by their bureaucratic mechanisms.

The Social Context of Police Lying

JENNIFER HUNT

AND PETER K. MANNING

INTRODUCTION

Police, like many people in official capacities, lie. We intend here to examine the culturally grounded bases for police lying using ethnographic materials.[1] Following the earlier work of Manning (1974), we define lies as speech acts which the speaker knows are misleading or false, and are intended to deceive. Evidence that proves the contrary must be known to the observer.[2] Lying is not an obvious matter: it is always socially and contextually defined with reference to what an audience will credit; thus, its meaning changes and its effects are often ambiguous (Goffman 1959, Pp. 58–66). The moral context of lying is very important insofar as its definition may be relative to membership status. The outsider may not appreciate distinctions held scrupulously within a group; indeed, differences between what is and is [not] said may constitute a lie to an outsider, but these distinctions may not be so easily made by an insider. In a sense, lies do not exist in the abstract; rather they are objects within a negotiated occupational order (Maines 1982). In analytic terms, acceptable or normal lies become one criterion for membership within a group, and inappropriate lying, contextually defined, sets a person on the margins of that order.

The structural sources of police lying are several. Lying is a useful way to manipulate the public when applying the law and other threats are of little use (see Bittner 1970; Westley 1970; Skolnick 1966; Klockars 1983, 1984; Wilson 1968; Stinchombe 1964). The police serve as gatherers and screeners of facts, shaping them within the legal realities and routines of court settings (Buckner 1978). The risks involved in establishing often problematic facts and the adversarial context of court narratives increase the value of secrecy and of concealing and controlling information generally (Reiss 1974). Police are protected for their lies by law under stipulated circumstances (see McBarnett 1981; Erikson and Shearing 1986).[3] The internal organization of policing as well as the occupational culture emphasize control, punishment, and secrecy (Westley 1970; Manning 1977). Some police tasks, especially those in specialized police units such as vice, narcotics and internal affairs, clearly require and reward lying skills more than others, and such units my be subject to periodic scandals and public outcry (Manning 1980). The unfilled and perhaps impossible

From: "The Social Context of Police Lying," Jennifer Hunt and Peter K. Manning, *Symbolic Interaction* 14(1), 1991. Reprinted by permission of JAI Press Inc.

expectations in drug enforcement may escalate the use of lies in the "war on drugs," further reducing public trust when officers' lies are exposed. Most police officers in large forces at one time or another participate in some form of illicit or illegal activity, from the violation of departmental morals codes to the use of extra-legal force. Perhaps more importantly, there is an accepted view that it is impossible to "police by the book"; that any good officer, in the course of a given day, will violate at least one of the myriad rules and regulations governing police conduct. This is certain; what is seen as contingent is when, how, . . . where [and by whom] detection . . . will take place.

Lying is a sanctioned practice, differentially rewarded and performed, judged by local occupationally-grounded standards of competence.[4] However, it is likely that these standards are changing; as police claims to professional competence and capacity to control crime and incivilities in cities are validated, and absent any changes in internal or external sources of control and accountability (Cf. Reiss 1974), police may encounter less external pressure and public support in routine tasks and are less likely to be called to account. Policing has emerged as a more "professional" occupation and may be less at risk generally to public outcry. One inference of this line of conjecture is that lying is perceived as less risky by police. The occupational culture in departments studied by researchers contains a rich set of stories told to both colleagues and criminals. However, like the routinely required application of violence, some lies are "normal," and acceptable to audiences, especially colleagues, whereas others are not.

Given the pervasiveness of police lies, it is surprising that no research has identified and provided examples of types of lies viewed from the officers' perspective. We focus here on patrol officers' lies and note the skill with which they cope with situations in which lies are produced. Some officers are more frequently in trouble than others, and some more inclined to lie. We suggest a distinction between lies that excuse from those that justify an action, between troublesome and non-troublesome lies, and between case and cover lies. Lies are troublesome when they arise in a context such as a courtroom or a report in which the individual is sworn to uphold the truth. In such a context, lying may risk legal and/or moral sanctions, resulting in punishment and a loss in status. *Case lies* and *cover stories* are routinely told types of troublesome lies. Case lies are stories an officer utilizes systematically in a courtroom or on paper to facilitate the conviction of a suspect. Cover stories are lies an officer tells in court, to supervisors, and to colleagues in order to provide a verbal shield or mitigation in the event of anticipated discipline.

METHODOLOGY

The senior author was funded to study police training in a large Metropolitan police department ("Metro City"). Continuous fieldwork, undertaken as a known observer-participant for eighteen months, focused on the differences and similarities in the socialization experiences of young female and male officers.[5] The fieldwork included observation, participation in training with an incoming class in the police academy, [and] tape-recording interviews in relaxing informal settings with key informants selected for their verbal skills and willingness to give lengthy interviews. The social milieu encouraged them to provide detailed and detached stories. The observer had access to the personnel files of the two hundred officers who entered the force during the research period. She attended a variety of off-duty events and activities ranging from meetings of the Fraternal Order of the Police, sporting events, parties, and funerals (for further details see Hunt 1984). The data presented here are drawn primarily from tape recorded interviews.

LEARNING TO LIE

In the police academy, instructors encouraged recruits to lie in some situations, while strongly discouraging it in others. Officers are told it is "good police work," and encouraged to lie, to substitute guile for force, in situations of crisis intervention, investigation and interrogation, and especially with the mentally ill (Harris 1973).[6] During classes on law and court testimony, on the other hand, students were taught that the use of deception in court was illegal, morally wrong, and unacceptable and would subject the officer to legal and departmental sanctions. Through films and discussions, recruits learned that the only appropriate means to win court cases was to undertake and complete a "solid," "by the books" investigation including displaying a professional demeanor while delivering a succinct but "factual" narrative in court testimony.

Job experience changes the rookies' beliefs about the circumstances under which it is appropriate to lie. Learning to lie is a key to membership. Rookies in Metro City learn on the job, for example, that police routinely participate in a variety of illicit activities which reduce the discomfort of the job such as drinking, sleeping on duty, and staying inside during inclement weather. As these patterns of work avoidance may result in discovery, they demand the learning of explanatory stories which rationalize informal behavior in ways that jeopardize neither colleagues nor supervisors (Cain 1973; Chatterton 1975). Rookies and veteran police who demonstrate little skill in constructing these routine lies were informally criticized. For example, veteran officers in Metro City commented sympathetically about rookies who froze on foot-beats because they were too green to know that "a good cop never gets cold or wet . . ." and too new to have attained expertise in explaining their whereabouts if they were to leave their beats. After a few months in the district, several veteran officers approvingly noted that most of the rookies had learned not only where to hide but what to say if questioned by supervisors.

Rookies also learn the situational utility of lying when they observe detectives changing reports to avoid unnecessary paperwork and maintain the clearance rate. In the Metro City police department, some cases defined initially as robberies, assaults, and burglaries were later reduced to less serious offenses (Cf Sudnow 1965). Police argued that this practice reduced the time and effort spent on "bullshit jobs" little likely to be cleared. Rookies who opposed this practice and insisted on filing cases as they saw fit were ridiculed and labelled troublemakers. As a result, most division detectives provided minimal cooperation to these "troublemakers." This added the task of reworking already time-consuming and tedious reports to the workload of young officers who were already given little prospective guidance and routine assistance in completing their paperwork.

Young police also observe veterans lying in court testimony regarding, for example, the presence of probable cause in situations of search, seizure and arrest (see, for example, McClure 1986, Pp. 230–232).

There are also counter-pressures. While learning to lie, rookies also recognize that the public and court officials disapprove of lying, and that if caught in a serious lie, they may be subject to either legal sanctioning and/or departmental punishment. But recognizing external standards and their relevance does not exhaust the learning required. There are also relevant tacit rules within the occupational culture about what constitutes a normal lie. Complexity and guile, and agile verbal constructions are appreciated, while lying that enmeshes or makes colleagues vulnerable or is "sloppy" is condemned. Lying is judged largely in pragmatic terms otherwise. Soon, some rookies are as skillful as veterans at lying.[7]

POLICE ACCOUNTS OF LYING

Lies are made normal or acceptable by means of socially approved vocabularies for relieving responsibility or neutralizing the consequences of an event. These accounts are provided *after* an act if and when conduct is called into question (see the classic, Mills 1940; Sykes and Matza 1957; Scott and Lyman 1968). Police routinely normalize lying by two types of accounts, excuses and justifications (Van Maanen 1980; Hunt 1985; Waegel 1984). These accounts are not mutually exclusive, and a combination is typically employed in practice. The greater the number of excuses and justifications condensed in a given account, the more the police officer is able to reduce personal and peer related conflicts. These accounts are typically tailored to an audience. A cover story directed to an "external" audience such as the district attorneys, courts, or the media is considered more problematic than a lie directed to supervisors or peers (Manning 1974). Lies are more troublesome also when the audience is perceived as less trustworthy (Goffman 1959, p. 58).

Excuses deny full responsibility for an act of lying but acknowledge its inappropriateness. Police distinguish passive lies which involve omission, or covering oneself, from active lies such as a "frame," of a person for a crime by, for example, planting a gun, or the construction of a sophisticated story. The latter are more often viewed as morally problematic.

Justifications accept responsibility for the illegal lie in question but deny that the act is wrongful or blameworthy. They socially construct a set of justifications, used with both public and other police, according to a number of principles. (These are analogous to the neutralizations found by Sykes and Matza 1957 in another context.) When lying, police may appeal to "higher" loyalties that justify the means used, deny[ing] that anyone is truly hurt or a victim of the lies. Police may also deny injury by claiming that court testimony has little consequence as it is merely an extension of the "cops and robbers" game. It is simply a tool in one's repertoire that requires a modicum of verbal skill (see Blumberg 1967). Finally, as seen in "cover stories," officers justify lies instrumentally and pragmatically (see Van Maanen 1980; Waegel 1984).

LYING IN ACTION

Case Stories and the Construction of Probable Cause

The most common form of case lying, used to gain a conviction in court, involves the construction of probable cause for arrest, or search and seizure in situations where the legally required basis in the street encounter is weak or absent.[8] Probable cause can be constructed by reorganizing the sequence of events, "shading" or adding to the facts, omitting embarrassing facts, or entering facts into a testimony that were not considered at the time of arrest or while writing the report.

The following is a typical case story-account chosen from a taped interview in which probable cause was socially constructed. The officer was called to a "burglary in progress" with no further details included and found a door forced open at the back of the factory.

> So I arrive at the scene, and I say, I know: I do have an open property and I'm going in to search it. And I'm looking around, and I hear noise. Then, I hear glass break. And I run to the window, and obviously something just jumped out the window and is running and I hear skirmishing. So I run, and I still don't see anyone yet. I just hear something. I still haven't seen anyone. You hear a window, you see a window and you hear footsteps running. Then you don't hear it anymore. I don't find anybody. So I say to myself "whoever it is is around here somewhere." So, fifteen minutes go by, twenty go by. The job resumes.
>
> One cop stays in the front and about a half an hour or forty-five minutes later, low and behold, I see someone half a block away coming out of a field. Now, the field is on the other side of the factory. It's the same field that I chased this noise into. So a half hour later I see this guy at quarter to four in the morning just happened to be walking out of this field. So I grab him. "Who are you? What are you doing?" Bla bla bla. . . . And I see that he has flour on him, like flour which is what's inside of this factory. So I say to myself, "you're the suspect under arrest for burglary." Well, I really, at this point it was iffy if I had probable cause or not. . . . a very conservative judge would say that was enough. . . . but probably not, because the courts are so jammed that the weak probable cause would be

enough to have it thrown out. So in order to make it stick, what I said was "As I went to the factory and I noticed the door open and I entered the factory to search to see if there was anybody inside. Inside, by the other side of the wall, I see a young black male, approximately twenty-two years old, wearing a blue shirt and khaki pants, jump out of the window, and I chase him and I lost him in the bushes. An hour later I saw this very same black male walking out of the field and I arrested him. He was the same one I saw inside." O.K.? . . .

What I did was to construct probable cause that would definitely stick in court and I knew he was guilty. So in order to make it stick. . . . That's the kind of lies that happen all the time. I would defend that.[9]

The officer's account of his activity during the arrest of a suspect and subsequent testimony in court reveals a combination of excuses and justifications which rationalize perjury. He clearly distinguishes the story he tells the interviewer from the lie he told in court. Near the end of the vignette, by saying "O.K.?" he seeks to emphasize phatic contact as well to establish whether the interviewer understood how and why he lied and how he justified it. Within the account, he excuses his lies with reference to organizational factors ("conservative judges"—those who adhere to procedural guarantees—and "overcrowded courts"), and implies that these are responsible for releasing guilty suspects who should be jailed. These factors force the officer to lie in court in order to sustain an ambiguous and weak probable cause. The officer further justifies his lies by claiming that he believes the suspect to be guilty, and is responsible for perpetrating crime more serious than the lie used to convict him. He ends by claiming that such lies are acceptable to his peers—they "happen all the time"—and implicitly appeals to the higher goal of justice. As in this case, officers can shape and combine observed and invented facts to form a complex, elaborate yet coherent, picture which may help solve a crime, clear a case, or convict a criminal.

Case Stories and the Manipulation of the Court as an Informal Entity

As a result of the community pressure in Metro City, a specialized unit was created to arrest juveniles who "hung out" on street corners and disturbed neighborhoods. The unit was considered a desirable assignment because officers worked steady shifts and were paid overtime for court appearances. They were to be judged by convictions obtained, not solely upon their arrests. An officer in this unit explains some of the enforcement constraints produced by the law and how they can be circumvented:

Legally . . . when there's any amount of kids over five there is noise, but it's not really defined legally being unruly even though the community complains that they are drunk and noisy. Anyway, you get there, you see five kids and there's noise. It's not really criminal, but you gotta lock them up, particularly if someone had called and complained. So you lock them up for disorderly conduct and you tell your story. If they plead not guilty then you have to actually tell a story.

   . . . It's almost like a game. The kids know that they can plead guilty and get a $12.50 fine or a harder judge will give them a $30.50 fine or they can plead not guilty and have the officer tell their story of what occurred which lead to the arrest.

   . . . The game is who manipulates better, the kid or the cop? The one who lies better wins.

   Well, the kids are really cocky. I had arrested this group of kids, and when we went to court the defense attorney for the kids was arguing that all of the kids, who I claimed were there the first time that I warned them to get offa the corner, weren't there at all. Now, you don't really have to warn them to get offa the corner before you arrest them, but the judge likes it if you warn them once.

   Meanwhile, one of the kids is laughing in the courtroom, and the judge asks why he's laughing in her courtroom and showing disrespect.

   At this point, the kid's attorney asks me what the kid was wearing when I arrested him. I couldn't remember exactly what he had on so I just gave the standard uniform; dungarees, shirt, sneakers . . . Then, the defense attorney turns around and asks the kid what he was wearing and he gives this description of white pants with a white sports jacket. Now, you just know the kid is lying because there ain't a kid in that neighborhood who dresses like that. But, anyway, I figure they got me on this one.

   But then I signal the District Attorney to ask me how I remembered this kid outta the whole bunch who was on the corner. So the District Attorney asks me, "Officer, what was it that made you remember this male the first time?" And I said, "Well, your honor, I referred this one here because the first time that I warned the group to get offa the corner, this male was the one that laughed the hardest." I knew this would get the judge because the kid has pissed her off in the first place by laughing in the courtroom. Well, the judge's eyes lit up like she knew what I was talking about.

   "Found guilty. . . . 60 dollars." (The Judge ruled.)

In this case, the officer believed the boy was guilty because he "hung out" regularly with the juvenile corner group. Although the officer forgot the boy's dress on the day he was arrested, he testifies in court that it was the "standard [juvenile] uniform." The boy, however, claims he wore pants and a sports jacket. The officer was in a potentially embarrassing and awkward spot. In order to affirm the identification of the suspect and win the case, the officer constructs another lie using the District Attorney's question. He manipulates the emotions of the judge whose authority was previously threatened by the boy's disrespectful courtroom demeanor. He claims that he knows this was the boy because he displayed arrogance by laughing when arrested just as he had in court.

The officer's account of the unit's organization, the arrest, and his court-room testimony reveals a combination of excuses and justifications. He justifies his lie to make the arrest without probable cause and to gain a conviction citing the organizational and community pressures. He also justifies perjury by denying the reality and potential injury to the suspect caused by his actions. He sees courtroom communications as a game, and argues that the penalty is minor in view of the offense and the age of the suspect. The officer argues instrumentally that the lie was a means to gain or regain control as well as a means to punish an offender who has not accepted the police definition of the situation. The latter is evident in the officer's assertion that the boys are "really cocky." Their attempts to question the police version of the story by presenting themselves as clean cut children with good families is apparently viewed as a demonstration of deliberate arrogance deserving of retaliation in the form of a lie which facilitates their conviction.

Another officer from the same unit describes a similar example of case construction to gain a conviction. The clumsy character of the lie suggests that the officer believes he is at little risk of perjury. According to a colleague's account:

> We arrested a group of kids in a park right across from the hospital. They all know us and we know them, so they are getting as good as we are at knowing which stories go over better on the judge. So the kids in this instance plead not guilty which is a real slap in the face because you know that they are going to come up with a story that you are going to have to top [that is, the case will go to court and require testimony].
>
> So, the kids' story was that they were just sitting in the park and waiting for someone and that they were only having a conversation. [The police officer's testimony was]: "The kids were making so much noise, the kids were so loud. . . . They had this enormous radio blasting and the people in the hospital were so disturbed that they were just hanging outta the windows. And some nuns, some of the nuns that work in the hospital, they were coming outside because it was so loud." And the thing that appalled the officer the most was that this was going on right in front of the entrance to the hospital. The kids were acting in such a manner that the officer immediately arrested them without even a warning.
>
> Well, the kid, when he hears this, likely drops dead. He kept saying "what radio, what radio?" The funny part of it was that the other police officers who were in the back of the courtroom watching the cop testify kept rolling their eyes at him. First of all, because when he said that the people were hanging outta the hospital windows, the windows in that hospital don't open. They're sealed. Another thing was that the cop said this occurred at the entrance of the hospital. Two years ago this was the entrance to the hospital. But it's not the entrance now. Another thing was

that there never was a radio. But when the officer testified regarding the radio, he got confused. He actually did think there was one but in fact, the radio blasting was from another job. The cop realized after he testified that the kid didn't have a radio blasting.

The lies are described as instrumental: they are designed to regain control in court and to punish the offender for violating the officer's authority by verbally "slapping him in the face." In addition, the court-as-a-game metaphor is evident in the notion that each participant must top the other's story in court. The amused reaction evidenced by peers listening to what they viewed as absurd testimony, rolling their eyes, also suggests their bemused approval. The informant's ironic identification of his colleague's factual errors points out the recognized and displayed limits and constraints upon lying. The officers recognized the difference between a rather sloppy or merely effective lie and an admired lie that artfully combines facts, observations, and subtle inferences. Perhaps it is not unimportant to note that the police engaged in the first instance in a kind of social construction of the required social order. The police lied in virtually every key facet of this situation because they believed that the juveniles should be controlled. What might be called the police ordering of a situation was the precondition for both of these court lies. Such decisions are potentially a factor in community policing when police define and then defend in court with lies their notions of public order (see Wilson and Kelling 1982).

Cover Stories

A cover story is the second kind of legal lie that police routinely tell on paper, in court, and to colleagues. Like most case stories, cover stories are constructed using sub-cultural nuances to make retelling the dynamics of encounters legally rational. Maintaining the capacity to produce a cover story is viewed as an essential skill required to protect against disciplinary action.

A cover story may involve the manipulation of legal and departmental rules, or taken-for-granted-knowledge regarding a neighborhood, actions of people and things. A common cover story involves failure to respond to a radio call. Every officer knows, for example, that some districts have radio "dead spots" where radio transmissions do not reach. If "radio" (central communications) calls an officer who doesn't respond or accept, "pick up the job," radio will usually recall. If he still doesn't respond, another officer typically takes the job to cover for him, or a friendly dispatcher may assign the job to another unit. However, if radio assigns the same job to the same car a third time and the unit still doesn't accept the job, the officer may be subject to formal disciplinary action. One acceptable account for temporary unavailability (for whatever reason) is to claim that one's radio malfunctioned or that one was in a "dead spot" (see Rubinstein 1973; Manning 1988).

The most common cover stories involving criminal matters are constructed to protect the officer against charges of brutality or homicide. Such cover stories serve to bridge the gap between the normal use of force which character-izes the informal world of the street and its legal use as defined by the court.

Self protection is the presumed justification for cover stories. Since officers often equate verbal challenges with actual physical violence, both of which are grounds for retaliatory violence, either may underlie a story.[10] Threats of harm to self, partner, or citizens, are especially powerful bases for rationalizations. Even an officer who is believed by colleagues to use brutal force and seen as a poor partner as a result, is expected to lie to protect himself (see Hunt 1985; Waegel 1984). He or she would be considered odd, or even untrustworthy, if he or she did not. There is an interaction between violence and lying under-stood by police standards.

In the following account, the officer who fired his weapon exceeded "normal" force, and committed a "bad shooting" (see Van Maanen 1980). Few officers would condone the shooting of an unarmed boy who they did not see commit a crime. Nevertheless, the officers participate in the construc-tion of a cover story to protect their colleague against disciplinary action and justify it on the basis of self-defense and loyalty. Officers arrived at a scene that had been described mistakenly in a radio call as a "burglary in progress" (in fact, boys were stripping a previously stolen car). Since this is a call with arrest potential, it drew several police vehicles and officers soon began to chase the suspect(s):

> Then they get into a back yard chasing one kid. The kid starts running up a rain spout like he's a spider man, and one of the cops took a shot at him. So now they're all panicky because the kid made it to the roof and he let out a scream and the cops thought that they hit him. And that was a bad shooting! What would you think if you was that kid's mother? Not only did they not have an open property, but they don't know if it's a stolen car at all. Well, when they shot the kid I gave them an excuse by mistake, inadvertently. I was on the other side of the place with my partner when I heard the one shot. [The officer telling the story is on one side of an iron gate, and another officer J.J. was on the other side. He kicks open the gate, thinking it is locked. It is not locked and swings wildly open, striking J.J. in the head]. . . .
>
> J.J. keeps stepping backward like he wanted to cry . . . like he was in a daze. "It's all right, it's not bleeding," I says to him . . . like he was stunned.
>
> So then the Sergeant gets to the scene and asks what happened, cause this shot has been fired, and the kid screamed, and you figure some kid's been hit and he's up on the roof.
>
> They gotta explain this dead kid and the shot to the sergeant when he gets there. So J.J. and Eddy discuss this. Eddy was the cop who'd fired the shot at the kid climbing up the rain spout, and all of a sudden they decide to claim he got hit with something in the head, and J.J. yells, "I'm hit, I'm hit." Then Eddy, thinking his partner's been shot, fires a shot at the kid.

So they reported this all to the Captain and J.J. gets reprimanded for yelling "I'm shot" when he said, "I'm hit."

J.J. was never involved at all, but he just says this to cover for Eddy.

Meanwhile, the fire department is out there looking on the roof for the kid and they never found him so you figure he never got hit.

Here, the officer telling the story demonstrates his solidarity with colleagues by passively validating (refusing to discredit) the construction of an episode created by collusion between two other officers. The moral ambiguity of participation in such a troublesome lie is recognized and indicated by the interviewed officer. He disclaims responsibility for his involvement in the lie by insisting that he was not at the scene of the shooting and only "by mistake . . . inadvertently . . . gave them [the other officers] an excuse" used to create the cover story. In such morally ambiguous situations, individual officers remain in some moral tension. Note the officers' role-taking capacity, empathy and concern for the generalized other when he asks in the vignette what the interviewer's thoughts would be if ". . . you was that kid's mother?" Such views may conflict with those of peers and supervisors. . . .

CONCLUSION

This ethnographic analysis relies principally upon the perspective of the officers observed and interviewed. It draws, however, on broad ethnographic accounts or general formulations of the police mandate and tasks. We attempt to integrate the pressures inherent in the inevitable negotiation within hierarchical systems between official expectations and roles, and one's individual sense of self. Officers learn how to define and control the public and other officers, and to negotiate meanings. The social constructions or lies which arise result from situational integration of organizational, political and moral pressures. These are not easily captured in rules, norms, or values. Repeatedly, officers must negotiate organizational realities *and* maintain self-worth. Police lies, serving in part to maintain a viable self, are surrounded by cultural assumptions and designations, a social context which defines normal or acceptable lies and distinguishes them from those deviant or marginal to good practice. The meanings imputed to the concepts "lie," "lying" and "truth" are negotiated and indicate or connote subtle intergroup relationships. In a crisis, ability to display solidarity by telling a proper and effective lie is highly valued and rewarded. The ironic epithet "police liar" is neutralized. Subtle redefinition of truth includes forms of group-based honesty that are unrecognized by legal standards or by the standards of outsiders. These findings have implications that might be further researched.

Lying is a feature of everyday life found in a variety of personal, occupational and political interactions. Although telling the full truth may be formally encouraged throughout life, it is not always admired or rewarded. Neither truth nor lies are simple and uniform; cultural variation exists in the idea of normal lying and its contrast conception. Those who continue to tell the truth and do not understand communications as complex negotiations of formal and informal behavioral norms, find themselves in social dilemmas, and are vulnerable to a variety of labels used in everyday life like "tactless," "undersocialized," "deviant," or "mentally ill." The application of the label is contingent upon taken-for-granted modes of deception that structure interpersonal relations. As the last few years have shown, given the impossible mandate of the police, certain police tasks are more highly visible e.g. drug enforcement, and even greater pressure to lie may emerge. Thus, the mandate is shaped and patterned by tasks as well as general social expectations, and the sources of lying may differ as well.

The cultural grounds explored here are features of any organization which lies as part of its routine activities, such as government agencies carrying out domestic intelligence operations and covert foreign activities. Standards of truth and falsehood drawn from everyday life do not hold here, and this shifting ground of fact and reality is often difficult to grasp and hold for both insiders and outsiders. As a result, organization members, like the police, develop sophisticated and culturally sanctioned mechanisms for neutralizing the guilt and responsibility that troublesome and even morally ambiguous lying may often entail. In time, accounts which retrospectively justify and excuse a lie may become techniques of neutralization which prospectively facilitate the construction of new lies with ready-made justifications. When grounds for lying are well-known in advance, it takes a self-reflective act to tell the truth, rather than to passively accept and use lies when they are taken-for-granted and expected. Police, like politicians, look to "internal standards" and practices to pin down the meaning of events that resonate with questions of public morality and propriety (Katz 1977). When closely examined in a public inquiry, the foreground of everyday internal standards may become merely the background for a public scandal. Normal lies, when revealed and subjected to public standards, can become the basis for scandals. This may be the first occasion on which members of the organization recognize their potential to be seen in such a fashion.

Finally, the extent to which an organization utilizing lies or heavily dependent upon them perceives that it is "under siege" varies. In attempts to shore up their mandate, organizations may tacitly justify lying. As a result, the organization may increase its isolation, lose public trust and credibility, and begin to believe its own lies. This differentially occurs within policing, across departments, and in agencies of control generally. Such dynamics are suggested by this analysis.

NOTES

1. The many social functions of lying, a necessary correlate of trust and symbolic communication generally, are noted elsewhere (Ekman 1985;

Simmel 1954; Manning 1977). Our focus is restricted. We do not discuss varieties of concealment, falsification and leakage (Ekman 1985, Pp. 28–29), nor interpersonal dynamics, such as the consequence of a sequence of lies and cover lies that often occurs. We omit the case in which the target, such as a theater audience or someone conned, is prepared in advance to accept lies (Ekman 1985, p. 28). Nor do we discuss in detail horizontal or vertical collusions within organizations that generate and sustain lying (e.g. Honeycombe 1974).

2. We do not distinguish "the lie" from the original event, since we are concerned with verbal rationalizations in the sense employed by Mills (1940), Lindesmith and Strauss (1956) and Scott and Lyman (1968). We cluster what might be called accounts for lies (lies about lies found in the interview material included here) with lies, and argue that the complexity of the formulations, and their embeddedness in any instance (the fact that a story may include several excuses and justifications, and may include how these, in turn, were presented to a judge) makes it misleading to adhere to a strict typology of lies such as routine vs. non–routine, case lies (both justifications and excuses) vs. cover stories (both justifications and excuses, and troublesome vs. not troublesome lies. If each distinction were worked out in a table, as one reader noted, omitting ambiguous lies, at least 16 categories of lies would result. After considering internal distinctions among lies in policing, we concluded that a typology would suggest a misleading degree of certainty and clarity. More ethnographic material is required to refine the categories outlined here.

3. Police organization, courts and the law permit sanctioned freedom to redefine the facts of a case, the origins of the case, the bases of the arrest and the charge, the number of offenders and the number of violations. Like many public officials, they are allowed to lie when public well-being is at issue (for example, posing as drug dealers, buying and selling drugs, lying about their personal biographies and so on; see Manning 1980). Officers are protected if they lie in order to enter homes, to encourage people to confess, and to facilitate people who would otherwise be committing crimes to commit them. They have warrant to misrepresent, dissemble, conceal and reveal as routine aspects of an investigation.

4. Evidence further suggests, in a point we do not examine here, that departments differ in the support given for lies. This may be related to legalistic aspects of the social organization of police departments (Cf Wilson 1968, ch. 6). Ironically for members of specialized units like "sting operations" or narcotics, the line between truth and lies becomes so blurred that according to Ekman's definition (the liar must know the truth and intend to lie), they are virtually always "telling the truth." Furthermore, as noted above, such units are more vulnerable to public criticism because they are held to unrealistic standards, and feel greater pressure to achieve illegally what cannot be accomplished legally. Marx

(1988) argues that increased use of covert deceptive operations leads to further penetration of private life, confusion of public standards, and reduced expectations of police morality.

5. She spent some 12 weeks in recruit classes at the academy. For fifteen months, she rode as a non-uniformed research observer, usually in the front seat of a one officer car, from 4–midnight and occasionally on midnight to eight shifts. Although she rode with veteran officers for the first few weeks in order to learn official procedures, the remainder of the time was spent with rookie officers. Follow-up interviews were conducted several years after the completion of the initial 18 months of observation.

6. Typically, recruits were successful in calming the "psychotic" actor when they demonstrated convincingly that they shared the psychotic's delusion and would rescue him/her from his/her persecutors by, for instance, threatening to shoot them. Such techniques were justified scientifically by trained psychologists who also stressed their practical use to avoid violence in potentially volatile situations.

7. Previous research has shown how detailed the knowledge is of officers of how and why to lie, and it demonstrates that trainees are taught to lie by specific instructions and examples (see Harris 1973; McClure 1984; Fielding 1988).

8. Technically, adding facts one recalls later, even in court, are [sic] not the basis for lies. Lies, in our view, must be intended.

9. This is taken verbatim from an interview, and thus several rather interesting linguistic turns (especially changes in perspective) are evidenced. Analysis of this sociolinguistically might suggest how this quote replicates in microcosm the problem officers have in maintaining a moral self. They dance repeatedly along the edges of at least two versions of the truth.

10. Waegel (1984) explores the retrospective and prospective accounts police use to excuse and justify the use of force. However, he does not distinguish accounts told by colleagues which are viewed as true by the speaker and those told to representatives of the legal order which are viewed as lies and fit the description of a cover story. For example, the account of accidental discharge which Waegel perceives as a denial of responsibility may also be a cover story which itself is justified as "self defense" against formal reprimand. In contrast, other police excuses and justifications invoked to account for the use of force are often renditions of events that present the officer in a morally favorable light rather than actual lies (see Van Maanen 1980; Hunt 1985; Waegel 1984). Whether the police categorize their use of force as "normal" or "brutal" (Hunt 1985) also structures the moral assessment of a lie, a point which Waegel also overlooks. Thus, acts of normal force which can be excused or justified with reference to routine accounting practices may necessitate the construction of cover stories which become morally neutral by virtue of the act they disguise. Other

acts of violence viewed as demonstrating incompetence or brutality may not be excused or justified according to routine accounting practices. Although cover stories in such cases are perceived as rational, they may not provide moral protection for the officer because the lie takes on aspects of the moral stigma associated with the act of violence which it conceals.

REFERENCES

Bittner, E. 1970. *Functions of the Police in an Urban Society*. Bethesda: NIMH.

———. 1974. "A Theory of Police: Florence Nightingale in Pursuit of Willie Sutton." In *The Potential for Reform of Criminal Justice,* edited by H. Jacob. Beverly Hills: Sage.

Blumberg, A. 1967. *Criminal Justice.* Chicago: Quadrangle Books.

Buckner, H. T. 1978. "Transformations of Reality in the Legal Process," *Social Research* 37:88–101.

Cain, M. 1973. *Society and the Policeman's Role*. London: Routledge, Kegan Paul.

Chatterton, M. 1975. "Organizational Relationships and Processes in Police Work: A Case Study of Urban Policing." Unpublished Ph.D. thesis, University of Manchester.

———. 1979. "The Supervision of Patrol Work Under the Fixed Points System." In *The British Police,* edited by S. Holdaway. London: Edward Arnold.

Ekman, P. 1985. *Telling Lies.* New York: W. W. Norton

Erikson, R., and C. Shearing. 1986. "The Scientification of the Police." In *The Knowledge Society,* edited by G. Bohme and N. Stehr. Dordrecht. Boston: D. Reidel.

Fielding, N. 1988. *Joining Forces.* London: Tavistock.

Goffman, E. 1959. *The Presentation of Self in Everyday Life*. New York: Doubleday Anchor Books.

Harris, R. 1973. *The Police Academy.* New York: Wiley.

Honeycombe, G. 1974. *Adam's Tale.* London: Arrow Books.

Hunt, J. C. 1984. "The Development of Rapport Through the Negotiation of Gender in Fieldwork among the Police." *Human Organization* 43:283–296.

———. 1985. "Police Accounts of Normal Force." *Urban Life* 13:315–342.

Katz, J. 1977. "Cover-up and Collective Integrity: On the Natural Antagonisms of Authority Internal and External to Organizations." *Social Problems* 25:3–17.

Klockars, C. 1983. "The Dirty Harry Problem." *Annals of the American Academy of Political and Social Science* 452 (November):33–47.

————. 1984. "Blue Lies and Police Placebos." *American Behavioral Scientist* 27:529–544.

Lindesmith, A., and A. Strauss. 1956. *Social Psychology*. New York: Holt, Dryden.

McBarnett, D. 1981. *Conviction*. London: MacMillan.

McClure, J. 1986. *Cop World*. New York: Laurel/Dell.

Maines, D. 1982. "In Search of Mesostructure: Studies in the Negotiated Order." *Urban Life* 11:267–279.

Manning, P. K. 1974. "Police Lying." *Urban Life* 3:283–306.

————. 1977. *Police Work*. Cambridge, MA: M.I.T. Press.

————. 1980. *Narc's Game*. Cambridge, MA: M.I.T. Press.

————. 1988. *Symbolic Communication: Signifying Calls and the Police Response*. Cambridge, MA: M.I.T. Press.

Marx, G. 1988. *Undercover Policework in America: Problems and Paradoxes of a Necessary Evil*. Berkeley: University of California Press.

Mills, C. W. 1940. "Situated Actions and Vocabularies of Motive." *ASR 6* (December):904–913.

Punch, M. 1985. *Conduct Unbecoming*. London: Tavistock.

Reiss, A. J., Jr. 1971. *The Police and the Public*. New Haven: Yale University Press.

————. 1974. "Discretionary Justice." Pp. 679–699 in *The Handbook of Criminal Justice,* edited by Daniel Glaser. Chicago: Rand-McNally.

Rubinstein, J. 1973. *City Police*. New York: Farrar, Straus and Giroux.

Scott, M. B., and S. Lyman. 1968. "Accounts." *American Sociological Review* 33:46–62.

Simmel, G. 1954. *The Society of George Simmel,* edited by Kurt Wolff. Glencoe: Free Press.

Skolnick, J. 1966. *Justice Without Trial*. New York: Wiley.

Stinchcombe, A. 1964. "Institutions of Privacy in the Determination of Police Administrative Practice." *American Journal of Sociology* 69:150–160.

Sykes, G. M., and D. Matza. 1957. "Techniques of Neutralization: A Theory of Delinquency." *American Sociological Review* 22:664–670.

Van Maanen, J. 1974. "Working the Street . . ." In *Prospects for Reform in Criminal Justice,* edited by H. Jacob. Newbury Park, CA: Sage.

————. 1975. "Police Socialization: A Longitudinal Examination of Job Attitudes in an Urban Police Department." *Administrative Science Quarterly* 20 (June):207–228.

————. 1978. "The Asshole." In *Policing: A View From the Street,* edited by P. K. Manning and J. Van Maanen. New York: Random House.

————. 1980. "Beyond Account: The Personal Impact of Police Shootings."
    Annals of the American Academy of Political and Science 342:145–156.

Waegel, W. 1984. "How Police Justify the Use of Deadly Force." *Social
    Problems* 32:144–155.

Westley, W. 1970. *Violence and the Police*. Cambridge, MA: M.I.T. Press.

Wilson, James, Q. 1968. *Varieties of Police Behavior*. Cambridge: Harvard
    University Press.

Wilson, J. Q., and G. Kelling. 1982. "The Police and Neighborhood Safety:
    Broken Windows." *Atlantic* 127 (March):29–38.

13

State-Organized Crime

WILLIAM J. CHAMBLISS

STATE-ORGANIZED CRIME DEFINED

The most important type of criminality organized by the state consists of acts defined by law as criminal and committed by state officials in the pursuit of their job as representatives of the state. Examples include a state's complicity in piracy, smuggling, assassinations, criminal conspiracies, acting as an accessory before or after the fact, and violating laws that limit their activities. In the latter category would be included the use of illegal methods of spying on citizens, diverting funds in ways prohibited by law (e.g., illegal campaign contributions, selling arms to countries prohibited by law, and supporting terrorist activities).

State-organized crime does not include criminal acts that benefit only individual officeholders, such as the acceptance of bribes or the illegal use of violence by the police against individuals, unless such acts violate existing criminal law and are official policy. For example, the current policies of torture and random violence by the police in South Africa are incorporated under the category of state-organized crime because, apparently, those practices are both state policy and in violation of existing South African law. On the other hand, the excessive use of violence by the police in urban ghettoes is not state-organized crime for it lacks the necessary institutionalized policy of the state. . . .

From: "State-Organized Crime," William J. Chambliss, *Criminology* 27(2), 1989.
Reprinted by permission of the American Society of Criminology and the author.

SMUGGLING

Smuggling occurs when a government has successfully cornered the market on some commodity or when it seeks to keep a commodity of another nation from crossing its borders. In the annals of crime, everything from sheep to people, wool to wine, gold to drugs, and even ideas, has been prohibited for either export or import. Paradoxically, whatever is prohibited, it is at the expense of one group of people for the benefit of another. Thus, the laws that prohibit the import or export of a commodity inevitably face a built-in resistance. Some part of the population will always want to either possess or to distribute the prohibited goods. At times, the state finds itself in the position of having its own interests served by violating precisely the same laws passed to prohibit the export or import of the goods it has defined as illegal.

Narcotics and the Vietnam War

Sometime around the eighth century, Turkish traders discovered a market for opium in Southeast Asia (Cahmbliss, 1977; McCoy, 1973). Portuguese traders several centuries later found a thriving business in opium trafficking conducted by small ships sailing between trading ports in the area. One of the prizes of Portuguese piracy was the opium that was taken from local traders and exchanged for tea, spices, and pottery. Several centuries later, when the French colonized Indochina, the traffic in opium was a thriving business. The French joined the drug traffickers and licensed opium dens throughout Indochina. With the profits from those licenses, the French supported 50% of the cost of their colonial government (McCoy, 1973: 27).

When the Communists began threatening French rule in Indochina, the French government used the opium profits to finance the war. It also used cooperation with the hill tribes who controlled opium production as a means of ensuring the allegiance of the hill tribes in the war against the Communists (McCoy, 1973).

The French were defeated in Vietnam and withdrew, only to be replaced by the United States. The United States inherited the dependence on opium profits and the cooperation of the hill tribes, who in turn depended on being allowed to continue growing and shipping opium. The CIA went a step further than the French and provided the opium-growing feudal lords in the mountains of Vietnam, Laos, Cambodia, and Thailand with transportation for their opium via Air America, the CIA airline in Vietnam.

Air America regularly transported bundles of opium from airstrips in Laos, Cambodia, and Burma to Saigon and Hong Kong (Chambliss, 1977: 56). An American stationed at Long Cheng, the secret CIA military base in northern Laos during the war, observed:

> . . . so long as the Meo leadership could keep their wards in the boon-
> docks fighting and dying in the name of, for these unfortunates anyway,
> some nebulous cause . . . the Meo leadership [was paid off] in the form of

a carte-blanche to exploit U.S.-supplied airplanes and communication gear to the end of greatly streamlining the opium operations. . . . (Chambliss, 1977: 56).

This report was confirmed by Laotian Army General Ouane Rattikone, who told me in an interview in 1974 that he was the principal overseer of the shipment of opium out of the Golden Triangle via Air America. U.S. law did not permit the CIA or any of its agents to engage in the smuggling of opium.

After France withdrew from Vietnam and left the protection of democracy to the United States, the French intelligence service that preceded the CIA in managing the opium smuggling in Asia continued to support part of its clandestine operations through drug trafficking (Kruger, 1980). Although those operations are shrouded in secrecy, the evidence is very strong that the French intelligence agencies helped to organize the movement of opium through the Middle East (especially Morocco) after their revenue from opium from Southeast Asia was cut off.

In 1969 Michael Hand, a former Green Beret and one of the CIA agents stationed at Long Cheng when Air America was shipping opium, moved to Australia, ostensibly as a private citizen. On arriving in Australia, Hand entered into a business partnership with an Australian national, Frank Nugan. In 1976 they established the Nugan Hand Bank in Sydney (Commonwealth of New South Wales, 1982a, 1982b). The Nugan Hand Bank began as a storefront operation with minimal capital investment, but almost immediately it boasted deposits of over $25 million. The rapid growth of the bank resulted from large deposits of secret funds made by narcotics and arms smugglers and large deposits from the CIA (Nihill, 1982).

In addition to the records from the bank that suggest the CIA was using the bank as a conduit for its funds, the bank's connection to the CIA and other U.S. intelligence agencies is evidenced by the people who formed the directors and principal officers of the bank, including the following:

- Admiral Earl F. Yates, president of the Nugan Hand Bank was, during the Vietnam War, chief of staff for strategic planning of U.S. forces in Asia and the Pacific.

- General Edwin F. Black, president of Nugan Hand's Hawaii branch, was commander of U.S. troops in Thailand during the Vietnam War and, after the war, assistant army chief of staff for the Pacific.

- General Erle Cocke, Jr., head of the Nugan Hand Washington, D.C., office.

- George Farris, worked in the Nugan Hand Hong Kong and Washington, D.C. offices. Farris was a military intelligence specialist who worked in a special forces training base in the Pacific.

- Bernie Houghton, Nugan Hand's representative in Saudi Arabia. Houghton was also a U.S. naval intelligence undercover agent.

- Thomas Clines, director of training in the CIA's clandestine service, was a London operative for Nugan Hand who helped in the takeover of a London-based bank and was stationed at Long Cheng with Michael Hand and Theodore S. Shackley during the Vietnam War.

- Dale Holmgreen, former flight service manager in Vietnam for Civil Air Transport, which became Air America. He was on the board of directors of Nugan Hand and ran the bank's Taiwan office.

- Walter McDonald, an economist and former deputy director of CIA for economic research, was a specialist in petroleum. He became a consultant to Nugan Hand and served as head of its Annapolis, Maryland, branch.

- General Roy Manor, who ran the Nugan Hand Philippine office, was a Vietnam veteran who helped coordinate the aborted attempt to rescue the Iranian hostages, chief of staff for the U.S. Pacific command, and the U.S. government's liaison officer to Philippine President Ferdinand Marcos.

On the board of directors of the parent company formed by Michael Hand that preceded the Nugan Hand Bank were Grant Walters, Robert Peterson, David M. Houton, and Spencer Smith, all of whom listed their address as c/o Air America, Army Post Office, San Francisco, California.

Also working through the Nugan Hand Bank was Edwin F. Wilson, a CIA agent involved in smuggling arms to the Middle East and later sentenced to prison by a U.S. court for smuggling illegal arms to Libya. Edwin Wilson's associate in Mideast arms shipments was Theodore Shackley, head of the Miami, Florida, CIA station.* In 1973, when William Colby was made director of Central Intelligence, Shackley replaced him as head of covert operations for the Far East; on his retirement from the CIA William Colby became Nugan Hand's lawyer.

In the late 1970s the bank experienced financial difficulties, which led to the death of Frank Nugan. He was found dead of a shotgun blast in his Mercedes Benz on a remote road outside Sydney. The official explanation was suicide, but some investigators speculated that he might have been murdered. In any event, Nugan's death created a major banking scandal and culminated in a government investigation. The investigation revealed that millions of dollars were unaccounted for in the bank's records and that the bank was serving as a money-laundering operation for narcotics smugglers and as a conduit through which the CIA was financing gun smuggling and other illegal operations throughout the world. These operations included illegally smuggling arms to South Africa and the Middle East. There was also evidence that the CIA used the Nugan Hand Bank to pay for political cam-

---

*It was Shackley who, along with Rafael "Chi Chi" Quintero, a Cuban-American, forged the plot to assassinate Fidel Castro by using organized-crime figures Santo Trafficante, Jr., John Roselli, and Sam Giancana.

paigns that slandered politicians, including Australia's Prime Minister Witham (Kwitny, 1987).

Michael Hand tried desperately to cover up the operations of the bank. Hundreds of documents were destroyed before investigators could get into the bank. Despite Hand's efforts, the scandal mushroomed and eventually Hand was forced to flee Australia. He managed this, while under indictment for a rash of felonies, with the aid of a CIA official who flew to Australia with a false passport and accompanied him out of the country. Hand's father, who lives in New York, denies knowing anything about his son's whereabouts.

Thus, the evidence uncovered by the government investigation in Australia linked high-level CIA officials to a bank in Sydney that was responsible for financing and laundering money for a significant part of the narcotics trafficking originating in Southeast Asia (Commonwealth of New South Wales, 1982b; Owen, 1983). It also linked the CIA to arms smuggling and illegal involvement in the democratic processes of a friendly nation. Other investigations reveal that the events in Australia were but part of a worldwide involvement in narcotics and arms smuggling by the CIA and French intelligence (Hougan, 1978; Kruger, 1980; Owen, 1983).

Arms Smuggling

One of the most important forms of state-organized crime today is arms smuggling. To a significant extent, U.S. involvement in narcotics smuggling after the Vietnam War can be understood as a means of funding the purchase of military weapons for nations and insurgent groups that could not be funded legally through congressional allocations or for which U.S. law prohibited support (NARMIC, 1984).

In violation of U.S. law, members of the National Security Council (NSC), the Department of Defense, and the CIA carried out a plan to sell millions of dollars worth of arms to Iran and use profits from those sales to support the Contras in Nicaragua (Senate Hearings, 1986). The Boland amendment, effective in 1985, prohibited any U.S. official from directly or indirectly assisting the Contras. To circumvent the law, a group of intelligence and military officials established a "secret team" of U.S. operatives, including Lt. Colonel Oliver North, Theodore Shackley, Thomas Clines, and Maj. General Richard Secord, among other (testimony before U.S. Senate, 1986). Shackley and Clines, as noted, were CIA agents in Long Cheng; along with Michael Hand they ran the secret war in Laos, which was financed in part from profits from opium smuggling. Shackley and Clines had also been involved in the 1961 invasion of Cuba and were instrumental in hiring organized-crime figures in an attempt to assassinate Fidel Castro.

Senator Daniel Inouye of Hawaii claims that this "secret government within our government" waging war in Third World countries was part of the Reagan doctrine (the *Guardian,* July 29, 1987). Whether President Reagan or then Vice President Bush was aware of the operations is yet to be established.

What cannot be doubted in the face of overwhelming evidence in testimony before the Senate and from court documents is that this group of officials of the state oversaw and coordinated the distribution and sale of weapons to Iran and to the Contras in Nicaragua. These acts were in direct violation of the Illegal Arms Export Control Act, which made the sale of arms to Iran unlawful, and the Boland amendment, which made it a criminal act to supply the Contras with arms or funds.

The weapons that were sold to Iran were obtained by the CIA through the Pentagon. Secretary of Defense Caspar Weinberger ordered the transfer of weapons from Army stocks to the CIA without the knowledge of Congress four times in 1986. The arms were then transferred to middlemen, such as Iranian arms dealer Yaacov Nimrodi, exiled Iranian arms dealer Manucher Ghorbanifar, and Saudi Arabian businessman Adnan Khashoggi. Weapons were also flown directly to the Contras, and funds from the sale of weapons were diverted to support Contra warfare. There is also considerable evidence that this "secret team," along with other military and CIA officials, cooperated with narcotics smuggling in Latin America in order to fund the Contras in Nicaragua.

In 1986, the Reagan administration admitted that Adolfo Chamorro's Contra group, which was supported by the CIA, was helping a Colombian drug trafficker transport drugs into the United States. Chamorro was arrested in April 1986 for his involvement (Potter and Bullington, 1987: 54). Testimony in several trials of major drug traffickers in the past 5 years has revealed innumerable instances in which drugs were flown from Central America into the United States with the cooperation of military and CIA personnel. These reports have also been confirmed by military personnel and private citizens who testified that they saw drugs being loaded on planes in Central America and unloaded at military bases in the United States. Pilots who flew planes with arms to the Contras report returning with planes carrying drugs.

At the same time that the United States was illegally supplying the Nicaraguan Contras with arms purchased, at least in part, with profits from the sale of illegal drugs, the administration launched a campaign against the Sandinistas for their alleged involvement in drug trafficking. Twice during his weekly radio shows in 1986, President Reagan accused the Sandinistas of smuggling drugs. Barry Seal, an informant and pilot for the Drug Enforcement Administration (DEA) was ordered by members of the CIA and DEA to photograph the Sandinistas loading a plane. During a televised speech in March 1986, Reagan showed the picture that Seal took and said that it showed Sandinista officials loading a plane with drugs for shipment to the United States. After the photo was displayed, Congress appropriated $100 million in aid for the Contras. Seal later admitted to reporters that the photograph he took was a plane being loaded with crates that did not contain drugs. He also told reporters that he was aware of the drug smuggling activities of the Contra network and a Colombian cocaine syndicate. For his candor, Seal was murdered in February 1987. Shortly after his murder, the DEA issued a "low key clarification" regarding the validity of the photograph, admitting that there was no evidence that the plane was being loaded with drugs.

Other testimony linking the CIA and U.S. military officials to complicity in drug trafficking includes the testimony of John Stockwell, a former high-ranking CIA official, who claims that drug smuggling and the CIA were essential components in the private campaign for the Contras. Corroboration for these assertions comes also from George Morales, one of the largest drug traffickers in South America, who testified that he was approached by the CIA in 1984 to fly weapons into Nicaragua. Morales claims that the CIA opened up an airstrip in Costa Rica and gave the pilots information on how to avoid radar traps. According to Morales, he flew 20 shipments of weapons into Costa Rica in 1984 and 1985. In return, the CIA helped him to smuggle thousands of kilos of cocaine into the United States. Morales alone channeled $250,000 quarterly to Contra leader Adolfo Chamorro from his trafficking activity. A pilot for Morales, Gary Betzner, substantiated Morales's claims and admitted flying 4,000 pounds of arms into Costa Rica and 500 kilos of cocaine to Lakeland, Florida, on his return trips. From 1985 to 1987, the CIA arranged 50 to 100 flights using U.S. airports that did not undergo inspection.

The destination of the flights by Morales and Betzner was a hidden airstrip on the ranch of John Hull. Hull, an admitted CIA agent, was a primary player in Oliver North's plan to aid the Contras. Hull's activities were closely monitored by Robert Owen, a key player in the Contra Supply network. Owen established the Institute for Democracy, Education, and Assistance, which raised money to buy arms for the Contras and which, in October 1985, was asked by Congress to distribute $50,000 in "humanitarian aid" to the Contras. Owen worked for Oliver North in coordinating illegal aid to the Contras and setting up the airstrip on the ranch of John Hull.

According to an article in the *Nation,* Oliver North's network of operatives and mercenaries had been linked to the largest drug cartel in South America since 1983. The DEA estimates that Colombian Jorge Ochoa Vasquez, the "kingpin" of the Medellin drug empire, is responsible for supplying 70% to 80% of the cocaine that enters the United States every year. Ochoa was taken into custody by Spanish police in October 1984 when a verbal order was sent by the U.S. Embassy in Madrid for his arrest. The embassy specified that Officer Cos-Gayon, who had undergone training with the DEA, should make the arrest. Other members of the Madrid Judicial Police were connected to the DEA and North's arms smuggling network. Ochoa's lawyers informed him that the United States would alter his extradition if he agreed to implicate the Sandanista government in drug trafficking. Ochoa refused and spent 20 months in jail before returning to Colombia. The Spanish courts ruled that the United States was trying to use Ochoa to discredit Nicaragua and released him (the *Nation,* September 5, 1987.)

There are other links between the U.S. government and the Medellin cartel. Jose Blandon, General Noriega's former chief advisor, claims that DEA operations have protected the drug empire in the past and that the DEA paid Noriega $4.7 million for his silence. Blandon also testified in Senate committee hearings that Panama's bases were used as training camps for the Contras in exchange for "economic" support from the United States. Finally, Blandon contends that the CIA gave Panamanian leaders intelligence documents about U.S. senators and aides; the CIA denies these charges (the *Christian Science Monitor,* February 11, 1988: 3).

Other evidence of the interrelationship among drug trafficking, the CIA, the NSC, and aid to the Contras includes the following:

- In January 1983, two Contra leaders in Costa Rica persuaded the Justice Department to return over $36,000 in drug profits to drug dealers Julio Zavala and Carlos Cabezas for aid to the Contras (Potter and Bullington, 1987: 22).

- Michael Palmer, a drug dealer in Miami, testified that the State Department's Nicaraguan humanitarian assistance office contracted with his company, Vortex Sales and Leasing, to take humanitarian aid to the Contras. Palmer claims that he smuggled $40 million in marijuana to the United States between 1977 and 1985 (the *Guardian,* March 20, 1988: 3).

- During House and Senate hearings in 1986, it was revealed that a major DEA investigation of the Medellin drug cartel of Colombia, which was expected to culminate in the arrest of several leaders of the cartel, was compromised when someone in the White House leaked the story of the investigation to the *Washington Times* (a conservative newspaper in Washington, D.C.), which published the story on July 17, 1984. According to DEA Administrator John Lawn, the leak destroyed what was "probably one of the most significant operations in DEA history" (Sharkey, 1988: 24).

- When Honduran General Jose Buseo, who was described by the Justice Department as an "international terrorist," was indicted for conspiring to murder the president of Honduras in a plot financed by profits from cocaine smuggling, Oliver North and officials from the Department of Defense and the CIA pressured the Justice Department to be lenient with General Buseo. In a memo disclosed by the Iran-Contra committee, North stated that if Buseo was not protected "he will break his long-standing silence about the Nic[araguan] resistance and other sensitive operations" (Sharkey, 1988: 27).

On first blush, it seems odd that government agencies and officials would engage in such wholesale disregard of the law. As a first step in building an explanation for these and other forms of state-organized crime, let us try to understand why officials of the CIA, the NSC, and the Department of Defense would be willing to commit criminal acts in pursuit of other goals.

WHY?

Why would government officials from the NSC, the Defense Department, the State Department, and the CIA become involved in smuggling arms and narcotics, money laundering, assassinations, and other criminal activities? The answer lies in the structural contradictions that inhere in nation-states (Chambliss, 1980).

As Weber, Marx, and Gramsci pointed out, no state can survive without establishing legitimacy. The law is a fundamental cornerstone in creating legitimacy and an illusion (at least) of social order. It claims universal principles that demand some behaviors and prohibit others. The protection of property and personal security are obligations assumed by states everywhere both as a means of legitimizing the state's franchise on violence and as a means of protecting commercial interests (Chambliss and Seidman, 1982).

The threat posed by smuggling to both personal security and property interests makes laws prohibiting smuggling essential. Under some circumstances, however, such laws contradict other interests of the state. This contradiction prepares the ground for state-organized crime as a solution to the conflicts and dilemmas posed by the simultaneous existence of contradictory "legitimate" goals.

The military-intelligence establishment in the United States is resolutely committed to fighting the spread of "communism" throughout the world. This mission is not new but has prevailed since the 1800s. Congress and the presidency are not consistent in their support for the money and policies thought by the front-line warriors to be necessary to accomplish their lofty goals. As a result, programs under way are sometimes undermined by a lack of funding and even by laws that prohibit their continuation (such as the passage of laws prohibiting support for the Contras). Officials of government agencies adversely affected by political changes are thus placed squarely in a dilemma: If they comply with the legal limitations on their activities they sacrifice their mission. The dilemma is heightened by the fact that they can anticipate future policy changes that will reinstate their resources and their freedom. When that time comes, however, programs adversely affected will be difficult if not impossible to re-create.

A number of events that occurred between 1960 and 1980 left the military and the CIA with badly tarnished images. Those events and political changes underscored their vulnerability. The CIA lost considerable political clout with elected officials when its planned invasion of Cuba (the infamous Bay of Pigs invasion) was a complete disaster. Perhaps as never before in its history, the United States showed itself vulnerable to the resistance of a small nation. The CIA was blamed for this fiasco even though it was President Kennedy's decision to go ahead with the plans that he inherited from the previous administration. To add to the agency's problems, the complicity between it and ITT to invade Chile and overthrow the Allende government was yet another scar (see below), as was the involvement of the CIA in narcotics smuggling in Vietnam.

These and other political realities led to a serious breach between Presidents Kennedy, Johnson, Nixon, and Carter and the CIA. During President Nixon's tenure in the White House, one of the CIA's top men, James Angleton, referred to Nixon's national security advisor, Henry Kissinger (who became secretary of state) as "objectively, a Soviet Agent" (Hougan, 1984: 75). Another top agent of the CIA, James McCord (later implicated in the Watergate burglary) wrote a secret letter to his superior, General Paul Gaynor, in January 1973 in which he said:

> When the hundreds of dedicated fine men and women of the CIA no longer write intelligence summaries and reports with integrity, without fear of political recrimination—when their fine Director [Richard Helms] is being summarily discharged in order to make way for a politician who will write or rewrite intelligence the way the politicians want them written, instead of the way truth and best judgment dictates, our nation is in the deepest of trouble and freedom itself was never so imperiled. Nazi Germany rose and fell under exactly the same philosophy of governmental operation. (Hougan, 1984: 26–27)

McCord (1974: 60) spoke for many of the top military and intelligence officers in the United States when he wrote in his autobiography: "I believed that the whole future of the nation was at stake." These views show the depth of feeling toward the dangers of political "interference" with what is generally accepted in the military–intelligence establishment as their mission (Goulden, 1984).

When Jimmy Carter was elected president, he appointed Admiral Stansfield Turner as director of Central Intelligence. At the outset, Turner made it clear that he and the president did not share the agency's view that they were conducting their mission properly (Goulden, 1984; Turner, 1985). Turner insisted on centralizing power in the director's office and on overseeing clandestine and covert operations. He met with a great deal of resistance. Against considerable opposition from within the agency, he reduced the size of the covert operation section from 1,200 to 400 agents. Agency people still refer to this as the "Halloween massacre."

Old hands at the CIA do not think their work is dispensable. They believe zealously, protectively, and one is tempted to say, with religious fervor, that the work they are doing is essential for the salvation of humankind. With threats from both Republicans and Democratic administrations, the agency sought alternative sources of revenue to carry out its mission. The alternative was already in place with the connections to the international narcotics traffic, arms smuggling, the existence of secret corporations incorporated in foreign countries (such as Panama), and the established links to banks for the laundering of money for covert operations.

STATE-ORGANIZED
ASSASSINATIONS AND MURDER

Assassination plots and political murders are usually associated in people's minds with military dictatorships and European monarchies. The practice of assassination, however, is not limited to unique historical events but has become a tool of international politics that involves modern nation-states of many different types.

In the 1960s a French intelligence agency hired Christian David to assassinate the Moroccan leader Ben Barka (Hougan, 1978: 204–207). Christian David was one of those international "spooks" with connections to the DEA, the CIA, and international arms smugglers, such as Robert Vesco.

In 1953, the CIA organized and supervised a coup d'etat in Iran that overthrew the democratically elected government of Mohammed Mossadegh, who had become unpopular with the United States when he nationalized foreign-owned oil companies. The CIA's coup replaced Mossadegh with Reza Shah Pahlevi, who denationalized the oil companies and with CIA guidance established one of the most vicious secret intelligence organizations in the world: SAVAK. In the years to follow, the shah and CIA-trained agents of SAVAK murdered thousands of Iranian citizens. They arrested almost 1,500 people monthly, most of whom were subjected to inhuman torture and punishments without trial. No only were SAVAK agents trained by the CIA, but there is evidence that they were instructed in techniques of torture (Hersh, 1979: 13).

In 1970 the CIA repeated the practice of overthrowing democratically elected governments that were not completely favorable to U.S. investments. When Salvador Allende was elected president of Chile, the CIA organized a coup that overthrew Allende, during which he was murdered, along with the head of the military, General Rene Schneider. Following Allende's overthrow, the CIA trained agents for the Chilean secret service (DINA). DINA set up a team of assassins who could "travel anywhere in the world . . . to carry out sanctions including assassinations" (Dinges and Landau, 1980: 239). One of the assassinations carried out by DINA was the murder of Orlando Letellier, Allende's ambassador to the United States and his former minister of defense. Letellier was killed when a car bomb blew up his car on Embassy Row in Washington, D.C. (Dinges and Landau, 1982).

Other bloody coups known to have been planned, organized, and executed by U.S. agents include coups in Guatemala, Nicaragua, the Dominican Republic, and Vietnam. American involvement in those coups was never legally authorized. The murders, assassinations, and terrorist acts that accompany coups are criminal acts by law, both in the United States and in the country in which they take place.

More recent examples of murder and assassination for which government officials are responsible include the death of 80 people in Beirut, Lebanon, when a car bomb exploded on May 8, 1985. The bomb was set by a Lebanese counterterrorist unit working with the CIA. Senator Daniel Moynihan has said that when he was vice president of the Senate Intelligence Committee, President Reagan ordered the CIA to form a small antiterrorist effort in the Mideast. Two sources said that the CIA was working with the group that planted the bomb to kill the Shiite leader Hussein Fadallah (the *New York Times,* May 13, 1985).

A host of terrorist plans and activities connected with the attempt to over-throw the Nicaraguan government, including several murders and assassinations, were exposed in an affidavit filed by free-lance reporters Tony Avirgan and Martha Honey. They began investigating Contra activities after Avirgan was injured in an attempt on the life of Contra leader Eden Pastora. In 1986, Honey and Avirgan filed a complaint with the U.S. District Court in Miami charging John Hull, Robert Owen, Theodore Shackley, Thomas Clines, Chi Chi Quintero, Maj. General Richard Secord, and others working for the CIA in Central America with criminal conspiracy and the smuggling of cocaine to aid the Nicaraguan rebels.

A criminal conspiracy in which the CIA admits participating is the publication of a manual, *Psychological Operation in Guerilla Warfare,* which was distributed to the people in Nicaragua. The manual describes how the people should proceed to commit murder, sabotage, vandalism, and violent acts in order to undermine the government. Encouraging or instigating such crimes is not only a violation of U.S. law, it was also prohibited by Reagan's executive order of 1981, which forbade any U.S. participation in foreign assassinations.

The CIA is not alone in hatching criminal conspiracies. The DEA organized a "Special Operations Group," which was responsible for working out plans to assassinate political and business leaders in foreign countries who were involved in drug trafficking. The head of this group was a former CIA agent, Lou Conein (also known as "Black Luigi"). George Crile wrote in the *Washington Post* (June 13, 1976):

> When you get down to it, Conein was organizing an assassination program. He was frustrated by the big-time operators who were just too insulated to get to. . . . Meetings were held to decide whom to target and what method of assassination to employ.

Crile's findings were also supported by the investigative journalist Jim Hougan (1878: 132).

It is a crime to conspire to commit murder. The official record, including testimony by participants in three conspiracies before the U.S. Congress and in court, make it abundantly clear that the crime of conspiring to commit murder is not infrequent in the intelligence agencies of the United States and other countries.

It is also a crime to cover up criminal acts, but there are innumerable examples of instances in which the CIA and the FBI conspired to interfere with the criminal prosecution of drug dealers, murderers, and assassins. In the death of Letellier, mentioned earlier, the FBI and the CIA refused to cooperate with the prosecution of the DINA agents who murdered Letellier (Dinges and Landau, 1980: 208–209). Those agencies were also involved in the cover-up of the criminal activities of a Cuban exile, Ricardo (Monkey) Morales. While an employee of the FBI and the CIA, Morales planted a bomb on an Air Cubana flight from Venezuela, which killed 73 people. The Miami police confirmed Morales's claim that he was acting under orders from the CIA (Lernoux, 1984: 188). In fact, Morales, who was arrested for overseeing the shipment of 10 tons of marijuana, admitted to being a CIA contract agent who conducted bombings, murders, and assassinations. He was himself killed in a bar after he made public his work with the CIA and the FBI.

Colonel Muammar Qaddafi, like Fidel Castro, has been the target of a number of assassination attempts and conspiracies by the U.S. government. One plot, the *Washington Post* reported, included an effort to "lure [Qaddafi] into some foreign adventure or terrorist exploit that would give a growing number of Qaddafi opponents in the Libyan military a chance to seize power, or such a foreign adventure might give one of Qaddafi's neighbors, such as Algeria or Egypt, a justification for responding to Qaddafi militarily" (the *Washington Post,* April 14, 1986). The CIA recommended "stimulating" Qaddafi's fall "by encouraging disaffected elements in the Libyan army who could be spurred to assassination attempts" (the *Guardian,* November 20, 1985: 6).

Opposition to government policies can be a very risky business, as the ecology group Greenpeace discovered when it opposed French nuclear testing in the Pacific. In the fall of 1985 the French government planned a series of atomic tests in the South Pacific. Greenpeace sent its flagship to New Zealand with instructions to sail into the area where the atomic testing was scheduled to occur. Before the ship could arrive at the scene, however, the French secret service located the ship in the harbor and blew it up. The blast from the bomb killed one of the crew.

OTHER STATE-ORGANIZED CRIMES

Every agency of government is restricted by law in certain fundamental ways. Yet structural pressures exist that can push agencies to go beyond their legal limits. The CIA, for example, is not permitted to engage in domestic intelligence. Despite this, the CIA has opened and photographed the mail of over 1 million private citizens (Rockefeller Report, 1975: 101–115), illegally entered people's homes, and conducted domestic surveillance through electronic devices (Parenti, 1983: 170–171).

Agencies of the government also cannot legally conduct experiments on human subjects that violate civil rights or endanger the lives of the subjects.

But the CIA conducted experiments on unknowing subjects by hiring prostitutes to administer drugs to their clients. CIA-trained medical doctors and
psychologists observed the effects of the drugs through a two-way mirror in
expensive apartments furnished to the prostitutes by the CIA. At least one of
the victims of these experiments died and others suffered considerable trauma
(Anderson and Whitten, 1976; Crewdson and Thomas, 1977; Jacobs 1977a,
1977b).

The most flagrant violation of civil rights by federal agencies is the FBI's
counterintelligence program, known as COINTELPRO. This program was
designed to disrupt, harass, and discredit groups that the FBI decided were in
some way "un-American." Such groups included the American Civil Liberties
Union, antiwar movements, civil rights organizations, and a host of other
legally constituted political groups whose views opposed some of the policies
of the United States (Church Committee, 1976). With the exposure of
COINTELPRO, the group was disbanded. There is evidence, however, that
the illegal surveillance of U.S. citizens did not stop with the abolition of
COINTELPRO but continues today (Klein, 1988). . . .

CONCLUSION

. . . We need to explore different political, economic, and social systems in
varying historical periods to discover why some forms of social organization
are more likely to create state-organized crimes than others. We need to
explore the possibility that some types of state agencies are more prone to
engaging in criminality than others. It seems likely, for example, that state
agencies whose activities can be hidden from scrutiny are more likely to
engage in criminal acts than those whose record is public. This principle may
also apply to whole nation-states: the more open the society, the less likely it
is that state-organized crime will become institutionalized.

There are also important parallels between state-organized criminality
and the criminality of police and law enforcement agencies generally. Local
police departments that find it more useful to cooperate with criminal syndi-
cates than to combat them are responding to their own particular contradictions, conflicts, and dilemmas (Chambliss, 1988). An exploration of the theoretical implications of these similarities could yield some important
findings.

The issue of state-organized crime raises again the question of how crime
should be defined to be scientifically useful. For the purposes of this analysis, I
have accepted the conventional criminological definition of crime as acts that
are in violation of the criminal law. This definition has obvious limitations
(see Schwendinger and Schwendinger, 1975), and the study of state-organized
crime may facilitate the development of a more useful definition by underlying the interrelationship between crime and the legal process. At the very
least, the study of state-organized crime serves as a reminder that crime is a
political phenomenon and must be analyzed accordingly.

REFERENCES

Anderson, Jack, and Lee Whitten. 1976. The CIA's "sex squad." The *Washington Post,* June 22:B13.

Chambliss, William J. 1977. Markets, profits, labor and smack. *Contemporary Crises* 1:53–57.

———. 1980. On lawmaking. *British Journal of Law and Society* 6:149–172.

———. 1988. *On the Take: From Petty Crooks to Presidents.* Revised ed. Bloomington: Indiana University Press.

Chambliss, William J., and Robert B. Seidman. 1982. *Law, Order and Power.* Rev. ed. Reading Mass.: Addison-Wesley.

Church Committee. 1976. *Intelligence Activities and the Rights of Americans.* Washington, D.C.: Government Printing Office.

Commonwealth of New South Wales. 1982a. New South Wales Joint Task Force on Drug Trafficking. Federal Parliament Report. Sydney: Government of New South Wales.

———. 1982b. Preliminary Report of the Royal Commission to Investigate the Nugan Hand Bank Failure. Federal Parliament Report. Sydney: Government of New South Wales.

Crewdson, John M., and Jo Thomas. 1977. Abuses in testing of drugs by CIA to be panel focus. The *New York Times,* September 20.

Dinges, John, and Saul Landau. 1980. *Assassination on Embassy Row.* New York: McGraw-Hill.

———. 1982. The CIA's link to Chile's plot. The *Nation,* June 12:712–713.

Goulden, Joseph C. 1984. *Death Merchant: The Brutal True Story of Edwin P. Wilson.* New York: Simon and Schuster.

Hersh, Seymour. 1979. Ex-analyst says CIA rejected warning on Shah. *The New York Times,* January 7:A10. Cited in Piers Beirne and James Messerschmidt, Criminology. New York: Harcourt Brace Jovanovich, forthcoming.

Hougan, Jim. 1978. *Spooks: The Haunting of America—The Private Use of Secret Agents.* New York: William Morrow.

———. 1984. *Secret Agenda: Watergate, Deep Throat, and the CIA.* New York: Random House.

Jacobs, John. 1977a. The diaries of a CIA operative. The *Washington Post,* September 5:1.

———. 1977b. Turner cites 149 drug-test projects. The *Washington Post,* August 4:1.

Klein, Lloyd. 1988. *Big Brother Is Still Watching You.* Paper presented at the annual meetings of the American Society of Criminology, Chicago, November 12.

Kruger, Henrik. 1980. *The Great Heroin Coup.* Boston: South End Press.

Kwitny, Jonathan. 1987. *The Crimes of Patriots.* New York: W. W. Norton.

Lernoux, Penny. 1984. The Miami connection. The *Nation,* February 18:186–198.

McCord, James W., Jr. 1974. *A Piece of Tape.* Rockville, Md: Washington Media Services.

McCoy, Alfred W. 1973. *The Politics of Heroin in Southeast Asia.* New York: Harper & Row.

NARMIC. 1984. Military Exports to South Africa: A Research Report on the Arms Embargo. Philadelphia: American Friends Service Committee.

Nihill, Grant. 1982. Bank links to spies, drugs. The *Advertiser,* November 10:1.

Owen, John. 1983. *Sleight of Hand: The $25 Million Nugan Hand Bank Scandal.* Sydney: Calporteur Press.

Parenti, Michael. 1983. *Democracy for the Few.* New York: St. Martin's.

Potter, Gary W., and Bruce Bullington. 1987. *Drug Trafficking and the Contras: A Case Study of State-Organized Crime.* Paper presented at annual meeting of the American Society of Criminology, Montreal.

Rockefeller Report. 1975. Report to the President by the Commission on CIA Activities within the United States. Washington, D.C.: Government Printing Office.

Schwendinger, Herman, and Julia Schwendinger. 1975. Defenders of order or guardians of human rights. Issue in *Criminology* 7:72–81.

Senate Hearings. 1986. Senate Select Committee on Assassination, Alleged Assassination Plots Involving Foreign Leaders. Interim Report of the Senate Select Committee to Study Governmental Operations with Respect to Intelligence Activities. 94th Cong., 1st sess., November 20. Washington, D.C.: Government Printing Office.

Sharkey, Jacqueline. 1988. The Contra-drug trade-off. *Common Cause Magazine,* September-October: 23–33.

Turner, Stansfield. 1985. *Secrecy and Democracy: The CIA in Transition.* New York: Houghton Miflin.

U.S. Department of State. 1985. Revolution Beyond Our Border: Information on Central America. State Department Report N 132. Washington, D.C.: U.S. Department of State.

14

The Moral Career
of the Mental Patient

ERVING GOFFMAN

Traditionally the term *career* has been reserved for those who expect to enjoy the rises laid out within a respectable profession. The term is coming to be used, however, in a broadened sense to refer to any social strand of any person's course through life. The perspective of natural history is taken: unique outcomes are neglected in favor of such changes over time as are basic and common to the members of a social category, although occurring independently to each of them. Such a career is not a thing that can be brilliant or disappointing; it can no more be a success than a failure. In this light, I want to consider the mental patient, drawing mainly upon data collected during a year's participant observation of patient social life in a public mental hospital,[1] wherein an attempt was made to take the patient's point of view.

One value of the concept of career is its two-sidedness. One side is linked to internal matters held dearly and closely, such as image of self and felt identity; the other side concerns official position, jural relations, and style of life, and is part of a publicly accessible institutional complex. The concept of career, then, allows one to move back and forth between the personal and the public, between the self and its significant society, without having overly to rely for data upon what the person says he thinks he imagines himself to be.

This paper, then, is an exercise in the institutional approach to the study of self. The main concern will be with the *moral* aspects of career—that is, the regular sequence of changes that career entails in the person's self and in his framework of imagery for judging himself and others.[2]

The category "mental patient" itself will be understood in one strictly sociological sense. In this perspective, the psychiatric view of a person becomes significant only in so far as this view itself alters his social fate—an alteration which seems to become fundamental in our society when, and only when, the person is put through the process of hospitalization.[3] I therefore exclude certain neighboring categories: the undiscovered candidates who would be judged "sick" by psychiatric standards but who never come to be viewed as such by themselves or others, although they may cause everyone a great deal of trouble;[4] the office patient whom a psychiatrist feels he can handle with drugs or shock on the outside; the mental client who engages in psychotherapeutic relationships. And I include anyone, however robust in tem-

From: "The Moral Career of the Mental Patient," Erving Goffman, *Psychiatry* 22, 1959.
Reprinted by permission of the Guilford Press.

perament, who somehow gets caught up in the heavy machinery of mental hospital servicing. In this way the effects of being treated as a mental patient can be kept quite distinct from the effects upon a person's life of traits a clinician would view as psychopathological.[5] Persons who become mental hospital patients vary widely in the kind and degree of illness that a psychiatrist would impute to them, and in the attributes by which laymen would describe them. But once started on the way, they are confronted by some importantly similar circumstances and respond to these in some importantly similar ways. Since these similarities do not come from mental illness, they would seem to occur in spite of it. It is thus a tribute to the power of social forces that the uniform status of mental patient cannot only assure an aggregate of persons a common fate and eventually, because of this, a common character, but that this social reworking can be done upon what is perhaps the most obstinate diversity of human materials that can be brought together by society. Here there lacks only the frequent forming of a protective group-life by ex-patients to illustrate in full the classic cycle of response by which deviant subgroupings are psychodynamically formed in society.

This general sociological perspective is heavily reinforced by one key finding of sociologically oriented students in mental hospital research. As has been repeatedly shown in the study of nonliterate societies, the awesomeness, distastefulness, and barbarity of a foreign culture can decrease in the degree that the student becomes familiar with the point of view to life that is taken by his subjects. Similarly, the student of mental hospitals can discover that the craziness or "sick behavior" claimed for the mental patient is by and large a product of the claimant's social distance from the situation that the patient is in, and is not primarily a product of mental illness. Whatever the refinements of the various patients' psychiatric diagnoses, and whatever the special ways in which social life on the "inside" is unique, the researcher can find that he is participating in a community not significantly different from any other he has studied.[6] Of course, while restricting himself to the off-ward grounds community of paroled patients, he may feel, as some patients do, that life in the locked wards is bizarre; and while on a locked admissions or convalescent ward, he may feel that chronic "back" wards are socially crazy places. But he need only move his sphere of sympathetic participation to the "worst" ward in the hospital, and this too can come into social focus as a place with a livable and continuously meaningful social world. This in no way denies that he will find a minority in any ward or patient group that continues to seem quite beyond the capacity to follow rules of social organization, or that the orderly fulfillment of normative expectations in patient society is partly made possible by strategic measures that have somehow come to be institutionalized in mental hospitals.

The career of the mental patient falls popularly and naturalistically into three main phases: the period prior to entering the hospital, which I shall call the *prepatient phase;* the period in the hospital, the *inpatient phase;* the period after discharge from the hospital, should this occur, namely, the *ex-patient phase.*[7] This paper will deal only with the first two phases.

THE PREPATIENT PHASE

A relatively small group of prepatients come into the mental hospital willingly, because of their own idea of what will be good for them, or because of whole-hearted agreement with the relevant members of their family. Presumably these recruits have found themselves acting in a way which is evidence to them that they are losing their minds or losing control of themselves. This view of oneself would seem to be one of the most pervasively threatening things that can happen to the self in our society, especially since that it is likely to occur at a time when the person is in any case sufficiently troubled to exhibit the kind of symptom which he himself can see. As Sullivan described it,

> What we discover in the self-system of a person undergoing schizophrenic changes or schizophrenic processes, is then, in its simplest form, an extremely fear-marked puzzlement, consisting of the use of rather generalized and anything but exquisitely refined referential processes in an attempt to cope with what is essentially a failure at being human—a failure at being anything that one could respect as worth being.[8]

Coupled with the person's disintegrative re-evaluation of himself will be the new, almost equally pervasive circumstance of attempting to conceal from others what he takes to be the new fundamental facts about himself, and attempting to discover whether others too have discovered them.[9] Here I want to stress that perception of losing one's mind is based on culturally derived and socially engrained stereotypes as to the significance of symptoms such as hearing voices, losing temporal and spatial orientation, and sensing that one is being followed, and that many of the most spectacular and convincing of these symptoms in some instances psychiatrically signify merely a temporary emotional upset in a stressful situation, however terrifying to the person at the time. Similarly, the anxiety consequent upon this perception of oneself, and the strategies devised to reduce this anxiety, are not a product of abnormal psychology, but would be exhibited by any person socialized into our culture who came to conceive of himself as someone losing his mind. Interestingly, subcultures in American society apparently differ in the amount of ready imagery and encouragement they supply for such self-views, leading to differential rates of *self*-referral; the capacity to take this disintegrative view of oneself without psychiatric prompting seems to be one of the questionable cultural privileges of the upper classes.[10]

For the person who has come to see himself—with whatever justification—as mentally unbalanced, entrance to the mental hospital can sometimes bring relief, perhaps in part because of the sudden transformation in the structure of his basic social situations; instead of being to himself a questionable person trying to maintain a role as a full one, he can become an officially questioned person known to himself to be not so questionable as that. In other cases, hospitalization can make matters worse for the willing patient, confirming by the objective situation what has theretofore been a matter of the private experience of self.

Once the willing prepatient enters the hospital, he may go through the same routine of experiences as do those who enter unwillingly. In any case, it is the latter that I mainly want to consider, since in America at present these are by far the more numerous kind.[11] Their approach to the institution takes one of three classic forms: they come because they have been implored by their family or threatened with the abrogation of family ties unless they go "willingly"; they come by force under police escort; they come under misapprehension purposely induced by others, this last restricted mainly to youthful prepatients.

The prepatient's career may be seen in terms of an extrusory model; he starts out with relationships and rights, and ends up, at the beginning of his hospital stay, with hardly any of either. The moral aspects of this career, then, typically begin with the experience of abandonment, disloyalty, and embitterment. This is the case even though to others it may be obvious that he was in need of treatment, and even though in the hospital he may soon come to agree.

The case histories of most mental patients document offense against some arrangement for face-to-face living—a domestic establishment, a work place, a semipublic organization such as a church or store, a public region such as a street or park. Often there is also a record of some *complainant,* some figure who takes that action against the offender which eventually leads to this hospitalization. This may not be the person who makes the first move, but it is the person who makes what turns out to be the first effective move. Here is the *social* beginning of the patient's career, regardless of where one might locate the psychological beginning of his mental illness.

The kinds of offenses which lead to hospitalization are felt to differ in nature from those which lead to other extrusory consequences—to imprisonment, divorce, loss of job, disownment, regional exile, noninstitutional psychiatric treatment, and so forth. But little seems known about these differentiating factors; and when one studies actual commitments, alternate outcomes frequently appear to have been possible. It seems true, moreover, that for every offense that leads to an effective complaint, there are many psychiatrically similar ones that never do. No action is taken; or action is taken which leads to other extrusory outcomes; or ineffective action is taken, leading to the mere pacifying or putting off of the person who complains. Thus, as Clausen and Yarrow have nicely shown, even offenders who are eventually hospitalized are likely to have had a long series of ineffective actions taken against them.[12]

Separating those offenses which could have been used as grounds for hospitalizing the offender from those that are so used, one finds a vast number of what students of occupation call career contingencies.[13] Some of these contingencies in the mental patient's career have been suggested, if not explored, such as socio-economic status, visibility of the offense, proximity to a mental hospital, amount of treatment facilities available, community regard for the type of treatment given in available hospitals, and so on.[14] For information about other contingencies one must rely on atrocity tales: a psychotic man is tolerated by his wife until she finds herself a boyfriend, or by his adult children until they move from a house to an apartment; an alcoholic is sent to a mental hospital because the jail is full, and a drug addict because he declines to avail himself of psychiatric treatment on the outside; a rebellious adolescent daughter can no longer be managed at home because she now threatens to have an open affair with an unsuitable companion; and so on. Correspondingly there is an equally important set of contingencies causing the person to bypass this fate. And should the person enter the hospital, still another set of contingencies will help determine when he is to obtain a discharge—such as the desire of his family for his return, the availability of a "manageable" job, and so on. The society's official view is that inmates of mental hospitals are there primarily because they are suffering from mental illness. However, in the degree that the "mentally ill" outside hospitals numerically approach or surpass those inside hospitals, one could say that mental patients *distinctively* suffer not from mental illness, but from contingencies.

Career contingencies occur in conjunction with a second feature of the prepatient's career—the *circuit of agents*—and agencies—that participate fatefully in his passage from civilian to patient status.[15] Here is an instance of that increasingly important class of social system whose elements are agents and agencies, which are brought into systemic connection through having to take up and send on the same persons. Some of these agent-roles will be cited now, with the understanding that in any concrete circuit a role may be filled more than once, and a single person may fill more than one of them.

First is the *next-of-relation*—the person whom the prepatient sees as the most available of those upon whom he should be able to most depend in times of trouble; in this instance the last to doubt his sanity and the first to have done everything to save him from the fate which, it transpires, he has been approaching. The patient's next-of-relation is usually his next of kin; the special term is introduced because he need not be. Second is the *complainant*, the person who retrospectively appears to have started the person on his way to the hospital. Third are the *mediators*—the sequence of agents and agencies to which the prepatient is referred and through which he is relayed and processed on his way to the hospital. Here are included police, clergy, general medical practitioners, office psychiatrists, personnel in public clinics, lawyers, social service workers, school teachers, and so on. One of these agents will have the legal mandate to sanction commitment and will exercise it, and so those agents who precede him in the process will be involved in something whose outcome is not yet settled. When the mediators retire from the scene, the prepatient has become an inpatient, and the significant agent has become the hospital administrator.

While the complainant usually takes action in a lay capacity as a citizen, an employer, a neighbor, or a kinsman, mediators tend to be specialists and differ from those they serve in significant ways. They have experience in handling trouble, and some professional distance from what they handle. Except in the case of policemen, and perhaps some clergy, they tend to be more psychiatrically oriented than the lay public, and will see the need for treatment at times when the public does not.[16]

An interesting feature of these roles is the functional effects of their inter-digitation. For example, the feelings of the patient will be influenced by whether or not the person who fills the role of complainant also has the role of next-of-relation—an embarrassing combination more prevalent, apparently, in the higher classes than in the lower.[17] Some of these emergent effects will be considered now.[18]

In the prepatient's progress from home to the hospital he may participate as a third person in what he may come to experience as a kind of *alienative coalition*. His next-of-relation presses him into coming to "talk things over" with a medical practitioner, an office psychiatrist, or some other counselor. Disinclination on his part may be met by threatening him with desertion, dis-ownment, or other legal action, or by stressing the joint and explorative nature of the interview. But typically the next-of-relation will have set the interview up, in the sense of selecting the professional, arranging for time, telling the professional something about the case, and so on. This move effec-tively tends to establish the next-of-relation as the responsible person to whom pertinent findings can be divulged, while effectively establishing the other as the patient. The prepatient often goes to the interview with the understanding that he is going as an equal of someone who is so bound together with him that a third person could not come between them in fun-damental matters; this, after all, is one way in which close relationships are defined in our society. Upon arrival at the office the prepatient suddenly finds that he and his next-of-relation have not been accorded the same roles, and apparently that a prior understanding between the professional and the next-of-relation has been put in operation against him. In the extreme but common case the professional first sees the prepatient alone, in the role of examiner and diagnostician, and then sees the next-of-relation alone, in the role of advisor, while carefully avoiding talking things over seriously with them both together.[19] And even in those nonconsultative cases where public officials must forcibly extract a person from a family that wants to tolerate him, the next-of-relation is likely to be induced to "go along" with the official action, so that even here the prepatient may feel that an alienative coalition has been formed against him.

The moral experience of being third man in such a coalition is likely to embitter the prepatient, especially since his troubles have already probably led to some estrangement from his next-of-relation. After he enters the hospital, continued visits by his next-of-relation can give the patient the "insight" that his own best interests were being served. But the initial visits may temporarily strengthen his feeling of abandonment; he is likely to beg his visitor to get him out or at least to get him more privileges and to sympathize with the mon-strousness of his plight—to which the visitor ordinarily can respond only by trying to maintain a hopeful note, by not "hearing" the requests, or by assur-ing the patient that the medical authorities know about these things and are doing what is medically best. The visitor then nonchalantly goes back into a world that the patient has learned is incredibly thick with freedom and privileges, causing the patient to feel that his next-of-relation is merely adding a pious gloss to a clear case of traitorous desertion.

The depth to which the patient may feel betrayed by his next-of-relation seems to be increased by the fact that another witnesses his betrayal—a factor which is apparently significant in many three-party situations. An offended person may well act forbearantly and accommodatively toward an offender when the two are alone, choosing peace ahead of justice. The presence of a witness, however, seems to add something to the implications of the offense. For then it is beyond the power of the offended and offender to forget about, erase, or suppress what has happened; the offense has become a public social fact.[20] When the witness is a mental health commission, as is sometimes the case, the witnessed betrayal can verge on a "degradation ceremony."[21] In such circumstances, the offended patient may feel that some kind of extensive reparative action is required before witnesses, if his honor and social weight are to be restored.

Two other aspects of sensed betrayal should be mentioned. First, those who suggest the possibility of another's entering a mental hospital are not likely to provide a realistic picture of how in fact it may strike him when he arrives. Often he is told that he will get required medical treatment and a rest, and may well be out in a few months or so. In some cases they may thus be concealing what they know, but I think, in general, they will be telling what they see as the truth. For here there is a quite relevant difference between patients and mediating professionals; mediators, more so than the public at large, may conceive of mental hospitals as short-term medical establishments where required rest and attention can be voluntarily obtained, and not as places of coerced exile. When the prepatient finally arrives he is likely to learn quite quickly, quite differently. He then finds that the information given him about life in the hospital has had the effect of his having put up less resistance to entering than he now sees he would have put up had he known the facts. Whatever the intentions of those who participated in his transition from person to patient, he may sense they have in effect "conned" him into his present predicament.

I am suggesting that the prepatient starts out with at least a portion of the rights, liberties, and satisfactions of the civilian and ends up on a psychiatric ward stripped of almost everything. The question here is *how* this stripping is managed. This is the second aspect of betrayal I want to consider.

As the prepatient may see it, the circuit of significant figures can function as a kind of *betrayal funnel*. Passage from person to patient may be effected through a series of linked stages, each managed by a different agent. While each stage tends to bring a sharp decrease in adult free status, each agent may try to maintain the fiction that no further decrease will occur. He may even manage to turn the prepatient over to the next agent while sustaining this note. Further, through words, cues, and gestures, the prepatient is implicitly asked by the current agent to join with him in sustaining a running line of polite small talk that tactfully avoids the administrative facts of the situation, becoming, with each stage, progressively more at odds with these facts. The spouse would rather not have to cry to get the prepatient to visit a psychiatrist; psychiatrists would rather not have a scene when the prepatient learns that he and his spouse are being seen separately and in different ways; the police infrequently bring a prepatient to the hospital in a strait jacket, finding it much easier all around to give him a cigarette, some kindly words, and freedom to relax in the back seat of the patrol car; and finally, the admitting psychiatrist finds he can do his work better in the relative quiet and luxury of the "admission suite" where, as an incidental consequence, the notion can survive that a mental hospital is indeed a comforting place. If the prepatient heeds all of these implied requests and is reasonably decent about the whole thing, he can travel the whole circuit from home to hospital without forcing anyone to look directly at what is happening or to deal with the raw emotion that his situation might well cause him to express. His showing consideration for those who are moving him toward the hospital allows them to show consideration for him, with the joint result that these interactions can be sustained with some of the protective harmony characteristic of ordinary face-to-face dealings. But should the new patient cast his mind back over the sequence of steps leading to hospitalization, he may feel that everyone's *current* comfort was being busily sustained while his long-range welfare was being undermined. This realization may constitute a moral experience that further separates him for the time from the people on the outside.[22]

I would now like to look at the circuit of career agents from the point of view of the agents themselves. Mediators in the person's transition from civil to patient status—as well as his keepers, once he is in the hospital—have an interest in establishing a responsible next-of-relation as the patient's deputy or *guardian;* should there be no obvious candidate for the role, someone may be sought out and pressed into it. Thus while a person is gradually being transformed into a patient, a next-of-relation is gradually being transformed into a guardian. With a guardian on the scene, the whole transition process can be kept tidy. He is likely to be familiar with the prepatient's civil involvements and business, and can tie up loose ends that might otherwise be left to entangle the hospital. Some of the prepatient's abrogated civil rights can be transferred to him, thus helping to sustain the legal fiction that while the prepatient does not actually have his rights he somehow actually has not lost them.

Inpatients commonly sense, at least for a time, that hospitalization is a massive unjust deprivation, and sometimes succeed in convincing a few persons on the outside that this is the case. It often turns out to be useful, then, for those identified with inflicting these deprivations, however justifiably, to be able to point to the cooperation and agreement of someone whose relationship to the patient places him above suspicion, firmly defining him as the person most likely to have the patient's personal interest at heart. If the guardian is satisfied with what is happening to the new inpatient, the world ought to be.[23]

Now it would seem that the greater the legitimate personal stake one party has in another, the better he can take the role of guardian to the other. But the structural arrangements in society which lead to the acknowledged merging of two persons' interests lead to additional consequences. For the person to whom the patient turns for help—for protection against such threats as involuntary commitment—is just the person to whom the mediators and hospital administrators logically turn for authorization. It is understandable, then, that some patients will come to sense, at least for a time, that the closeness of a relationship tells nothing of its trustworthiness.

There are still other functional effects emerging from this complement of roles. If and when the next-of-relation appeals to mediators for help in the trouble he is having with the prepatient, hospitalization may not, in fact, be in his mind. He may not even perceive the prepatient as mentally sick, or, if he does, he may not consistently hold to this view.[24] It is the circuit of mediators, with their greater psychiatric sophistication and their belief in the medical character of mental hospitals, that will often define the situation for the next-of-relation, assuring him that hospitalization is a possible solution and a good one, that it involves no betrayal, but is rather a medical action taken in the best interests of the prepatient. Here the next-of-relation may learn that doing his duty to the prepatient may cause the prepatient to distrust and even hate him for the time. But the fact that this course of action may have had to be pointed out and prescribed by professionals, and be defined by them as a moral duty, relieves the next-of-relation of some of the guilt he may feel.[25] It is a poignant fact that an adult son or daughter may be pressed into the role of mediator, so that the hostility that might otherwise be directed against the spouse is passed on to the child.[26]

Once the prepatient is in the hospital, the same guilt-carrying function may become a significant part of the staff's job in regard to the next-of-relation.[27] These reasons for feeling that he himself has not betrayed the patient, even though the patient may then think so, can later provide the next-of-relation with a defensible line to take when visiting the patient in the hospital and a basis for hoping that the relationship can be reestablished after its hospital moratorium. And of course this position, when sensed by the patient, can provide him with excuses for the next-of-relation, when and if he comes to look for them.[28]

Thus while the next-of-relation can perform important functions for the mediators and hospital administrators, they in turn can perform important functions for him. One finds, then, an emergent unintended exchange or reciprocation of functions, these functions themselves being often unintended.

The final point I want to consider about the prepatient's moral career is its peculiarly *retroactive* character. Until a person actually arrives at the hospital there usually seems no way of knowing for sure that he is destined to do so, given the determinative role of career contingencies. And until the point of hospitalization is reached, he or others may not conceive of him as a person who is becoming a mental patient. However, since he will be held against his will in the hospital, his next-of-relation and the hospital staff will be in great need of a rationale for the hardships they are sponsoring. The medical elements of the staff will also need evidence that they are still in the trade they were trained for. These problems are eased, no doubt unintentionally, by the case-history construction that is placed on the patient's past life, this having the effect of demonstrating that all along he had been becoming sick, that he finally became very sick, and that if he had not been hospitalized much worse things would have happened to him—all of which, of course, may be true. Incidentally, if the patient wants to make sense out of his stay in the hospital, and, as already suggested, keep alive the possibility of once again conceiving of his next-of-relation as a decent, well-meaning person, then he too will have reason to believe some of this psychiatric work-up of his past.

Here is a very ticklish point for the sociology of careers. An important aspect of every career is the view the person constructs when he looks backward over his progress; in a sense, however, the whole of the prepatient career derives from this reconstruction. The fact of having had a prepatient career, starting with an effective complaint, becomes an important part of the mental patient's orientation, but this part can begin to be played only after hospitalization proves that what he had been having, but no longer has, is a career as a prepatient.

THE INPATIENT PHASE

The last step in the prepatient's career can involve his realization—justified or not—that he has been deserted by society and turned out of relationships by those closest to him. Interestingly enough, the patient, especially a first admission, may manage to keep himself from coming to the end of this trail, even though in fact he is now in a locked mental hospital ward. On entering the hospital, he may very strongly feel the desire not to be known to anyone as a person who could possibly be reduced to these present circumstances, or as a person who conducted himself in the way he did prior to commitment. Consequently, he may avoid talking to anyone, may stay by himself when possible, and may even be "out of contact" or "manic" so as to avoid ratifying any interaction that presses a politely reciprocal role upon him and opens him up to what he has become in the eyes of others. When the next-of-relation makes an effort to visit, he may be rejected by mutism, or by the patient's refusal to enter the visiting room, these strategies sometimes suggesting that the patient still clings to a remnant of relatedness to those who made up his past, and is protecting this remnant from the final destructiveness of dealing with the new people that they have become.[29]

Usually the patient comes to give up this taxing effort at anonymity, at not-hereness, and begins to present himself for conventional social interaction to the hospital community. Thereafter he withdraws only in special ways—by always using his nickname, by signing his contribution to the patient weekly with his initial only, or by using the innocuous "cover" address tactfully provided by some hospitals; or he withdraws only at special times, when, say, a flock of nursing students makes a passing tour of the ward, or when, paroled to the hospital grounds, he suddenly sees he is about to cross the path of a civilian he happens to know from home. Sometimes this making of oneself available is called "settling down" by the attendants. It marks a new stand openly taken and supported by the patient, and resembles the "coming out" process that occurs in other groupings.[30]

Once the prepatient begins to settle down, the main outlines of his fate tend to follow those of a whole class of segregated establishments—jails, concentration camps, monasteries, work camps, and so on—in which the inmate spends the whole round of life on the grounds, and marches through his regimented day in the immediate company of a group of persons of his own institutional status.[31]

Like the neophyte in many of these "total institutions," the new inpatient finds himself cleanly stripped of many of his accustomed affirmations, satisfactions, and defenses, and is subjected to a rather full set of mortifying experiences: restriction of free movement; communal living; diffuse authority of a whole echelon of people; and so on. Here one begins to learn about the limited extent to which a conception of oneself can be sustained when the usual setting of supports for it are suddenly removed.

While undergoing these humbling moral experiences, the inpatient learns to orient himself in terms of the "ward system."[32] In public mental hospitals this usually consists of a series of graded living arrangements built around wards, administrative units called services, and parole statuses. The "worst" level involves often nothing but wooden benches to sit on, some quite indifferent food, and a small piece of room to sleep in. The "best" level may involve a room of one's own, ground and town privileges, contacts with staff that are relatively undamaging, and what is seen as good food and ample recreational facilities. For disobeying the pervasive house rules, the inmate will receive stringent punishments expressed in terms of loss of privileges; for obedience he will eventually be allowed to reacquire some of the minor satisfactions he took for granted on the outside.

The institutionalization of these radically different levels of living throws light on the implications for self of social settings. And this in turn affirms that the self arises not merely out of its possessor's interactions with significant others, but also out of the arrangements that are evolved in an organization for its members.

There are some settings which the person easily discounts as an expression or extension of him. When a tourist goes slumming, he may take pleasure in the situation not because it is a reflection of him but because it so assuredly is not. There are other settings, such as living rooms, which the person manages on his own and employs to influence in a favorable direction other persons' views of him. And there are still other settings, such as a work place, which express the employee's occupational status, but over which he has no final control, this being exerted, however tactfully, by his employer. Mental hospitals provide an extreme instance of this latter possibility. And this is due not merely to their uniquely degraded living levels, but also to the unique way in which significance for self is made explicit to the patient, piercingly, persistently, and thoroughly. Once lodged on a given ward, the patient is firmly instructed that the restrictions and deprivations he encounters are not due to such things as tradition or economy—and hence dissociable from self—but are intentional parts of his treatment, part of his need at the time, and therefore an expression of the state that his self has fallen to. Having every reason to initiate requests for better conditions, he is told that when the staff feels he is "able to manage" or will be "comfortable with" a higher ward level, then appropriate action will be taken. In short, assignment to a given ward is presented not as a reward or punishment, but as an expression of his general level of social functioning, his status as a person. Given the fact that the worst ward levels provide a round of life that inpatients with organic brain damage can easily manage, and that these quite limited human beings are present to prove it, one can appreciate some of the mirroring effects of the hospital.[33]

The ward system, then, is an extreme instance of how the physical facts of an establishment can be explicitly employed to frame the conception a person takes of himself. In addition, the official psychiatric mandate of mental hospitals gives rise to even more direct, even more blatant, attacks upon the inmate's view of himself. The more "medical" and the more progressive a mental hospital is—the more it attempts to be therapeutic and not merely custodial—the more he may be confronted by high-ranking staff arguing that his past has been a failure, that the cause of this has been within himself, that his attitude to life is wrong, and that if he wants to be a person he will have to change his way of dealing with people and his conceptions of himself. Often the moral value of these verbal assaults will be brought home to him by requiring him to practice taking this psychiatric view of himself in arranged confessional periods, whether in private sessions or group psychotherapy.

Now a general point may be made about the moral career of inpatients which has bearing on many moral careers. Given the stage that any person has reached in a career, one typically finds that he constructs an image of his life course—past, present, and future—which selects, abstracts, and distorts in such a way as to provide him with a view of himself that he can usefully expound in current situations. Quite generally, the person's line concerning self defensively brings him into appropriate alignment with the basic values of his society, and so may be called an *apologia*. If the person can manage to present a view of his current situation which shows the operation of favorable personal qualities in the past and a favorable destiny awaiting him, it may be called a *success story*. If the facts of a person's past and present are extremely dismal, then about the best he can do is to show that he is not responsible for what has become of him, and the term *sad tale* is appropriate. Interestingly enough, the more the person's past forces him out of apparent alignment with central moral values, the more often he seems compelled to tell his sad tale in any company in which he finds himself. Perhaps he partly responds to the need he feels in others of not having their sense of proper life courses affronted. In any case, it is among convicts, "wino's," and prostitutes that one seems to obtain sad tales the most readily.[34] It is the vicissitudes of the mental patient's sad tale that I want to consider now.

In the mental hospital, the setting and the house rules press home to the patient that he is, after all, a mental case who has suffered some kind of social collapse on the outside, having failed in some over-all way, and that here he is of little social weight, being hardly capable of acting like a full-fledged person at all. These humiliations are likely to be most keenly felt by middle-class patients, since their previous condition of life little immunizes them against such affronts; but all patients feel some downgrading. Just as any normal member of his outside subculture would do, the patient often responds to this situation by attempting to assert a sad tale proving that he is not "sick," that the "little trouble" he did get into was really somebody else's fault, that his past life course had some honor and rectitude, and that the hospital is therefore unjust in forcing the status of mental patient upon him. This self-respecting tendency is heavily institutionalized within the patient society where opening social contacts typically involve the participants' volunteering information about their current ward location and length of stay so far, but not the reasons for their stay—such interaction being conducted in the manner of small talk on the outside.[35] With greater familiarity, each patient usually volunteers relatively acceptable reasons for his hospitalization, at the same time accepting without open immediate question the lines offered by other patients. Such stories as the following are given and overtly accepted.

> I was going to night school to get a M.A. degree, and holding down a job in addition, and the load got too much for me.

> The others here are sick mentally but I'm suffering from a bad nervous system and that is what is giving me these phobias.

> I got here by mistake because of a diabetes diagnosis, and I'll leave in a couple of days. [The patient had been in seven seeks.]

> I failed as a child, and later with my wife I reached out for dependency.

> My trouble is that I can't work. That's what I'm in for. I had two jobs with a good home and all the money I wanted.[36]

The patient sometimes reinforces these stories by an optimistic definition of his occupational status: A man who managed to obtain an audition as a radio announcer styles himself a radio announcer; another who worked for some months as a copy boy and was then given a job as a reporter on a large trade journal, but fired after three weeks, defines himself as a reporter.

A whole social role in the patient community may be constructed on the basis of these reciprocally sustained fictions. For these face-to-face niceties tend to be qualified by behind-the-back gossip that comes only a degree closer to the "objective" facts. Here, of course, one can see a classic social function of informal networks of equals: they serve as one another's audience for self-supporting tales—tales that are somewhat more solid than pure fantasy and somewhat thinner than the facts.

But the patient's *apologia* is called forth in a unique setting, for few settings could be so destructive of self-stories except, of course, those stories already constructed along psychiatric lines. And this destructiveness rests on more than the official sheet of paper which attests that the patient is of unsound mind, a danger to himself and others—an attestation, incidentally, which seems to cut deeply into the patient's pride, and into the possibility of his having any.

Certainly the degrading conditions of the hospital setting belie many of the self-stories that are presented by patients; and the very fact of being in the mental hospital is evidence against these tales. And of course, there is not always sufficient patient solidarity to prevent patient discrediting patient, just as there is not always a sufficient number of "professionalized" attendants to prevent attendant discrediting patient. As one patient informant repeatedly suggested to a fellow patient:

If you're so smart, how come you got your ass in here?

The mental hospital setting, however, is more treacherous still. Staff has much to gain through discreditings of the patient's story—whatever the felt reason for such discreditings. If the custodial faction in the hospital is to succeed in managing his daily round without complaint or trouble from him, then it will prove useful to be able to point out to him that the claims about himself upon which he rationalizes his demands are false, that he is not what he is claiming to be, and that in fact he is a failure as a person. If the psychiatric faction is to impress upon him its views about his personal make-up, then they must be able to show in detail how their version of his past and their version of his character hold up much better than his own.[37] If both the custodial and psychiatric factions are to get him to cooperate in the various psychiatric treatments, then it will prove useful to disabuse him of *his* view of their purposes, and cause him to appreciate that they know what they are doing, and are doing what is best for him. In brief, the difficulties caused by a patient are closely tied to his version of what has been happening to him, and if cooperation is to be secured, it helps if this version is discredited. The patient must "insightfully" come to take, or affect to take, the hospital's view of himself.

NOTES

1. The study was conducted during 1955–56 under the auspices of the Laboratory of Social-environmental Studies of the National Institute of Mental Health. I am grateful to the Laboratory Chief, John A. Clausen, and to Dr. Winfred Overholser, Superintendent, and the late Dr. Jay Hoffman, then First Assistant Physician of Saint Elizabeths Hospital, Washington, D.C., for the ideal cooperation they freely provided. A preliminary report is contained in Goffman, "Interpersonal Persuasion," pp. 117–193; in *Group Processes: Transactions of the Third Conference*, edited

by Bertram Schaffner: New York, Josiah Macy, Jr. Foundation, 1957. A shorter version of this paper was presented at the Annual Meeting of the American Sociological Society, Washington, D.C., August 1957.

2. Material on moral career can be found in early social anthropological work on ceremonies of status transition, and in classic social psychological descriptions of those spectacular changes in one's view of self that can accompany participation in social movements and sects. Recently new kinds of relevant data have been suggested by psychiatric interest in the problem of "identity" and sociological studies of work careers and "adult socialization."

3. This point has recently been made by Elaine and John Cumming, *Closed Ranks*; Cambridge, Commonwealth Fund, Harvard Univ. Press, 1957; pp. 101–102. "Clinical experience supports the impression that many people define mental illness as 'That condition for which a person is treated in a mental hospital.' . . . Mental illness, it seems, is a condition which afflicts people who must go to a mental institution, but until they do almost anything they do is normal." Leila Deasy has pointed out to me the correspondence here with the situation in white collar crime. Of those who are detected in this activity, only the ones who do not manage to avoid going to prison find themselves accorded the social role of the criminal.

4. Case records in mental hospitals are just now coming to be exploited to show the incredible amount of trouble a person may cause for himself and others before anyone begins to think about him psychiatrically, let alone take psychiatric action against him. See John A. Clausen and Marian Radke Yarrow, "Paths to the Mental Hospital," *J. Social Issues* (1955) 11:25–32; August B. Hollingshead and Frederick C. Redlich, *Social Class and Mental Illness;* New York, Wiley, 1958: pp. 173–174.

5. An illustration of how this perspective may be taken to all forms of deviancy may be found in Edwin Lemert, *Social Pathology;* New York, McGraw-Hill, 1951; see especially pp. 74–76. A specific application to mental defectives may be found in Stewart E. Perry, "Some Theoretic Problems of Mental Deficiency and Their Action Implications," *Psychiatry* (1954) 17:45–73; see especially p. 68.

6. Conscientious objectors who voluntarily went to jail sometimes arrived at the same conclusion regarding criminal inmates. See, for example, Alfred Hassler, *Diary of a Self-made Convict;* Chicago, Regnery, 1954; p. 74.

7. This simple picture is complicated by the somewhat special experience of roughly a third of ex-patients—namely, readmission to the hospital, this being the recidivist or "repatient" phase.

8. Harry Stack Sullivan, *Clinical Studies in Psychiatry;* edited by Helen Swick Perry, Mary Ladd Gawel, and Martha Gibbon; New York, Norton, 1956; pp. 184–185.

9. This moral experience can be contrasted with that of a person learning to become a marihuana addict, whose discovery that he can be "high" and still "op" effectively without being detected apparently leads to a new level of use. See Howard S. Becker, "Marihuana Use and Social Control," *Social Problems* (1955) 3:35–44; see especially pp. 40–41.

10. See note 4: Hollingshead and Redlich, p. 187, Table 6, where relative frequency is given of self-referral by social class grouping.

11. The distinction employed here between willing and unwilling patients cuts across the legal one, of voluntary and committed, since some persons who are glad to come to the mental hospital may be legally committed, and of those who come only because of strong familial pressure, some may sign themselves in as voluntary patients.

12. Clausen and Yarrow; see note 4.

13. An explicit application of this notion to the field of mental health may be found in Edwin M. Lemert, "Legal Commitment and Social Control," *Sociology and Social Research* (1946) 30:370–378.

14. For example, Jerome K. Meyers and Leslie Schaffer, "Social Stratification and Psychiatric Practice: A Study of an Outpatient Clinc," *Amer. Sociological Rev.* (1954) 19:307–310, Lemert, see note 5; pp. 402–403. *Patients in Mental Institutions,* 1941; Washington, D.C., Department of Commerce, Bureau of Census, 1941; p. 2.

15. For one circuit of agents and its bearing on career contingencies, see Oswald Hall, "The Stages of a Medical Career," *Amer. J. Sociology* (1948) 53:227–336.

16. See Cumming, note 3; p. 92.

17. Hollingshead and Redlich, note 4; p. 187.

18. For an analysis of some of these circuit implications for the inpatient, see Leila C. Deasy and Olive W. Quinn, "The Wife of the Mental Patient and the Hospital Psychiatrist,"*J. Social Issues* (1955) 11:49–60. An interesting illustration of this kind of analysis may also be found in Alan G. Gowman, "Blindness and the Role of Companion," *Social Problems* (1956) 4:68–75. A general statement may be found in Robert Merton, "The Role Set: Problems in Sociological Theory," *British J. Sociology* (1957) 8:106–120.

19. I have one case record of a man who claims he thought *he* was taking his wife to see the psychiatrist, not realizing until too late that his wife had made the arrangements.

20. A paraphrase from Kurt Riezler, "The Social Psychology of Shame," *Amer. J. Sociology* (1943) 48:458.

21. See Harold Garfinkel, "Conditions of Successful Degradation Ceremonies," *Amer. J. Sociology* (1956) 61:420–424.

22. Concentration camp practices provide a good example of the function of the betrayal funnel in inducing cooperation and reducing struggle and fuss, although here the mediators could not be said to be acting in the best interests of the inmates. Police picking up persons from their homes would sometimes joke good-naturedly and to offer to wait while coffee was being served. Gas chambers were fitted out like delousing rooms, and victims taking off their clothes were told to note where they were leaving them. The sick, aged, weak, or insane who were selected for extermination were sometimes driven away in Red Cross ambulances to camps referred to by terms such as "observation hospital." See David Boder, *I Did Not Interview the Dead;* Urbana, Univ. of Illinois Press, 1949; p. 81; and Elie A. Cohen, *Human Behavior in the Concentration Camp;* London, Cape, 1954; pp. 32, 37, 107.

23. Interviews collected by the Clausen group at NIMH suggest that when a wife comes to be a guardian, the responsibility may disrupt previous distance from in-laws, leading either to a new supportive coalition with them or to a marked withdrawal from them.

24. For an analysis of these nonpsychiatric kinds of perception, see Marian Radke Yarrow, Charlotte Green Schwartz, Harriet S. Murphy, and Leila Calhoun Deasy, "The Psychological Meaning of Mental Illness in the Family," *J. Social Issues* (1955) 11:12–24; Charlotte Green Schwartz, "Perspectives on Deviance: Wives' Definitions of their Husbands' Mental Illness," *Psychiatry* (1957) 20:275–291.

25. This guilt-carrying function is found, of course, in other role-complexes. Thus, when a middle-class couple engages in the process of legal separation or divorce, each of their lawyers usually takes the position that his job is to acquaint his client with all of the potential claims and rights, pressing his client into demanding these, in spite of any nicety of feelings about the rights and honorableness of the ex-partner. The client, in all good faith, can then say to self and to the ex-partner that the demands are being made only because the lawyer insists it is best to do so.

26. Recorded in the Clausen data.

27. This point is made by Cumming, see note 3; p. 129.

28. There is an interesting contrast here with the moral career of the tuberculosis patient. I am told by Julius Roth that tuberculosis patients are likely to come to the hospital willingly, agreeing with their next-of-relation about treatment. Later in their hospital career, when they learn how long they yet have to stay and how depriving and irrational some of the hospital rulings are, they may seek to leave, be advised against this by the staff and by relatives, and only then begin to feel betrayed.

29. The inmate's initial strategy of holding himself aloof from ratifying contact may partly account for the relative lack of group-information among

inmates in public mental hospitals, a connection that has been suggested to me by William R. Smith. The desire to avoid personal bonds that would give license to the asking of biographical questions could also be a factor. In mental hospitals, of course, as in prisoner camps, the staff may consciously break up incipient group-formation in order to avoid collective rebellious action and other ward disturbances.

30. A comparable coming out occurs in the homosexual world, when a person finally comes frankly to present himself to a "gay" gathering not as a tourist but as someone who is "available." See Evelyn Hooker, "A Preliminary Examination of Group Behavior of Homosexuals," *J. Psychology* (1956) 42:217–225; especially p. 221. A good fictionalized treatment may be found in James Baldwin's *Giovanni's Room*; New York, Dial, 1956; pp. 41–63. A familiar instance of the coming out process is no doubt to be found among prepubertal children at the moment one of these actors sidles *back* into a room that had been left in an angered huff and injured *amour-propre*. The phrase itself presumably derives from a *rite-de-passage* ceremony once arranged by upper-class mothers for their daughters. Interestingly enough, in large mental hospitals the patient sometimes symbolizes a complete coming out by his first active participation in the hospital wide patient dance.

31. See Goffman, "Characteristics of Total Institutions," pp. 43–84; in *Proceedings of the Symposium of Preventive and Social Psychiatry*; Washington, D.C., Walter Reed Army Institute of Research, 1958.

32. A good description of the ward system may be found in Ivan Belknap, *Human Problems of a State Mental Hospital;* New York, McGraw-Hill, 1956; see especially p. 164.

33. Here is one way in which mental hospitals can be worse than concentration camps and prisons as places in which to "do" time; in the latter, self-insulation from the symbolic implications of the settings may be easier. In fact, self-insulation from hospital settings may be so difficult that patients have to employ devices for this which staff interpret as psychotic symptoms.

34. In regard to convicts, see Anthony Heckstall-Smith, *Eighteen Months;* London, Wingate, 1954; pp. 52–53. For "winos's" see the discussion in Howard G. Bain, "A Sociological Analysis of the Chicago Skid-Row Lifeway;" unpublished M.A. thesis, Dept. of Sociology, pp. 141–146. Bain's neglected thesis is a useful source of material on moral careers.

　　Apparently one of the occupational hazards of prostitution is that clients and other professional contacts sometimes persist in expressing sympathy by asking for a defensible dramatic explanation for the fall from grace. In having to bother to have a sad tale ready, perhaps the prostitute is more to be pitied than damned. Good examples of prostitute sad tales may be found in Sir Henry Mayhew, "Those that Will Not Work," pp. 210–272; in his *London Labour and the London Poor,* Vol. 4; London,

Griffin, Bohn, and Cox, 1862. For a contemporary source, see *Women of the Streets,* edited by C. H. Rolph; London, Zecker and Warburg, 1955; especially p. 6. "Almost always, however, after a few comments on the police, the girl would begin to explain how it was that she was in the life, usually in terms of self-justification." Lately, of course, the psychological expert has helped out the profession in the construction of wholly remarkable sad tales. See, for example, Harold Greenwald, *Call Girl;* New York, Ballantine, 1958.

35. A similar self-protecting rule has been observed in prisons. Thus, Hassler, see note 6, in describing a conversation with a fellow-prisoner; "He didn't say much about why he was sentenced, and I didn't ask him, that being the accepted behavior in prison" (p. 76). A novelistic version for the mental hospital may be found in J. Kerkhoff, *How Thin the Veil: A Newspaperman's Story of His Own Mental Crack-up and Recovery;* New York, Greenberg, 1952; p. 27.

36. From the writer's field notes of informal interaction with patients, transcribed as near verbatim as he was able.

37. The process of examining a person psychiatrically and then altering or reducing his status in consequence is known in hospital and prison parlance as *bugging,* the assumption being that once you come to the attention of the testers you either will automatically be labeled crazy or the process of testing itself will make you crazy. Thus psychiatric staff are sometimes seen not as *discovering* whether you are sick, but as *making* you sick; and "Don't bug me, man," can mean, "Don't pester me to the point where I'll get upset." Sheldon Messenger has suggested to me that this meaning of bugging is related to the other colloquial meaning, of wiring a room with a secret microphone to collect information usable for discrediting the speaker.

15

The Making of Blind Men

ROBERT A. SCOTT

When a blind person first comes to an organization for the blind, he usually has some specific ideas about what his primary problems are and how they can be solved. Most new clients request services that they feel will solve or ameliorate the specific problems they experience because of their visual impairment. Many want only to be able to read better, and therefore request optical aids. Others desire help with mobility problems, or with special problems of dressing, eating, or housekeeping. Some need money or medical care. A few contact agencies for the blind in search of scientific discoveries that will restore their vision. Although the exact type of help sought varies considerably, many clients feel that the substance of their problems is contained in their specific requests. . . .

The personal conceptions that blinded persons have about the nature of their problems are in sharp contrast with beliefs that workers for the blind share about the problems of blindness. The latter regard blindness as one of the most severe of all handicaps, the effects of which are long-lasting, pervasive, and extremely difficult to ameliorate. They believe that if these problems are to be solved, blind persons must understand them and all their manifestations and willingly submit themselves to a prolonged, intensive, and comprehensive program of psychological and restorative services. *Effective socialization of the client largely depends upon changing his views about his problem.* In order to do this, the client's views about the problems of blindness must be discredited. Workers must convince him that simplistic ideas about solving the problems of blindness by means of one or a few services are unrealistic. Workers regard the client's initial definition of his problems as akin to the visible portion of an iceberg. Beneath the surface of awareness lies a tremendously complicated mass of problems that must be dealt with before the surface problems can ever be successfully solved.

Discrediting the client's personal ideas about his problems is achieved in several ways. His initial statements about why he has come to the organization and what he hopes to receive from it are euphemistically termed "the presenting problem," a phrase that implies superficiality in the client's views. During the intake interview and then later with the caseworker or psychologist, the client is encouraged to discuss his feelings and aspirations. . . . However, when concrete plans are formulated, the client learns that his personal views about his problems are largely ignored. A client's request for help with a reading

From: Robert A. Scott, *The Making of Blind Men: A Study of Adult Socialization* (New York: Russell Sage Foundation, 1969). Reprinted by permission of the publisher.

problem produces a recommendation by the worker for a comprehensive psychological work-up. A client's inquiries regarding the availability of financial or medical aid may elicit the suggestion that he enroll in a complicated long-term program of testing, evaluation, and training. In short, blind persons who are acceptable to the agency for the blind will often find that intake workers listen attentively to their views but then dismiss them as superficial or inaccurate. . . . For most persons who have come this far in the process, however, dropping out is not a particularly realistic alternative, since it implies that the blind person has other resources open to him. For the most part, such resources are not available.

. . . [The] experiences a blind person has before being inducted into an agency make him vulnerable to the wishes and intentions of the workers who deal with them. The ability to withstand the pressure to act, think, and feel in conformity with the workers' concept of a model blind person is further reduced by the fact that the workers have a virtual monopoly on the rewards and punishments in the system. By manipulating these rewards and punishments, workers are able to pressure the client into rejecting personal conceptions of problems in favor of the worker's own definition of them. Much evaluative work, in fact, involves attempts to get the client to understand and accept the agency's conception of the problems of blindness. . . . In face-to-face situations, the blind person is rewarded for showing insight and subtly reprimanded for continuing to adhere to earlier notions about his problems. He is led to think that he "really" understands past and present experiences when he couches them in terms acceptable to his therapist. . . .

Psychological rewards are not the only rewards at stake in this process. A fundamental tenet of work for the blind is that a client must accept the fact of his blindness and everything implied by it before he can be effectively rehabilitated. As a result, a client must show signs of understanding his problem in the therapist's terms before he will be permitted to progress any further in the program. Since most blind persons are anxious to move along in the program as rapidly as possible, the implications of being labeled "uncooperative" are serious. Such a label prevents him from receiving basic restorative services. The uncooperative client is assigned low priority for entering preferred job programs. Workers for the blind are less willing to extend themselves on his behalf. As a result, the alert client quickly learns to become "insightful," to behave as workers expect him to.

Under these circumstances, the assumptions and theories of workers for the blind concerning blindness and rehabilitation take on new significance, for what they do is to create, shape, and mold the attitudes and behavior of the client in his role as a blind person. . . . [It] is in organizations for the blind that theories and explicit and implicit assumptions about blindness and rehabilitation become actualized in the clients' attitudes and behavior. We can therefore gain an understanding about the behavior of clients as blind people by examining the theories and assumptions about blindness and rehabilitation held by workers for the blind.

THE PRACTICE THEORIES
OF BLINDNESS WORKERS

The beliefs, ideologies, and assumptions about blindness and rehabilitation that make up practice theories of work for the blind are legion. They include global and limited theories about blindness, ethical principles, commonsense ideas, and an array of specific beliefs that are unrelated, and often contradictory, to one another. Contained in this total array of ideas are two basically different approaches to the problems of blindness. The first I will call the "restorative approach"; the most complete and explicit version of this approach is contained in the writings of Father Thomas Carroll.[1] The second I will call the "accommodative approach." This approach has never been formulated into a codified practice theory; rather, it is only apparent in the programs and policies of more orthodox agencies for the blind.

The Restorative Approach

The basic premise of the restorative approach to blindness is that most blind people can be restored to a high level of independence enabling them to lead a reasonably normal life. However, these goals are attainable only if the person accepts completely the fact that he is blind, and only after he has received competent professional counseling and training. . . .

Seven basic kinds of losses resulting from blindness are identified: (1) the losses to psychological security—the losses of physical integrity, confidence in the remaining senses, reality contact with the environment, visual background, and light security; (2) the losses of the skills of mobility and techniques of daily living; (3) the communication losses, such as the loss of ease of written and spoken communication, and of information about daily events in the world; (4) the losses of appreciation, which include the loss of the visual perception of the pleasurable and of the beautiful; (5) the losses of occupational and financial status, which consist of financial security, career, vocational goals, job opportunities, and ordinary recreational activities; (6) the resulting losses to the whole personality, including the loss of personal independence, social adequacy, self-esteem, and total personality organization; and (7) the concomitant losses of sleep, of physical tone of the body, and of decision, and the sense of control over one's life.[2]

Rehabilitation, in this scheme, is the process "whereby adults in varying stages of helplessness, emotional disturbance, and dependence come to gain new understanding of themselves and their handicap, the new skills necessary for their state, and a new control of their emotions and their environment."[3] This process is not a simple one; it involves the pain and recurrent crises that accompany the acceptance of the many "deaths" to sighted life. It consists of "restorations" for each of the losses involved in blindness. The final objective of total rehabilitation involves returning and integrating the blinded person in his society.

. . . The various restorations in each of these phases correspond to the losses the person has encountered. The loss of confidence in the remaining senses is restored through deliberate training of these senses; the loss of mobility is restored through training in the use of a long cane or a guide dog; the loss of ease of written communication is restored through learning braille; and so on. The goal of this process is to reintegrate the components of the restored personality into an effectively functioning whole. . . .

[In] several rehabilitation centers and general agencies . . . the ideas contained in . . . [Father Carroll's] book are used as the basis for a formal course taught to blind people while they are obtaining services. The purpose of this course is to clarify for them what they have lost because they are blind, how they must change through the course of rehabilitation, and what their lives will be like when rehabilitation has been completed. These ideas are given added weight by the fact that they are shared by all staff members who deal directly with the client and, in some agencies at least, by other nonservice personnel who have occasional contacts with clients. . . .

We cannot assume that there is a necessary correspondence between these beliefs regarding the limits and potentialities imposed by blindness and the blind client's self-image. The question of the full impact of the former on the latter is an empirical one on which there are no hard data. Our analysis of the client's "set" when he enters an agency for the blind does suggest, however, that such beliefs probably have a profound impact on his self-image. . . . [When] the client comes to an agency, he is often seeking direction and guidance and, more often than not, he is in a state of crisis. Consequently, the authority of the system makes the client highly suggestible to the attitudes of those whose help he seeks.

There is evidence that some blind people resist the pressures of the environment of agencies and centers that adopt this philosophy by feigning belief in the workers' ideas for the sake of "making out" in the system.[4] In such cases, the impact of workers on the client's self-image will be attenuated. Despite this, he will learn only those skills made available to him by the agency or center. These skills, which the workers regard as opportunities for individual fulfillment, act also as limits. The choice of compensatory skills around which the theory revolves means the exclusion of a spectrum of other possibilities.

The Accommodative Approach

A basic premise of the restorative approach is that most blind people possess the capacity to function independently enough to lead normal lives. Rehabilitation centers and general service agencies that have embraced this approach therefore gear their entire service programs toward achieving this goal. In other agencies for the blind, no disagreement is voiced about the desirability of blind people's attaining independence, but there is considerable skepticism as to whether this is a feasible goal for more than a small fraction of the client population.[5] According to this view, blindness poses enormous obstacles to independence—obstacles seen as insurmountable by a majority of people. . . . Settings and programs are designed to accommodate the helpless, dependent blind person.

The physical environment in such agencies is often contrived specifically to suit certain limitations inherent in blindness. In some agencies, for example, the elevators have tape recorders that report the floor at which the elevator is stopping and the direction in which it is going, and panels of braille numbers for each floor as well. Other agencies have mounted over their front doors special bells that ring at regular intervals to indicate to blind people that they are approaching the building. Many agencies maintain fleets of cars to pick up clients at their homes and bring them to the agency for services. In the cafeterias of many agencies, special precautions are taken to serve only food that blind people can eat without awkwardness. In one agency cafeteria, for example, the food is cut before it is served, and only spoons are provided.

Recreation programs in agencies that have adopted the accommodative approach consist of games and activities tailored to the disability. For example, bingo, a common activity in many programs, is played with the aid of a corps of volunteers who oversee the game, attending to anything the blind person is unable to do himself.

Employment training for clients in accommodative agencies involves instruction in the use of equipment specifically adapted to the disability. Work tasks, and even the entire method of production, are engineered with this disability in mind, so that there is little resemblance between an average commercial industrial setting and a sheltered workshop. Indeed, the blind person who has been taught to do industrial work in a training facility of an agency for the blind will acquire skills and methods of production that may be unknown in most commercial industries.

The general environment of such agencies is also accommodative in character. Clients are rewarded for trivial things and praised for performing tasks in a mediocre fashion. This superficial and overgenerous reward system makes it impossible for most clients to assess their accomplishments accurately. Eventually, since anything they do is praised as outstanding, many of them come to believe that the underlying assumption must be that blindness makes them incompetent.

The unstated assumption of accommodative agencies is that most of their clients will end up organizing their lives around the agency. Most will become regular participants in the agency's recreation programs, and those who can work will obtain employment in a sheltered workshop or other agency-sponsored employment program. The accommodative approach therefore produces a blind person who can function effectively only within the confines of the agency's contrived environment. He learns skills and behavior that are necessary for participating in activities and programs of the agency, which make it more difficult to cope with the environment of the larger community. A blind person who has been fully socialized in an accommodative agency will be maladjusted to the larger community. In most cases, he does not have the resources, the skills, the means, or the opportunity to overcome the maladaptive patterns of behavior he has learned. He has little choice but to remain a part of the environment that has been designed and engineered to accommodate him.

This portrayal of accommodative agencies suggests that the workers in them, like those in restorative agencies, make certain assumptions about the limitations that blindness imposes, and that these assumptions are manifested in expectations about attitudes and behavior that people ought to have because they are blind. . . .

Unfortunately, no hard data are available on socialization outcomes in agencies that adopt either of the two approaches I have described. However, the materials I collected from interviews with blind people suggest that a number of discernably patterned reactions occur.[6] Some clients and trainees behave according to workers' expectations of them deliberately and consciously in order to extract from the system whatever rewards it may have. Others behave according to expectations because they have accepted and internalized them as genuine qualities of character. The former are the "expedient" blind people, and the latter are the "true believers."

Expedient blind people consciously play a part, acting convincingly the way they sense their counselors and instructors want them to act. They develop a keen sense of timing that enables them to be at their best when circumstances call for it. When the circumstances change, the façade is discarded, much as the Negro discards his "Uncle Tomisms" in the absence of whites. As a rule, the expedient blind person is one who recognizes that few alternatives are open to him in the community; his response is an understandable effort to maximize his gains in a bad situation.

True believers are blind people for whom workers' beliefs and assumptions about blindness are unquestioned ideals toward which they feel impelled earnestly to strive. While this pattern is probably found in all agencies for the blind, it is most obvious in those which embrace the accommodative approach to blindness. Clients who become true believers in such agencies actually experience the emotions that workers believe they must feel. They experience and spontaneously verbalize the proper degree of gratitude, they genuinely believe themselves to be helpless, and they feel that their world must be one of darkness and dependency.

NOTES

1. Thomas J. Carroll, *Blindness: What It Is, What It Does, and How to Live with It,* Little, Brown & Company, Boston, 1961.

2. *Ibid.,* pp. 14–79.

3. *Ibid.,* pp. 96.

4. *Information Bulletin No. 59,* University of Utah, Regional Rehabilitation Research Institute, Salt Lake City, 1968.

5. Roger G. Barker et al., *Adjustment to Physical Handicap and Illness: A Survey of the Social Psychology of Physique and Disability,* Social Science Research Council, New York, 1953.

6. Most of this discussion applies to blind people who have been exposed to agencies that adopt an accommodative approach to rehabilitation. Little information could be gathered on those who have been trainees in restorative agencies, primarily because such agencies are comparatively few in number and recent in origin.

16

Discrediting Victims' Allegations of Sexual Assault: Prosecutorial Accounts of Case Rejections

LISA FROHMANN

Case screening is the gateway to the criminal court system. Prosecutors, acting as gatekeepers, decide which instances of alleged victimization will be passed on for adjudication by the courts. A recent study by the Department of Justice (Boland et al. 1990) suggests that a significant percentage of felony cases never get beyond this point, with only cases characterized as "solid" or "convictable" being filed (Stanko 1981, 1982; Mather 1979). This paper will examine how prosecutors account for the decision to reject sexual assault cases for prosecution and looks at the centrality of discrediting victims' rape allegations in this justification.

A number of studies on sexual assault have found that victim credibility is important in police decisions to investigate and make arrests in sexual assault cases (LaFree 1981; Rose and Randall 1982; Kerstetter 1990; Kerstetter and Van Winkle 1990). Similarly, victim credibility has been shown to influence prosecutors' decisions at a number of stages in the handling of sexual assault cases (LaFree 1980, 1989; Chandler and Torney 1981; Kerstetter 1990).

Much of this prior research has assumed, to varying degrees, that victim credibility is a phenomenon that exists independently of prosecutors' interpre-

From: "Discrediting Victims' Allegations of Sexual Assault," Lisa Frohmann, *Social Problems,* Vol. 38, No. 2, May 1991. © 1991 Society for the Study of Social Problems. Reprinted by permission of University of California Press Journals and the author.

tations and assessments of such credibility. Particularly when operationalized in terms of quantitative variables, victim credibility is treated statistically as a series of fixed, objective features of cases. Such approaches neglect the processes whereby prosecutors actively assess and negotiate victim credibility in actual, ongoing case processing.

An alternative view examines victim credibility as a phenomenon constructed and maintained through interaction (Stanko 1980). Several qualitative studies have begun to identify and analyze these processes. For example, Holmstrom and Burgess's (1983) analysis of a victim's experience with the institutional handling of sexual assault cases discusses the importance of victim credibility through the prosecutor's evaluation of a complainant as a "good witness." A "good witness" is someone who, through her appearance and demeanor, can convince a jury to accept her account of "what happened." Her testimony is "consistent," her behavior "sincere," and she cooperates in case preparation. Stanko's (1981, 1982) study of felony case filing decisions similarly emphasizes prosecutors' reliance on the notion of the "stand-up" witness—someone who can appear to the judge and jury as articulate and credible. Her work emphasizes the centrality of victim credibility in complaint-filing decisions.

In this article I extend these approaches by systematically analyzing the kinds of accounts prosecutors offer in sexual assault cases to support their complaint-filing decisions. Examining the justifications for decisions provides an understanding of how these decisions appear irrational, necessary, and appropriate to decision-makers as they do the work of a case screening. It allows us to uncover the inner, indigenous logic of prosecutors' decisions and the organizational structures in which those decisions are embedded (Garfinkel 1984).

I focus on prosecutorial accounting for case rejection for three reasons. First, since a significant percentage of cases are not filed, an important component of the case-screening process involves case rejection. Second, the organization of case filing requires prosecutors to justify case rejection, not case acceptance, to superiors and fellow deputies. By examining deputy district attorneys' (DDAs') reasons for case rejection, we can gain access to what they consider "solid" cases, providing further insight into the case-filing process. Third, in case screening, prosecutors orient to the rule—when in doubt, reject. Their behavior is organized more to avoiding the error of filing cases that are not likely to result in conviction than to avoiding the error of rejecting cases that will probably end in conviction (Scheff 1966). Thus, I suggest that prosecutors are actively looking for "holes" or problems that will make the victim's version of "what happened" unbelievable or not convincing beyond a reasonable doubt, hence unconvictable (see Miller [1970], Neubauer [1974], and Stanko [1980, 1981] for the importance of conviction in prosecutors' decisions to file cases). This bias is grounded within the organizational context of complaint filing.

DATA AND METHODS

The research was part of an ethnographic field study of the prosecution of sexual assault crimes by deputy district attorneys in the sexual assault units of two branch offices of the district attorney's offices in a metropolitan area on the West Coast.* Research was conducted on a full-time basis in 1989 for nine months in Bay City and on a full-time basis in 1990 for eight months in Center Heights. Three prosecutors were assigned to the unit in Bay City, and four prosecutors to the unit in Center Heights. The data came from 17 months of observation of more than three hundred case screenings. These screenings involved the presentation and assessment of a police report by a sexual assault detective to a prosecutor, conversations between detectives and deputies regarding the "filability"/reject status of a police report, interviews of victims by deputies about the alleged sexual assault, and discussions between deputies regarding the file/reject status of a report. Since tape recordings were prohibited, I took extensive field notes and tried to record as accurately as possible conversation between the parties. In addition, I also conducted open-ended interviews with prosecutors in the sexual assault units and with investigating officers who handled these cases. The accounts presented in the data below include both those offered in the course of negotiating a decision to reject or file a case (usually to the investigating officer [IO] but sometimes with other prosecutors or to me as an insider), and the more or less fixed accounts offered for a decision already made (usually to me). Although I will indicate the context in which the account occurs, I will not emphasize the differences between accounts in the analysis.

The data were analyzed using the constant comparison method of grounded theory (Glaser and Strauss 1967.) I collected all accounts of case rejection from both offices. Through constant comparison of the data, I developed coding schema which provide the analytic framework of the paper.

The two branches of the district attorney's office I studied cover two communities differing in socioeconomic and racial composition. Bay City is primarily a white middle-to-upper-class community, and Center Heights is primarily a black and Latino lower-class community. Center Heights has heavy gang-drug activity, and most of the cases brought to the district attorney were assumed to involve gang members (both the complainant and the assailant) or a sex-drug or sex-money transaction. Because of the activities that occur in this community, the prior relationships between the parties are often the result of gang affiliation. This tendency, in connection with the sex-drug and sex-money transactions, gives a twist to the "consent defense" in "acquaintance" rapes. In Bay City, in contrast, the gang activity is much more limited and the majority of acquaintance situations that came to the prosecutors' attention could be categorized as "date rape."

---

*To protect the confidentiality of people and places studied, pseudonyms are used throughout this article.

THE ORGANIZATIONAL CONTEXT
OF COMPLAINT FILING

Several features of the court setting that I studied provided the context for prosecutors' decisions. These features are prosecutorial concern with maintaining a high conviction rate to promote an image of the "community's legal protector," and prosecutorial and court procedures for processing sexual assault cases.

The promotion policy of the county district attorney's (DA) office encourages prosecutors to accept only "strong" or "winnable" cases for prosecution by using conviction rates as a measure of prosecutorial performance. In the DA's office, guilty verdicts carry more weight than a conviction by case settlement. The stronger the case, the greater likelihood of a guilty verdict, the better the "stats" for promotion considerations. The inducement to take risks—to take cases to court that might not result in conviction—is tempered in three ways: First, a pattern of not-guilty verdicts is used by the DA's office as an indicator of prosecutorial incompetency. Second, prosecutors are given credit for the number of cases they reject as a recognition of their commitment to the organizational concern of reducing the case load of an already overcrowded court system. Third, to continually pursue cases that should have been rejected outright may lead judges to question the prosecutor's competence as a member of the court.

Sexual assault cases are among those crimes that have been deemed by the state legislature to be priority prosecution cases. That is, in instances where both "sex" and "nonsex" cases are trailing (waiting for a court date to open), sexual assault cases are given priority for court time. Judges become annoyed when they feel that court time is being "wasted" with cases that "should" have been negotiated or rejected in the first place, especially when those cases have been given priority over other cases. Procedurally, the prosecutor's office handles sexual assault crimes differently from other felony crimes. Other felonies are handled by a referral system; they are handed from one DDA to another at each stage in the prosecution of the case. But sexual assault cases are vertically prosecuted; the deputy who files the case remains with it until its disposition, and therefore is closely connected with the case outcome.

ACCOUNTING FOR REJECTION BECAUSE
OF "DISCREPANCIES"

Within this organizational context, a central feature of prosecutorial accounts of case rejection is the discrediting of victims' allegations of sexual assault. Below I examine two techniques used by prosecutors to discredit victim's complaints: discrepant accounts and ulterior motives.

Using Official Reports and Records
to Detect Discrepancies

In the course of reporting a rape, victims recount their story to several criminal justice officials. Prosecutors treat consistent accounts of the incident over time as an indicator of a victim's credibility. In the first example two prosecutors are discussing a case brought in for filing the previous day.

> DDA Tamara Jacobs: In the police report she said all three men were kissing the victim. Later in the interview she said that was wrong. It seems strange because there are things wrong on major events like oral copulation and intercourse . . . , for example whether she had John's penis in her mouth. Another thing wrong is whether he forced her into the bedroom immediately after they got to his room or, as the police report said, they all sat on the couch and watched TV. This is something a cop isn't going to get wrong, how the report started. (Bay City)

The prosecutor questions the credibility of the victim's allegation by finding "inconsistencies" between the complainant's account given to the police and the account given to the prosecutor. The prosecutor formulates differences in these accounts as "discrepancies" by noting that they involve "major events"—events so significant no one would confuse them, forget them, or get them wrong. This is in contrast to some differences that may involve acceptable, "normal inconsistencies" in victims' accounts of sexual assault. By "normal inconsistencies," I mean those that are expected and explainable because the victim is confused, upset, or shaken after the assault.

The DDA also discredited the victim's account by referring to a typification of police work. She assumes that the inconsistencies in the accounts could not be attributed to the incorrect writing of the report by the police officer on the grounds that they "wouldn't get wrong how the report started." Similarly, in the following example, a typification of police work is invoked to discredit the victim's interview.

> DDA Sabrina Johnson: [T]he police report doesn't say anything about her face being swollen, only her hand. If they took pictures of her hand, wouldn't the police have taken a picture of her face if it was swollen? (Bay City)

The prosecutor calls the credibility of the victim's complaint into question by pointing to a discrepancy between her subsequent account of injuries received during the incident and the notation of injuries on the police reports taken at the time the incident was reported. Suspicion of the complainant's account is also expressed in the prosecutor's inference that if the police went to the trouble of photographing the victim's injured hand they would have taken pictures of her face had it also shown signs of injury.

In the next case the prosecutor cites two types of inconsistencies between accounts. The first set of inconsistencies is the victim's accounts to the prosecutor and to the police. The second set is between the account the victim gave to the prosecutor and the statements the defendants gave to the police. This excerpt was obtained during an interview.

DDA Tracy Timmerton: The reason I did not believe her [the victim] was, I get the police report first and I'll read that, so I have read the police report which recounts her version of the facts but it also has the statement of both defendants. Both defendants were arrested at separate times and gave separate independent statements that were virtually the same. Her story when I had her recount it to me in the DA's office, the number of acts changed, the chronological order of how they happened has changed. (Bay City)

When the prosecutor compared the suspects' accounts with the victim's account, she interpreted the suspects' accounts as credible because both of their accounts, given separately to police, were similar. This rests on the assumption that if suspects give similar accounts when arrested together, they are presumed to have colluded on the story, but if they give similar accounts independent of the knowledge of the other's arrest, there is presumed to be a degree of truth to the story. This stands in contrast to the discrepant accounts the complainant gave to law enforcement officials and the prosecutor.

Using Official Typifications of
Rape-Relevant Behavior

In the routine handling of sexual assault cases prosecutors develop a repertoire of knowledge about the features of these crimes.* This knowledge includes how particular kinds of rape are committed, post-incident interaction between the parties in an acquaintance situation, and victims' emotional and psychological reactions to rape and their effects on victims' behavior. The typifications of rape-relevant behavior are another resource for discrediting a victim's account of "what happened."

Typifications of Rape Scenarios   Prosecutors distinguish between different types of sexual assault. They characterize these types by the sex acts that occur, the situation in which the incident occurred, and the relationship between the parties. In the following excerpt the prosecutor discredits the victim's version of events by focusing on incongruities between the victim's description of the sex acts and the prosecutor's knowledge of the typical features of kidnap-rape. During an interview a DDA described the following:

DDA Tracy Timmerton: [T]he only acts she complained of was intercourse, and my experience has been that when a rapist has a victim cornered

---

*The use of practitioners' knowledge to inform decision making is not unique to prosecutors. For example, such practices are found among police (Bittner 1967; Rubinstein 1973), public defenders (Sudnow 1965), and juvenile court officials (Emerson 1969).

for a long period of time, they engage in multiple acts and different types of sexual acts and very rarely do just intercourse. (Bay City)

The victim's account is questioned by noting that she did not complain about or describe other sex acts considered "typical" of kidnap-rape situations. She only complained of intercourse. In the next example the DDA and IO are talking about a case involving the molestation of a teenage girl.

DDA William Nelson: Something bothers me, all three acts are the same. She's on her stomach and has her clothes on and he has a "hard and long penis." All three.times he is grinding his penis into her butt. It seems to me he should be trying to do more than that by the third time. (Center Heights)

Here the prosecutor is challenging the credibility of the victim's account by comparing her version of "what happened" with his typification of the way these crimes usually occur. His experience suggests there should be an escalation of sex acts over time, not repetition of the same act.

Often the typification invoked by the prosecutor is highly situational and local. In discussing a drug-sex-related rape in Center Heights, for example, the prosecutor draws on his knowledge of street activity in the community and the types of rapes that occur there to question whether the victim's version of events is what "really" happened. The prosecutor is describing a case he received the day before to an investigating officer there on another matter.

DDA Kent Fernome: I really feel guilty about this case I got yesterday. The girl is 20 going on 65. She is real skinny and gangly. Looks like a cluckhead [crack addict]—they cut off her hair. She went to her uncle's house, left her clothes there, drinks some beers and said she was going to visit a friend in Center Heights who she said she met at a drug rehab program. She is not sure where this friend Cathy lives. Why she went to Center Heights after midnight, God Knows? It isn't clear what she was doing between 12 and 4 A.M. Some gang bangers came by and offered her a ride. They picked her up on the corner of Main and Lincoln. I think she was turning a trick, or looking for a rock, but she wouldn't budge from her story. . . . There are lots of conflicts between what she told the police and what she told me. The sequence of events, the sex acts performed, who ejaculates. She doesn't say who is who. . . . She's beat up, bruises on face and a laceration on her neck. The cop and doctor say there is no trauma—she's done by six guys. That concerns me. There is no semen that they see. It looks like this to me—maybe she is a strawberry, she's hooking or looking for a rock, but somewhere along the line it is not consensual. . . . She is [a] real street-worn woman. She's not leveling with me— visiting a woman with an unknown address on a bus in Center Heights—I don't buy it. . . . (Center Heights)

The prosecutor questioned the complainant's reason for being in Center Heights because, based on his knowledge of the area, he found it unlikely that a woman would come to this community at midnight to visit a friend at an unknown address. The deputy proposed an alternative account of the victim's action based on his knowledge of activities in the community—specifically, prostitution and drug dealing—and questioned elements of the victim's account, particularly her insufficiently accounted for activity between 12 and 4 A.M., coming to Center Heights late at night to visit a friend at an unknown address, and "hanging out" on the corner.

The DDA uses "person-descriptions" (Maynard 1984) to construct part of the account, describing the complainant's appearance as a "cluckhead" and "street-worn." These descriptions suggested she was a drug user, did not have a "stable" residence or employment, and was probably in Center Heights in search of drugs. This description is filled in by her previous "participation in a drug rehab program," the description of her activity as "hanging out" and being "picked up" by gang bangers, and a medical report which states that no trauma or semen was found when she was "done by six guys." Each of these features of the account suggests that the complainant is a prostitute or "strawberry" who came to Center Heights to trade sex or money for drugs. This alternative scenario combined with "conflicts between what she told the police and what she told me" justify case rejection because it is unlikely that the prosecutor could get a conviction.

The prosecutor acknowledges the distinction between the violation of women's sexual/physical integrity—"somewhere along the line it wasn't consensual"—and prosecutable actions. The organizational concern with "downstream consequences" (Emerson and Paley, forthcoming) mitigate against the case being filed.

Typifications of Post-Incident Interaction   In an acquaintance rape, the interaction between the parties after the incident is a critical element in assessing the validity of a rape complaint. As implied below by the prosecutors, the typical interaction pattern between victim and suspect after a rape incident is not to see one another. In the following cases the prosecutor challenges the validity of the victims' allegations by suggesting that the complainants' behavior runs counter to a typical rape victim's behavior. In the first instance the parties involved in the incident had a previous relationship and were planning to live together. The DDA is talking to me about the case prior to her decision to reject.

DDA Sabrina Johnson: I am going to reject the case. She is making it very difficult to try the case. She told me she let him into her apartment last night because she is easily influenced. The week before this happened [the alleged rape] she agreed to have sex with him. Also, first she says "he raped me" and then she lets him into her apartment. (Bay City)

Here the prosecutor raises doubt about the veracity of the victim's rape allegation by contrasting it to her willingness to allow the suspect into her apartment after the incident. This "atypical" behavior is used to discredit the complainant's allegation.

In the next excerpt the prosecutor was talking about two cases. In both instances the parties knew each other prior to the rape incident as well as having had sexual relations after the incident. As in the previous instance, the victims' allegations are discredited by referring to their atypical behavior.

DDA Sabrina Johnson: I can't take either case because of the women's behavior after the fact. By seeing these guys again and having sex with them they are absolving them of their guilt. (Bay City)

In each instance the "downstream" concern with convictability is indicated in the prosecutor's talk—"She is making it very difficult to try the case" and "By seeing these guys again and having sex with them they are absolving them of their guilt." This concern is informed by a series of common-sense assumptions about normal heterosexual relations that the prosecutors assume judges and juries use to assess the believability of the victim: First, appropriate behavior within ongoing relationships is noncoercive and nonviolent. Second, sex that occurs within the context of ongoing relationships is consensual. Third, if coercion or violence occurs, the appropriate response is to sever the relationship, at least for a time. When complainants allege they have been raped by their partner within a continuing relationship, they challenge the taken-for-granted assumptions of normal heterosexual relationships. The prosecutors anticipate that this challenge will create problems for the successful prosecution of a case because they think that judges and jurors will use this typification to question the credibility of the victim's allegation. They assume that the triers of fact will assume that if there is "evidence" of ongoing normal heterosexual relations— she didn't leave and the sexual relationship continued—then there was no coercive sex. Thus the certitude that a crime originally occurred can be retrospectively undermined by the interaction between complainant and suspect after the alleged incident. Implicit in this is the assumed primacy of the normal heterosexual relations typification as the standard on which to assess the victim's credibility even though an allegation of rape has been made.

Typifications of Rape Reporting An important feature of sexual assault cases is the timeliness in which they are reported to the police (see Torrey, forthcoming). Prosecutors expect rape victims to report the incident relatively promptly: "She didn't call the police until four hours later. That isn't consistent with someone who has been raped." If a woman reports "late," her motives for reporting and the sincerity of her allegation are questioned if they fall outside the typification of officially recognizable/explainable reasons for late reporting. The typification is characterized by the features that can be explained by Rape Trauma Syndrome (RTS). In the first excerpt the victim's credibility is not challenged as a result of her delayed reporting. The prosecutor describes her behavior and motives as characteristic of RTS. The DDA is describing a case to me that came in that morning.

DDA Tamara Jacobs: Charlene was in the car with her three assailants after the rape. John (the driver) was pulled over by the CHP [California Highway Patrol] for erratic driving behavior. The victim did not tell the officers that she had just been raped by these three men. When she arrived home, she didn't tell anyone what happened for approximately 24 hours. When her best friend found out from the assailants (who were mutual friends) and confronted the victim, Charlene told her what happened. She then reported it to the police. When asked why she didn't report the crime earlier, she said that she was embarrassed and afraid they would hurt her more if she reported it to the police. The DDA went on to say that the victim's behavior and reasons for delayed reporting were symptomatic of RTS. During the trial an expert in Rape Trauma Syndrome was called by the prosecution to explain the "normality" and commonness of the victim's reaction. (Bay City)

Other typical motives include "wanting to return home first and get family support" or "wanting to talk the decision to report over with family and friends." In all these examples, the victims sustained injuries of varying degrees in addition to the trauma of the rape itself, and they reported the crime within 24 hours. At the time the victims reported the incident, their injuries were still visible, providing corroboration for their accounts of what happened.

In the next excerpt we see the connection between atypical motives for delayed reporting and ulterior motives for reporting a rape allegation. At this point I focus on the prosecutors' use of typification as a resource for discrediting the victim's account. I will examine ulterior motives as a technique of discrediting in a later section. The deputy is telling me about a case she recently rejected.

DDA Sabrina Johnson: She doesn't tell anyone after the rape. Soon after this happened she met him in a public place to talk business. Her car doesn't start, he drives her home and starts to attack her. She jumps from the car and runs home. Again she doesn't tell anyone. She said she didn't tell anyone because she didn't want to lose his business. Then the check bounces, and she ends up with VD. She has to tell her fiance so he can be treated. He insists she tell the police. It is three weeks after the incident. I have to look at what the defense would say about the cases. Looks like she consented, and told only when she had to because of the infection and because he made a fool out of her by having the check bounce. (Bay City)

The victim's account is discredited because her motives for delayed reporting—not wanting to jeopardize a business deal—fall outside those considered officially recognizable and explicable.

Typifications of Victim's Demeanor   In the course of interviewing hundreds of victims, prosecutors develop a notion of a victim's comportment when she tells what happened. They distinguish between behavior that signifies "lying" versus "discomfort." In the first two exchanges the DDA and IO cite the victim's behavior as an indication of lying. Below, the deputy and IO are discussing the case immediately after the intake interview.

IO Nancy Fauteck: I think something happened. There was an exchange of body language that makes me question what she was doing. She was yawning, hedging, fudging something.

DDA Sabrina Johnson: Yawning is a sign of stress and nervousness.

IO Nancy Fauteck: She started yawning when I talked to her about her record earlier, and she stopped when we finished talking about it. (Bay City)

The prosecutor and the investigating officer collaboratively draw on their common-sense knowledge and practical work experience to interpret the yawns, nervousness, and demeanor of the complainant as running counter to behavior they expect from one who is "telling the whole truth." They interpret the victim's behavior as a continuum of interaction first with the investigating officer and then with the district attorney. The investigating officer refers to the victim's recurrent behavior (yawning) as an indication that something other than what the victim is reporting actually occurred.

In the next excerpt the prosecutor and IO discredit the victim's account by referencing two typifications—demeanor and appropriate rape-victim behavior. The IO and prosecutor are telling me about the case immediately after they finished the screening interview.

IO Dina Alvarez: One on one, no corroboration.

DDA William Nelson: She's a poor witness, though that doesn't mean she wasn't raped. I won't file a one-on-one case.

IO Dina Alvarez: I don't like her body language.

DDA William Nelson: She's timid, shy, naive, virginal, and she didn't do all the right things. I'm not convinced she is even telling the truth. She's not even angry about what happened to her. . . .

DDA William Nelson: Before a jury if we have a one on one, he denies it, no witnesses, no physical evidence or medical corroboration, they won't vote guilty.

IO Dina Alvarez: I agree, and I didn't believe her because of her body language. She looks down, mumbles, crosses her arms, and twists her hands.

DDA William Nelson: . . . She has the same mannerisms and demeanor as a person who is lying. A jury just won't believe her. She has low self-esteem and self-confidence. . . . (Center Heights)

The prosecutor and IO account for case rejection by characterizing the victim as unbelievable and the case as unconvictable. They establish their disbelief in the victim's account by citing the victim's actions that fall outside the typified notions of believable and expected behavior—"she has the same mannerisms and demeanor as a person who is lying," and "I'm not convinced she is even telling the truth. She isn't even angry about what happened." They assume that potential jurors will also find the victim's demeanor and post-incident behavior problematic. They demonstrate the unconvictability of the case by citing the "holes" in the case—a combination of a "poor witness" whom "the jury just won't believe" and "one on one, [with] no corroboration" and a defense in which the defendant denies anything happened or denies it was nonconsensual sex.

Prosecutors and investigating officers do not routinely provide explicit accounts of "expected/honest" demeanor. Explicit accounts of victim demeanor tend to occur when DDAs are providing grounds for discrediting a rape allegation. When as a researcher I pushed for an account of expected behavior, the following exchange occurred. The DDA had just concluded the interview and asked the victim to wait in the lobby.*

IO Nancy Fauteck: Don't you think he's credible?

DDA Sabrina Johnson: Yes

LF: What seems funny to me is that someone who said he was so unwilling to do this talked about it pretty easily.

IO Nancy Fauteck: Didn't you see his eyes, they were like saucers.

DDA Sabrina Johnson: And [he] was shaking too. (Bay City)

This provides evidence that DDAs and IOs are orienting to victims' comportment and could provide accounts of "expected/honest" demeanor if necessary. Other behavior that might be included in this typification are the switch from looking at to looking away from the prosecutor when the victim begins to discuss the specific details of the rape itself; a stiffening of the body and tightening of the face as though to hold in tears when the victim begins to tell about the particulars of the incident; shaking of the body and crying when describing the details of the incident; and a lowering of the voice and long pauses when the victim tells the specifics of the sexual assault incident.

Prosecutors have a number of resources they call on to develop typification related to rape scenarios and reporting. These include how sexual assaults are committed, community residents and activities, interactions between suspect and defendants after a rape incident, and the way victims' emotional and psychological responses to rape influence their behavior. These typifications highlight discrepancies between prosecutors' knowledge and victims'

---

*Unlike the majority of rape cases I observed, this case had a male victim. Due to lack of data, I am unable to tell if this made him more or less credible in the eyes of the prosecutor and police.

accounts. They are used to discredit the victims' allegation of events, justifying case rejection.

As we have seen, one technique used by prosecutors to discredit a victim's allegations of rape as a justification of case rejection is the detection of discrepancies. The resources for this are official documents and records and typifications of rape scenarios and rape reporting. A second technique prosecutors use is the identification of ulterior motives for the victim's rape allegation.

ACCOUNTING FOR REJECTION BY "ULTERIOR MOTIVE"

Ulterior motives rest on the assumption that a woman consented to sexual activity and for some reason needed to deny it afterwards. These motives are drawn from the prosecutor's knowledge of the victim's personal history and the community in which the incident occurred. They are elaborated and supported by other techniques and knowledge prosecutors use in the accounting process.

I identify two types of ulterior motives prosecutors use to justify rejection: The first type suggests the victim has a reason to file a false rape complaint. The second type acknowledges the legitimacy of the rape allegation, framing the motives as an organizational concern with convictability.

Knowledge of Victim's Current Circumstances

Prosecutors accumulate the details of the victims' lives from police interviews, official documents, and filing interviews. They may identify ulterior motives by drawing on this information. Note that unlike the court trial itself, where the rape incident is often taken out of the context of the victim's life, here the DDAs call on the texture of a victim's life to justify case rejection. In an excerpt previously discussed, the DDA uses her knowledge of the victim's personal relationship and business transactions as a resource for formulating ulterior motive. Drawing on the victim's current circumstances, the prosecutor suggests two ulterior motives for the rape allegation—disclosure to her fiance about the need to treat a sexually transmitted disease, and anger and embarrassment about the bounced check. Both of these are motives for making a false complaint. The ulterior motives are supported by the typification for case reporting. Twice unreported sexual assault incidents with the same suspect, a three-week delay in reporting, and reporting only after the fiance insisted she do so are not within the typified behavior and reasons for late reporting. Her atypical behavior provides plausibility to the alternative version of the events—the interaction was consensual and only reported as a rape because the victim needed to explain a potentially explosive matter (how she contracted venereal disease) to her fiance. In addition she felt duped on a business deal.

Resources for imputing ulterior motives also come from the specifics of the rape incident. Below, the prosecutor's knowledge of the residents and activities in Center Heights supply the reason: the type of activity the victim wanted to cover up from her boyfriend. The justification for rejection is strengthened by conflicting accounts between the victim and witness on the purpose for being in Center Heights. The DDA and IO are talking about the case before they interview the complainant.

DDA William Nelson: A white girl from Addison comes to buy dope. She gets kidnapped and raped.

IO Brandon Palmer: She tells her boyfriend and he beats her up for being so stupid for going to Center Heights. . . . The drug dealer positively ID'd the two suspects, but she's got a credibility problem because she said she wasn't selling dope, but the other two witnesses say they bought dope from her. . . .

LF: I see you have a blue sheet [a sheet used to write up case rejections] already written up.

IO Brandon Palmer: Oh yes. But there was no doubt in my mind that she was raped. But do you see the problems?

DDA William Nelson: Too bad because these guys really messed her up. . . . She has a credibility problem. I don't think she is telling the truth about the drugs. It would be better if she said she did come to buy drugs. The defense is going to rip her up because of the drugs. He is going to say, isn't it true you had sex with these guys but didn't want to tell your boyfriend, so you lied about the rape like you did about the drugs, or that she had sex for drugs. . . . (Center Heights)

The prosecutor expresses doubt about the victim's account because it conflicts with his knowledge of the community. He uses this knowledge to formulate the ulterior motive for the victim's complaint—to hide from her boyfriend the "fact" that she traded sex for drugs. The victim, "a white woman from Addison," alleges she drove to Center Heights "in the middle of the night" as a favor to a friend. She asserted that she did not come to purchase drugs. The DDA "knows" that white people don't live in Center Heights. He assumes that whites who come to Center Heights, especially in the middle of the night, are there to buy drugs or trade sex for drugs. The prosecutor's scenario is strengthened by the statements of the victim's two friends who accompanied her to Center Heights, were present at the scene, and admitted buying drugs. The prosecutor frames the ulterior motives as an organizational concern with defense arguments and convictability. This concern is reinforced by citing conflicting accounts between witnesses and the victim. He does not suggest that the victim's allegation was false—"there is no doubt in my mind she was raped"; rather, the case isn't convictable—"she has a credibility problem" and "the defense is going to rip her up."

Criminal Connections

The presence of criminal connections can also be used as a resource for identifying ulterior motives. Knowledge of a victim's criminal activity enables prosecutors to "find" ulterior motives for her allegation. In the first excerpt the complainant's presence in an area known by police as "where prostitutes bring their clients" is used to formulate an ulterior motive for her rape complaint. This excerpt is from an exchange in which the DDA was telling me about a case he had just rejected.

DDA William Nelson: Young female is raped under questionable circum-
    stances. One on one. The guy states it is consensual sex. There is no
    corroboration, no medicals. We ran the woman's rap sheet, and she had
    a series of prostitution arrests. She's with this guy in the car in a dark
    alley having sex. The police know this is where prostitutes bring their
    customers, so she knew she had better do something fast unless she is
    going to be busted for prostitution, so, lo and behold, she comes run-
    ning out of the car yelling "he's raped me." He says no. He picked her
    up on Long Beach Boulevard, paid her $25 and this is "where she
    brought me." He's real scared, he has no record. (Center Heights)

Above, the prosecutor, relying on police knowledge of a particular location, assumes the woman is a prostitute. Her presence in the location places her in a "suspicious" category, triggering a check on her criminal history. Her record of prostitution arrests is used as the resource for developing an ulterior motive for her complaint: To avoid being busted for prostitution again, she made a false allegation of rape. Here the woman's record of prostitution and the imminent possibility of arrest are used to provide the ulterior motive to discredit her account. The woman's account is further discredited by comparing her criminal history—"a series of prostitution arrests" with that of the suspect, who "has no record," thus suggesting that he is the more credible of the two parties.

Prosecutors and investigating officers often decide to run a rap sheet (a chronicle of a person's arrests and convictions) on a rape victim. These decisions are triggered when a victim falls into certain "suspicious" categories, categories that have a class/race bias. Rap sheets are not run on women who live in the wealthier parts of town (the majority of whom are white) or have professional careers. They are run on women who live in Center Heights (who are black and Latina), who are homeless, or who are involved in illegal activities that could be related to the incident.

In the next case the prosecutor's knowledge of the victim's criminal conviction for narcotics is the resource for formulating an ulterior motive. This excerpt was obtained during an interview.

DDA Tracy Timmerton: I had one woman who had claimed that she had been
    kidnapped off the street after she had car trouble by these two gentle-
    men who locked her in a room all night and had repeated intercourse

with her. Now she was on a cocaine diversion [a drug treatment program where the court places persons convicted of cocaine possession instead of prison], and these two guys' stories essentially were that the one guy picked her up, they went down and got some cocaine, had sex in exchange for the cocaine, and the other guy comes along and they are all having sex and all doing cocaine. She has real reason to lie, she was doing cocaine, and because she has then violated the terms of her diversion and is now subject to criminal prosecution for her possession of cocaine charge. She is also supposed to be in a drug program which she has really violated, so this is her excuse and her explanation to explain why she has fallen off her program. (Bay City)

The prosecutor used the victim's previous criminal conviction for cocaine and her probation conditions to provide ulterior motives for her rape allegation— the need to avoid being violated on probation for the possession of cocaine and her absence from a drug diversion program. She suggests that the allegation made by the victim was false.

Prosecutors develop the basis for ulterior motives from the knowledge they have of the victim's personal life and criminal connections. They create two types of ulterior motives, those that suggest the victim made a false rape complaint and those that acknowledge the legitimacy of the complaint but discredit the account because of its unconvictability. In the accounts prosecutors give, ulterior motives for case rejection are supported with discrepancies in victims' accounts and other practitioners' knowledge.

CONCLUSION

Case filing is a critical stage in the prosecutorial process. It is here that prosecutors decide which instances of alleged victimization will be forwarded for adjudication by the courts. A significant percentage of sexual assault cases are rejected at this stage. This research has examined prosecutorial accounts for case rejection and the centrality of victim discreditability in those accounts. I have elucidated the techniques of case rejection (discrepant accounts and ulterior motives), the resources prosecutors use to develop these techniques (official reports and records, typifications of rape-relevant behavior, criminal connections and knowledge of a victim's personal life), and how these resources are used to discredit victims' allegations of sexual assault.

This examination has also provided the beginnings of an investigation into the logic and organization of prosecutors' decisions to reject/accept cases for prosecution. The research suggests that prosecutors are orienting a "downstream" concern with convictability. They are constantly "in dialogue with" anticipated defense arguments and anticipated judge and juror responses to case testimony. These dialogues illustrate the intricacy of prosecutorial decision-making. They make visible how prosecutors rely on assumptions about relationships, gender, and sexuality (implicit in this analysis, but critical and requiring of specific and explicit attention) in complaint filing of sexual assault cases. They also make evident how the processes of distinguishing truths from untruths and the practical concerns of trying cases are central to these decisions. Each of these issues, in all its complexity, needs to be examined if we are to understand the logic and organization of filing sexual assault cases.

The organizational logic unveiled by these accounts has political implications for the prosecution of sexual assault crimes. These implications are particularly acute for acquaintance rape situations. As I have shown, the typification of normal heterosexual relations plays an important role in assessing these cases, and case conviction is key to filing cases. As noted by DDA William Nelson: "There is a difference between believing a woman was assaulted and being able to get a conviction in court." Unless we are able to challenge the assumptions on which these typifications are based, many cases of rape will never get beyond the filing process because of unconvictability.

REFERENCES

Bittner, Egon A. 1967. "The police on skid-row: A study of peace keeping." American Sociological Review 32:699–715.

Boland, Barbara, Catherine H. Conly, Paul Mahanna, Lynn Warner, and Ronald Sones. 1990. The Prosecution of Felony Arrests, 1987. Washington, D.C.: Bureau of Justice Statistics, U.S. Department of Justice.

Chandler, Susan M., and Martha Torney. 1981. "The decision and the processing of rape victims through the criminal justice system." California Sociologist 4:155–69.

Emerson, Robert M. 1969. Judging Delinquents: Context and Process in Juvenile Court. Chicago: Aldine Publishing Co.

Emerson, Robert M., and Blair Paley. Forthcoming. "Organizational horizons and complaint-filing." In The Uses of Discretion, ed. Keith Hawkins. Oxford: Oxford University Press.

Garfinkel, Harold. 1984. Studies in Ethnomethodology. Cambridge, Eng.: Polity Press.

Glaser, Barney, and Anselm Strauss. 1967. The Discovery of Grounded Theory. Chicago: Aldine Publishing Co.

Holmstrom, Lynda Lytle, and Ann Wolbert Burgess. 1983. The Victim of Rape: Institutional Reactions. New Brunswick, N.J.: Transaction Books.

Kerstetter, Wayne A. 1990. "Gateway to justice: Police and prosecutorial response to sexual assaults against women." Journal of Criminal Law and Criminology 81:267–313.

Kerstetter, Wayne A., and Barrik Van Winkle. 1990. "Who decides? A study of the complainant's decision to prosecute in rape cases." Criminal Justice and Behavior 17:268–83.

LaFree, Gary D. 1980. "Variables affecting guilty pleas and convictions in rape cases: Toward a social theory of rape processing." Social Forces 58:833–50.

———. 1981. "Official reactions to social problems: Police decisions in sexual assault cases." Social Problems 28:582–94.

———. 1989. Rape and Criminal Justice: The Social Construction of Sexual Assault. Belmont, Calif.: Wadsworth Publishing Co.

Mather, Lynn M. 1979. Plea Bargaining or Trial? The Process of Criminal-Case Disposition. Lexington, Mass.: Lexington Books.

Maynard, Douglas W. 1984. Inside Plea Bargaining: The Language of Negotiation. New York: Plenum Press.

Miller, Frank. 1970. Prosecution: The Decision to Charge a Suspect with a Crime. Boston: Little, Brown.

Neubauer, David. 1974. Criminal Justice in Middle America. Morristown, N.J.: General Learning Press.

Rose, Vicki M., and Susan C. Randall. 1982. "The impact of investigator perceptions of victim legitimacy on the processing of rape/sexual assault cases." Symbolic Interaction 5:23–36.

Rubinstein, Jonathan. 1973. City Police. New York: Farrar, Straus & Giroux.

Scheff, Thomas. 1966. Being Mentally Ill: A Sociological Theory. Chicago: Aldine Publishing Co.

Stanko, Elizabeth A. 1980. "These are the cases that try themselves: An examination of extra-legal criteria in felony case processing." Presented at the Annual Meetings of the North Central Sociological Association, December. Buffalo, N.Y.

———. 1981. "The impact of victim assessment on prosecutor's screening decisions: The case of the New York District Attorney's Office." Law and Society Review 16:225–39.

———. 1982. "Would you believe this woman? Prosecutorial screening for "credible" witnesses and a problem of justice." In Judge, Lawyer, Victim, Thief, ed. Nicole Hahn Rafter and Elizabeth A. Stanko, 63–82. Boston: Northeastern University Press.

Sudnow, David. 1965. "Normal crimes: Sociological features of the penal code in a public defenders office." Social Problems 12:255–76.

Torrey, Morrison. Forthcoming. "When will we be believed? Rape myths and the idea coming of a fair trial in rape prosecutions." U.C. Davis Law Review.

Waegel, William B. 1981. "Case routinization in investigative police work." Social Problems 28:263–75.

Williams, Kristen M. 1978a. The Role of the Victim in the Prosecution of Violent Crimes. Washington, D.C.: Institute for Law and Social Research.

————. 1978b. The Prosecution of Sexual Assaults. Washington, D.C.: Institute for Law and Social Research.

PART V

■

Deviant Identity

Bureaucratic organizations are not the only contexts in which people are labeled as deviant; this labeling can also occur in interpersonal situations. Becoming deviant does not only entail having a definition of deviance and an environment in which it can occur; it also requires that people accept the identity and make it their own. In Part V we will examine this process: how the concept of deviance becomes applied to individuals and how it affects their self-conception.

IDENTITY DEVELOPMENT

We mentioned earlier that although many people engage in deviance, the label is applied to only a small percentage of them. Such labeling is tied to their formerly secret deviance becoming exposed or to an abstract status coming to bear on their personal experience. Thus, Jews may not feel stigmatized unless they experience anti-Semitism, and embezzlers may not think of themselves as thieves until they are caught. When this happens, they enter the pathway to the deviant identity, a pathway that follows a certain trajectory. Howard Becker (1963) has suggested that we can think of this path as a "*deviant career.*"

Once people are publicly identified as deviant, their life changes in several ways. People start to think of them differently. For example, suppose there has been a rash of thefts in a college dormitory, and Jessica, a freshman, is finally caught and identified as the culprit. She may or may not be reported to authorities and charged with theft, but regardless, she will experience an informal labeling process. People will probably change their attitudes toward her, as they find themselves talking about her behind her back. They may look back on her behavior and engage in *"retrospective interpretation"* (Kitsuse 1962) as they think about her differently in light of their new information. Where did she say she was when the last theft occurred, where did she say she got the money to buy that new sweater? She may develop what Goffman (1963) has called a *"spoiled identity,"* one with a damaged reputation.

Jessica's dorm mates and former friends may then engage in what Lemert (1951) has called *"the dynamics of exclusion,"* deriding her and ostracizing her from their social group. When she enters the room, she may notice that a sudden hush falls over the conversation. People may not feel comfortable leaving her alone in their room. They may exclude her from their meal plans and study groups. She may become progressively shut out from non-deviant activities and circles. At the same time, others may welcome, or *include,* her in their deviant circles or activities. She may find that she has developed a reputation that, though repelling to some groups, is attractive to others. They may welcome her as "cool" and invite her into their circles. The more people start to treat her and interact with her as a deviant, the more she is likely to *internalize the label* and regard herself as a deviant.

Once people are labeled as deviant and accept that label into their self-conception, a variety of outcomes may ensue. Everett Hughes (1945) has suggested that we all have a series of statuses, or identities, through which we relate to people, including those of sibling, child, friend, student, neighbor, and customer. These identities also derive from some of our demographic or occupational features, like race, gender, age, religion, or social class. Hughes asserts that some statuses are very dominant, overpowering others and coloring the way in which people are viewed. Having a deviant identity may become one of these *master statuses*, rising to the top of the hierarchy and infusing people's self-concept and others' reactions. People who are labeled as deviants, like heroin addicts, cult members, or homosexuals, may be viewed by others through this lens no matter what the situation or setting. In addition, every master status has a set of *auxiliary traits* that accompany it, and once people label someone with a deviant master status, they will expect to see the relevant auxiliary traits. A heroin addict may be suspected of being a prostitute or thief, and a homosexual may be suspected of being sexually promiscuous or AIDS-infected. This type of identification spreads the image of deviance to cover the person as a whole and not just one part of him or her.

The process of developing a deviant master status and auxiliary traits helps explain the move from primary deviance to secondary deviance. As Lemert (1967) defined it, *primary deviance* refers to the initial type of deviance in which people engage, one that has not necessarily yielded them the master status of deviance. Getting caught and labeled reinforces the deviant identity, spoils one's reputation, and leads to an altered self-concept. People accept the deviant view of themselves, and, stripped of their fear of losing their good reputations, they may go on to engage in more and different kinds of deviant behavior. *Secondary deviance* thus refers both to the diffusion of involvement in deviance and the change in self-conception.

Some of these processes are illustrated in the two readings in this first section of Part V. Naomi Gerstel's "Divorce and Stigma" highlights the dynamics of exclusion associated with divorcees' shift from the status of married to that of divorced. Within a short time after the divorce, they progressively find themselves excluded from the company of their former couple friends and cast into the company of other divorcees. Treated as outsiders and social misfits, they develop feelings of devaluation and demoralization. In "Anorexia Nervosa and Bulimia: The Development of Deviant Identities," Penelope McLorg and Diane Taub describe and analyze women's progression from socially conforming eating behavior through primary deviance and on to secondary deviance. Along the way, the women they studied moved through stages of more common fixations about dieting; to frustration with dieting and movement toward more radical solutions such as binging, purging, compulsive exercising, and lack of eating (devoid of reconception of their selves—primary deviance); and to recognition of serious eating disorders accompanied by the internalization of these labels into their self-identities. By interacting with others through the vehicle of their eating disorders, these women reinforced their anorexic and bulimic behaviors.

ACCOUNTS

Marvin Scott and Stanford Lyman (1968) have suggested that we all engage in instances of deviant behavior but that we desire to maintain a positive self-image in both our own eyes and the eyes of others. To do so, we offer *accounts* intended to explain and *normalize* our deviant behavior. Three kinds of accounts predominate: excuses, justifications, and techniques of neutralization.

In offering *excuses*, individuals admit the wrongfulness of their actions but distance themselves from the blame. These excuses may precede or follow the act of deviance. They are often fairly standard phrases or ideas designed to soften the deviance and relieve individuals of their accountability. These may include appeals to accidents ("My computer malfunctioned and lost my file"), appeals to defeasibility ("I thought my roommate turned my paper in"), appeals to biological drives ("I couldn't stop myself"), and scapegoating ("She borrowed my notes, and I couldn't get them back in time to study for the test").

In offering *justifications*, individuals accept responsibility for their actions but seek to have specific instances accepted. In so doing, they try to legitimate the act or its consequences. In drawing on justifications, individuals may invoke sad tales ("I turn tricks because I was sexually abused as a child") or the need for self-fulfillment ("Taking hallucinogenic drugs expands my consciousness").

Techniques of neutralization attempt to resolve the contradictions between what people say and what they do. Even people who engage in deviance want to feel good about themselves, so they offer neutralizations to allow a conception of themselves as "moral" (Sykes and Matza 1957). These techniques include the denial of responsibility ("It wasn't my fault"), denial of injury ("No harm done, it was just a prank"), denial of the victim ("Nobody was hurt"; "he deserved it"), condemnation of the condemners ("Society is unfair"), and the appeal to higher loyalties ("Some things are more important").

Diana Scully and Joseph Marolla's "Convicted Rapists' Vocabularies of Motive: Excuses and Justifications" offers a fascinating glimpse into the rationalizations offered by a group of incarcerated rapists. Representing the most hard-core segment of the rapist population, those sentenced to prison time, these men gave a variety of accounts to legitimate their violent crimes. In dividing these accounts into the categories of excuses and justifications, Scully and Marolla both illustrate these disavowal techniques and shed light on the repertoire of culturally available excuses men use to neutralize this behavior.

Donald McCabe then presents some rationalizations for students' deviant behavior in "The Influence of Situational Ethics on Cheating among College Students." Based on a survey of six thousand students, McCabe shows how Sykes and Matza's neutralization techniques can be used in a familiar situation. Readers will no doubt recognize some of these familiar rationales, as they tie their everyday life surroundings to deviance concepts.

Moving from the disrespectability of rapists, the ranks of future upstanding citizens, and society's business elite, we focus on how people in positions of power and trust use logic to disavow their deviant acts. Michael Benson's "Denying the Guilty Mind: Accounting for Involvement in a White-Collar Crime" sheds insight into the thoughts and statements of people who have been caught in practices such as price rigging, income tax violation, embezzlement, stock swindling, fraudulent loan scams, and check kiting. It is interesting to compare the excuses of people working in the legitimate business world with those of students and violent criminals to see the correspondences and differences in ideology among members of these diverse segments of society. Benson's article also shows us the fine line dividing aggressive business practice and crime, as his subjects legitimate their acts as nondeviant.

INDIVIDUAL STIGMA MANAGEMENT

The label of deviant marks people with a stigma in the eyes of society. As we have seen, this label may lead to devaluation and exclusion. Consequently, people with deviant features learn how to "manage" their stigma so that they are not shamed or ostracized. This effort requires considerable social skills.

Goffman (1963) has suggested that people with a potential deviant stigma fall into two categories: "*the discreditable*" and "*the discredited.*" The former are those with concealable deviant traits (ex-convicts, secret homosexuals) who may manage themselves so as to avoid the deviant stigma. The latter are either members of the former category who have revealed their deviance or those who cannot hide their deviance (the obese, the physically handicapped). These people's lives are characterized by a constant focus on secrecy and information control. Goffman observed that most discreditables engage in "*passing*" as "normals" in their everyday lives, concealing their deviance and fitting in with regular people. They may do this by avoiding contacts with "stigma symbols," those objects or behaviors that would tip people off to their deviant condition (an unwed father avoiding his pregnant girlfriend, AIDS patients avoiding their medication). Another technique for passing includes using "disidentifiers" such as props, actions, or verbal expressions to distract and fool people into thinking that they do not have the deviant stigma (homosexuals bragging about sexual conquests or taking a date to the company picnic). Finally, they may "lead a double life," maintaining two different lifestyles with two distinct groups of people, one that knows about their deviance and one that does not.

In this endeavor, people may employ the aid of others to help conceal their deviance by "*covering*" for them. In these team performances, friends and family members may assist the deviants by concealing their identities, their whereabouts, their deficiencies, or their pasts. They may even coach the deviants on how to construct stories designed to hide their deviance.

Another form of stigma management, sometimes adopted when concealment fails, involves disclosing the deviance. People may do this for cathartic reasons (alleviating their burden of secrecy), therapeutic reasons (casting it in a positive light), or preventive reasons (so others don't find out in negative ways later, so that efforts at developing relationships are not totally wasted). Although many people find that their disclosures lead to rejection, others are more fortunate.

Disclosures of deviance can follow two courses. In observing the interactions between deviants and normals, Fred Davis (1961) noted that some nondeviant people engage in "*deviance disavowal*." That is, they go through a process of normalizing their relationship with the deviant person largely through failing to acknowledge the deviant trait. Although this disavowal usually begins with a conspicuous and stilted ignoring of the individual's deviance, it progresses through a stage where more relaxed interaction begins, directed at features of the person other than his or her deviant stigma, and finally gets to the point where the deviant stigma is overlooked and almost forgotten.

In contrast, deviant people can strive to normalize their relationships with nondeviants through "*deviance avowal*" (Turner 1972), in which they openly acknowledge their stigma and try to present themselves in a positive light. This avowal often takes the form of humor, "breaking the ice" by joking about their deviant attribute. In this way they show others that they can take the perspective of the normal and see themselves as deviant too, thus forming a bridge to others. This action further asserts that they have nondeviant aspects and they can see the world as others do.

Nancy Herman's "Return to Sender: Reintegrative Stigma-Management Strategies of Ex-Psychiatric Patients" discusses individuals' efforts to manage stigma by addressing some of the problems encountered by former mental patients after their release from treatment. She poignantly describes individuals' painful decisions about whether to conceal or disclose their stigmatizing past and some of the factors on which they base their decisions. She then explores the ways in which they try to manage their disclosures, including some of the difficulties they encounter. While some of their confidants prove sympathetic, others react poorly, rejecting and restigmatizing the former patients. In a foreshadowing of the next section, Herman also discusses ways in which the individuals she studied collectively organized to fight and manage their deviant stigma.

Kent Sandstrom's reading, "Confronting Deadly Disease: The Drama of Identity Construction among Gay Men with AIDS," illustrates some of the dynamics associated with the processes of interpersonal labeling and stigma management. AIDS patients progressively "personalize the illness" and realize the extent of their social stigma through their interactions with doctors, potential sexual partners, friends, and co-workers. From being firmly embedded in previous social anchorages, they suddenly find themselves *liminal* people, floating in between groups and statuses but rejected by all. This status leads to feelings of alienation, forcing them to creatively reshape their new identities. Sandstrom also addresses the way in which AIDS patients try to conceal their deviant stigma through passing, covering, isolation, and insulation and some of the ways in which they seek normalization through disclosure and deviance avowal.

COLLECTIVE STIGMA MANAGEMENT

Thus far we have considered individual modes of adaptation to a deviant stigma. Yet these stigmata can also be managed through a group, or collective, effort. Many voluntary associations of stigmatized individuals exist, from the early organization of prostitutes (COYOTE—Call Off Your Old Tired Ethics), to more recent ones such as the Gay Liberation Front, the Little People of America, the National Stuttering Project, and the Gray Panthers. Most well known are the 12-step programs modeled after the tremendous success of Alcoholics Anonymous (AA), including such groups as Overeaters Anonymous, Narcotics Anonymous, and Gamblers Anonymous.

These groups vary in character. Some are organized as what Stanford Lyman (1970) has called *expressive* groups, whose primary function is to provide support for their members. This support can take the form of organizing social and recreational activities, dispersing legal or medical information, or offering services such as shopping, meals, or transportation. Expressive groups tend to be apolitical, helping their members adapt to their social stigma rather than evade it. They also serve as places where members can come together in the company of other deviants, avoid the censure of nonstigmatized normals, and seek collective solutions to their common problems. It is here that they can make disclosures to others without fear of rejection.

Lyman has also described *instrumental* groups, where members come together not only to accomplish the expressive functions but also to organize for political activism. Building on Lemert's (1967) terms, Kitsuse (1980) has called this activity *tertiary deviation*, in which individuals reject the societal conception and treatment of their stigma and organize to change social definitions. They fight to get others to modify their views of the status or behavior in question so that society, like them, will no longer regard it as deviant. Examples of such groups include ACT UP, an AIDS organization whose members have tried to change social attitudes toward AIDS patients, the National Organization for Women, and the Disabled in Action.

David Karp's "Illness Ambiguity and the Search for Meaning: A Case Study of a Self-Help Group for Affective Disorders" offers us an excellent example of an expressive, collective stigma management group. Based on the AA model, this group of manic-depressives met weekly to discuss ways of coping with their illness and its accompanying stigma. Through concrete suggestion and emotional empathy, they helped one another navigate the complex and often painful pathway through diagnosis, forging relationships with physicians and psychiatrists, and dealing with the positive and negative aspects of their experiences with therapeutic drugs. Renee Anspach looks at more instrumental groups in "From Stigma to Identity Politics: Political Activism among the Physically Disabled and Former Mental Patients." She discusses how her subjects rejected both societal values and the negative self-concept associated with the deviant stigma through their collective organization, and she examines the role of political activism among a range of other options for deviant groups.

17

Divorce and Stigma

NAOMI GERSTEL

By most accounts, tolerance of variation in family life has increased dramatically in the United States. Public opinion polls over the last two decades reveal declining disapproval of extended singlehood (Veroff et al., 1981), premarital sex and pregnancy (Gerstel, 1982), employment of mothers with young children (Cherlin, 1981), and voluntary childlessness (Huber and Spitze, 1983). Divorce resembles these other situations; in fact public tolerance of divorce appears to have increased especially dramatically over the last few decades (Veroff et al., 1981).

In the mid-1950's, Goode (1956:10) could still observe: "We know that in our own society, divorce has been a possible, but disapproved, solution for marital conflict." However, comparing attitudes in 1958 and 1971, McRae (1978) found an increasing proportion of adults believing that divorce was only "sometimes wrong" while a decreasing proportion felt that it was "always wrong." These data, he claimed, indicated attitudes toward divorce had shifted "from moral absolutism to situational ethics" (1978:228). In an analysis of panel data collected between 1960 and 1980, Thornton (1985) found that changes in attitudes toward divorce were not only large but pervasive: all subgroups—whether defined by age, class, or even religion—showed substantial declines in disapproval of marital separation.

What are the implications of declining disapproval of divorce? In historical perspective, it is clear that the divorced are no longer subject to the moral outrage they encountered centuries, or even decades, ago. Certainly, divorce is no longer treated as a sin calling for repressive punishment, as it was in theological doctrine and practice (be it Catholic or Protestant) until the beginning of the twentieth century (Halem, 1980; O'Neil, 1967). In electing a divorced president and many divorced senators and governors, U.S. citizens seem to have repudiated the idea that divorce is grounds for exclusion from public life. With the recent passage of no-fault divorce laws in every state, U.S. courts no longer insist on attributing wrongdoing to one party to a divorce (Weitzman, 1985).

Most recent commentators on a divorce even argue that it is no longer stigmatized. For example, Spanier and Thompson (1984:15) claim that "the social stigma associated with divorce has disappeared" and Weitzman (1981:146) suggests that "the decline in the social stigma traditionally attached

From: "Divorce and Stigma," Naomi Gerstel, *Social Problems*, Vol. 34, No. 2, April 1987.

to divorce is one of the most striking changes in the social climate surrounding divorce."

However, I argue in this paper that the stigma attached to divorce has disappeared in only two very limited senses. First, although other studies have
shown a clear decline in disapproval of divorce as a general category, disapproval of divorced individuals persists contingent on the specific conditions of
their divorce. Thus, as I show below, some divorced people experience disapproval and at least one party to a divorce often feels blamed.

Second, while many of the formal, institutional controls on divorce—
imposed in the public realm of church or state—have weakened, the individual who divorces suffers informal, relational sanctions. These are the interpersonal controls that emerge more or less spontaneously in social life. I will
present evidence indicating that the divorced believe the married often
exclude them and that the divorced themselves frequently pull toward, yet
devalue, others who divorce.

In these two senses, I argue that the divorced are still subject to the same
social processes and evaluations associated with stigmatization more generally.
As in Goffman's (1963:3) classic formulation—which stresses both the condi-
tional and relational aspects of stigma—my findings suggest that the divorced
come to be seen (and to see themselves) as "of a less desired kind . . . reduced
in our minds from a whole and usual person to a tainted, discounted one."

METHODS

My data come from interviews with 104 separated and divorced respondents:
52 women and 52 men. Based on a conception of marital dissolution as a
process rather than a static life event, the research team sampled respondents in
different stages of divorce: one-third of the respondents were separated less
than one year; one-third separated one to two years; one-third separated two
or more years. To obtain respondents, we could not rely on court records
alone, for most couples who have filed for a divorce have already been living
apart for at least a year. Thus, 61 percent of the respondents were selected from
probate court records in two counties in the Northeast; the others came from
referrals.[1] Comparisons between the court cases and referred respondents show
no statistically significant differences on demographic characteristics.

Sample Characteristics

In contrast to the samples in most previous research on separation and divorce,
the respondents are a heterogeneous group. They include people in the working class as well as in the middle class whose household incomes ranged from
under $4,000 to over $50,000, with a median of $18,000 (with women's sig-
nificantly lower than men's.)[2] Levels of education varied widely: about one-
fourth had less than a high school degree, and slightly less than one-fourth had
four or more years of college. The sample also includes significant numbers
whose primary source of income came from public assistance and from manu-

al, clerical, and professional jobs. Only 11 percent were not currently employed while another 9 percent were working part-time. The median age of the respondents was 33 years, and the mean number of years married was nine. Finally, 30 percent of the sample had no children, 19 percent had one child, and 51 percent had more than one child.

FINDINGS

Social Exclusion: Rejection by the Married

Partners to a divorce not only split friends; they are often excluded from social interaction with the married more generally.[3] Many ex-husbands and ex-wives found they could not maintain friendships with married couples: about one-half of both men (43 percent) and women (58 percent) agreed with the statement: "Married couples don't want to see me now." Moreover, less than one-fourth (23 percent) of the women and men agreed: "I am as close to my married friends as when I was married."[4] By getting a divorce, then, they became marginal to at least part of the community on which they had previously relied.

One man summed up the views of many when he spoke of the "normal life" of the married.

One of the things I recognized not long after I was separated is that this is a couple's world. People do things in couples, normally (C043, male).

Remembering his own marriage, he now recognized its impact:

We mostly went out with couples. I now have little or no contact with them.

Discovering "they don't invite me anymore" or "they never call," many of the divorced felt rejected:

The couples we shared our life with, uh, I'm an outsider now. They stay away. Not being invited to a lot of parties that we was always invited to. It's with males and females. It sucks (C030, male).

The divorced developed explanations for their exclusion. Finding themselves outsiders, some simply thought that their very presence destabilized the social life of couples: "I guess I threaten the balance" (N004, female). They found themselves social misfits in that world, using terms like "a third wheel" (N010, female; N027, male) and "odd person out" (N006, male) to describe their newly precarious relationship with the married.

Some went further, suggesting that those still in couples felt threatened by the divorce or were afraid it would harm their own marriages. "They say, 'My God, it's happening all over.' It scares them" (N019, male). Men and women expressed this form of rejection in terms of "contagion" (C027, male) and "a fear it's going to rub off on them" (N010, female).[5] Because the difficulties of marriage are often concealed, others found their divorce came as a surprise to married friends. That surprise reinforced the idea that "it can happen to any-one" and "so they tend to stay away" (C027, male).

A few turned the explanation around, suspecting that married couples rejected them out of jealousy rather than fear. One woman, speaking of a friend who no longer called, explained: "It was like me living out her fan-tasies" (N008, female). A salesman in his mid-40s, who had an affair before getting a divorce, believed:

> I get a kick out of it because . . . I am the envy of both men and women because, some of it is courage, others look upon it as freedom. Both words have been used a number of times. People become very envious, and a spouse of the envious person will feel extremely threatened (N043, male).

These few could turn an unpleasant experience into an enviable one. For a small minority, then, the experience of exclusion did not produce a sense of devaluation.

But more of the divorced, men as well as women, were troubled by the thought that old friends now defined them in terms of their sexual availability and, as a result, avoided them. One woman, a teacher's aide with a very young daughter and son at home, felt insulted that friends misconstrued her situation:

> Well, I now have no married friends. It's as if I all of a sudden became single and I'm going to chase after their husbands (N011, female).

And a plumbing contractor, unusual because he had custody of his five chil-dren, was particularly hurt by the image of sexual availability because he had resisted any sexual entanglements. Describing one woman who "couldn't understand why I didn't want to hop into bed with her," he noted "she told me there must be something wrong with me" (C047, male). He went on:

> I would say couples in general, there seems to be a, well, they are nice to me, but distant. I think the men don't want a single man around their wives.

And when asked, "Can you tell me about that?," he associated his seeming sexual availability with a threat to the cohesion of the community:

> (They) don't really want to involve me in things that are going on . . . neighborhood picnics . . . couples' things. . . . I have had men say that they figure that I'm out chasing women all over. So I'm considered some-what of an unstable person.

He added that he was not the only person who had reached this conclusion:

> And this is quite common with divorced people. In group discussions, everyone seems to experience the same thing.

As his final comment indicates, the divorced talk to each other about this experience and generate a shared explanation for it—that they are viewed as somewhat "unstable." Thus, some divorced people come to believe that married acquaintances saw them as "misfits"—unstable individuals who could not maintain a stable marriage, a threat to the routines of a community made up of the "normal" married.[6]

The exclusion of the divorced from the social life they had enjoyed while married constitutes a negative sanction on divorce. This is not simply a functional process of friendship formation based on homogamy (cf. Lazarsfeld and Merton, 1964): it involves conflict, producing a sense of devaluation on the part of one group (the divorced) who feel rejected by another group still considered normal (the married).[7]

The divorced try to come to terms with their experience by talking to others who share it. Together they develop a shared understanding similar to what Goffman (1963:5) calls a "stigma theory": the married feel uncomfortable, even threatened by them, and act as if divorce, as a "social disease," is contagious. Or divorce poses a threat because of the desired freedom and sexuality it (perhaps falsely) represents. Finally, divorced people mutually develop a broader explanation for the modern response to them: they are avoided because the dissolution of marriage is so common, so possible, that it becomes a real threat both to any given couple and to the social world built on, and routinized by, groups consisting of couples.

Colleagues and Demoralization

The separation of the divorced from the married is even more clearly apparent in the social life developed by the divorced themselves. The divorced pull away from the married and into the lives of others like them. Goffman (1963:18) argues that the stigmatized turn to others like them in anticipation that "mixed social contact will make for anxious, unanchored interaction." Accordingly, many of the divorced said they felt "uncomfortable" (C042, female), "strained" (C003, male; N014 and C029, females), "strange," (C034, female), and "awkward" (C019, male) in a world composed of couples. And some abandoned the married: "I've been pulling away from my coupled friends" (C026, female).

Drawing together with other divorced, they develop as well as share their "sad tales" (Goffman, 1963:19) and learn how to behave. In fact, over half of the people with whom these divorced men (52 percent) and women (62 percent) socialized were other divorced individuals, a far higher proportion than is found in the general population (U.S. Bureau of the Census, 1983). The divorced used many well-worn phrases to talk about others who shared their marital status: "birds of a feather flock together" (N021, male) and "likes attract likes" (N025, female).

Many discovered their interests and concerns, at least for a time, were based on their newfound marital status. When asked to respond to the statement, "I have more in common with singles now," over half of the men (55 percent) and women (58 percent) agreed. One 28-year-old working-class man sought out divorced people for the same reason many respondents avoided the married. He said of others who shared his marital status:

> We have something in common. We have almost right off the bat something to talk about. It makes it easier to talk because you have gone through it (C018, male).

Equally important, respondents often felt they could turn to other divorced people as experienced "veterans" (Caplan, 1974) and "colleagues" (Best and Luckenbill, 1980) for their newfound marital state. These others served as role models, showing them ways to cope as spouseless adults and, in doing so, bolstered their new identities. A man referred to other divorced as providing an "experience bank" and elaborated by saying:

> I solicited assistance, guidance from people who had gone through or who were going through similar experiences. So I might get a better understanding of how they reacted to it (C043, male).

So, too, other divorced people could help them do what Hochschild (1983:254) has called "feeling work" or "the shaping, modulating, or inducing of feelings." That is, colleagues encouraged them to manage and change their feelings, and to realize how their marriage (or many marriages) were not as good as they had thought. That helped them disengage from their ex-spouses.

But these colleagues could do something more. They showed them that the life of a divorced adult was not all anguish and pain. One 30-year-old middle-class woman found she felt closest to her old friends who had been through divorce because:

> I could talk to them and they helped me to talk about my problems. What went wrong. I felt they could understand. And it was interesting to hear what they had been through, you know (C016, female).

Importantly, she added:

> I wanted to hear what it was like. It sorta helped me to think that they had made a go of their life again. They were happy afterwards.

The divorced needed reassurance that divorce did not imply a serious character flaw. They needed to find those who, after getting a divorce, had made a successful transition. A 32-year-old man, who in the first month "isolated" himself "for fear that people might think I'm doing something wrong," found a few months later that:

> It's nice to hear people who have gone through the same experience, talking to me about it. It's nice to hear a lot of these things because you realize: "Hey, I'm not so bad. I'm not the only one this happens to." And looking at the person and seeing that they made it okay. And that I will, too (N009, male).

What Goffman (1963:20) wrote more generally about the stigmatized, then, characterizes the modern divorced: "They can provide instruction for tricks of the trade and a circle of lament to which he can withdraw for moral support and comfort of feeling at home, at ease, accepted as a person who is really like any normal person." The divorced turn to others like themselves to get reassurance, advice, and encouragement, and to make sense of their often dislocated lives. Telling their story to those like themselves is therapeutic (Conrad and Schneider, 1980).

Yet, there are also pitfalls in this attraction to others like them. As time passes, the divorced may find themselves bored by constant discussion of divorce, that "the whole matter of focusing on atrocity tales . . . on the 'problem' is one of the largest penalties for having one" (Goffman, 1963:21). As a man divorced close to two years put it:

> You see, I've been locked in too much with divorced people. You're relating to them relative to the separation. And what happened to you, when you did it, you know (N019, male).

The divorced, especially those who had been separated more than a year, spoke of how they were getting tired of "problems dominating the conversation" (C042, female, divorced three years) as they felt "dissipated" (C009, male, divorced two years) and "wanted to talk to people about something different" (N016, female, divorced a year-and-a-half) because "the less you talk about the problems, the less you think about them, and the less you feel about them" (N028, female, divorced a year-and-a-half). One man described how, in the first months of divorce, he had "learned a lot about divorce" from others who had the same experience because "they understood me." But then he went on to complain about the problems with this association: "You feel up and then someone drags you down with their problems" (C021, male). . . .

The Devaluation of Self

Perhaps the most striking evidence that the divorced devalue their own condition is found in their assessment of organizations established for the divorced. Only 10 percent of the respondents were in such groups.[8] In fact, most of the divorced—male as well as female—explicitly rejected such formal mechanisms of integration set up by and for others like them.

For the relatively small number of people who did join, such groups provided both a source of entertainment for their children as well as an opportunity to meet other adults. However, in explaining why they joined, the divorced typically stressed child care. Thus, children were not simply a reason for joining; they provided legitimation for membership. By explaining membership in instrumental, rather than expressive terms, and in terms of children rather than themselves, the divorced distanced themselves from the potentially damaging implications of membership for their own identity. In this sense, children provide a "face-saving device," much like those inventoried by Berk (1977) among people who attended single dances.

The notion that groups for the divorced—and therefore those who join them—are stigmatized is substantiated still further by the comments of those who did not join. They gave a number of reasons for their reluctance. Some attributed their lack of participation to a lack of knowledge. Others simply felt they did not have the time or energy. When asked why she had not joined any divorce group, one 25-year-old saleswoman said:

> I've thought about it, but I have just never done anything about it. I know it is not getting me anywhere by not doing anything. Basically I am a lazy person (C024, female).

But while she first blamed herself for non-participation in these groups, she then went on to add a more critical note: "I think I would feel funny walking into a place like that." Her second thought reiterated a common theme—an attitude toward divorce and membership in organizations for them—which came through with compelling force. Many imagined that people who joined such groups were unacceptable in a variety of ways, or even that to join them was somehow a sign of weakness. For example:

> These people really don't have somebody to turn to. I guess that's the main reason for them belonging and I do have someone to turn to, matter of fact, more than one. They're really not sure of themselves; they're insecure (N013, female).

Such comments reveal that respondents saw divorce as a discredit, at least insofar as it became the axis of one's social life. Consequently, to join such groups was to reinforce the very devaluation they hoped to avoid. One welfare mother with a young child had been told by her social worker that joining a singles' group might alleviate the enormous loneliness she experienced. But she resisted:

> It's kind of degrading to me or something. Not that I'm putting these other people down. I could join something but I couldn't join something that was actually called a singles' group (N016, female).

Others reiterated the same theme. To them, groups of the divorced were "rejects looking, you know, going after rejects. They need a crutch" (N011, female). Or they asked rhetorically, "Is that for the very, very lonely?" (C016, female). These to them were "people with as many, if not worse, problems than I have" (C040, male) or "weirdos" (N029, female). As these comments show, the divorced were quick to put a pejorative label on groups consisting of other divorced.

Or, the divorced we spoke to felt that such groups were unacceptable because they were sexual marketplaces. In the words of a 28-year-old plant supervisor, who (like most others) had never actually been to any organization for the divorced:

> I refuse to go to a place where I'm looked at as a side of beef and women are looked at as sides of beef. It disgusts me (C040, male).

. Association with such groups would reinforce the very view that so many of the divorced work so hard to dispel. While their rejection of such groups is a way to separate themselves from a stigmatized status (Berk, 1977:542), the very character and strength of the rejection confirms that the status is stigmatized. Thus, in distancing themselves, the divorced reveal that they share the belief that individuals who divorce, especially if they use that divorce to organize their social worlds, continue to be somehow tainted.

Despite their negative reaction to divorce groups, respondents did not reject all organized routes to friendship formation. The majority (82 percent) were members of at least one group—including, for example, sports, cultural, religious and service groups. Women participated in a median of 2.24 groups; men, a median of 3.56. In fact, many spoke of joining these other groups as a way to "make friends" and to cope with the loneliness they felt. Such groups may provide access to others who are divorced, but only coincidentally. These organizational memberships—and the relationships they allow—are legitimized by their *dissociation* from marital status. It is in this context that respondents' resistance to joining divorce groups becomes especially compelling as evidence for their devaluation of the status of divorce.

CONCLUSION

To argue that the divorced are no longer stigmatized is to misunderstand their experience. To be sure, divorce is now less deviant in a statistical sense than it was a decade ago. As a group, the divorced are not categorized as sinful, criminal, or even wrong. Moreover, even though the divorced lose married friends and have smaller networks than the married, they do not become completely ghettoized into subcultures of the divorced (Gerstel et al., 1985; Weiss, 1979). Finally, . . . the divorced themselves do not think that most of their kin and friends disapprove.

However, a decrease in statistical deviance, a relaxation of institutional controls by church or state, or a decline in categorical disapproval is not the same as the absence of stigmatization. Although a majority of Americans claim they are indifferent in principle to those who make a "personal decision" to leave a "bad" marriage, this indifference does not carry over into the social construction of private lives. The divorced believe they are the targets of informal relational sanctions—exclusion, blame, and devaluation. If we understand stigma as referring not simply to the realm of public sanctions but rather see it as emerging out of everyday experience, then we can see that the divorced continue to be stigmatized. . . .

NOTES

1. A small number of these referrals were located through a "snowball" strategy: various people who heard about our study told us about individuals who had just separated. But the majority of referrals were located

through respondents. At the end of each interview, we asked for the names of other people who had been separated less than a year and interviewed a maximum of one person named by each respondent.

2. Women's mean household income was $14,000 while the men's mean income was $22,000.

3. Using a variety of methods and samples, a number of other studies also find that the divorced lose married friends (e.g., Spanier and Thompson, 1984; Wallerstein and Kelly, 1980; Weiss, 1979).

4. In contrast to my finding that divorced men and women were equally likely to experience a certain distance from the married, Hetherington et. al. (1976:422) found "dissociation from married friends was greater for women than for men." However, they studied only divorced parents of children in nursery school. Fischer's (1982) findings would lead us to believe that, of any group, mothers—married or divorced—of young children are most isolated. In fact, my data suggest that the gender differences may well characterize only this very special group. Among male respondents, the presence of children is not significantly associated with their belief that "married couples don't want to see them" or their feeling that they "are not as close to those couples." In contrast, for women, the presence of any children, especially young children, increases disassociation from the married. Among female respondents, the correlation between "married couples don't want to see me now" and the presence of any children is .25 (p < .05) and with having children less than 12-years-old is .39 (p < .01). This difference between women and men may well be a result of the fact that women obtain custody of the children far more often than do men. While Hetherington's findings are often cited as evidence for general gender differences in the social life of the divorced, this implication probably should be limited to this special group—the parents of young children.

5. This belief that others see their "condition" as "contagious" is similar to the experience of others who are stigmatized, like the mentally ill. See Foucault (1967) for a discussion of the development of the belief that "unreason" is contagious, that anyone could "catch it," and the consequent movement to isolate the insane.

6. Wallerstein and Kelly (1980:33) hypothesize that still another factor may explain why the married move away from the divorced: the married "feel uncomfortable and inadequate in providing solace." While this is certainly a possible (and generous) explanation, the divorced nonetheless experience the loss of friendship as rejection and exclusion.

7. To be sure, research on "single individuals"—be they widows (Lopata, 1979), never married (Stein, 1981), or divorced—suggests there is a general pattern of friendship based on homogeneity of marital status. Indeed,

as Simmel (1950) points out in his classic work, the triad is a more unstable group than the dyad. Hence, a "third party" is likely to be excluded. Here, I am suggesting that such third parties, especially the divorced, are likely to interpret the separation of marital groups as exclusion and hence as devaluation.

8. These findings suggest, of course, that those studies which draw entirely on members of singles' groups are seriously flawed: they represent a small and atypical population of the divorced.

REFERENCES

Berk, Bernard. 1977. "Face saving at the singles dance." *Social Problems* 24:530–44.

Best, Joel, and David Luckenbill. 1980. "The social organization of deviants." *Social Problems* 28:14–31.

Caplan, Gerald. 1974. *Social Supports and Community Mental Health.* New York: Behavioral Publications.

Cherlin, Andrew. 1981. *Marriage, Divorce and Remarriage.* Cambridge, MA: Harvard University Press.

Conrad, Peter, and Joseph W. Schneider. 1980. *Deviance and Medicalization: From Badness to Sickness.* St. Louis: C.V. Mosby.

Fischer, Claude. 1982. *To Dwell Among Friends.* Chicago: University of Chicago Press.

Foucault, Michel. 1967. *Madness and Civilization.* London: Tavistock.

Gerstel, Naomi. 1982. "The new right and the family." Pp. 6–20 in Barbara Haber (ed.), *The Woman's Annual.* New York: G.K. Hall.

Gerstel, Naomi, Catherine Kohler Riessman, and Sarah Rosenfield. 1985. "Explaining the symptomatology of separated and divorced women and men: the role of material resources and social networks." *Social Forces* 64:84–101.

Goffman, Erving. 1963. *Stigma.* Englewood Cliffs, NJ: Prentice-Hall.

Goode, William. 1956. *Women in Divorce.* New York: Free Press.

Halem, Lynne Carol. 1980. *Divorce Reform.* New York: Free Press.

Hetherington, E. M., M. Cox, and R. Cox. 1976. "Divorced fathers." *The Family Coordinator* 25:417–28.

Hochschild, Arlie Russell. 1983. "Attending to, codifying and managing feelings: sex differences in love." Pp. 250–62 in Laurel Richardson and Verta Taylor (eds.). *Feminist Frontiers.* Reading, MA: Addison–Wesley.

Huber, Joan, and Glenna Spitze. 1983. *Stratification: Children, Housework, and Jobs.* New York: Academic Press.

Lopata, Helena. 1979. *Women as Widows*. New York: Elsevier.

McRae, James A. 1978. "The secularization of divorce." Pp. 227–42 in Beverly Duncan and Otis Dudley Duncan (eds.), *Sex Typing and Sex Roles*. New York: Academic Press.

O'Neil, William L. 1967. *Divorce in the Progressive Era*. New Haven: Yale University Press.

Simmel, Georg. 1950. *The Sociology of Georg Simmel*. Kurt H. Wolff, translator. New York: Free Press.

Spanier, Graham, and Linda Thompson. 1984. *Parting*. Beverly Hills, CA: Sage.

Stein, Peter (ed.). 1981. *Single Life*. Englewood Cliffs, NJ: Prentice–Hall.

Thornton, Arland. 1985. "Changing attitudes toward separation and divorce: causes and consequences." *American Journal of Sociology* 90:856–72.

U.S. Bureau of Census. 1983. "Marital status and living arrangements: March, 1983." Current Population Reports, Series P-20, #389. Washington, DC: U.S. Government Printing Office.

Veroff, Joseph, Elizabeth Douvan, and Richard A. Kulka. 1981. *The Inner American: A Self-Portrait from 1957–1976*. New York: Basic Books.

Wallerstein, Judith S., and Joan B. Kelly. 1980. *Surviving the Breakup*. New York: Basic Books.

Weiss, Robert. 1979. *Going It Alone*. New York: Basic Books.

Weitzman, Lenore. 1981. *The Marriage Contract*. New York: Free Press.

———. 1985. *The Divorce Revolution: The Unexpected Social and Economic Consequences for Women and Children in America*. New York: Free Press.

18

Anorexia Nervosa and Bulimia:
The Development of Deviant Identities

PENELOPE A. McLORG

AND DIANE E. TAUB

INTRODUCTION

Current appearance norms stipulate thinness for women and muscularity for men; these expectations, like any norms, entail rewards for compliance and negative sanctions for violations. Fear of being overweight—of being visually deviant—has led to a striving for thinness, especially among women. In the extreme, this avoidance of overweight engenders eating disorders, which themselves constitute deviance. Anorexia nervosa, or purposeful starvation, embodies visual as well as behavioral deviation; bulimia, binge-eating followed by vomiting and/or laxative abuse, is primarily behaviorally deviant.

Besides a fear of fatness, anorexics and bulimics exhibit distorted body images. In anorexia nervosa, a 20-25 percent loss of initial body weight occurs, resulting from self-starvation alone or in combination with excessive exercising, occasional binge-eating, vomiting and/or laxative abuse. Bulimia denotes cyclical (daily, weekly, for example) binge-eating followed by vomiting or laxative abuse; weight is normal or close to normal (Humphries et al., 1982). Common physical manifestations of these eating disorders include menstrual cessation or irregularities and electrolyte imbalances; among behavioral traits are depression, obsessions/compulsions, and anxiety (Russell, 1979; Thompson and Schwartz, 1982).

Increasingly prevalent in the past two decades, anorexia nervosa and bulimia have emerged as major health and social problems. Termed an epidemic on college campuses (Brody, as quoted in Schur, 1984:76), bulimia affects 13% of college students (Halmi et al., 1981). Less prevalent, anorexia nervosa was diagnosed in 0.6% of students utilizing a university health center (Stangler and Printz, 1980). However, the overall mortality rate of anorexia nervosa is 6% (Schwartz and Thompson, 1981) to 20% (Humphries et al., 1982); bulimia appears to be less life-threatening (Russell, 1979).

From: "Anorexia Nervosa and Bulimia: The Development of Deviant Identities,"
Penelope A. McLorg and Diane E. Taub, *Deviant Behavior,* Vol. 8, 1987. Reprinted by
permission of Taylor and Francis Inc. All rights reserved.

Particularly affecting certain demographic groups, eating disorders are most prevalent among young, white, affluent (upper-middle to upper class) women in modern, industrialized countries (Crisp, 1977; Willi and Grossman, 1983). Combining all of these risk factors (female sex, youth, high socioeconomic status, and residence in an industrialized country), prevalence of anorexia nervosa in upper class English girls' schools is reported at 1 in 100 (Crisp et al., 1976). The age of onset for anorexia nervosa is bimodal at 14.5 and 18 years (Humphries et al., 1982); the most frequent age of onset for bulimia is 18 (Russell, 1979).

Eating disorders have primarily been studied from psychological and medical perspectives.[1] Theories of etiology have generally fallen into three categories: the ego psychological (involving an impaired child-maternal environment); the family systems (implicating enmeshed, rigid families); and the endocrinological (involving a precipating hormonal defect). Although relatively ignored in previous studies, the sociocultural components of anorexia nervosa and bulimia (the slimness norm and its agents of reinforcement, such as role models) have been postulated as accounting for the recent, dramatic increases in these disorders (Schwartz et al., 1982; Boskind-White, 1985).[2]

Medical and psychological approaches to anorexia nervosa and bulimia obscure the social facets of the disorders and neglect the individuals' own definitions of their situations. Among the social processes involved in the development of an eating disorder is the sequence of conforming behavior, primary deviance, and secondary deviance. Societal reaction is the critical mediator affecting the movement through the deviant career (Becker, 1973). Within a framework of labeling theory, this study focuses on the emergence of anorexic and bulimic identities, as well as on the consequences of being career deviants.

METHODOLOGY

Sampling and Procedures

Most research on eating disorders has utilized clinical subjects or non-clinical respondents completing questionnaires. Such studies can be criticized for simply counting and describing behaviors and/or neglecting the social construction of the disorders. Moreover, the work of clinicians is often limited by therapeutic orientation. Previous research may also have included individuals who were not in therapy on their own volition and who resisted admitting that they had an eating disorder.

Past studies thus disregard the intersubjective meanings respondents attach to their behavior and emphasize researchers' criteria for definition as anorexic or bulimic. In order to supplement these sampling and procedural designs, the present study utilizes participant observation of a group of self-defined anorexics and bulimics.[3] As the individuals had acknowledged their eating disorders, frank discussion and disclosure were facilitated.

Data are derived from a self-help group, BANISH, Bulimics/Anorexics In Self-Help, which met at a university in an urban center of the mid-South. Founded by one of the researchers (D.E.T.), BANISH was advertised in local newspapers as offering a group experience for individuals who were anorexic or bulimic. Despite the local advertisements, the campus location of the meetings may have selectively encouraged university students to attend. Nonetheless, in view of the modal age of onset and socioeconomic status of individuals with eating disorders, college students have been considered target populations (Crisp et al., 1976; Halmi et al., 1981).

The group's weekly two-hour meetings were observed for two years. During the course of this study, thirty individuals attended at least one of the meetings. Attendance at meetings was varied: ten individuals came nearly every Sunday; five attended approximately twice a month; and the remaining fifteen participated once a month or less frequently, often when their eating problems were "more severe" or "bizarre." The modal number of members at meetings was twelve. The diversity in attendance was to be expected in self-help groups of anorexics and bulimics.

> . . . most people's involvement will not be forever or even a long time. Most people get the support they need and drop out. Some take the time to help others after they themselves have been helped but even they may withdraw after a time. It is a natural and in many cases *necessary* process (emphasis in original) (American Anorexia/Bulimia Association, 1983).

Modeled after Alcoholics Anonymous, BANISH allowed participants to discuss their backgrounds and experiences with others who empathized. For many members, the group constituted their only source of help; these respondents were reluctant to contact health professionals because of shame, embarrassment, or financial difficulties.

In addition to field notes from group meetings, records of other encounters with all members were maintained. Participants visited the office of one of the researchers (D.E.T.), called both researchers by phone, and invited them to their homes or out for a cup of coffee. Such interaction facilitated genuine communication and mutual trust. Even among the fifteen individuals who did not attend the meetings regularly, contact was maintained with ten members on a monthly basis.

Supplementing field notes were informal interviews with fifteen group members, lasting from two to four hours. Because they appeared to represent more extensive experience with eating disorders, these interviewees were chosen to amplify their comments about the labeling process, made during group meetings. Conducted near the end of the two-year observation period, the interviews focused on what the respondents thought antedated and maintained their eating disorders. In addition, participants described others' reactions to their behaviors as well as their own interpretations of these reactions.

To protect the confidentiality of individuals quoted in the study, pseudonyms are employed.

Description of Members

The demographic composite of the sample typifies what has been found in other studies (Fox and James, 1976; Crisp, 1977; Herzog, 1982; Schlesier-Stropp, 1984). Group members' ages ranged from nineteen to thirty-six, with the modal age being twenty-one. The respondents were white, and all but one were female. The sole male and three of the females were anorexic; the remaining females were bulimic.[4]

Primarily composed of college students, the group included four non-students, three of whom had college degrees. Nearly all members derived from upper-middle or lower-upper class households. Eighteen students and two non-students were never-marrieds and uninvolved in serious relationships; two non-students were married (one with two children); two students were divorced (one with two children); and six students were involved in serious relationships. The duration of eating disorders ranged from three to fifteen years.

CONFORMING BEHAVIOR

In the backgrounds of most anorexics and bulimics, dieting figures prominently, beginning in the teen years (Crisp, 1977; Johnson et al., 1982; Lacey et al., 1986). As dieters, these individuals are conformist in their adherence to the cultural norms emphasizing thinness (Garner et al., 1980; Schwartz et al., 1982). In our society, slim bodies are regarded as the most worthy and attractive; overweight is viewed as physically and morally unhealthy—"obscene," "lazy," "slothful," and "gluttonous" (DeJong, 1980; Ritenbaugh, 1982; Schwartz et al., 1982).

Among the agents of socialization promoting the slimness norm is advertising. Female models in newspaper, magazine, and television advertisements are uniformly slender. In addition, product names and slogans exploit the thin orientation; examples include "Ultra Slim Lipstick," "Miller Lite," and "Virginia Slims." While retaining pressures toward thinness, an Ayds commercial attempts a compromise for those wanting to savor food: "Ayds . . . so you can taste, chew, and enjoy, while you lose weight." Appealing particularly to women, a nationwide fast-food restaurant chain offers low-calorie selections, so individuals can have a "license to eat." In the latter two examples, the notion of enjoying food is combined with the message to be slim. Food and restaurant advertisements overall convey the pleasures of eating, whereas advertisements for other products, such as fashions and diet aids, reinforce the idea that fatness is undesirable.

Emphasis on being slim affects everyone in our culture, but it influences women especially because of society's traditional emphasis on women's appearance. The slimness norm and its concomitant narrow beauty standards exacerbate the objectification of women (Schur, 1984). Women view themselves as visual entities and recognize that conforming to appearance expectations and "becoming attractive object[s] [are] role obligation[s]" (Laws, as quoted in Schur, 1984:66). Demonstrating the beauty motivation behind dieting, a recent Nielson survey indicated that of the 56 percent of all women aged 24 to 54 who dieted during the previous year, 76 percent did so for cosmetic, rather than health, reasons (Schwartz et al., 1982). For most female group members, dieting was viewed as a means of gaining attractiveness and appeal to the opposite sex. The male respondent, as well, indicated that "when I was fat, girls didn't look at me, but when I got thinner, I was suddenly popular."

In addition to responding to the specter of obesity, individuals who develop anorexia nervosa and bulimia are conformist in their strong commitment to other conventional norms and goals. They consistently excel at school and work (Russell, 1979; Bruch, 1981; Humphries et al., 1982), maintaining high aspirations in both areas (Theander, 1970; Lacey et al., 1986). Group members generally completed college-preparatory courses in high school, aware from an early age that they would strive for a college degree. Also, in college as well as high school, respondents joined honor societies and academic clubs.

Moreover, pre-anorexics and -bulimics display notable conventionality as "model children" (Humphries et al., 1982:199), "the pride and joy" of their parents (Bruch, 1981:215), accommodating themselves to the wishes of others. Parents of these individuals emphasize conformity and value achievement (Bruch, 1981). Respondents felt that perfect or near-perfect grades were expected of them; however, good grades were not rewarded by parents, because "A's" were common for these children. In addition, their parents suppressed conflicts, to preserve the image of the "all-American family" (Humphries et al., 1982). Group members reported that they seldom, if ever, heard their parents argue or raise their voices.

Also conformist in their affective ties, individuals who develop anorexia nervosa and bulimia are strongly, even excessively, attached to their parents. Respondents' families appeared close-knit, demonstrating palpable emotional ties. Several group members, for example, reported habitually calling home at prescribed times, whether or not they had any news. Such families have been termed "enmeshed" and "overprotective," displaying intense interaction and concern for members' welfare (Minuchin et al., 1978; Selvini-Palazzoli, 1978). These qualities could be viewed as marked conformity to the norm of familial closeness.[5]

Another element of notable conformity in the family milieu of pre-anorexics and -bulimics concerns eating, body weight/shape, and exercising (Kalucy et al., 1977; Humphries et al., 1982). Respondents reported their fathers' preoccupation with exercising and their mothers' engrossment in food preparation. When group members dieted and lost weight, they received an extraordinary amount of approval. Among the family, body size became a matter of "friendly rivalry." One bulimic informant recalled that she, her mother, and her coed sister all strived to wear a size 5, regardless of their heights and body frames. Subsequent to this study, the researchers learned that both the mother and sister had become bulimic.

As pre-anorexics and -bulimics, group members thus exhibited marked conformity to cultural norms of thinness, achievement, compliance, and parental attachment. Their families reinforced their conformity by adherence to norms of family closeness and weight/body shape consciousness.

PRIMARY DEVIANCE

Even with familial encouragement, respondents, like nearly all dieters (Chernin, 1981), failed to maintain their lowered weights. Many cited their lack of willpower to eat only restricted foods. For the emerging anorexics and bulimics, extremes such as purposeful starvation or binging accompanied by vomiting and/or laxative abuse appeared as "obvious solutions" to the problem of retaining weight loss. Associated with these behaviors was a regained feeling of control in lives that had been disrupted by a major crisis. Group members' extreme weight-loss efforts operated as coping mechanisms for entering college, leaving home, or feeling rejected by the opposite sex.

The primary inducement for both eating adaptations was the drive for slimness: with slimness came more self-respect and a feeling of superiority over "unsuccessful dieters." Brian, for example, experienced a "power trip" upon consistent weight loss through starvation. Binges allowed the purging respondents to cope with stress through eating while maintaining a slim appearance. As former strict dieters, Teresa and Jennifer used binging/purging as an alternative to the constant self-denial of starvation. Acknowledging their parents' desires for them to be slim, most respondents still felt it was a conscious choice on their part to continue extreme weight-loss efforts. Being thin became the "most important thing" in their lives—their "greatest ambition."

In explaining the development of an anorexic or bulimic identity, Lemert's (1951; 1967) concept of primary deviance is salient. Primary deviance refers to a transitory period of norm violations which do not affect an individual's self-concept or performance of social roles. Although respondents were exhibiting anorexic or bulimic behavior, they did not consider themselves to be anorexic or bulimic.

At first, anorexics' significant others complimented their weight loss, expounding on their new "sleekness" and "good looks." Branch and Eurman (1980:631) also found anorexics' families and friends describing them as "well-groomed," "neat," "fashionable," and "victorious." Not until the respondents approached emaciation did some parents or friends become concerned and withdraw their praise. Significant others also became increasingly aware of the anorexics' compulsive exercising, preoccupation with food preparation (but not consumption), and ritualistic eating patterns (such as cutting food into minute pieces and eating only certain foods at prescribed times).

For bulimics, friends or family members began to question how the respondents could eat such large amounts of food (often in excess of 10,000 calories a day) and stay slim. Significant others also noticed calluses across the bulimics' hands, which were caused by repeated inducement of vomiting.

Several bulimics were "caught in the act," bent over commodes. Generally, friends and family required substantial evidence before believing that the respondents' binging or purging was no longer sporadic.

SECONDARY DEVIANCE

Heightened awareness of group members' eating behavior ultimately led others to label the respondents "anorexic" or "bulimic." Respondents differed in their histories of being labeled and accepting the labels. Generally first termed anorexic by friends, family, or medical personnel, the anorexics initially vigorously denied the label. They felt they were not "anorexic enough," not skinny enough; Robin did not regard herself as having the "skeletal" appearance she associated with anorexia nervosa. These group members found it difficult to differentiate between socially approved modes of weight loss—eating less and exercising more—and the extremes of those behaviors. In fact, many of their activities—cheerleading, modeling, gymnastics, aerobics—reinforced their pursuit of thinness. Like other anorexics, Chris felt she was being "ultra-healthy," with "total control" over her body.

For several respondents, admitting they were anorexic followed the realization that their lives were disrupted by their eating disorder. Anorexics' inflexible eating patterns unsettled family meals and holiday gatherings. Their regimented lifestyle of compulsively scheduled activities—exercising, school, and meals—precluded any spontaneous social interactions. Realization of their adverse behaviors preceded the anorexics' acknowledgment of their subnormal body weight and size.

Contrasting with anorexics, the binge/purgers, when confronted, more readily admitted that they were bulimic and that their means of weight loss was "abnormal." Teresa, for example, knew "very well" that her bulimic behavior was "wrong and unhealthy," although "worth the physical risks." While the bulimics initially maintained that their purging was only a temporary weight-loss method, they eventually realized that their disorder represented a "loss of control." Although these respondents regretted the self-indulgence, "shame," and "wasted time," they acknowledged their growing dependence on binging/purging for weight management and stress regulation.

The application of anorexic or bulimic labels precipitated secondary deviance, wherein group members internalized these identities. Secondary deviance refers to norm violations which are a response to society's labeling: "secondary deviation . . . becomes a means of social defense, attack or adaptation to the overt and covert problems created by the societal reaction to primary deviance" (Lemert, 1967:17). In contrast to primary deviance, secondary deviance is generally prolonged, alters the individual's self-concept, and affects the performance of his/her social roles.

As secondary deviants, respondents felt that their disorders "gave a purpose" to their lives. Nicole resisted attaining a normal weight because it was not "her"—she accepted her anorexic weight as her "true" weight. For

Teresa, bulimia became a "companion"; and Julie felt "every aspect of her life," including time management and social activities, was affected by her bulimia. Group members' eating disorders became the salient element of their self-concepts, so that they related to familiar people and new acquaintances as anorexics or bulimics. For example, respondents regularly compared their body shapes and sizes with those of others. They also became sensitized to comments about their appearance, whether or not the remarks were made by someone aware of their eating disorder.

With their behavior increasingly attuned to their eating disorders, group members exhibited role engulfment (Schur, 1971). Through accepting anorexic or bulimic identities, individuals centered activities around their deviant role, downgrading other social roles. Their obligations as students, family members, and friends became subordinate to their eating and exercising rituals. Socializing, for example, was gradually curtailed because it interfered with compulsive exercising, binging, or purging.

Labeled anorexic or bulimic, respondents were ascribed a new status with a different set of role expectations. Regardless of other positions the individuals occupied, their deviant status, or master status (Hughes, 1958; Becker, 1973), was identified before all others. Among group members, Nicole, who was known as the "school's brain," became known as the "school's anorexic." No longer viewed as conforming model individuals, some respondents were termed "starving waifs" or "pigs."

Because of their identities as deviants, anorexics' and bulimics' interactions with others were altered. Group members' eating habits were scrutinized by friends and family and used as a "catch-all" for everything negative that happened to them. Respondents felt self-conscious around individuals who knew of their disorders; for example, Robin imagined people "watching and whispering" behind her. In addition, group members believed others expected them to "act" anorexic or bulimic. Friends of some anorexic group members never offered them food or drink, assuming continued disinterest on the respondents' part. While being hospitalized, Denise felt she had to prove to others she was not still vomiting, by keeping her bathroom door open. Other bulimics, who lived in dormitories, were hesitant to use the restroom for normal purposes lest several friends be huddling at the door, listening for vomiting. In general, individuals interacted with the respondents largely on the basis of their eating disorder; in doing so, they reinforced anorexic and bulimic behaviors.

Bulimic respondents, whose weight-loss behavior was not generally detectable from their appearance, tried earnestly to hide their bulimia by binging and purging in secret. Their main purpose in concealment was to avoid the negative consequences of being known as a bulimic. For these individuals, bulimia connoted a "cop-out": like "weak anorexics," bulimics pursued thinness but yielded to urges to eat. Respondents felt other people regarded bulimia as "gross" and had little sympathy for the sufferer. To avoid these stigmas or "spoiled identities," the bulimics shrouded their behaviors.

Distinguishing types of stigma, Goffman (1963) describes discredited (visible) stigmas and discreditable (invisible) stigmas. Bulimics, whose weight was approximately normal or even slightly elevated, harbored discreditable stigmas. Anorexics, on the other hand, suffered both discreditable and discredited stigmas—the latter due to their emaciated appearance. Certain anorexics were more reconciled than the bulimics to their stigmas: for Brian, the "stigma of anorexia was better than the stigma of being fat." Common to the stigmatized individuals was an inability to interact spontaneously with others. Respondents were constantly on guard against topics of eating and body size.

Both anorexics and bulimics were held responsible by others for their behavior and presumed able to "get out of it if they tried." Many anorexics reported being told to "just eat more," while bulimics were enjoined to simply "stop eating so much." Such appeals were made without regard for the complexities of the problem. Ostracized by certain friends and family members, anorexics and bulimics felt increasingly isolated. For respondents, the self-help group presented a non-threatening forum for discussing their disorders. Here, they found mutual understanding, empathy, and support. Many participants viewed BANISH as a haven from stigmatization by "others."

Group members, as secondary deviants, thus endured negative consequences, such as stigmatization, from being labeled. As they internalized the labels anorexic or bulimic, individuals' self-concepts were significantly influenced. When others interacted with the respondents on the basis of their eating disorders, anorexic or bulimic identities were encouraged. Moreover, group members' efforts to counteract the deviant labels were thwarted by their master statuses.

DISCUSSION

Previous research on eating disorders has dwelt almost exclusively on medical and psychological facets. Although necessary for a comprehensive understanding of anorexia nervosa and bulimia, these approaches neglect the social processes involved. The phenomena of eating disorders transcend concrete disease entities and clinical diagnoses. Multifaceted and complex, anorexia nervosa and bulimia require a holistic research design, in which sociological insights must be included.

A limitation of medical/psychiatric studies, in particular, is researchers' use of a priori criteria in establishing salient variables. Rather than utilizing predetermined standards of inclusion, the present study allows respondents to construct their own reality. Concomitant to this innovative approach to eating disorders is the selection of a sample of self-admitted anorexics and bulimics. Individuals' perceptions of what it means to become anorexic or bulimic are explored. Although based on a small sample, findings can be used to guide researchers in other settings.

With only five to ten percent of reported cases appearing in males (Crisp, 1977; Stangler and Printz, 1980), eating disorders are primarily a women's aberrance. The deviance of anorexia nervosa and bulimia is rooted in the visual objectification of women and attendant slimness norm. Indeed, purposeful starvation and binging/purging reinforce the notion that "a society gets the deviance it deserves" (Schur, 1979:71). As recently noted (Schur, 1984), the sociology of deviance has generally bypassed systematic studies of women's norm violations. Like male deviants, females endure label applications, internalizations, and fulfillments.

The social processes involved in developing anorexic or bulimic identities comprise the sequence of conforming behavior, primary deviance, and secondary deviance. With a background of exceptional adherence to conventional norms, especially the striving for thinness, respondents subsequently exhibit the primary deviance of starving or binging/purging. Societal reaction to these behaviors leads to secondary deviance, wherein respondents' self-concepts and master statuses become anorexic or bulimic. Within this framework of labeling theory, the persistence of eating disorders, as well as the effects of stigmatization, are elucidated.

Although during the course of this research some respondents alleviated their symptoms through psychiatric help or hospital treatment programs, no one was labeled "cured." An anorexic is considered recovered when weight is normal for two years; a bulimic is termed recovered after being symptom-free for one and one-half years (American Anorexia/Bulimia Association Newsletter, 1985). Thus deviance disavowal (Schur, 1971), or efforts after normalization to counteract the deviant labels, remains a topic for future exploration.

NOTES

1. Although instructive, an integration of the medical, psychological, and sociocultural perspectives on eating disorders is beyond the scope of this paper.

2. Exceptions to the neglect of sociocultural factors are discussions of sex-role socialization in the development of eating disorders. Anorexics' girlish appearance has been interpreted as a rejection of femininity and womanhood (Orbach, 1979; Bruch, 1981; Orbach, 1985). In contrast, bulimics have been characterized as over-conforming to traditional female sex roles (Boskind-Lodahl, 1976).

3. Although a group experience for self-defined bulimics has been reported (Boskind-Lodahl, 1976), the researcher, from the outset, focused on Gestalt and behaviorist techniques within a feminist orientation.

4. One explanation for fewer anorexics than bulimics in the sample is that, in the general population, anorexics are outnumbered by bulimics at 8 or 10 to 1 (Lawson, as reprinted in American Anorexia/Bulimia Association Newsletter, 1985:1). The proportion of bulimics to anorexics in the sample is 6.5 to 1. In addition, compared to bulimics, anorexics may be less likely to attend a self-help group as they have a greater tendency to deny

the existence of an eating problem (Humphries et al., 1982). However, the four anorexics in the present study were among the members who attended the meetings most often.

5. Interactions in the families of anorexics and bulimics might seem deviant in being inordinately close. However, in the larger societal context, the family members epitomize the norms of family cohesiveness. Perhaps unusual in their occurrence, these families are still within the realm of conformity. Humphries and colleagues (1982:202) refer to the "highly enmeshed and protective" family as part of the "idealized family myth."

REFERENCES

American Anorexia/Bulimia Association. 1983. Correspondence. April.

American Anorexia/Bulimia Association Newsletter. 1985. 8(3).

Becker, Howard S. 1973. *Outsiders*. New York: Free Press.

Boskind-Lodahl, Marlene. 1976. "Cinderella's stepsisters: A feminist perspective on anorexia nervosa and bulimia." *Signs, Journal of Women in Culture and Society* 2:342–56.

Boskind-White, Marlene. 1985. "Bulimarexia: A sociocultural perspective." Pp. 113–26 in S. W. Emmett (ed.), *Theory and Treatment of Anorexia Nervosa and Bulimia: Biomedical, Sociocultural and Psychological Perspectives*. New York: Brunner/Mazel.

Branch, C. H. Hardin, and Linda J. Eurman. 1980. "Social attitudes toward patients with anorexia nervosa." *American Journal of Psychiatry* 137:631–32.

Bruch, Hilde. 1981. "Developmental considerations of anorexia nervosa and obesity." *Canadian Journal of Psychiatry* 26:212–16.

Chernin, Kim. 1981. *The Obsession: Reflections on the Tyranny of Slenderness*. New York: Harper and Row.

Crisp, A. H. 1977. "The prevalence of anorexia nervosa and some of its associations in the general population." *Advances in Psychosomatic Medicine* 9:38–47.

Crisp, A. H., R. L. Palmer, and R. S. Kalucy. 1976. "How common is anorexia nervosa? A prevalence study." *British Journal of Psychiatry* 128:549–54.

DeJong, William. 1980. "The stigma of obesity: The consequences of naive assumptions concerning the causes of physical deviance." *Journal of Health and Social Behavior* 21:75–87.

Fox, K. C. and N. McI. James. 1976. "Anorexia nervosa: A study of 44 strictly defined cases." *New Zealand Medical Journal* 84:309–12.

Garner, David M., Paul E. Garfinkel, Donald Schwartz, and Michael Thompson. 1980. "Cultural expectations of thinness in women." *Psychological Reports* 47:483–91.

Goffman, Erving. 1963. *Stigma*. Englewood Cliffs, NJ: Prentice-Hall.

Halmi, Katherine A., James R. Falk, and Estelle Schwartz. 1981. "Binge-eating and vomiting: A survey of a college population." *Psychological Medicine* 11:697–706.

Herzog, David B. 1982. "Bulimia: The secretive syndrome." *Psychosomatics* 23:481–83.

Hughes, Everett C. 1958. *Men and Their Work*. New York: Free Press.

Humphries, Laurie L., Sylvia Wrobel, and H. Thomas Wiegert. 1982. "Anorexia nervosa." *American Family Physician* 26:199–204.

Johnson, Craig L., Marilyn K. Stuckey, Linda D. Lewis, and Donald M. Schwartz. 1982. "Bulimia: A descriptive survey of 316 cases." *International Journal of Eating Disorders* 2(1):3–16.

Kalucy, R. S., A. H. Crisp, and Britta Harding. 1977. "A study of 56 families with anorexia nervosa." *British Journal of Medical Psychology* 50:381–95.

Lacey, Hubert J., Sian Coker, and S. A. Birtchnell. 1986. "Bulimia: Factors associated with its etiology and maintenance." *International Journal of Eating Disorders* 5:475–87.

Lemert, Edwin M. 1951. *Social Pathology*. New York: McGraw-Hill.

———. 1967. *Human Deviance, Social Problems and Social Control*. Englewood Cliffs, NJ: Prentice-Hall.

Minuchin, Salvador, Bernice L. Rosman, and Lester Baker. 1978. *Psychosomatic Families: Anorexia Nervosa in Context*. Cambridge, MA: Harvard University Press.

Orbach, Susie. 1979. *Fat is a Feminist Issue*. New York: Berkeley.

———. 1985. "Visibility/invisibility: Social considerations in anorexia nervosa—a feminist perspective." Pp. 127–38 in S. W. Emmett (ed.), *Theory and Treatment of Anorexia Nervosa and Bulimia: Biomedical, Sociocultural, and Psychological Perspectives*. New York: Brunner/Mazel.

Ritenbaugh, Cheryl. 1982. "Obesity as a culture-bound syndrome." *Culture, Medicine and Psychiatry* 6:347–61.

Russell, Gerald. 1979. "Bulimia nervosa: An ominous variant of anorexia nervosa." *Psychological Medicine* 9:429–48.

Schlesier-Stropp, Barbara. 1984. "Bulimia: A review of the literature." *Psychological Bulletin* 95:247–57.

Schur, Edwin M. 1971. *Labeling Deviant Behavior*. New York: Harper and Row.

———. 1979. *Interpreting Deviance: A Sociological Introduction*. New York: Harper and Row.

———. 1984. *Labeling Women Deviant: Gender, Stigma, and Social Control*. New York: Random House.

Schwartz, Donald M., and Michael G. Thompson. 1981. "Do anorectics get well? Current research and future needs." *American Journal of Psychiatry* 138:319–23.

Schwartz, Donald M., Michael G. Thompson, and Craig L. Johnson. 1982. "Anorexia nervosa and bulimia: The socio-cultural context." *International Journal of Eating Disorders* 1(3):20–36.

Selvini-Palazzoli, Mara. 1978. *Self-Starvation: From Individual to Family Therapy in the Treatment of Anorexia Nervosa.* New York: Jason Aronson.

Stangler, Ronnie S., and Adolph M. Printz. 1980. "DSM-III: Psychiatric diagnosis in a university population." *American Journal of Psychiatry* 137:937–40.

Theander, Sten. 1970. "Anorexia nervosa." *Acta Psychiatrica Scandinavica Supplement* 214:24–31.

Thompson, Michael G., and Donald M. Schwartz. 1982. "Life adjustment of women with anorexia nervosa and anorexic-like behavior." *International Journal of Eating Disorders* 1(2):47–60.

Willi, Jurg, and Samuel Grossmann. 1983. "Epidemiology of anorexia nervosa in a defined region of Switzerland." *American Journal of Psychiatry* 140:564–67.

19

Convicted Rapists' Vocabulary of Motive: Excuses and Justifications

DIANA SCULLY

AND JOSEPH MAROLLA

Psychiatry has dominated the literature on rapists since "irresistible impulse" (Glueck, 1925:243) and "disease of the mind" (Glueck, 1925:243) were introduced as the causes of rape. Research has been based on small samples of men, frequently the clinicians' own patient population. Not surprisingly, the medical model has predominated: rape is viewed as

From: "Convicted Rapists' Vocabulary of Motive: Excuses and Justifications," Diana Scully and Joseph Marolla, *Social Problems,* Vol. 31, No. 5, 1984. © 1984 Society for the Study of Social Problems. Reprinted by permission of University of California Press Journals and Diana Scully.

an individualistic, idiosyncratic symptom of a disordered personality. That is, rape is assumed to be a psychopathologic problem and individual rapists are assumed to be "sick." However, advocates of this model have been unable to isolate a typical or even predictable pattern of symptoms that are causally liked to rape. Additionally, research has demonstrated that fewer than 5 percent of rapists were psychotic at the time of their rape (Abel et al., 1980).

We view rape as behavior learned socially through interaction with others; convicted rapists have learned the attitudes and actions consistent with sexual aggression against women. Learning also includes the acquisition of culturally derived vocabularies of motive, which can be used to diminish responsibility and to negotiate a non-deviant identity.

Sociologists have long noted that people can, and do, commit acts they define as wrong and, having done so, engage various techniques to disavow deviance and present themselves as normal. Through the concept of "vocabulary of motive," Mills (1940:904) was among the first to shed light on this seemingly perplexing contradiction. Wrong-doers attempt to reinterpret their actions through the use of a linguistic device by which norm-breaking conduct is socially interpreted. That is, anticipating the negative consequences of their behavior, wrong-doers attempt to present the act in terms that are both culturally appropriate and acceptable.

Following Mills, a number of sociologists have focused on the types of techniques employed by actors in problematic situations (Hall and Hewitt, 1970; Hewitt and Hall, 1973; Hewitt and Stokes, 1975; Sykes and Matza, 1957). Scott and Lyman (1968) describe excuses and justifications, linguistic "accounts" that explain and remove culpability for an untoward act after it has been committed. *Excuses* admit the act was bad or inappropriate but deny full responsibility, often through appeals to accident, or biological drive, or through scapegoating. In contrast, *justifications* accept responsibility for the act but deny that it was wrong—that is, they show in this situation the act was appropriate. *Accounts* are socially approved vocabularies that neutralize an act or its consequences and are always a manifestation of an underlying negotiation of identity.

Stokes and Hewitt (1976:837) use the term "aligning actions" to refer to those tactics and techniques used by actors when some feature of a situation is problematic. Stated simply, the concept refers to an actor's attempt, through various means, to bring his or her conduct into alignment with culture. Culture in this sense is conceptualized as a "set of cognitive constraints—objects—to which people must relate as they form lines of conduct" (1976:837), and includes physical constraints, expectations and definitions of others, and personal biography. Carrying out aligning actions implies both awareness of those elements of normative culture that are applicable to the deviant act and, in addition, an actual effort to bring the act into line with this awareness. The result is that deviant behavior is legitimized.

This paper presents an analysis of interviews we conducted with a sample of 114 convicted, incarcerated rapists. We use the concept of accounts (Scott and Lyman, 1968) as a tool to organize and analyze the vocabularies of motive which this group of rapists used to explain themselves and their actions. An analysis of their accounts demonstrates how it was possible for 83 percent (n = 114)[1] of these convicted rapists to view themselves as non-rapists.

When rapists' accounts are examined, a typology emerges that consists of admitters and deniers. Admitters (n = 47) acknowledged that they had forced sexual acts on their victims and defined the behavior as rape. In contrast, deniers[2] either eschewed sexual contact or all association with the victim (n = 35),[3] or admitted to sexual acts but did not define their behavior as rape (n = 32). . . . By and large, the deniers used justifications while the admitters used excuses. In some cases, both groups relied on the same themes, stereotypes, and images: some admitters, like most deniers, claimed that women enjoyed being raped. Some deniers excused their behavior by referring to alcohol or drug use, although they did so quite differently than admitters. Through these narrative accounts, we explore convicted rapists' own perceptions of their crimes. . . .

JUSTIFYING RAPE

Deniers attempted to justify their behavior by presenting the victim in a light that made her appear culpable, regardless of their own actions. Five themes run through attempts to justify their rapes: (1) women as seductresses; (2) women mean "yes" when they say "no"; (3) most women eventually relax and enjoy it; (4) nice girls don't get raped; and (5) guilty of a minor wrongdoing.

1) Women as Seductresses

Men who rape need not search far for cultural language which supports the premise that women provoke or are responsible for rape. In addition to common cultural stereotypes, the fields of psychiatry and criminology (particularly the subfield of victimology) have traditionally provided justifications for rape, often by portraying raped women as the victims of their own seduction (Albin, 1977; Marolla and Scully, 1979). For example, Hollander (1924:130) argues:

> Considering the amount of illicit intercourse, rape of women is very rare indeed. Flirtation and provocative conduct, i.e. tacit (if not actual) consent is generally the prelude to intercourse.

Since women are supposed to be coy about their sexual availability, refusal to comply with a man's sexual demands lacks meaning and rape appears normal. The fact that violence and, often, a weapon are used to accomplish the rape is not considered. As an example, Abrahamsen (1960:61) writes:

> The conscious or unconscious biological or psychological attraction between man and woman does not exist only on the part of the offender toward the woman but, also, on her part toward him, which in many instances may, to some extent, be the impetus for his sexual attack. Often a women [sic] unconsciously wishes to be taken by force—consider the theft of the bride in Peer Gynt.

Like Peer Gynt, the deniers we interviewed tried to demonstrate that their victims were willing and, in some cases, enthusiastic participants. In these accounts, the rape became more dependent upon the victim's behavior than upon their own actions.

Thirty-one percent (n = 10) of the deniers presented an extreme view of the victim. Not only willing, she was the aggressor, a seductress who lured them, unsuspecting, into sexual action. Typical was a denier convicted of his first rape and accompanying crimes of burglary, sodomy, and abduction. According to the pre-sentence reports, he had broken into the victim's house and raped her at knife point. While he admitted to the breaking and entry, which he claimed was for altruistic purposes ("to pay for the prenatal care of a friend's girlfriend"), he also argued that when the victim discovered him, he had tried to leave but she had asked him to stay. Telling him that she cheated on her husband, she had voluntarily removed her clothes and seduced him. She was, according to him, an exemplary sex partner who "enjoyed it very much and asked for oral sex.[4] Can I have it now?" he reported her as saying. He claimed they had spent hours in bed, after which the victim had told him he was good looking and asked to see him again. "Who would believe I'd meet a fellow like this?" he reported her as saying.

In addition to this extreme group, 25 percent (n = 8) of the deniers said the victim was willing and had made some sexual advances. An additional 9 percent (n = 3) said the victim was willing to have sex for money or drugs. In two of these three cases, the victim had been either an acquaintance or picked up, which the rapists said led them to expect sex.

2) Women Mean "Yes" When They Say "No"

Thirty-four percent (n = 11) of the deniers described their victim as unwilling, at least initially, indicating either that she had resisted or that she had said no. Despite this, and even though (according to pre-sentence reports) a weapon had been present in 64 percent (n = 7) of these 11 cases, the rapists justified their behavior by arguing that either the victim had not resisted enough or that her "no" had really meant "yes." For example, one denier who was serving time for a previous rape was subsequently convicted of attempting to rape a prison hospital nurse. He insisted he had actually completed the second rape, and said of his victim: "She semi-struggled but deep down inside I think she felt it was a fantasy come true." The nurse, according to him, had asked a question about his conviction for rape, which he interpreted as teasing. "It was like she was saying, 'rape me'." Further, he stated that she had helped him along with oral sex and "from her actions, she was enjoying it." In another case, a 34-year-old man convicted of abducting and raping a 15-year-old teenager at knife point as she walked on the beach, claimed it was a pickup. This rapist said women like to be overpowered before sex, but to dominate after it begins.

A man's body is like a Coke bottle, shake it up, put your thumb over the opening and feel the tension. When you take a woman out, woo her, then she says "no, I'm a nice girl," you have to use force. All men do this. She said "no" but it was a societal no, she wanted to be coaxed. All women say "no" when they mean "yes" but it's a societal no, so they won't have to feel responsible later.

Claims that the victim didn't resist or, if she did, didn't resist enough, were also used by 24 percent (n = 11) of admitters to explain why, during the incident, they believed the victim was willing and that they were not raping. These rapists didn't redefine their acts until some time after the crime. For example, an admitter who used a bayonet to threaten his victim, an employee of the store he had been robbing, stated:

> At the time I didn't think it was rape. I just asked her nicely and she didn't resist. I never considered prison. I just felt like I had met a friend. It took about five years of reading and going to school to change my mind about whether it was rape. I became familiar with the subtlety of violence. But at the time, I believed that as long as I didn't hurt anyone it wasn't wrong. At the time, I didn't think I would go to prison, I thought I would beat it.

Another typical case involved a gang rape in which the victim was abducted at knife point as she walked home about midnight. According to two of the rapists, both of whom were interviewed, at the time they had thought the victim had willingly accepted a ride from the third rapist (who was not interviewed). They claimed the victim didn't resist and one reported her as saying she would do anything if they would take her home. In this rapist's view, "She acted like she enjoyed it, but maybe she was just acting. She wasn't crying, she was engaging in it." He reported that she had been friendly to the rapist who abducted her and, claiming not to have a home phone, she gave him her office number—a tactic eventually used to catch the three. In retrospect, this young man had decided, "She was scared and just relaxed and enjoyed it to avoid getting hurt." Note, however, that while he had redefined the act as rape, he continued to believe she enjoyed it.

Men who claimed to have been unaware that they were raping viewed sexual aggression as a man's prerogative at the time of the rape. Thus they regarded their act as little more than a minor wrongdoing even though most possessed or used a weapon. As long as the victim survived without major physical injury, from their perspective, a rape had not taken place. Indeed, even U.S. courts have often taken the position that physical injury is a necessary ingredient for a rape conviction.

3) Most Women Eventually Relax and Enjoy It

Many of the rapists expected us to accept the image, drawn from cultural stereotype, that once the rape began, the victim relaxed and enjoyed it.[5] Indeed, 69 percent (n = 22) of deniers justified their behavior by claiming not only that the victim was willing, but also that she enjoyed herself, in some cases to an immense degree. Several men suggested that they had fulfilled their victims' dreams. Additionally, while most admitters used adjectives such as "dirty," "humiliated," and "disgusted" to describe how they thought rape made women feel, 20 percent (n = 9) believed that their victim enjoyed herself. For example, one denier had posed as a salesman to gain entry to his victim's house. But he claimed he had had a previous sexual relationship with the victim, that she agreed to have sex for drugs, and that the opportunity to have sex with him produced "a glow, because she was really into oral stuff and fascinated by the idea of sex with a black man. She felt satisfied, fulfilled, wanted me to stay, but I didn't want her." In another case, a denier who had broken into his victim's house but who insisted the victim was his lover and let him in voluntarily, declared "She felt good, kept kissing me and wanted me to stay the night. She felt proud after sex with me." And another denier, who had hid in his victim's closet and later attacked her while she slept, argued that while she was scared at first, "once we got into it, she was ok." He continued to believe he hadn't committed rape because "she enjoyed it and it was like she consented."

4) Nice Girls Don't Get Raped

The belief that "nice girls don't get raped" affects perception of fault. The victim's reputation, as well as characteristics or behavior which violate normative sex role expectations, are perceived as contributing to the commission of the crime. For example, Nelson and Amir (1975) defined hitchhike rape as a victim-precipitated offense.

In our study, 69 percent (n = 22) of deniers and 22 percent (n = 10) of admitters referred to their victims' sexual reputation, thereby evoking the stereotype that "nice girls don't get raped." They claimed that the victim was known to have been a prostitute, or a "loose" woman, or to have had a lot of affairs, or to have given birth to a child out of wedlock. For example, a denier who claimed he had picked up his victim while she was hitchhiking stated, "To be honest, we [his family] knew she was a damn whore and whether she screwed one or 50 guys didn't matter." According to pre-sentence reports this victim didn't know her attacker and he abducted her at knife point from the street. In another case, a denier who claimed to have known his victim by reputation stated:

> If you wanted drugs or a quick piece of ass, she would do it. In court she said she was a virgin, but I could tell during sex [rape] that she was very experienced.

When other types of discrediting biographical information were added to these sexual slurs, a total of 78 percent (n = 25) of the deniers used the victim's reputation to substantiate their accounts. Most frequently, they referred to the victim's emotional state or drug use. For example, one denier claimed his victim had been known to be loose and, additionally, had turned state's evidence against her husband to put him in prison and save herself from a burglary conviction. Further, he asserted that she had met her current boyfriend, who was himself in and out of prison, in a drug rehabilitation center where they were both clients.

Evoking the stereotype that women provoke rape by the way they dress, a description of the victim as seductively attired appeared in the accounts of 22 percent (n = 7) of deniers and 17 percent (n = 8) of admitters. Typically, these descriptions were used to substantiate their claims about the victim's reputation. Some men went to extremes to paint a tarnished picture of the victim, describing her as dressed in tight black clothes and without a bra; in one case, the victim was portrayed as sexually provocative in dress and carriage. Not only did she wear short skirts, but she was observed to "spread her legs while getting out of cars." Not all of the men attempted to assassinate their victim's reputation with equal vengeance. Numerous times they made subtle and off-hand remarks like, "She was a waitress and you know how they are."

The intent of these discrediting statements is clear. Deniers argued that the woman was a "legitimate" victim who got what she deserved. For example, one denier stated that all of his victims had been prostitutes; pre-sentence reports indicated they were not. Several times during his interview, he referred to them as "dirty sluts," and argued "anything I did to them was justified." Deniers also claimed their victim had wrongly accused them and was the type of woman who would perjure herself in court.

5) Only a Minor Wrongdoing

The majority of deniers did not claim to be completely innocent and they also accepted some accountability for their actions. Only 16 percent (n = 5) of deniers argued that they were totally free of blame. Instead, the majority of deniers pleaded guilty to a lesser charge. That is, they obfuscated the rape by pleading guilty to a less serious, more acceptable charge. They accepted being over-sexed, accused of poor judgment or trickery, even some violence, or guilty of adultery or contributing to the delinquency of a minor, charges that are hardly the equivalent of rape.

Typical of this reasoning is a denier who met his victim in a bar when the bartender asked him if he would try to repair her stalled car. After attempting unsuccessfully, he claimed the victim drank with him and later accepted a ride. Out riding, he pulled into a deserted area "to see how my luck would go." When the victim resisted his advances, he beat her and he stated:

> I did something stupid. I pulled a knife on her and I hit her as hard as I would hit a man. But I shouldn't be in prison for what I did. I shouldn't have all this time [sentence] for going to bed with a broad.

This rapist continued to believe that while the knife was wrong, his sexual behavior was justified.

In another case, the denier claimed he picked up his under-age victim at a party and that she voluntarily went with him to a motel. According to pre-sentence reports, the victim had been abducted at knife point from a party. He explained:

After I paid for a motel, she would have to have sex but I wouldn't use a weapon. I would have explained. I spent money and, if she still said no, I would have forced her. If it had happened that way, it would have been rape to some people but not to my way of thinking. I've done that kind of thing before. I'm guilty of sex and contributing to the delinquency of a minor, but not rape.

In sum, deniers argued that, while their behavior may not have been completely proper, it should not have been considered rape. To accomplish this, they attempted to discredit and blame the victim while presenting their own actions as justified in the context. Not surprisingly, none of the deniers thought of himself as a rapist. A minority of the admitters attempted to lessen the impact of their crime by claiming the victim enjoyed being raped. But despite this similarity, the nature and tone of admitters' and deniers' accounts were essentially different.

EXCUSING RAPE

In stark contrast to deniers, admitters regarded their behavior as morally wrong and beyond justification. They blamed themselves rather than the victim, although some continued to cling to the belief that the victim had contributed to the crime somewhat, for example, by not resisting enough.

Several of the admitters expressed the view that rape was an act of such moral outrage that it was unforgivable. Several admitters broke into tears at intervals during their interviews. A typical sentiment was,

> I equate rape with someone throwing you up against a wall and tearing your liver and guts out of you. . . . Rape is worse than murder . . . and I'm disgusting.

Another young admitter frequently referred to himself as repulsive and confided:

> I'm in here for rape and in my own mind, it's the most disgusting crime, sickening. When people see me and know, I get sick.

Admitters tried to explain their crime in a way that allowed them to retain a semblance of moral integrity. Thus, in contrast to deniers' justifications, admitters used excuses to explain how they were compelled to rape. These excuses appealed to the existence of forces outside of the rapists' control. Through the use of excuses, they attempted to demonstrate that either intent was absent or responsibility was diminished. This allowed them to admit rape while reducing the threat to their identity as a moral person. Excuses also permitted them to view their behavior as idiosyncratic rather than typical and, thus, to believe they were not "really" rapists. Three themes run through these accounts: (1) the use of alcohol and drugs; (2) emotional problems; and (3) nice guy image.

1) The Use of Alcohol and Drugs

A number of studies have noted a high incidence of alcohol and drug consumption by convicted rapists prior to their crime (Groth, 1979; Queen's Bench Foundation, 1976). However, more recent research has tentatively concluded that the connection between substance use and crime is not as direct as previously thought (Ladouceur, 1983). Another facet of alcohol and drug use mentioned in the literature is its utility in disavowing deviance. McCaghy (1968) found that child molesters used alcohol as a technique for neutralizing their deviant identity. Marolla and Scully (1979), in a review of psychiatric literature, demonstrated how alcohol consumption is applied differently as a vocabulary of motive. Rapists can use alcohol both as an excuse for their behavior and to discredit the victim and make her more responsible. We found the former common among admitters and the latter common among deniers.

Alcohol and/or drugs were mentioned in the accounts of 77 percent (n = 30) of admitters and 84 percent (n = 21) of deniers and both groups were equally likely to have acknowledged consuming a substance—admitters, 77 percent (n = 30); deniers, 72 percent (n = 18). However, admitters said they had been affected by the substance; if not the cause of their behavior, it was at least a contributing factor. For example, an admitter who estimated his consumption to have been eight beers and four "hits of acid" reported:

> Straight, I don't have the guts to rape. I could fight a man but not that. To say, "I'm going to do it to a woman," knowing it will scare and hurt her, takes guts or you have to be sick.

Another admitter believed that his alcohol and drug use,

> . . . brought out what was already there but in such intensity it was uncontrollable. Feelings of being dominant, powerful, using someone for my own gratification, all rose to the surface.

In contrast, deniers' justifications required that they not be substantially impaired. To say that they had been drunk or high would cast doubt on their ability to control themselves or to remember events as they actually happened. Consistent with this, when we asked if the alcohol and/or drugs had had an effect on their behavior, 69 percent (n = 27) of admitters, but only 40 percent (n = 10) of deniers, said they had been affected.

Even more interesting were references to the victim's alcohol and/or drug use. Since admitters had already relieved themselves of responsibility through claims of being drunk or high, they had nothing to gain from the assertion that the victim had used or been affected by alcohol and/or drugs. On the other hand, it was very much in the interest of deniers to declare that their victim had been intoxicated or high: that fact lessened her credibility and made her more responsible for the act. Reflecting these observations, 72 percent (n = 18) of deniers and 26 percent (n = 10) of admitters maintained that

TABLE 1 *Rapists' Accounts of Own and Victims' Alcohol and/or Drug (A/D) Use and Effect*

	Admitters n = 39 %	Deniers n = 25 %
Neither Self nor Victim Used A/D	23	16
Self Used A/D	77	72
Of Self Used, No Victim Use	51	12
Self Affected By A/D	69	40
Of Self Affected, No Victim Use or Affect	54	24
Self A/D Users Who Were Affected	90	56
Victim Used A/D	26	72
Of Victim Used, No Self Use	0	0
Victim Affected By A/D	15	56
Of Victim Affected, No Self Use or Affect	0	40
Victim A/D Users Who Were Affected	60	78
Both Self and Victim Used and Affected By A/D	15	16

alcohol or drugs had been consumed by the victim. Further, while 56 percent (n = 14) of deniers declared she had been affected by this use, only 15 percent (n = 6) of admitters made a similar claim. Typically deniers argued that the alcohol and drugs had sexually aroused their victim or rendered her out of control. For example, one denier insisted that his victim had become hysterical from drugs, not from being raped, and it was because of the drugs that she had reported him to the police. In addition, 40 percent (n = 10) of deniers argued that while the victim had been drunk or high, they themselves either hadn't ingested or weren't affected by alcohol and/or drugs. None of the admitters made this claim. In fact, in all of the 15 percent (n = 6) of cases where an admitter said the victim was drunk or high, he also admitted to being similarly affected.

These data strongly suggest that whatever role alcohol and drugs play in sexual and other types of violent crime, rapists have learned the advantage to be gained from using alcohol and drugs as an account. Our sample were aware that their victim would be discredited and their own behavior excused or justified by referring to alcohol and/or drugs.

2) Emotional Problems

Admitters frequently attributed their acts to emotional problems. Forty percent (n = 19) of admitters said they believed an emotional problem had been at the root of their rape behavior, and 33 percent (n = 15) specifically related the problem to an unhappy, unstable childhood or a marital-domestic situation. Still others claimed to have been in a general state of unease. For example, one admitter said that at the time of the rape he had been depressed, feeling he couldn't do anything right, and that something had been missing from his life. But he also added, "being a rapist is not part of my personality." Even admitters who could locate no source for an emotional problem evoked the popular image of rapists as the product of disordered personalities to argue they also must have problems:

> The fact that I'm a rapist makes me different. Rapists aren't all there. They have problems. It was wrong so there must be a reason why I did it. I must have a problem.

Our data do indicate that a precipitating event, involving an upsetting problem of everyday living, appeared in the accounts of 80 percent (n = 38) of admitters and 25 percent (n = 8) of deniers. Of those experiencing a precipitating event, including deniers, 76 percent (n = 35) involved a wife or girlfriend. Over and over, these men described themselves as having been in a rage because of an incident involving a woman with whom they believed they were in love.

Frequently, the upsetting event was related to a rigid and unrealistic double standard for sexual conduct and virtue which they applied to "their" woman but which they didn't expect from men, didn't apply to themselves, and, obviously, didn't honor in other women. To discover that the "pedestal" didn't apply to their wife or girlfriend sent them into a fury. One especially articulate and typical admitter described his feeling as follows. After serving a short prison term for auto theft, he married his "childhood sweetheart" and secured a well-paying job. Between his job and the volunteer work he was doing with an ex-offender group, he was spending long hours away from home, a situation that had bothered his wife. In response to her request, he gave up his volunteer work, though it was clearly meaningful to him. Then, one day, he discovered his wife with her former boyfriend "and my life fell apart." During the next several days, he said his anger had made him withdraw into himself and, after three days of drinking in a motel room, he abducted and raped a stranger. He stated:

> My parents have been married for many years and I had high expectations about marriage. I put my wife on a pedestal. When I walked in on her, I felt like my life had been destroyed, it was such a shock. I was bitter and angry about the fact that I hadn't done anything to my wife for cheating. I didn't want to hurt her [victim], only to scare and degrade her.

It is clear that many admitters, and a minority of deniers, were under stress at the time of their rapes. However, their problems were ordinary—the types of upsetting events that everyone experiences at some point in life. The overwhelming majority of the men were not clinically defined as mentally ill in court-ordered psychiatric examinations prior to their trials. Indeed, our sample is consistent with Abel et al. (1980) who found fewer than 5 percent of rapists were psychotic at the time of their offense.

As with alcohol and drug intoxication, a claim of emotional problems works differently depending upon whether the behavior in question is being justified or excused. It would have been counter-productive for deniers to have claimed to have had emotional problems at the time of the rape. Admitters used psychological explanations to portray themselves as having been temporarily "sick" at the time of the rape. Sick people are usually blamed for neither the cause of their illness nor for acts committed while in that state of diminished capacity. Thus, adopting the sick role removed responsibility by excusing the behavior as having been beyond the ability of the individual to control. Since the rapists were not "themselves," the rape was idiosyncratic rather than typical behavior. Admitters asserted a non-deviant identity despite their self-proclaimed disgust with what they had done. Although admitters were willing to assume the sick role, they did not view their problem as a chronic condition, nor did they believe themselves to be insane or permanently impaired. Said one admitter, who believed that he needed psychological counseling: "I have a mental disorder, but I'm not crazy." Instead, admitters viewed their "problem" as mild, transient, and curable. Indeed, part of the appeal of this excuse was that not only did it relieve responsibility, but, as with alcohol and drug addiction, it allowed the rapist to "recover." Thus, at the time of their interviews, only 31 percent (n = 14) of admitters indicated that "being a rapist" was part of their self-concept. Twenty-eight percent (n = 13) of admitters stated they had never thought of themselves as rapists, 8 percent (n = 4) said they were unsure, and 33 percent (n = 16) asserted they had been a rapist at one time but now were recovered. A multiple "ex-rapist," who believed his "problem" was due to "something buried in my subconscious" that was triggered when his girlfriend broke up with him, expressed a typical opinion:

> I was a rapist, but not now. I've grown up, had to live with it. I've hit the bottom of the well and it can't get worse. I feel born again to deal with my problems.

3) Nice Guy Image

Admitters attempted to further neutralize their crime and negotiate a non-rapist identity by painting an image of themselves as a "nice guy." Admitters projected the image of someone who had made a serious mistake but, in every other respect, was a decent person. Fifty-seven percent (n = 27) expressed regret and sorrow for their victim indicating that they wished there were a way to apologize for or amend their behavior. For example, a participant in a rape-murder, who insisted his partner did the murder, confided, "I wish there was something I could do besides saying 'I'm sorry, I'm sorry.' I live with it 24 hours a day and, sometimes, I wake up crying in the middle of the night because of it."

Schlenker and Darby (1981) explain the significance of apologies beyond the obvious expression of regret. An apology allows a person to admit guilt while at the same time seeking a pardon by signaling that the event should not be considered a fair representation of what the person is really like. An apology separates the bad self from the good self, and promises more acceptable behavior in the future. When apologizing, an individual is attempting to say: "I have repented and should be forgiven," thus making it appear that no further rehabilitation is required.

The "nice guy" statements of the admitters reflected an attempt to communicate a message consistent with Schlenker's and Darby's analysis of apologies. It was an attempt to convey that rape was not a representation of their "true" self. For example,

> It's different from anything else I've ever done. I feel more guilt about this. It's not consistent with me. When I talk about it, it's like being assaulted myself. I don't know why I did it, but once I started, I got into it. Armed robbery was a way of life for me, but not rape. I feel like I wasn't being myself.

Admitters also used "nice guy" statements to register their moral opposition to violence and harming women, even though, in some cases, they had seriously injured their victims. Such was the case of an admitter convicted of gang rape:

> I'm against hurting women. She should have resisted. None of us were the type of person that would use force on a woman. I never positioned myself on a woman unless she showed an interest in me. They would play to me, not me to them. My weakness is to follow. I never would have stopped, let alone pick her up without the others. I never would have let anyone beat her. I never bothered women who didn't want sex; never had a problem with sex or getting it. I loved her—like all women.

Finally, a number of admitters attempted to improve their self-image by demonstrating that, while they had raped, it could have been worse if they had not been a "nice guy." For example, one admitter professed to being especially gentle with his victim after she told him she had just had a baby. Others claimed to have given the victim money to get home or make a phone call, or to have made sure the victim's children were not in the room. A multiple rapist, whose pattern was to break in and attack sleeping victims in their homes, stated:

> I never beat any of my victims and I told them I wouldn't hurt them if they cooperated. I'm a professional thief. But I never robbed the women I raped because I felt so bad about what I had already done to them.

Even a young man, who raped his five victims at gun point and then stabbed them to death, attempted to improve his image by stating:

> Physically they enjoyed the sex [rape]. Once they got involved, it would be difficult to resist. I was always gentle and kind until I started to kill them. And the killing was always sudden, so they wouldn't know it was coming.

SUMMARY AND CONCLUSIONS

Convicted rapists' accounts of their crimes include both excuses and justifications. Those who deny what they did was rape justify their actions; those who admit it was rape attempt to excuse it or themselves. This study does not address why some men admit while others deny, but future research might address this question. This paper does provide insight on how men who are sexually aggressive or violent construct reality, describing the different strategies of admitters and deniers.

Admitters expressed the belief that rape was morally reprehensible. But they explained themselves and their acts by appealing to forces beyond their control, forces which reduced their capacity to act rationally and thus compelled them to rape. Two types of excuses predominated: alcohol/drug intoxication and emotional problems. Admitters used these excuses to negotiate a moral identity for themselves by viewing rape as idiosyncratic rather than typical behavior. This allowed them to reconceptualize themselves as recovered or "ex-rapists," someone who had made a serious mistake which did not represent their "true" self.

In contrast, deniers' accounts indicate that these men raped because their value system provided no compelling reason not to do so. When sex is viewed as a male entitlement, rape is no longer seen as criminal. However, the deniers had been convicted of rape, and like the admitters, they attempted to negotiate an identity. Through justifications, they constructed a "controversial" rape and attempted to demonstrate how their behavior, even if not quite right, was appropriate in the situation. Their denials, drawn from common cultural rape stereotypes, took two forms, both of which ultimately denied the existence of a victim.

The first form of denial was buttressed by the cultural view of men as sexually masterful and women as coy but seductive. Injury was denied by portraying the victim as willing, even enthusiastic, or as politely resistant at first but eventually yielding to "relax and enjoy it." In these accounts, force appeared merely as a seductive technique. Rape was disclaimed: rather than harm the woman, the rapist had fulfilled her dreams. In the second form of denial, the victim was portrayed as the type of woman who "got what she deserved." Through attacks on the victim's sexual reputation and, to a lesser degree, her emotional state, deniers attempted to demonstrate that since the victim wasn't a "nice girl," they were not rapists. Consistent with both forms of denial was the self-interested use of alcohol and drugs as a justification. Thus, in contrast to admitters, who accentuated their own use as an excuse, deniers emphasized the victim's consumption in an effort to both discredit her and make her appear more responsible for the rape. It is important to remember that deniers did not invent these justifications. Rather, they reflect a belief system which has historically victimized women by promulgating the myth that women both enjoy and are responsible for their own rape.

While admitters and deniers present an essentially contrasting view of men who rape, there were some shared characteristics. Justifications particularly, but also excuses, are buttressed by the cultural view of women as sexual commodities, dehumanized and devoid of autonomy and dignity. In this sense, the sexual objectification of women must be understood as an important factor contributing to an environment that trivializes, neutralizes, and, perhaps, facilitates rape.

Finally, we must comment on the consequences of allowing one perspective to dominate thought on a social problem. Rape, like any complex continuum of behavior, has multiple causes and is influenced by a number of social factors. Yet, dominated by psychiatry and the medical model, the underlying assumption that rapists are "sick" has pervaded research. Although methodologically unsound, conclusions have been based almost exclusively on small clinical populations of rapists—that extreme group of rapists who seek counseling in prison and are the most likely to exhibit psychopathology. From this small, atypical group of men, psychiatric findings have been generalized to all men who rape. Our research, however, based on volunteers from the entire prison population, indicates that some rapists, like deniers, viewed and understood their behavior from a popular cultural perspective. This strongly suggests that cultural perspectives, and not an idiosyncratic illness, motivated their behavior. Indeed, we can argue that the psychiatric perspective has contributed to the vocabulary of motive that rapists use to excuse and justify their behavior (Scully and Marolla, 1984).

Efforts to arrive at a general explanation for rape have been retarded by the narrow focus of the medical model and the preoccupation with clinical populations. The continued reduction of such complex behavior to a singular cause hinders, rather than enhances, our understanding of rape.

NOTES

1. These numbers include pretest interviews. When the analysis involves either questions that were not asked in the pretest or that were changed, they are excluded and thus the number changes.

2. There is, of course, the possibility that some of these men really were innocent of rape. However, while the U.S. criminal justice system is not without flaw, we assume that it is highly unlikely that this many men could have been unjustly convicted of rape, especially since rape is a crime with traditionally low conviction rates. Instead, for purposes of this research, we assume that these men were guilty as charged and that their attempt to maintain an image of non-rapist springs from some psychologically or sociologically interpretable mechanism.

3. Because of their outright denial, interviews with this group of rapists did not contain the data being analyzed here and, consequently, they are not included in this paper.

4. It is worth noting that a number of deniers specifically mentioned the victim's alleged interest in oral sex. Since our interview questions about sexual history indicated that the rapists themselves found oral sex

marginally acceptable, the frequent mention is probably another attempt to discredit the victim. However, since a tape recorder could not be used for the interviews and the importance of these claims didn't emerge until the data was being coded and analyzed, it is possible that it was mentioned even more frequently but not recorded.

5. Research shows clearly that women do not enjoy rape. Holmstrom and Burgess (1978) asked 93 adult rape victims, "How did it feel sexually?" Not one said they enjoyed it. Further, the trauma of rape is so great that it disrupts sexual functioning (both frequency and satisfaction) for the overwhelming majority of victims, at least during the period immediately following the rape and, in fewer cases, for an extended period of time (Burgess and Holmstrom, 1979; Feldman-Summers et al., 1979). In addition, a number of studies have shown that rape victims experience adverse consequences prompting some to move, change jobs, or drop out of school (Burgess and Holmstrom, 1974; Kilpatrick et al., 1979; Ruch et al., 1980; Shore, 1979).

REFERENCES

Abel, Gene, Judith Becker, and Linda Skinner. 1980. "Aggressive behavior and sex." *Psychiatric Clinics of North America* 3(2):133–151.

Abrahamsen, David. 1960. *The Psychology of Crime.* New York: John Wiley.

Albin, Rochelle. 1977. "Psychological studies of rape." *Signs* 3(2):423–435.

Burgess, Ann Wolbert, and Lynda Lytle Holmstrom. 1974. *Rape: Victims of Crisis.* Bowie: Robert J. Brady.

———. 1979. "Rape: Sexual disruption and recovery." *American Journal of Orthopsychiatry* 49(4):648–657.

Feldman-Summers, Shirley, Patricia E. Gordon, and Jeanette R. Meagher. 1979. "The impact of rape on sexual satisfaction." *Journal of Abnormal Psychology* 88(1):101–105.

Glueck, Sheldon. 1925. *Mental Disorders and the Criminal Law.* New York: Little Brown.

Groth, Nicholas A. 1979. *Men Who Rape.* New York: Plenum Press.

Hall, Peter M., and John P. Hewitt. 1970. "The quasi-theory of communication and the management of dissent." *Social Problems* 18(1):17–27.

Hewitt, John P., and Peter M. Hall. 1973. "Social problems, problematic situations, and quasi-theories." *American Journal of Sociology* 38(3):367–374.

Hewitt, John P., and Randall Stokes. 1975. "Disclaimers." *American Sociological Review* 40(1):1–11.

Hollander, Bernard. 1924. *The Psychology of Misconduct, Vice and Crime.* New York: Macmillan.

Holmstrom, Lynda Lytle, and Ann Wolbert Burgess. 1978. "Sexual behavior of assailant and victim during rape." Paper presented at the annual meetings of the American Sociological Association, San Francisco, September 2–8.

Kilpatrick, Dean G., Lois Veronen, and Patricia A. Resnick. 1979. "The aftermath of rape: Recent empirical findings." *American Journal of Orthopsychiatry* 49(4):658–669.

Ladouceur, Patricia. 1983. "The relative impact of drugs and alcohol on serious felons." Paper presented at the annual meetings of the American Society of Criminology, Denver, November 9–12.

McCaghy, Charles. 1968. "Drinking and deviance disavowal: The case of child molesters." *Social Problems* 16(1):43–49.

Marolla, Joseph and Diana Scully. 1979. "Rape and psychiatric vocabularies of motive." Pp. 301-318 in Edith S. Gomberg and Violet Franks (eds.), *Gender and Disordered Behavior: Sex Differences in Psychopathology.* New York: Brunner/Mazel.

Mills, C. Write. 1940. "Situated actions and vocabularies of motive." *American Sociological Review* 5(6):904–913.

Nelson, Steve, and Menachem, Amir. 1975. "The hitchhike victim of rape: A research report." Pp. 47–65 in Israel Drapkin and Emilio Viano (eds.), *Victimology: A New Focus.* Lexington, KY: Lexington Books.

Queen's Bench Foundation. 1976. *Rape: Prevention and Resistance.* San Francisco: Queen's Bench Foundation.

Ruch, Libby O., Susan Meyers Chandler, and Richard A. Harter. 1980. "Life change and rape impact." *Journal of Health and Social Behavior* 21(3):248–260.

Schlenker, Barry R., and Bruce W. Darby. 1981. "The use of apologies in social predicaments." *Social Psychology Quarterly* 44(3):271–278.

Scott, Marvin, and Stanford Lyman. 1968. "Accounts." *American Sociological Review* 33(1):46–62.

Scully, Diana, and Joseph Marolla. 1984. "Rape and psychiatric vocabularies of motive: Alternative perspectives." In Ann Wolbert Burgess (ed.), *Handbook on Rape and Sexual Assault.* New York: Garland Publishing.

Shore, Barbara K. 1979. "An Examination of Critical Process and Outcome Factors in Rape." Rockville, MD: National Institute of Mental Health.

Stokes, Randall, and John P. Hewitt. 1976. "Aligning actions." *American Sociological Review* 41(5):837–849.

Sykes, Gresham M., and David Matza. 1957. "Techniques of neutralization." *American Sociological Review* 22(6):664–670.

20

The Influence of Situational Ethics on
Cheating among College Students

DONALD L. McCABE

INTRODUCTION

Numerous studies have demonstrated the pervasive nature of cheating among college students (Baird 1980; Haines, Diekhoff, LaBeff, and Clark 1986; Michaels and Miethe 1989; Davis, et al. 1992). This research has examined a variety of factors that help explain cheating behavior, but the strength of the relationships between individual factors and cheating has varied considerably from study to study (Tittle and Rowe 1973; Baird 1980; Eisenberger and Shank 1985; Haines, et al. 1986; Ward 1986; Michaels and Miethe 1989; Perry, Kane, Bernesser, and Spicker 1990; Ward and Beck 1990).

Although the factors examined in these studies (for example, personal work ethic, gender, self-esteem, rational choice, social learning, deterrence) are clearly important, the work of LaBeff, Clark, Haines, and Diekhoff (1990) suggests that the concept of situational ethics may be particularly helpful in understanding student rationalizations for cheating. Extending the arguments of Norris and Dodder (1979), LaBeff et al. conclude

> that students hold qualified guidelines for behavior which are situationally determined. As such, the concept of situational ethics might well describe . . . college cheating [as having] rules for behavior [that] may not be considered rigid but depend on the circumstances involved (1990, p. 191).

LaBeff et al. believe a utilitarian calculus of "the ends justifies the means" underlies this reasoning process and "what is wrong in most situations might be considered right or acceptable if the end is defined as appropriate" (1990, p. 191). As argued by Edwards (1967), the situation determines what is right or wrong in this decision-making calculus and also dictates the appropriate principles to be used in guiding and judging behavior.

Sykes and Matza (1957) hypothesize that such rationalizations, that is, "justifications for deviance that are seen as valid by the delinquent but not by the legal system or society at large" (p. 666), are common. However, they

From: "The Influence of Situational Ethics on Cheating among College Students," Donald L. McCabe, *Sociological Inquiry,* Vol. 62:3 (Summer 1992). Reprinted by permission of the University of Texas Press and the author.

challenge conventional wisdom that such rationalizations typically follow deviant behavior as a means of protecting "the individual from self-blame and the blame of others after the act" (p. 666). They develop convincing arguments that these rationalizations may logically precede the deviant behavior and "[d]isapproval from internalized norms and conforming others in the social environment is neutralized, turned back, or deflated in advance. Social controls that serve to check or inhibit deviant motivational patterns are rendered inoperative, and the individual is freed to engage in delinquency without serious damage to his self-image" (pp. 666–667).

Using a sample of 380 undergraduate students at a small southwestern university, LaBeff et al. (1990) attempted to classify techniques employed by students in the neutralization of cheating behavior into the five categories of neutralization proposed by Sykes and Matza (1957): (1) denial of responsibility, (2) condemnation of condemners, (3) appeal to higher loyalties, (4) denial of victim, and (5) denial of injury. Although student responses could easily be classified into three of these techniques, denial of responsibility, appeal to higher loyalties, and condemnation of condemners, LaBeff et al. conclude that "[i]t is unlikely that students will either deny injury or deny the victim since there are no real targets in cheating" (1990, p. 196).

The research described here responds to LaBeff et al. in two ways; first, it answers their call to "test the salience of neutralization . . . in more diverse university environments" (p. 197) and second, it challenges their dismissal of denial of injury and denial of victim as neutralization techniques employed by students in their justification of cheating behavior.

METHODOLOGY

The data discussed here were gathered as part of a study of college cheating conducted during the 1990–1991 academic year. A seventy-two-item questionnaire concerning cheating behavior was administered to students at thirty-one highly selective colleges across the country. Surveys were mailed to a minimum of five hundred students at each school and a total of 6,096 completed surveys were returned (38.3 percent response rate). Eighty-eight percent of the respondents were seniors, nine percent were juniors, and the remaining three percent could not be classified. Survey administration emphasized voluntary participation and assurances of anonymity to help combat issues of non-response bias and the need to accept responses without the chance to question or contest them.

The final sample included 61.2 percent females (which reflects the inclusion of five all-female schools in the sample and a slightly higher return rate among female students) and 95.4 percent U.S. citizens. The sample paralleled the ethnic diversity of the participating schools (85.5 percent Anglo, 7.2 percent Asian, 2.6 percent African American, 2.2 percent Hispanic and 2.5 percent other); their religious diversity (including a large percentage of students who claimed no religious preference, 27.1 percent); and their mix of undergraduate majors (36.0 percent humanities, 28.8 percent social sciences, 26.8 percent natural sciences and engineering, 4.5 percent business, and 3.9 percent other).

RESULTS

Of the 6,096 students participating in this research, over two-thirds (67.4 percent) indicated that they had cheated on a test or major assignment at least once while an undergraduate. This cheating took a variety of different forms, but among the most popular (listed in decreasing order of mention) were: (1) a failure to footnote sources in written work, (2) collaboration on assignments when the instructor specifically asked for individual work, (3) copying from other students on tests and examinations, (4) fabrication of bibliographies, (5) helping someone else cheat on a test, and (6) using unfair methods to learn the content of a test ahead of time. Almost one in five students (19.1 percent) could be classified as active cheaters (five or more self-reported incidents of cheating). This is double the rate reported by LaBeff et al. (1990), but they asked students to report only cheating incidents that had taken place in the last six months. Students in this research were asked to report all cheating in which they had engaged while an undergraduate—a period of three years for most respondents at the time of this survey.

Students admitting to any cheating activity were asked to rate the importance of several specific factors that might have influenced their decisions to cheat. These data establish the importance of denial of responsibility and condemnation of condemners as neutralization techniques. For example, 52.4 percent of the respondents who admitted to cheating rated the pressure to get good grades as an important influence in their decision to cheat with parental pressures and competition to gain admission into professional schools singled out as the primary grade pressures. Forty-six percent of those who had engaged in cheating cited excessive workloads and an inability to keep up with assignments as important factors in their decisions to cheat.

In addition to rating the importance of such preselected factors, 426 respondents (11.0 percent of the admitted cheaters) offered their own justifications for cheating in response to an open-ended question on motivations for cheating. These responses confirm the importance of denial of responsibility and condemnation of condemners as neutralization techniques. They also support LaBeff et al.'s (1990) claim that appeal to higher loyalties is an important neutralization technique. However, these responses also suggest that LaBeff et al.'s dismissal of denial of injury as a justification for student cheating is arguable.

As shown in Table 1, denial of responsibility was the technique most frequently cited (216 responses, 61.0 percent of the total) in the 354 responses classified into one of Sykes and Matza's five categories of neutralization. The most common responses in this category were mind block, no understanding of the material, a fear of failing, and unclear explanations of assignments.

TABLE 1 *Neutralization Strategies: Self-Admitted Cheaters*

Strategy	Number	Percent
Denial of responsibility	216	61.0
Mind block	90	25.4
No understanding of material	31	8.8
Other	95	26.8
Condemnation of condemners	99	28.0
Pointless assignment	35	9.9
No respect for professor	28	7.9
Other	36	10.2
Appeal to higher loyalties	24	6.8
Help a friend	10	2.8
Peer pressure	9	2.5
Other	5	1.5
Denial of injury	15	4.2
Cheating is harmless	9	2.5
Does not matter	6	1.7

(Although it is possible that some instances of mind block and a fear of failing included in this summary would be more accurately classified as rationalization, the wording of all responses included here suggests that rationalization preceded the cheating incident. Responses that seem to involve post hoc rationalizations were excluded from this summary). Condemnation of condemners was the second most popular neutralization technique observed (99 responses, 28.0 percent) and included such explanations as pointless assignments, lack of respect for individual professors, unfair tests, parents' expectations, and unfair professors. Twenty-four respondents (6.8 percent) appealed to higher loyalties to explain their behavior. In particular, helping a friend and responding to peer pressures were influences some students could not ignore. Finally fifteen students (4.2 percent) provided responses that clearly fit into the category of denial of injury. These students dismissed their cheating as harmless since it did not hurt anyone or they felt cheating did not matter in some cases (for example, where an assignment counted for a small percentage of the total course grade).

Detailed examination of selected student responses provides additional insight into the neutralization strategies they employ.

Denial of Responsibility

Denial of responsibility invokes the claim that the act was "due to forces outside of the individual and beyond his control such as unloving parents" (Sykes and Matza 1957, p. 667). For example, many students cite an unreasonable workload and the difficulty of keeping up as ample justification for cheating.

Here at . . . , you must cheat to stay alive. There's so much work and the quality of materials from which to learn, books, professors, is so bad that there's no other choice.

It's the only way to keep up.

I couldn't do the work myself.

The following descriptions of student cheating confirm fear of failure is also an important form of denial of responsibility:

. . . a take-home exam in a class I was failing.

. . . was near failing.

Some justified their cheating by citing the behavior of peers:

Everyone has test files in fraternities, etc. If you don't, you're at a great disadvantage.

When most of the class is cheating on a difficult exam and they will ruin the curve, it influences you to cheat so your grade won't be affected.

All of these responses contain the essence of denial of responsibility: the cheater has deflected blame to others or to a specific situational context.

Denial of Injury

As noted in Table 1, denial of injury was identified as a neutralization technique employed by some respondents. A key element in denial of injury is whether one feels "anyone has clearly been hurt by [the] deviance." In invoking this defense, a cheater would argue "that his behavior does not really cause any great harm despite the fact that it runs counter to the law" (Sykes and Matza 1957, pp. 667–668). For example, a number of students argued that the assignment or test on which they cheated was so trivial that no one was really hurt by their cheating.

These grades aren't worth much therefore my copying doesn't mean very much. I am ashamed, but I'd probably do it the same way again.

If I extend the time on a take-home it is because I feel everyone does and the teacher kind of expects it. No one gets hurt.

As suggested earlier, these responses suggest the conclusion of LaBeff et al. that "[i]t is unlikely that students will . . . deny injury" (1990, p. 196) must be re-evaluated.

The Denial of the Victim

LaBeff et al. failed to find any evidence of denial of the victim in their student accounts. Although the student motivations for cheating summarized in Table 1 support this conclusion, at least four students (0.1% on the self-admitted cheaters in this study) provided comments elsewhere on the survey instrument which involved denial of the victim. The common element in these responses was a victim deserving of the consequences of the cheating behavior and cheating was viewed as "a form of rightful retaliation or punishment" (Sykes and Matza 1957, p. 668).

This feeling was extreme in one case, as suggested by the following student who felt her cheating was justified by the

> realization that this school is a manifestation of the bureaucratic capitalist system that systematically keeps the lower classes down, and that adhering to their rules was simply perpetuating the institution.

This "we" versus "they" mentality was raised by many students, but typically in comments about the policing of academic honesty rather than as justification for one's own cheating behavior. When used to justify cheating, the target was almost always an individual teacher rather than the institution and could be more accurately classified as a strategy of condemnation of condemners rather than denial of the victim.

The Condemnation of Condemners

Sykes and Matza describe the condemnation of condemners as an attempt to shift "the focus of attention from [one's] own deviant acts to the motives and behavior of those who disapprove of [the] violations. [B]y attacking others, the wrongfulness of [one's] own behavior is more easily repressed or lost to view" (1957, p. 668). The logic of this strategy for student cheaters focused on issues of favoritism and fairness. Students invoking this rationale describe "uncaring, unprofessional instructors with negative attitudes who were negligent in their behavior" (LaBeff et al 1990, p. 195). For example:

> In one instance, nothing was done by a professor because the student was a hockey player.

> The TAs who graded essays were unduly harsh.

> It is known by students that certain professors are more lenient to certain types, e.g., blondes or hockey players.

> I would guess that 90% of the students here have seen athletes and/or fraternity members cheating on an exam or papers. If you turn in one of these culprits, and I have, the penalty is a five-minute lecture from a coach and/or administrator. All these add up to a "who cares, they'll never do anything to you anyway" attitude here about cheating.

Concerns about the larger society were an important issue for some students:

> When community frowns upon dishonesty, then people will change.

If our leaders can commit heinous acts and then lie before Senate commit-tees about their total ignorance and innocence, *then why can't I cheat a little?*

In today's world you do anything to be above the competition.

In general, students found ready targets on which to blame their behavior and condemnation of the condemners was a popular neutralization strategy.

The Appeal to Higher Loyalties

The appeal to higher loyalties involves neutralizing "internal and external con-trols . . . by sacrificing the demands of the larger society for the demands of the smaller social groups to which the [offender] belongs. [D]eviation from certain norms may occur not because the norms are rejected but because other norms, held to be more pressing or involving a higher loyalty, are accorded precedence" (Sykes and Matza 1957, p. 669). For example, a difficult conflict for some students is balancing the desire to help a friend against the institu-tion's rules on cheating. The student may not challenge the rules, but rather views the need to help a friend, fellow fraternity/sorority member, or room-mate to be a greater obligation which justifies the cheating behavior.

Fraternities and sororities were singled out as a network where such behavior occurs with some frequency. For example, a female student at a small university in New England observed:

> There's a lot of cheating within the Greek system. Of all the cheating I've seen, it's often been men and women in fraternities and sororities who exchange information or cheat.

The appeal to higher loyalties was particularly evident in student reactions concerning the reporting of cheating violations. Although fourteen of the thirty-one schools participating in this research had explicit honor codes that generally require students to report cheating violations they observe, less than one-third (32.3 percent) indicated that they were likely to do so. When asked if they would report a friend, only 4 percent said they would and most stu-dents felt that they should not be expected to do so. Typical student com-ments included:

> Students should not be sitting in judgment of their own peers.

> The university is not a police state.

For some this decision was very practical:

> A lot of students, 50 percent, wouldn't because they know they will prob-ably cheat at some time themselves.

For others, the decision would depend on the severity of the violation they observed and many would not report what they consider to be minor violations, even those explicitly covered by the school's honor code or policies on academic honesty. Explicit examination or test cheating was one of the few violations where students exhibited any consensus concerning the need to report violations. Yet even in this case many students felt other factors must be considered. For example, a senior at a woman's college in the Northeast commented:

> It would depend on the circumstances. If someone was hurt, *very likely*. If there was no single victim in the case, if the victim was [the] institution . . . , then *very unlikely*.

Additional evidence of the strength of the appeal to higher loyalties as a neutralization technique is found in the fact that almost one in five respondents (17.8 percent) reported that they had helped someone cheat on an examination or major test. The percentage who have helped others cheat on papers and other assignments is likely much higher. Twenty-six percent of those students who helped someone else cheat on a test reported that they had never cheated on a test themselves, adding support to the argument that peer pressure to help friends is quite strong.

CONCLUSIONS

From this research it is clear that college students use a variety of neutralization techniques to rationalize their cheating behavior, deflecting blame to others and/or the situational context, and the framework of Sykes and Matza (1957) seems well-supported when student explanations of cheating behavior are analyzed. Unlike prior research (LaBeff et al. 1990), however, the present findings suggest that students employ all of the techniques described by Sykes and Matza, including denial of injury and denial of victim. Although there was very limited evidence of the use of denial of victim, denial of injury was not uncommon. Many students felt that some forms of cheating were victimless crimes, particularly on assignments that accounted for a small percentage of the total course grade. The present research does affirm LaBeff et al.'s finding that denial of responsibility and condemnation of condemners are the neutralization techniques most frequently utilized by college students. Appeal to higher loyalties is particularly evident in neutralizing institutional expectations that students report cheating violations they observe.

The present results clearly extend the findings of LaBeff et al. into a much wider range of contexts as this research ultimately involved 6,096 students at thirty-one geographically dispersed institutions ranging from small liberal arts colleges in the Northeast to nationally prominent research universities in the South and West. Fourteen of the thirty-one institutions have long-standing honor-code traditions. The code tradition at five of these schools dates to the late 1800s and all fourteen have codes that survived the student unrest of the 1960s. In such a context, the strength of the appeal to higher loyalties and the denial of responsibility as justifications for cheating is a very persuasive argument that neutralization techniques are salient to today's college student.

More importantly, it may suggest fruitful areas of future discourse between faculty, administrators, and students on the question of academic honesty.*

REFERENCES

Baird, John S. 1980. "Current Trends in College Cheating." *Psychology in Schools* 17:512–522.

Davis, Stephen F., Cathy A. Grover, Angela H. Becker, and Loretta N. McGregor. 1992. "Academic Dishonesty: Prevalence, Determinants, Techniques, and Punishments." *Teaching of Psychology*. In press.

Edwards, Paul. 1967. *The Encyclopedia of Philosophy,* no. 3, Paul Edwards (ed.). New York: Macmillan Company and Free Press.

Eisenberger, Robert, and Dolores M. Shank. 1985. "Personal Work Ethic and Effort Training Affect Cheating." *Journal of Personality and Social Psychology* 49:520–528.

Haines, Valerie J., George Diekhoff, Emily LaBeff, and Robert Clark. 1986. "College Cheating: Immaturity, Lack of Commitment, and the Neutralizing Attitude." *Research in Higher Education* 25:342–354.

LaBeff, Emily E., Robert E. Clark, Valerie J. Haines, and George M. Diekhoff. 1990. "Situational Ethics and College Student Cheating." *Sociological Inquiry* 60:190–198.

Michaels, James W., and Terance Miethe. 1989. "Applying Theories of Deviance to Academic Cheating." *Social Science Quarterly* 70:870–885.

Norris, Terry D., and Richard A. Dodder. 1979. "A Behavioral Continuum Synthesizing Neutralization Theory, Situational Ethics and Juvenile Delinquency." *Adolescence* 55:545–555.

Perry, Anthony R., Kevin M. Kane, Kevin J. Bernesser, and Paul T. Spicker. 1990. "Type A Behavior, Competitive Achievement-Striving, and Cheating Among College Students." *Psychological Reports* 66:459–465.

Sykes, Gresham M., and David Matza. 1957. "Techniques of Neutralization: A Theory of Delinquency." *American Sociological Review* 22:664–670.

Tittle, Charles, and Alan Rowe. 1973. "Moral Appeal, Sanction Threat, and Deviance: An Experimental Test." *Social Problems* 20:488–498.

Ward, David. 1986. "Self-Esteem and Dishonest Behavior Revisited." *Journal of Social Psychology* 123:709–713.

Ward, David, and Wendy L. Beck. 1990. "Gender and Dishonesty." *Journal of Social Psychology* 130:333–339.

---

*The author would like to acknowledge the support of the Rutgers Graduate School of Management Research Resources Committee, Exxon Corporation and First Fidelity Bancorporation.

21

Denying the. Guilty Mind:
Accounting for Involvement
in a White-Collar Crime

MICHAEL L. BENSON

The subjective experiences of those who pass through the criminal justice system have seldom been given explicit attention by the criminologist.[1] In particular, little attention has been paid to the offender's account of involvement in the offense.[2] In the case of white-collar criminals, the failure to analyze the offender's explanation for involvement in the offense is especially noteworthy. Although white-collar offenders are assumed to suffer subjectively as a result of the public humiliation of adjudication as criminals, they are also assumed, paradoxically, to be able to maintain a non-criminal self-concept and to successfully deny the criminality of their actions (Conklin, 1977).

The present study treats the accounts given by a sample of convicted white-collar offenders, focusing specifically on the techniques they use to deny their own criminality. The emphasis is on general patterns and regularities in the data. The central research question is: How do convicted white-collar offenders account for their adjudication as criminals?[3] While researchers have frequently expressed outrage at the denial of criminality that is thought to be typical of white-collar criminals, few attempts have been made to understand how this process occurs or to relate it to general deviance theory. Rather, researchers have all too often concentrated on morally condemning offenders (Clinard and Yeager, 1978).

Over 30 years ago, Sutherland (1949: 222, 225) wrote,

> Businessmen develop rationalizations which conceal the fact of crime. . . .
> Even when they violate the law, they do not conceive of themselves as
> criminals. . . . Businessmen fight whenever words that tend to break down
> this rationalization are used.

This view of white-collar offenders has continued to the present day (Geis, 1982; Meier and Geis, 1982). Indeed, failure to confront and penetrate the rationalizations used by white-collar offenders and to get beyond a sympathetic

From: "Denying the Guilty Mind: Accounting for Involvement in a White-Collar
Crime," Michael L. Benson, *Criminology,* Vol. 23, No. 4, 1985. Reprinted by permission
of the American Society of Criminology.

view of the individual offender is considered by some to be one of the reasons for the continued widespread prevalence of white-collar crimes (Geis, 1982: 55–57; Meier and Geis, 1982: 98). In addition, others have argued that the leniency with which white-collar criminals are treated by the justice system derives in part from their ability to evoke sympathy from judges (Conklin, 1977).

These widely held beliefs regarding the rationalization processes used by white-collar offenders to justify their crimes and their ability to provoke sympathetic consideration rely, however, on anecdotal evidence. The typical offender studied is a person of extremely high status who has been convicted of a particularly egregious offense. Examples include the studies of former Vice President Agnew (Naughton, Crewdson, Franklin, Lydon, and Solpukas, 1977) and the executives involved in the heavy electrical equipment antitrust case of 1961 (Geis, 1967). These cases are important for symbolic reasons, but the individuals involved most likely do not represent typical white-collar offenders.

High-status offenders tend to receive a great deal of media attention. The cases in which they are involved become public morality plays, and the protagonists become representatives of larger institutional or political constituencies: business versus consumers, governmental regulation versus free enterprise, or employers versus employees. A recent work by Fisse and Braithwaite (1983) gives many examples and is an important advance in the understanding of how corporate actors attempt to manage the impact of adverse publicity. However, although it is assumed that white-collar offenders are able to maintain a noncriminal self-concept and avoid stigmatization as criminals, the subtle interactional processes whereby individual white-collar offenders attempt to avoid the transformation of their personal and public identities have for the most part been ignored.

Rothman and Gandossy (1982), using quantitative methods, looked at the role the defendant's version of the offense plays in the decision to sanction, finding that the sufficiency of the defendant's story can influence the probation officer's evaluation of the defendant and indirectly the sanction imposed by the judge. The defendant's written version of the offense as it appears in the presentence investigation report was treated as an account.

It is likely that accounts vary depending upon the audience to which they are presented. In preparing his version of the offense, the defendant is talking to the probation officer and to the judge involved in his case. His explanation is private, and through it the defendant hopes to minimize the sanction to be imposed by the judge. But how offenders account for their behavior in nonjudicial settings is an open question, the answer to which may shed light on the perplexing ability of white-collar offenders to avoid being characterized as criminals even though they have been publicly convicted.

The major theme of the following discussion is that accounting for involvement in a white-collar offense is intimately involved with the social organization of the offense. The accounts developed by white-collar offenders are delimited by the type of offense committed, its mechanics, and its organizational context. They are further structured in that they must be constructed so as to defeat the conditions required for a successful degradation ceremony.

THE STUDY

This study is based primarily on interviews conducted with a sample of 30 convicted white-collar offenders. The interviews were supplemented by an examination of the files maintained on 80 white-collar offenders and by further interviews with federal probation officers, federal judges, Assistant U.S. Attorneys, and defense attorneys specializing in white-collar cases.

The sample of interviewed offenders was essentially self-selected. A letter which introduced the researcher and described the nature of the study was sent to most of the 80 offenders in the sample.[4] The letter indicated that the researcher was interested in the subject's impressions of the way in which his case was handled and in the effect that conviction had on his self-image and life prospects.[5] Offenders were assured that their remarks would not be attributable to them as individuals. The proposed interviews would be open-ended and unstructured.

In light of the small and non-random nature of the sample, the results reported here must be viewed as provisional. There are no systematic differences between the offenders who agreed to be interviewed and those who did not in terms of their social and offense characteristics, but there is some likelihood that the interviewees differed psychologically or experientially from those who refused to participate.

The letter inviting participation in the study was sent from the Probation Office, and it is possible that some offenders viewed their participation as a way of ingratiating themselves with their respective probation officers. They also may have felt under some coercion to participate in the study. While these potential sources of bias cannot be completely ruled out, it is the researcher's impression that most of the interviewees agreed to participate because they welcomed an opportunity to express their views on the criminal justice system in a confidential and nonjudgmental forum.

In the interviews no attempt was made to challenge the explanations or rationalizations given by offenders regarding their offenses. Rather, offenders were encouraged to talk of themselves and their feelings regarding the case and they were allowed to focus on the aspects they considered to be most important. This approach was followed for two reasons: first, the sensitive nature of the subject matter under discussion did not permit the use of an interrogatory or inquisitorial style. The emotional trauma wrought by conviction was, indeed, evident in many of the interviews and, considering the voluntary nature of the interviews, to challenge the subjects seemed insensitive and unnecessary. Second, the goal of the study was not to determine how strong the rationalizations were, nor was it to bring about a "rehabilitative" awareness in the offender of the criminality of past acts. Rather, it was to determine how offenders account for their actions to themselves and to significant others, who it is assumed are unlikely to challenge or refute their explanations.

The Offenders

For the purposes of this study, white-collar offenders were those convicted of economic offenses committed through the use of indirection, fraud, or collusion (Shapiro, 1980). The offenses represented in the sample are those that are usually thought of as presumptively white-collar offenses, such as securities and exchange fraud, antitrust violations, embezzlement, false claims and statements, and tax violations. In terms of socioeconomic status, the sample ranges from a formerly successful practitioner of international law to a man currently self-employed as a seller of jewelry trinkets. For some offenders, particularly licensed professionals and those employed in the public sector, conviction was accompanied by loss of occupation and other major changes in lifestyle. For others, such as businessmen and those employed in the private sector, conviction was not accompanied by collateral disabilities other than the expense and trauma of criminal justice processing (Benson, 1984).

Accounts

An account is a statement made to explain unanticipated or untoward behavior (Scott and Lyman, 1968). There are two general forms of accounts: (1) justifications, and (2) excuses. In a justification the actor admits responsibility for the act in question but denies its pejorative content. In an excuse the actor admits the act in question is wrong, but denies having full responsibility for it. Accounts are used to narrow the gap between expectation and behavior and to present the actor in a favorable light. They serve, that is, an exculpatory function. Related to the justification and the excuse is the apology. In an apology, the individual admits violating a rule, accepts the validity of the rule, and expresses embarrassment and anger at himself. The individual "splits himself into two parts, the part that is guilty of an offense and the part that disassociates itself from the delict and affirms a belief in the offended rule" (Goffman, 1972: 113).

Two perspectives are possible on the psychological status of accounts. They can be viewed as impression management techniques used by the offender to exonerate himself and to avoid being labeled a criminal. Alternatively, they can be viewed as indicators of the offender's cognitive structure—that is, of the way he interprets what he did. The present study cannot demonstrate conclusively which of these perspectives is more accurate. Given the obvious interest that offenders have in maintaining a noncriminal identity, the impression management perspective is perhaps more plausible in this instance. Nonetheless, it should be recognized that offenders probably engage in a process of self-persuasion in defining their situations. Accounts intended to shore up the offender's public identity may evolve into the offender's own view of himself and his offense. Regardless of which perspective is more appropriate, it can be suggested that the study of accounts has important implications for the field of criminology.

On the one hand, accounts relate to the techniques of neutralization identified by Sykes and Matza (1957) in their seminal article on juvenile delinquency. According to Sykes and Matza, before committing offenses, juvenile delinquents use techniques of neutralization to relieve themselves of the duty to behave according to the norms. The study of neutralizations is important, therefore, in helping us to understand the individual and situational causes of crime. Whether white-collar offenders engage in a similar process before their offenses is not clear in all instances. One prominent theory of embezzlement hinges on this assumption (Cressey, 1953). Considering that almost by definition white-collar offenders are more strongly committed to the central normative structure, it is reasonable to assume that many of them go through elaborate neutralization processes prior to their offenses. The accounts developed afterwards may describe the process employed by the offender. In this manner, the study of accounts can provide a guide to micro-level analysis of the causal factors involved in a white-collar crime.

An account, however, is not a foolproof guide to an actor's intentions (Nettler, 1982: 15–17). It is important to distinguish between neutralizations that cause or allow an offense to be committed and accounts that are developed afterwards to excuse or justify it. By definition, an account is a linguistic act presented to, or more correctly, performed before an audience. It is an example of what Goffman (1959) has called impression management.

In giving an account, the offender is not talking to himself for the purpose of identifying reasons that allow the commission of an offense. Rather, he is addressing an audience and attempting to explain the offense while at the same time demonstrating an essential lack of personal criminality. To accomplish this feat, he must bring his actions into correspondence with the class of actions that is implicitly acceptable in his society. For this reason, accounts should not be thought of as solely individual inventions. Rather, they are invented by, as well as in, an historical and institutional context. This context delimits the range of superficially plausible justifications and excuses.

Corporate capitalism creates opportunities for particular forms of crime and at the same time creates, or makes plausible, certain types of justifications and mitigations for engaging in crime. The diffusion of responsibility in corporations, for example, makes it plausible for an actor to either deny responsibility altogether or to partially excuse actions by claiming to have been working at the request of supervisors. The widespread acceptance of such concepts as profit, growth, and free enterprise make it plausible for an actor to argue that governmental regulations run counter to more basic societal values and goals. Criminal behavior can then be characterized as being in line with other higher laws of free enterprise (Denzin, 1977: 919). More generally, the idea, which seems to be at the core of capitalist economies, that society benefits most through the individual competitive strivings of its members, as opposed to the opposite notion that society benefits most through the submerging of individual goals in favor of group needs, would seem to provide a moral environment which facilitates the rationalization of criminal behavior. The idea of "just trying to get ahead" becomes an understandable and perhaps acceptable motive even when it occasionally leads to behavior that violates the law. A similar defense (or motive) would not seem to be possible in societies that do not promote individual material success as a desirable goal.

The study of accounts, therefore, can reveal how history and social structure make possible certain characterizations of events. This creates a moral environment conducive to crime in two ways. Convicted offenders can rationalize their behavior to themselves prior to engaging in crimes. After the discovery of an offender's involvement in criminal activity, he may be able to avoid the stigma of being labeled a criminal through the use of proper accounting practices.

Accounts are important, therefore, for three reasons. First, they shed light on neutralizations and vocabularies of motive and help one to understand the causes of crime at the individual and situational level. Second, as modes of impression management, they illuminate underlying assumptions about what constitutes culpable criminality versus acceptable illegalities. Finally, as Rothman and Gandossy (1982) have demonstrated, some white-collar offenders are more successful than others in minimizing the sanctions they receive by virtue of the sufficiency of their accounts. Thus, accounts may play an important and poorly understood role in the judicial process.

DENYING THE GUILTY MIND

Adjudication as a criminal is, to use Garfinkel's (1956) classic term, a degradation ceremony. The focus of this article is on how offenders attempt to defeat the success of this ceremony and deny their own criminality through the use of accounts. However, in the interest of showing in as much detail as possible all sides of the experience undergone by these offenders, it is necessary to treat first the guilt and inner anguish that is felt by many white-collar offenders even though they deny being criminals. This is best accomplished by beginning with a description of a unique feature of the prosecution of white-collar crimes.

In white-collar criminal cases, the issue is likely to be *why* something was done, rather than *who* did it (Edelhertz, 1970: 47). There is often relatively little disagreement as to what happened. In the words of one Assistant U.S. Attorney interviewed for the study:

> If you actually had a movie playing, neither side would dispute that a person moved in this way and handled this piece of paper, etc. What it comes down to is, did they have the criminal intent?

If the prosecution is to proceed past the investigatory stages, the prosecutor must infer from the pattern of events that conscious criminal intent was present and believe that sufficient evidence exists to convince a jury of this interpretation of the situation. As Katz (1979: 445–446) has noted, making this inference can be difficult because of the way in which white-collar illegalities are integrated into ordinary occupational routines. Thus, prosecutors in conducting trials, grand jury hearings, or plea negotiations spend a great deal of effort establishing that the defendant did indeed have the necessary criminal intent. By concentrating on the offender's motives, the prosecutor attacks the very essence of the white-collar offender's public and personal image as an upstanding member of the community. The offender is portrayed as someone with a guilty mind.

Not surprisingly, therefore, the most consistent and recurrent pattern in the interviews, though not present in all of them, was denial of criminal intent, as opposed to the outright denial of any criminal behavior whatsoever. Most offenders acknowledged that their behavior probably could be construed as falling within the conduct proscribed by statute, but they uniformly denied that their actions were motivated by a guilty mind. This is not to say, however, that offenders *felt* no guilt or shame as a result of conviction. On the contrary, indictment, prosecution, and conviction provoke a variety of emotions among offenders.

The enormous reality of the offender's lived emotion (Denzin, 1984) in admitting guilt is perhaps best illustrated by one offender's description of his feelings during the hearing at which he pled guilty.

> You know [the plea's] what really hurt. I didn't even know I had feet. I felt numb. My head was just floating. There was no feeling, except a state of suspended animation. . . . For a brief moment, I almost hesitated. I almost said not guilty. If I had been alone, I would have fought, but my family. . . .

The traumatic nature of this moment lies, in part, in the offender's feeling that only one aspect of his life is being considered. From the offender's point of view his crime represents only one small part of his life. It does not typify his inner self, and to judge him solely on the basis of this one event seems an atrocious injustice to the offender.

For some the memory of the event is so painful that they want to obliterate it entirely, as the two following quotations illustrate.

> I want quiet. I want to forget. I want to cut with the past.

> I've already divorced myself from the problem. I don't even want to hear the names of certain people ever again. It brings me pain.

For others, rage rather than embarrassment seemed to be the dominant emotion.

> I never really felt any embarrassment over the whole thing. I felt rage and it wasn't false or self-serving. It was really [something] to see this thing in action and recognize what the whole legal system has come to through its

development, and the abuse of the grand jury system and the abuse of the indictment system. . . .

The role of the news media in the process of punishment and stigmatization should not be overlooked. All offenders whose cases were reported on by the news media were either embarrassed or embittered or both by the public exposure.

> The only one I am bitter at is the newspapers, as many people are. They are unfair because you can't get even. They can say things that are untrue, and let me say this to you. They wrote an article on me that was so blasphemous, that was so horrible. They painted me as an insidious, miserable creature, wringing out the last penny. . . .

Offenders whose cases were not reported on by the news media expressed relief at having avoided that kind of embarrassment, sometimes saying that greater publicity would have been worse than any sentence they could have received.

In court, defense lawyers are fond of presenting white-collar offenders as having suffered enough by virtue of the humiliation of public adjudication as criminals. On the other hand, prosecutors present them as cavalier individuals who arrogantly ignore the law and brush off its weak efforts to stigmatize them as criminals. Neither of these stereotypes is entirely accurate. The subjective effects of conviction on white-collar offenders are varied and complex. One suspects that this is true of all offenders, not only white-collar offenders.

The emotional responses of offenders to conviction have not been the subject of extensive research. However, insofar as an individual's emotional response to adjudication may influence the deterrent or crime-reinforcing impact of punishment on him or her, further study might reveal why some offenders stop their criminal behavior while others go on to careers in crime (Casper, 1978: 80).

Although the offenders displayed a variety of different emotions with respect to their experiences, they were nearly unanimous in denying basic criminality. To see how white-collar offenders justify and excuse their crimes, we turn to their accounts. The small number of cases rules out the use of any elaborate classification techniques. Nonetheless, it is useful to group offenders by offense when presenting their interpretations.

Antitrust Violators

Four of the offenders had been convicted of antitrust violations, all in the same case involving the building and contracting industry. Four major themes characterized their accounts. First, antitrust offenders focused on the everyday character and historical continuity of their offenses.

> It was a way of doing business before we even got into the business. So it was like why do you brush your teeth in the morning or something. . . . It was part of the everyday. . . . It was a method of survival.

The offenders argued that they were merely following established and necessary industry practices. These practices were presented as being necessary for the well-being of the industry as a whole, not to mention their own companies. Further, they argued that cooperation among competitors was either allowed or actively promoted by the government in other industries or professions.

The second theme emphasized by the offenders was the characterization of their actions as blameless. They admitted talking to competitors and admitted submitting intentionally noncompetitive bids. However, they presented these practices as being done not for the purpose of rigging prices nor to make exorbitant profits. Rather, the everyday practices of the industry required them to occasionally submit bids on projects they really did not want to have. To avoid the effort and expense of preparing full-fledged bids, they would call a competitor to get a price to use. Such a situation might arise, for example, when a company already had enough work for the time being, but was asked by a valued customer to submit a bid anyway.

> All you want to do is show a bid, so that in some cases it was for as small a reason as getting your deposit back on the plans and specs. So you just simply have no interest in getting the job and just call to see if you can find someone to give you a price to use, so that you didn't have to go through the expense of an entire bid preparation. Now that is looked on very unfavorably, and it is a technical violation, but it was strictly an opportunity to keep your name in front of a desired customer. Or you may find yourself in a situation where somebody is doing work for a customer, has done work for many, many years and is totally acceptable, totally fair. There is no problem. But suddenly they (the customer) get an idea that they ought to have a few tentative figures, and you're called in, and you are in a moral dilemma. There's really no reason for you to attempt to compete in that circumstance. And so there was a way to back out.

Managed in this way, an action that appears on the surface to be a straightforward and conscious violation of antitrust regulations becomes merely a harmless business practice that happens to be a "technical violation." The offender can then refer to his personal history to verify his claim that, despite technical violations, he is in reality a law-abiding person. In the words of one offender, "Having been in the business for 33 years, you don't just automatically become a criminal overnight."

Third, offenders were very critical of the motives and tactics of prosecutors. Prosecutors were accused of being motivated solely by the opportunity for personal advancement presented by winning a big case. Further, they were accused of employing prosecution selectively and using tactics that allowed the most culpable offenders to go free. The Department of Justice was painted as using antitrust prosecutions for political purposes.

The fourth theme emphasized by the antitrust offenders involved a comparison between their crimes and the crimes of street criminals. Antitrust offenses differ in their mechanics from street crimes in that they are not committed in one place and at one time. Rather, they are spatially and temporally diffuse and are intermingled with legitimate behavior. In addition, the victims of antitrust offenses tend not to be identifiable individuals, as is the case with most street crimes. These characteristics are used by antitrust violators to contrast their own behavior with that of common stereotypes of criminality. Real crimes are pictured as discrete events that have beginnings and ends and involve individuals who directly and purposely victimize someone else in a particular place and at a particular time.

> It certainly wasn't a premeditated type of thing in our cases as far as I can see. . . . To me it's different than __ and I sitting down and we plan, well we're going to rob this bank tomorrow and premeditatedly go in there. . . . That wasn't the case at all. . . . It wasn't like sitting down and planning I'm going to rob this bank type of thing. . . . It was just a common everyday way of doing business and surviving.

A consistent threat running through all of the interviews was the necessity for antitrust-like practices, given the realities of the business world. Offenders seemed to define the situation in such a manner that two sets of rules could be seen to apply. On the one hand, there are the legislatively determined rules—laws—which govern how one is to conduct one's business affairs. On the other hand, there is a higher set of rules based on the concepts of profit and survival, which are taken to define what it means to be in business in a capitalistic society. These rules do not just regulate behavior; rather, they constitute or create the behavior in question. If one is not trying to make a profit or trying to keep one's business going, then one is not really "in business." Following Searle (1969: 33–41), the former type of rule can be called a regulative rule and the latter type a constitutive rule. In certain situations, one may have to violate a regulative rule in order to conform to the more basic constitutive rule of the activity in which one is engaged.

This point can best be illustrated through the use of an analogy involving competitive games. Trying to win is a constitutive rule of competitive games in the sense that if one is not trying to win, one is not really playing the game. In competitive games, situations may arise where a player deliberately breaks the rules even though he knows or expects he will be caught. In the game of basketball, for example, a player may deliberately foul an opponent to prevent him from making a sure basket. In this instance, one would understand that the fouler was trying to win by gambling that the opponent would not make the free throws. The player violates the rule against fouling in order to follow the higher rule of trying to win.

Trying to make a profit or survive in business can be thought of as a constitutive rule of capitalist economies. The laws that govern *how* one is allowed to make a profit are regulative rules, which can understandably be subordinated to the rules of trying to survive and profit. From the offender's point of view, he is doing what businessmen in our society are supposed to do—that is, stay in business and make a profit. Thus, an individual who violates society's laws or regulations in certain situations may actually conceive of himself as thereby acting more in accord with the central ethos of his society than if he had been a strict observer of its law. One might suggest, following Denzin (1977), that for businessmen in the building and contracting industry, an informal structure exists below the articulated legal structure, one which frequently supersedes the legal structure. The informal structure may define as moral and "legal" certain actions that the formal legal structure defines as immoral and "illegal."

Tax Violators

Six of the offenders interviewed were convicted of income tax violations. Like antitrust violators, tax violators can rely upon the complexity of the tax laws and an historical tradition in which cheating on taxes is not really criminal. Tax offenders would claim that everybody cheats somehow on their taxes and present themselves as victims of an unlucky break, because they got caught.

> Everybody cheats on their income tax, 95% of the people. Even if it's for $10 it's the same principle. I didn't cheat. I just didn't know how to report it.

The widespread belief that cheating on taxes is endemic helps to lend credence to the offender's claim to have been singled out and to be no more guilty than most people.

Tax offenders were more likely to have acted as individuals rather than as part of a group and, as a result, were more prone to account for their offenses by referring to them as either mistakes or the product of special circumstances. Violations were presented as simple errors which resulted from ignorance and poor recordkeeping. Deliberate intention to steal from the government for personal benefit was denied.

> I didn't take the money. I have no bank account to show for all this money, where all this money is at that I was supposed to have. They never found the money, ever. There is no Swiss bank account, believe me.

> My records were strictly one big mess. That's all it was. If only I had an accountant, this wouldn't even of happened. No way in God's creation would this ever have happened.

Other offenders would justify their actions by admitting that they were wrong while painting their motives as altruistic rather than criminal. Criminality was denied because they did not set out to deliberately cheat the government for their own personal gain. Like the antitrust offenders discussed above, one tax violator distinguished between his own crime and the crimes of real criminals.

> I'm not a criminal. That is, I'm not a criminal from the standpoint of taking a gun and doing this and that. I'm a criminal from the standpoint of making a mistake, a serious mistake. . . . The thing that really got me

involved in it is my feeling for the employees here, certain employees that are my right hand. In order to save them a certain amount of taxes and things like that, I'd extend money to them in cash, and the money came from these sources that I took it from. You know, cash sales and things of that nature, but practically all of it was turned over to the employees, because of my feeling for them.

All of the tax violators pointed out that they had no intention of deliberately victimizing the government. None of them denied the legitimacy of the tax laws, nor did they claim that they cheated because the government is not representative of the people (Conklin, 1977: 99). Rather, as a result of ignorance or for altruistic reasons, they made decisions which turned out to be criminal when viewed from the perspective of the law. While they acknowledged the technical criminality of their actions, they tried to show that what they did was not criminally motivated.

Violations of Financial Trust

Four offenders were involved in violations of financial trust. Three were bank officers who embezzled or misapplied funds, and the fourth was a union official who embezzled from a union pension fund. Perhaps because embezzlement is one crime in this sample that can be considered *mala in se*, these offenders were much more forthright about their crimes. Like the other offenders, the embezzlers would not go so far as to say "I am a criminal," but they did say "What I did was wrong, was criminal, and I knew it was." Thus, the embezzlers were unusual in that they explicitly admitted responsibility for their crimes. Two of the offenders clearly fit Cressey's scheme as persons with financial problems who used their positions to convert other people's money to their own use.

Unlike tax evasion, which can be excused by reference to the complex nature of tax regulations or antitrust violations, which can be justified as for the good of the organization as a whole, embezzlement requires deliberate action on the part of the offender and is almost inevitably committed for personal reasons. The crime of embezzlement, therefore, cannot be accounted for by using the same techniques that tax violators or antitrust violators do. The act itself can only be explained by showing that one was under extraordinary circumstances which explain one's uncharacteristic behavior. Three of the offenders referred explicitly to extraordinary circumstances and presented the offense as an aberration in their life history. For example, one offender described his situation in this manner:

As a kid, I never even—you know kids will sometimes shoplift from the dime store—I never even did that. I had never stolen a thing in my life and that was what was so unbelievable about the whole thing, but there were some psychological and personal questions that I wasn't dealing with very well. I wasn't terribly happily married. I was married to a very strong-willed woman and it just wasn't working out.

The offender in this instance goes on to explain how, in an effort to impress his wife, he lived beyond his means and fell into debt.

A structural characteristic of embezzlement also helps the offender demonstrate his essential lack of criminality. Embezzlement is integrated into ordinary occupational routines. The illegal action does not stand out clearly against the surrounding set of legal actions. Rather, there is a high degree of surface correspondence between legal and illegal behavior. To maintain this correspondence, the offender must exercise some restraint when committing his crime. The embezzler must be discrete in his stealing; he cannot take all of the money available to him without at the same time revealing the crime. Once exposed, the offender can point to this restraint on his part as evidence that he is not really a criminal. That is, he can compare what happened with what could have happened in order to show how much more serious the offense could have been if he was really a criminal at heart.

> What I could have done if I had truly had a devious criminal mind and perhaps if I had been a little smarter—and I am not saying that with any degree of pride or any degree of modesty whatever, [as] it's being smarter in a bad, an evil way—I could have pulled this off on a grander scale and I might still be doing it.

Even though the offender is forthright about admitting his guilt, he makes a distinction between himself and someone with a truly "devious criminal mind."

Contrary to Cressey's (1953: 57–66) findings, none of the embezzlers claimed that their offenses were justified because they were underpaid or badly treated by their employers. Rather, attention was focused on the unusual circumstances surrounding the offense and its atypical character when compared to the rest of the offender's life. This strategy is for the most part determined by the mechanics and organizational format of the offense itself. Embezzlement occurs within the organization but not for the organization. It cannot be committed accidentally or out of ignorance. It can be accounted for only by showing that the actor "was not himself" at the time of the offense or was under such extraordinary circumstances that embezzlement was an understandable response to an unfortunate situation. This may explain the finding that embezzlers tend to produce accounts that are viewed as more sufficient by the justice system than those produced by other offenders (Rothman and Gandossy, 1982). The only plausible option open to a convicted embezzler trying to explain his offense is to admit responsibility while justifying the action, an approach that apparently strikes a responsive chord with judges.

Fraud and False Statements

Ten offenders were convicted of some form of fraud or false statements charge. Unlike embezzlers, tax violators, or antitrust violators, these offenders were much more likely to deny committing any crime at all. Seven of the ten claimed that they, personally, were innocent of any crime, although each admitted that fraud had occurred. Typically, they claimed to have been set up by associates and to have been wrongfully convicted by the U.S. Attorney handling the case. One might call this the scapegoat strategy. Rather than admitting technical wrongdoing and then justifying or excusing it, the offender attempts to paint himself as a victim by shifting the blame entirely to another party. Prosecutors were presented as being either ignorant or politically motivated.

The outright denial of any crime whatsoever is unusual compared to the other types of offenders studied here. It may result from the nature of the crime of fraud. By definition, fraud involves a conscious attempt on the part of one or more persons to mislead others. While it is theoretically possible to accidentally violate the antitrust and tax laws, or to violate them for altruistic reasons, it is difficult to imagine how one could accidentally mislead someone else for his or her own good. Furthermore, in many instances, fraud is an aggressively acquisitive crime. The offender develops a scheme to bilk other people out of money or property, and does this not because of some personal problem but because the scheme is an easy way to get rich. Stock swindles, fraudulent loan scams, and so on are often so large and complicated that they cannot possibly be excused as foolish and desperate solutions to personal problems. Thus, those involved in large-scale frauds do not have the options open to most embezzlers of presenting themselves as persons responding defensively to difficult personal circumstances.

Furthermore, because fraud involves a deliberate attempt to mislead another, the offender who fails to remove himself from the scheme runs the risk of being shown to have a guilty mind. That is, he is shown to possess the most essential element of modern conceptions of criminality: an intent to harm another. His inner self would in this case be exposed as something other than what it has been presented as, and all of his previous actions would be subject to reinterpretation in light of this new perspective. For this reason, defrauders are most prone to denying any crime at all. The cooperative and conspiratorial nature of many fraudulent schemes makes it possible to put the blame on someone else and to present oneself as a scapegoat. Typically, this is done by claiming to have been duped by others.

Two illustrations of this strategy are presented below.

> I figured I wasn't guilty, so it wouldn't be that hard to disprove it, until, as I say, I went to court and all of a sudden they start bringing in these guys out of the woodwork implicating me that I never saw. Lot of it could be proved that I never saw.

> Inwardly, I personally felt that the only crime that I committed was not telling on these guys. Not that I deliberately, intentionally committed a crime against the system. My only crime was that I should have had the guts to tell on these guys, what they were doing, rather than putting up with it and then trying to gradually get out of the system without hurting them or without them thinking I was going to snitch on them.

Of the three offenders who admitted committing crimes, two acted alone and the third acted with only one other person. Their accounts were similar to the others presented earlier and tended to focus on either the harmless nature of their violations or on the unusual circumstances that drove them to commit their crimes. One claimed that his violations were only technical and that no one besides himself had been harmed.

> First of all, no money was stolen or anything of that nature. The bank didn't lose any money. . . . What I did was a technical violation. I made a mistake. There's no question about that, but the bank lost no money.

Another offender who directly admitted his guilt was involved in a check-kiting scheme. In a manner similar to embezzlers, he argued that his actions were motivated by exceptional circumstances.

> I was faced with the choice of all of a sudden, and I mean now, closing the doors or doing something else to keep that business open. . . . I'm not going to tell you that this wouldn't have happened if I'd had time to think it over, because I think it probably would have. You're sitting there with a dying patient. You are going to try to keep him alive.

In the other fraud cases more individuals were involved, and it was possible and perhaps necessary for each offender to claim that he was not really the culprit.

DISCUSSION: OFFENSES, ACCOUNTS, AND DEGRADATION CEREMONIES

The investigation, prosecution, and conviction of a white-collar offender involves him in a very undesirable status passage (Glaser and Strauss, 1971). The entire process can be viewed as a long and drawn-out degradation ceremony with the prosecutor as the chief denouncer and the offender's family and friends as the chief witnesses. The offender is moved from the status of law-abiding citizen to that of convicted felon. Accounts are developed to defeat the process of identity transformation that is the object of a degradation ceremony. They represent the offender's attempt to diminish the effect of his legal transformation and to prevent its becoming a publicly validated label. It can be suggested that the accounts developed by white-collar offenders take the forms that they do for two reasons: (1) the forms are required to defeat the success of the degradation ceremony, and (2) the specific forms used are the ones available given the mechanics, history, and organizational context of the offenses.

Three general patterns in accounting strategies stand out in the data. Each can be characterized by the subject matter on which it focuses: the event (offense), the perpetrator (offender), or the denouncer (prosecutor). These are the natural subjects of accounts in that to be successful, a degradation ceremony requires each of these elements to be presented in a particular manner (Garfinkel, 1956). If an account giver can undermine the presentation of one or more of the elements, then the effect of the ceremony can be reduced. Although there are overlaps in the accounting strategies used by the various types of offenders, and while any given offender may use more than one strategy, it appears that accounting strategies and offenses correlate. . . .

NOTES

1. There is a well-developed literature which treats the subjective experiences of deviants and which has obvious associations with the subject at hand. For examples, see Matza (1969) and Becker (1963).

2. Cressey's (1953) study of embezzlers is an exception to this generalization.

3. The term "account" is used here in a special sense developed by Scott and Lyman (1968). It is described in greater detail in a later section of the paper.

4. Some offenders whose files were examined were not available to be interviewed because they were incarcerated at the time of the study.

5. All of the offenders who consented to being interviewed were men.

REFERENCES

Becker, Howard S. 1963. Outsiders. New York: Free Press.

Benson, Michael. 1984. The fall from grace: Loss of occupational status among white-collar offenders. Criminology 22: 573–593.

Casper, Jonathan D. 1978. Criminal Courts: The Defendant's Perspective. Washington, DC: U.S. Department of Justice.

Clinard, Marshall B., and Peter C. Yeager. 1978. Corporate crime: Issues in research. Criminology 16: 255–272.

Conklin, John E. 1977. Illegal But Not Criminal: Business Crime in America. Englewood Cliffs, NJ: Prentice-Hall.

Cressey, Donald. 1953. Other People's Money. New York: Free Press.

Denzin, Norman K. 1977. Notes on the criminogenic hypothesis: A case study of the American liquor industry. American Sociological Review 42: 905–920.

———. 1984. On Understanding Emotion. San Francisco: Jossey-Bass.

Edelhertz, Herbert. 1970. The Nature, Impact, and Prosecution of White Collar Crime. Washington, DC: U.S. Government Printing Office.

Fisse, Brent, and John Braithwaite. 1983. The Impact of Publicity on Corporate Offenders. Albany, NY: State University of New York.

Garfinkel, Harold. 1956. Conditions of successful degradation ceremonies. American Journal of Sociology 61: 420–424.

Geis, Gilbert. 1967. The heavy electrical equipment antitrust cases of 1961. In Gilbert Geis and Robert Meier (eds.), White-Collar Crime. New York: Free Press.

———. 1982. On White-Collar Crime. Lexington, MA: Lexington.

Glaser, Barney G., and Anselm L. Strauss. 1971. Status Passage. Chicago: Aldine.

Goffman, Erving. 1959. The Presentation of Self in Everyday Life. Garden City, NY: Anchor.

———. 1972. Relations in Public. New York: Harper Colophon.

Katz, Jack. 1979. Legality and equality: Plea bargaining in the prosecution of white-collar crimes. Law and Society Review 13: 431–460.

Matza, David. 1969. Becoming Deviant. Englewood Cliffs, NJ: Prentice-Hall.

Meier, Robert, and Gilbert Geis. 1982. The psychology of the white-collar offender. In Gilbert Geis (ed.), On White-Collar Crime. Lexington, MA: Lexington.

Naughton, James M., John Crewdson, Ben Franklin, Christopher Lydon, and Agis Solpukas. 1977. How Agnew bartered his office to keep from going to jail. In Gilbert Geis and Robert F. Meier (eds.), White-Collar Crime. New York: Free Press.

Nettler, Gwynn. 1982. Explaining Criminals. Cincinnati, OH: Anderson.

Rothman, Martin, and Robert F. Gandossy. 1982. Sad tales: The accounts of white-collar defendants and the decision to sanction. Pacific Sociological Review 4: 449–473.

Scott, Marvin B., and Stanford M. Lyman. 1968. Accounts. American Sociological Review 33:46–62.

Searle, John R. 1969. Speech Acts. Cambridge: Cambridge University Press.

Shapiro, Susan P. 1980. Thinking about White-Collar Crime: Matters of Conceptualization and Research. Washington, DC: U.S. Government Printing Office.

Sutherland, Edwin H. 1949. White Collar Crime. New York: Dryden.

Sykes, Gresham, and David Matza. 1957. Techniques of neutralization: A theory of delinquency. American Sociological Review 22: 664–670.

22

Return to Sender: Reintegrative

Stigma-Management Strategies

of Ex-Psychiatric Patients

NANCY J. HERMAN

Although scholars have addressed the exit phase of deviant careers (cf. Adler and Adler 1983; Faupel 1991; Frazier 1976; Glassner et al. 1983; Harris 1973; Inciardi 1970; Irwin 1970; Luckenbill and Best 1981; Meisenhelder 1977; and Ray 1961), the issue of reintegrating deviants into society has received little sociological attention. So too has little attention been given to the wide array of factors affecting role exit and reintegration. . . .

Despite the preponderance of sociological research on the mentally ill, there is a dearth of ethnographically-based studies dealing with the post-hospital lives of ex-psychiatric patients. Such studies, as they do exist, deal largely with chronic ex-patients living in halfway houses or boarding homes, or involved in specific aftercare treatment programs (Cheadle et al. 1978; Estroff 1981; Lamb and Goertzel, 1977; Reynolds and Farberow 1977). Little systematic attention has been given to ex-patients' perceptions of mental illness as a stigmatizable/stigmatizing attribute,[1] the numerous problems they face on the outside, the ways such persons manage discreditable information about themselves in the context of social interaction with others, and the consequences of employing these strategies for altering their deviant identities and social reintegration. In this paper I address these deficits in the sociological literature by presenting ethnographic evidence from a study of 146 non-chronic,[2] ex-psychiatric patients. First, I begin with a discussion of the setting and methods used in this study. Second, I illustrate how ex-patients came to perceive their attribute as potentially stigmatizing. Third, I analyze the five strategies these persons developed and employed in their "management work." Finally, I address the implications of adopting such stratagems for identity transformation and social reintegration. In this paper I hope to contribute to the existing literature on stigma, deviant career exit, management work, and the reintegration of deviants.

From: "Return to Sender: Reintegrative Stigma-Management Strategies of Ex-Psychiatric Patients," Nancy J. Herman, *Journal of Contemporary Ethnography,* Vol. 22, No. 3, October 1993. Reprinted by permission of Sage Publications, Inc.

SETTING AND METHODS

My interest in mental illness and psychiatric patients has been a long-standing one. My father was employed as an occupational therapist for twenty-five years at a large psychiatric institution in a metropolitan city in Ontario, Canada. Throughout my childhood and adolescent years, I made frequent trips to the institution, interacting with many of the patients on the "admission" and "back" wards, in the hospital canteen, in the occupational therapy workshop, and on the grounds. Moreover, during those years, my father often brought a number of his patients into our home (those, in particular, whose families had abandoned them or who did not have any relatives) to spend Easter, Thanksgiving, and Christmas holidays with our family.

My childhood interest in mental patients sparked my initial involvement in this topic, and I began to study mental patients in 1980 as a graduate student at McMaster University. Having a father who was highly esteemed by the institutional gatekeepers greatly facilitated my access. . . . I spent eight months studying the institutionalization of psychiatric patients ethnographically (Herman 1981). After completing this study on the pre- and in-patient phases of the patients' careers, I became interested in learning about the post-patient phase of their careers. As a result of the movement toward deinstitutionalization, chronic patients, once institutionalized for periods of years, were being released into the community. Moreover, newly diagnosed patients were being hospitalized for brief periods and shortly released. I became interested in examining the post-hospital social worlds and experiences of these ex-psychiatric clients.

I began a four-year research project on this topic in January 1981 (see Herman 1986). In contrast to my earlier study on institutionalized mental patients, where I encountered numerous problems with institutional gatekeepers, this time I encountered relatively few such problems. Rather than dealing with the same provincial institution to obtain a list of discharged patients, I contacted the Director of Psychiatric Services and Professor of Psychiatry at the Medical School affiliated with my university. After hearing about my research proposal, he granted his support for the project and served as my sponsor to the Ethics Committee of the hospital. Approximately one month later, I was given access to a listing of discharged (chronic[3] and non-chronic) ex-psychiatric patients who had been released from the psychiatric wards of seven general hospitals and from two psychiatric institutions in the Southern Ontario area between 1975 and including 1981. In order to protect the identities of those ex-patients not wishing to participate in my study, and to avoid litigation against the hospital for violating rights of confidentiality, I agreed to only be given access to the names of those individuals who agreed to participate. A stratified random sample of 300 ex-patients was formed from the overall discharge list. Upon drawing this sample, the hospitals sent out a letter to each potential subject on my behalf, outlining the general nature of the study, my identity, and my affiliation. Two weeks later, hospital officials made follow-up telephone calls asking potential subjects if they were willing to participate in the study. I was subsequently given a list containing the names of 285 willing participants, 146 non-chronic and 139 chronic ex-patients.[4]

I initially conducted informal interviews with each of the ex-patients in coffee shops, in their homes, in malls, or at their places of work. The interviews lasted from three to five and one-half hours. These interviews not only provided me with a wealth of information about the social worlds of ex-patients, but many subjects invited me to subsequently attend and participate with them in various social settings, including self-help group meetings, activist group meetings and protest marches, and therapy sessions. In addition, I frequently met ex-patients where they worked during coffee and lunch breaks and was able to observe them interacting with co-workers. I ate lunches and dinners in their homes (as they did in mine) and watched them interacting with family members, friends, and neighbors. Each Wednesday afternoon, I met a group of six ex-patients at a local doughnut shop where they would discuss the problems they were facing "on the outside" and collectively search for possible remedies.

PERCEPTIONS OF MENTAL ILLNESS AS STIGMA

In his classic work *Stigma*, Goffman (1963:4) distinguishes between stigmata that are "discrediting" and those that are "discreditable"; the former refer to attributes that are immediately apparent to others, such as obesity, physical abnormalities, and blindness; the latter refer to attributes that are not visible or readily apparent to others, such as being a secret homosexual or ex-prisoner. Mental illness is conceptualized, for the most part, as a discreditable or potentially stigmatizing attribute in that it is not readily apparent to others. It is equally important to note however, that for some ex-psychiatric patients, mental illness is a deeply discrediting attribute. Many ex-patients, especially chronic ones, were rendered discredited by inappropriate patterns of social interaction and the side effects of medication, which took the forms of twitching, swaying, jerking, and other bizarre mannerisms (see Herman 1986).[5] When ex-patients made voluntary disclosures of personal information about their psychiatric histories or when other people somehow became aware of their histories, ex-patients were categorized in negative terms, as representing some sort of personal failure to "measure up" to the rest of society. Tom, age 45, summed it up for most of the ex-patients:

> Having been diagnosed as a psychiatric patient with psychotic tendencies is the worst thing that has ever happened to me. It's shitty to be mentally ill; it's not something to be proud of. It makes you realize just how different you are from everybody else—they're normal and you're not. Things are easy for them; things are hard for you. Life's a ball for them; life's a bitch for you! I'm like a mental cripple! I'm a failure at life! . . .

MENTAL ILLNESS AND STRATEGIES OF
INFORMATION MANAGEMENT

Some studies on the discreditable (Edgerton 1967; Humphreys 1972; Ponse 1976) have suggested that individuals either disclose their attribute to others *or* make attempts to actively conceal such information about their selves. Other studies (Bell and Weinberg 1978; Miall 1986; Schneider and Conrad 1980; Veevers 1980) have suggested that being a "secret deviant" is far more complex than either choosing to disclose or not disclose one's "failing." These studies suggest that individuals *selectively* conceal such information about themselves at certain times, in certain situations, and with certain individuals and freely disclose the same information at other times, in other situations, and with other individuals. Concealment and disclosure, then, are contingent upon a "complex interaction of one's learned perceptions of the stigma [of their attribute], actual 'test' experiences with others before/or after disclosure, and the nature of the particular relationship involved" (Schneider and Conrad 1980:39).

The complex reality of how individuals selectively conceal and disclose information was evident in the case of non-chronic ex-psychiatric patients. Examination of their post-hospital worlds revealed not only that many ex-patients faced economic hardships, had problems coping in the community, and experienced adverse side effects from their "meds"[6] but also that their perception of mental illness as a potentially stigmatizing attribute presented severe problems in their lives. Many lived their lives in states of emotional turmoil, afraid and frustrated—deciding who to tell or not tell, when to tell and when not to tell, and how to tell. Joan, a 56-year-old waitress, aptly summed it up for most non-chronics:

> It's a very difficult thing. It's not easy to distinguish the good ones from the bad ones. . . . You've gotta figure out who you can tell about your illness and who you better not tell. It is a tremendous stress and strain that you have to live with 24 hours a day!

Ex-psychiatric patients learned how, with whom, and under which circumstances to disclose or conceal their discreditable aspects of self, largely through a process of trial and error, committing numerous *faux pas* along the way. Frank, a 60-year-old factory worker, spoke of the number of mistakes he made in his "management work":

> I was released over two years now. And since then, I've developed an ulcer trying to figure out how to deal with my "sickness"—that is, how or whether others could handle it or not. I screwed up things a few times when I told a couple of guys on the bowling team. I made a mistake and thought that they were my buddies and would accept it.

In fact, even if no *faux pas* were committed, there was no guarantee that others would accept preferred meanings and definitions of self. As Charlie, a 29-year-old graduate student, hospitalized on three occasions, remarked:

I'm not a stupid person. I learned how to handle effectively the negative aspects of my sickness—I mean how others view it. I've been doing OK now since my discharge, but still, each time I'm entering a new situation, I get anxious; I'm not always a hundred percent sure of whether to tell or not to—especially in the case of dating relationships. Even if you've had success in telling certain types of people, there's always the chance—and it happens more than you think, that people will just not "buy" what you're trying so desperately to "sell" them.

Nearly 80 percent of the non-chronics in this study engaged in some form of information control about their illnesses and past hospitalizations. Specifically, the stratagems adopted and employed by the ex-patients, resembling those observed in other deviant groups (cf. Davis 1961; Hewitt and Stokes 1975; Levitin 1975; Miall 1986; Schneider and Conrad 1980), included (1) selective concealment, (2) therapeutic disclosure, (3) preventive disclosure, and (4) political activism—stratagems adopted by ex-patients in their effort to lessen or avoid the stigma potential of mental illness, elevate self-esteem, renegotiate societal conceptions of mental illness as a discreditable attribute, and alter deviant identities.

Selective Concealment

Selective concealment can be defined as the selective withholding or disclosure of information about self perceived as discreditable in cases where secrecy is the major stratagem for handling information about an attribute. Especially during the period directly following their psychiatric treatment, the majority of non-chronics had a marked desire to conceal such information about their selves from all others. Decisions about disclosure and concealment were made on the basis of their perceptions of others—that is, whether they were "safe others" or "risky others." So too were decisions based on prior, negative experiences with "certain types" of others. Speaking of her classification of others into "trustworthy" and "untrustworthy" others, Dawina, a 46-year-old secretary, institutionalized on seven occasions, said:

It's like this. There are two types of people out there, "trustworthy" ones—the people who will be understanding and supportive—and "untrustworthy" ones. Out of all of my friends and relations and even the people I work with at the company, I only decided to tell my friend Sue.

Moreover, there was a hierarchical pattern of selective disclosure based upon the individual's perceived degree of closeness and the ex-patient's revealing his or her discreditable attribute. In general, such information was most frequently revealed to family members, followed by close friends, and then acquaintances, a pattern also reflected in the literature on epileptics (Schneider and Conrad 1980) and involuntary childless women (Miall 1986). As Sarah, a 36-year-old mother of two, put it:

When I was discharged, I didn't automatically hide from everyone the fact that I was hospitalized for a nervous breakdown again. But I didn't go and tell everyone either. I phoned and told my relatives in "Logenport," and I confided in two of my close, good friends here in town.

Further, selective disclosures to normal others were frequently made to test reactions. Similar to Schneider and Conrad's (1980) epileptics, the ex-patients continued to disclose mental illness contingent upon responses they had received to previous disclosures. Rudy, a 39-year-old man hospitalized on ten occasions, stated:

> You learn through trial and error. When I was let out back in 1976, I was still naive, you know. I decided to tell a few people. Boy, was that a mistake. They acted as if I had AIDS. Nobody wanted anything further to do with me. . . . Since then, I've pretty much dummied up and not told anyone!

In those cases where concealment was the dominant strategy of information management, ex-patients usually disclosed only to one or two individuals. As Simon, a 25-year-old ex-patient, aptly expressed:

> I decided from the moment that my treatment ended, I would tell as few people as possible about my stay in the psychiatric hospital. I figured that it would be for the best to "keep it under a lid" for the most part. So, to this day, I've only confided in my friend Paul and a neighbor who had a similar illness a while back.

The employment of concealment as a stratagem of information management took the following forms: avoidance of selected "normals," redirection of conversations, withdrawal, the use of disidentifiers, and the avoidance of stigma symbols. Speaking on his efforts to redirect conversations, Mark, a 34-year-old non-chronic, explained:

> Look, you've got to remain on your toes at all times. More often than not, somebody brings up the topic about my past and starts probing around. Sometimes these people won't let up. . . . I use the tactic where I change the subject, answer their question with a question. . . . I try to manipulate the conversation so it works out in my favor.

For still others, concealment of their discreditable attribute was achieved through withdrawal. Over two-thirds of the ex-patients in this study engaged in withdrawal as a form of concealment, especially during the early months following discharge. Speaking of his use of this technique, Harry, a college junior, remarked:

> Sometimes when I'm at a party or some type of gathering with a number of people, I just remain pretty reticent. I don't participate too much in the conversations. . . . I'm really unsure how much to tell other people. For the most part, I just keep pretty quiet and remain a wallflower. People

may think I'm shy or stuck-up, but I'd rather deal with that than with the consequences of others finding out that I'm a mental patient.

A third technique, employed by over one-third of the ex-patients to conceal their discreditable aspects from others, involved the use of disidentifiers (Goffman 1963:44). That is, ex-patients utilized misleading physical or verbal symbols in an effort to prevent normal others from discovering their "failing." Similar to homosexuals (Carrier 1976; Delph 1978), unwed parents (Christensen 1953; Pfuhl 1978), and lesbians (Ponse 1978), who frequently made use of disidentifiers in their management work, non-chronic ex-patients also employed such techniques. Specifically, disidentifiers took the form of making jokes about psychiatric patients while in the presence of "normal" others and participating in protests *against* the integration of ex-patients into the community. Mike, a 26-year-old ex-patient, recently released after three hospitalizations, remarked (with some remorse) on his use of this tactic:

> They wanted to use this house down the street for a group home for discharged patients. All the neighbors on the street were up in arms over it. It didn't upset me personally, but the neighbors made up this petition, and to protect myself, I not only signed it, but I also went door-to-door convincing other neighbors to sign it and "keep those mentals out". . . . I felt sort of bad afterwards, but what else could I do?

In a similar vein, Morgan, a 49-year-old history professor, explained how his joke-telling aided in the concealment of his attribute:

> I conceal this information about myself—my psychiatric past—by frequently telling jokes about mental patients to my colleagues in the elevator, and sometimes even my lectures, and in everyday conversation. It's really a great ploy to use. People may think the jokes are in bad taste, but at the very least, it helps me to keep secret my illness.

A final form of concealing information on the part of ex-patients was through the avoidance of stigma symbols (cf. Goffman 1963:43)—signs that would bring into the forefront or disclose their discreditable attribute. It is interesting to note that the data presented here on non-chronics and their avoidance of stigma symbols supports observations made of other deviant groups, for example, transsexuals (Bogdan 1974; Kando 1973) and unwed fathers (Pfuhl 1978). . . . Among the 146 ex-patients studied, over two-thirds avoided contact with such stigma symbols as other ex-mental patients with whom they had become friends while institutionalized and self-help groups for ex-patients. So, too, did they avoid frequenting drop-in centers, attending dances and bingo games for ex-patients, and, in general, placing themselves where other "patients and ex-patients hung out." For still others, avoidance of stigma symbols entailed not attending post-hospital therapy sessions. Margarette, a stocky, middle-aged woman of German descent, explained her avoidance of post-discharge therapy sessions in the following manner:

After I was released, my psychiatrist asked that I make appointments and see him every two weeks for follow-up maintenance treatments. But I never did go because I didn't want someone to see me going into the psychiatric department of "Meadowbrook Hospital" and sitting in the waiting room of the "Nut Wing." Two of my nosy neighbors are employed at that hospital, and I just couldn't take the chance of them seeing me there one day.

In sum, as a strategy of information management, selective concealment of their attribute and past hospitalizations was done to protect themselves from the perceived negative consequences that might result from the revelation of their illness—an "offensive tactical maneuver" through which ex-patients attempted (although often unsuccessfully) to mitigate the stigma potential of mental illness on their daily lives. Notably, employing concealment as a strategy of information management was a *temporal* process. The majority of ex-patients employed this strategy primarily during the first eight months following their discharge. During this time, in particular, they expressed feelings of anxiety, fear, and trepidation. As time passed, however, ex-patients began to test reactions. They encountered both positive and negative responses from certain "normals," and their strong initial desires for secrecy were replaced by alternative strategies.

Therapeutic Disclosures

Therapeutic disclosure can be defined as the selective disclosure of a discreditable attribute to certain "trusted," "empathetic" supportive others in an effort to renegotiate personal perceptions of the stigma of "failing." . . . Thirty-six percent of the ex-patients felt that discussing their mental illnesses and past hospitalizations, getting it off their chests in a cathartic fashion, functioned to alleviate much of the burden of their loads. Attesting to the cathartic function disclosing served, Vincent, a 29-year-old ex-patient, remarked:

> Finally, letting it all out, after so many secrets, lies, it was so therapeutic for me. Keeping something like this all bottled up inside is self-destructive. When I came clean, this great burden was lifted from me!

Therapeutic disclosure was most often carried out with family members, close friends, and with other ex-psychiatric patients—individuals "sharing the same fate." Ida, age 52, discussing the circumstances surrounding her disclosure to a neighbor who had also been hospitalized in a psychiatric facility at one time, said:

> At first, I was apprehensive to talk about it. But keeping it inside of you all bottled up is no good either. One day, I walked down the street to a neighbor of mine and she invited me in to have tea. I knew what had happened to her years ago (her deceased husband confided in my husband). I let out all my anxieties and fears to her that afternoon. . . . I told

her everything and she was so sympathetic. . . . She knew exactly what I was going through. Once I let it all out, I felt so much better.

Even in cases where ex-patients disclosed to individuals who turned out to be unsympathetic and unsupportive, some considered this therapeutic:

When I came out of hiding and told people about my sickness, not everyone embraced me. A lot of people are shocked and just tense up. Some just stare. . . . A few never call you after that time or make up excuses not to meet with you. . . . But I don't care, because overall, telling made me feel better.

Just as therapeutic disclosure functioned to relieve ex-patients' anxieties and frustrations, it also allowed for the renegotiation of personal perceptions of mental illness as a discreditable attribute. Speaking of the manner by which she came to redefine mental illness in her own mind as a less stigmatizing attribute, Edith explained:

When I finally opened up and started talking about it, it really wasn't so bad after all. My Uncle John was very supportive and helped me to put my mind at rest, to realize that having mental illness isn't so bad; it's not like having cancer. He told me that thousands of people go into the hospital each year for psychiatric treatment and probably every third person I meet has had treatment. . . . After much talking, I no longer think of myself as less human, but more normally. . . . Having mental illness isn't the blight I thought it was.

In short, then, ex-patients employed therapeutic disclosure in order to relieve feelings of frustration and anxiety, to elevate their self-esteem, and to renegotiate (in their own minds) personal perceptions of mental illness as stigmatizing.

Preventive Disclosure

Preventive disclosure can be described as the selective disclosure to "normals" of a discreditable attribute in an effort to influence others' actions or perceptions about the ex-patient or about mental illness in general (cf. Miall 1986; Schneider and Conrad 1980). Preventive disclosure of their mental illness and past hospitalizations occurred in situations where ex-patients anticipated future rejection by "normal" others. In order to minimize the pain of subsequent rejection, 34 percent of the sample decided that the best strategy to employ with certain people was preventive disclosure *early* in their relationships. As Hector, a 40-year-old janitor, said:

I figured out that, for me, it is best to inform people right off the bat about my mental illness. Why? Because you don't waste a lot of time developing relationships and then are rejected later. That hurts too much. Tell them early and if they can't deal with it, and run away, you don't get hurt as much!

Preventive disclosure, then, represented a way ex-patients attempted to prevent a drop in their status at a later date, or a way of testing acquaintances in an effort to establish friendship boundaries.

Just as non-chronics used preventive disclosure to avoid future stigma and rejection, so too did they employ this strategy to influence normals' attitudes about themselves and about mental illness in general. Specifically, ex-patients used the following devices: (1) medical disclaimers (cf. Hewitt and Stokes 1975; Miall 1986; Schneider and Conrad 1980); (2) deception/coaching (cf. Goffman 1963; Miall 1986; Schneider and Conrad 1980); (3) education (cf. Schneider and Conrad 1980); and (4) normalization (cf. Cogswell 1967; Davis 1961, 1963; Levitin 1975; McCaghy 1968; Scott 1969).

Medical Disclaimers

Fifty-two percent of the ex-patients frequently used medical disclaimers in their management work—"blameless, beyond-my-control medical interpretation(s)" developed in order to "reduce the risk that more morally disreputable interpretations might be applied by naive others" discovering their failing (Schneider and Conrad 1980:41). Such interpretations were often used by ex-patients to evoke sympathy from others and to ensure that they would be treated in a charitable manner. As Dick, an unemployed laborer, put it:

> When I tell people about my hospitalization in a psychiatric hospital, I immediately emphasize that the problem isn't anything I did, it's a biological one. I didn't ask to get sick; it was just plain biology; or my genes that fucked me up. I try to tell people in a nice way so that they see mental illness just like other diseases—you know, cancer or the mumps. It's not my parents' fault or my own. . . . I just tell them, "Don't blame me, blame my genes!"

In a similar vein, Anna, a 29-year-old waitress, explained her use of medical disclaimers:

> Talking about it is quite tricky. When I tell them about it, I'm careful to emphasize that the three times I was admitted, was due to a biochemical imbalance—something that millions of people get. I couldn't do anything to help myself—I ate properly, didn't drink or screw around. It's not something I deserved. When you give people the facts and do it in a clinical fashion, you can sway many of them to sympathize with you.

Further, eleven ex-patients revealed their mental illness and past hospitalizations as a side effect of another medical problem or disease, such as childbirth, stroke, or heart disease, thereby legitimizing what otherwise might be considered a potentially stigmatizing condition. As Rebecca, age 36, confessed:

> I have had heart problems since birth. I was a very sick baby. I've had four operations since that time and I've been on all kinds of medications. The stress of dealing with such an illness led to my depression and subsequent

breakdowns. . . . When my friends hear about mental illness in this light, they are very empathetic.

While Sue spoke of her successes in influencing others' perceptions about her attribute and mental illness in general, Lenny lamented about his failure with the same strategy:

> Life's not easy for ex-nuts, you know. I tried telling two of my drinking buddies about my schizophrenia problem one night at the bar. I thought if I told 'em that it's a "disease" like having a heart problem that they would understand and pat me on the butt and say it didn't matter to them and that I was OK. Shit, it didn't work out like I planned; they flipped out on me. Sid couldn't handle it at all and just let out of there in a hurry; Jack stayed around me for about twenty minutes and then made some excuse and left.

In sum, through the use of medical disclaimers, ex-patients hoped to elevate their self-esteem and to renegotiate personal perceptions of mental illness as a non-stigmatizing attribute.

Deception/Coaching

Deception differed from strategies of concealment in that with the former, ex-patients readily disclosed their illness and past hospitalizations but explicitly distorted the conditions or circumstances surrounding it. Similar to Miall's (1986) involuntary childless women and Schneider and Conrad's (1980) epileptics, about one-third of the ex-patients employed deceptive practices developed with the assistance of coaches. Coaches included parents, close friends, spouses, and other ex-patients sharing the same stigma. Coaches actively provided ex-patients with practical suggestions on how to disclose their attribute in the least stigmatizing manner and present themselves in a favorable light. Maureen, age 32, explained of her "coaching sessions" with relatives:

> My parents and grandma really helped me out in terms of what I should say or tell others. They were so afraid I'd be hurt that they advised me what to tell my school mates, the manager at Wooldo where I got hired. We had numerous practice exercises where we'd role-play and I'd rehearse what I would say to others. . . . After a while I became quite convincing.

Moreover, it is interesting to note that about one quarter of the ex-patients employed deceptive practices together with medical disclaimers. As Benjamin, age 62, aptly expressed:

> To survive in this cruel, cold world, you've got to be sneaky. I mean, that you've got to try to win people over to your side. Whoever you decide to tell about your illness, you've got to make it clear that you had nothing to

do with getting sick; nobody can place blame on anyone. . . . And you've got to color the truth about how you ended up in the hospital by telling heart-sob stories to get people sympathetic to you. You never tell them the whole truth or they'll shun you like the plague!

Education

A third form of preventive disclosure used by ex-patients to influence others' perceptions of them and their ideas about mental illness was education. . . . Twenty-eight percent of the ex-patients revealed their attribute in an effort to educate others. Marge, age 39, speaking on her efforts to educate friends and neighbors, said:

> I have this urge inside of me to teach people out there, to let them know that they've been misinformed about mental illness and mental patients. We're not the way the media has portrayed us. That's why people are afraid of us. I feel very strongly that someone has to tell people the truth . . . give them the facts. . . . And when they hear it, they're amazed sometimes and begin to treat me without apprehension. . . . Each time I make a breakthrough, I think more highly of myself too.

Ex-patients did not automatically attempt to educate everyone they encountered but, rather, based on subjective typification of normals, made value judgments about whom to "educate." Brenda, speaking on this matter, explained:

> You just can't go ahead and tell everyone. You ponder who it is, what are the circumstances, and whether you think that they can be educated about it. There are some people that these efforts would be fatal and fruitless. Others however, you deem as a potential. And these are the people you work with.

While education proved successful for some ex-patients in their management activities with certain individuals, others found it less successful. Jim, recalling one disastrous experience with a former poker buddy, said:

> I really thought he would learn something from my discussion of the facts. I really misjudged Fred. I thought him to be an open-minded kind of guy but perhaps just naive, so I sat him down one afternoon and made him my personal "mission." I laid out my past and then talked to him about all the kinds of mental illnesses that are out there. He reacted terrible. All his biases came out, and he told me that all those people should be locked up and the key thrown away—that they were a danger to society. He was probably thinking the same thing about me too!

Following Goffman (1963:101), medical disclaimers, deception/coaching, and education are forms of "disclosure etiquette"—they are formulas for revealing a stigmatizing attribute "in a matter of fact way, supporting the assumption that those present are above such concerns while preventing them from trapping themselves into showing that they are not."

Normalization

A final form of preventive disclosure employed by ex-psychiatric patients to manage stigma was normalization. This concept is drawn from Davis's (1963) study on children with polio and is akin to deviance disavowal (cf. Davis 1961). Normalization is a strategy individuals use to deny that their behavior or attribute is deviant. It "seeks to render normal and morally acceptable that which has heretofore been regarded as abnormal and immoral" (Pfuhl 1986:163). Similar to observations made on pedophiles (McCaghy 1968), the obese (Millman 1980), the visibly handicapped (Levitin 1975), and paraplegics (Cogswell 1967), about one quarter of the ex-psychiatric patients I studied also employed this same strategy. Such persons were firmly committed to societal conceptions of normalcy and were aware that according to these standards, they were disqualified—they would never "measure up." Yet ex-patients made active attempts at rationalizing and downplaying the stigma attached to their failing. So, for example, they participated in a full round of normal activities and aspired to normal attainments. They participated in amateur theater groups, played competitive sports such as hockey and tennis, enrolled in college, and the like. Ex-patients whose stigma could be considered "discreditable," that is, not readily or visibly apparent to others, would disclose such information for preventive reasons, thereby rendering them "discredited" in the eyes of others. They would then attempt to negotiate with normals for preferred images, attitudes, roles, and non-deviant conceptions of self and definitions of mental illness as less stigmatizing. Discussing his utilization of this technique, "Weird Old Larry," age 59 said:

> The third time I got out [of the hospital], I tried to fit right in. I told some of my buddies and a couple of others about my sickness. It was easier to get it out in the open. But what I tried to show 'em was that I could do the same things they could, some of them, even better. I beat them at pool, at darts; I could outdrink them, I was holding down two jobs—one at the gas station and at K-mart. I tried to show them I was normal. I was cured! The key to success is being up-front and making them believe you're just as normal as them. . . . You can really change how they see and treat you.

If successfully carried out, this avowal normalized relations between ex-patients and others.

This is not to imply, however, that the strategy of normalization worked for all patients in all situations. Similar to Millman's (1980:78) overweight females who were accepted in certain roles but treated as deviant in others, many ex-patients expressed similar problems. Frederick, speaking on this problem with respect to co-workers, said:

It's really tragic, you know. When I told the other people at work that I was a manic-depressive but was treated and released, I emphasized that I was completely normal in every way . . . but they only accepted me normally part of the time, like when we were in the office. . . . But they never really accepted me as their friend, as one of "the boys"; and they never invited me over to dinner with their wife and family—they still saw me as an ex-crazy, not as an equal to be worthy being invited to dinner, or playing with their kids.

It is interesting to note that just as ex-patients whose attribute was discreditable employed the strategy of normalization, so, too, did other ex-patients with discrediting attributes (conditions *visibly apparent* or *known* to others) employ this same technique. Explaining how medication side effects rendered him discredited and how he attempted to reduce the stigma of mental illness through normalization, Ross said:

Taking all that dope the shrinks dish out makes my hands tremble. Look at my shaking legs too. I never used to have these twitches in my face either, but that's just the side effects, a bonus you get. It really fucks things up though. If I wanted to hide my illness, I couldn't; everyone just looks at me and knows. . . . So, what I do is to try to get people's attention and get them to see my positive side—that I can be quite normal, you know. I emphasize all the things that I can do!

In short, by presenting themselves as normals, ex-patients hoped to elicit positive responses from others whose reactions were deemed to be important. From a social-psychological perspective, others accepted and reinforced a non-deviant image of self through this process of negotiation, allowing ex-patients to achieve more positive, non-deviant identities.

In many cases, ex-psychiatric patients progressed from one strategy to another as they managed information about themselves. Specifically, they moved from a strategy of initial selective concealment to disclosure for therapeutic and preventive reasons. According to the ex-patients, such a progression was linked to their increased adjustment to their attribute as well as the result of positive responses from others to the revelation of their mental illness.

Political Activism

Just as ex-psychiatric patients developed and employed a number of individualized forms of information management to deal with the stigma potential of mental illness, enhance self-images, and alter deviant identities, they also employed one collective management strategy[7] to achieve the same ends, namely, joining and participating in ex-mental patient activist groups (cf. Anspach, 1979). Such groups, with their goal of self-affirmation, represent what Kitsuse (1980:9) terms "tertiary deviation"—referring to the deviant's confirmation, assessment, and rejection of the negative identity embedded in secondary deviation, and the transformation of that identity into a positive and viable self-conception."

Political activism served a three-fold function for ex-patients: (1) it repudiated standards of normalcy (standards to which they couldn't measure up) and the deviant labels placed upon these individuals; (2) it provided them with a new, positive, non-deviant identity, enhanced their self-respect, and afforded them a new sense of purpose; and (3) it served to propagate this new, positive image of ex-mental patient to individuals, groups, and organizations in society. The payoff from political activism, was, then, personal as well as social.

Similar to such activist groups as the Gay Liberation Front, the Disabled in Action, the Gray Panthers, or the Radical Feminist Movement, ex-mental patient activists rejected prevailing societal values of normalcy through participation in their groups. They repudiated the deviant identities, roles, and statuses placed on them. Moreover, these individuals flatly rejected the stigma associated with their identities. Steve, a 51-year-old electrician, aptly summed it up for most ex-patient activists:

> The whole way society had conceived of right and wrong, normal and abnormal is all wrong. They somehow have made us believe that to be mentally ill is to be *ashamed* of something; that these people are to be feared, that they are to blame for their sickness. Well I don't accept this vein anymore.

Upon repudiating prevailing cultural values and deviant identities, ex-patient activists collectively redefined themselves in a more positive, nondeviant light according to their *own* newly constructed set of standards. Speaking of her embracing a new non-deviant identity, Susan, age 39, who recently returned to teaching school, said:

> I no longer agree to accept what society says is normal and what is not. It's been so unfair to psychiatric patients. Who are they to say, just because we don't conform, that we're rejects of humanity. . . . The labels they've given us are degrading and make us feel sick. . . . [The labels] have a negative connotation to them. . . . So, we've gotten together and liberated ourselves. We've thrown away the old labels and negative images of self-worth, and we give ourselves new labels and images of self-worth—as human beings who should be treated with decency and respect.

In contrast to other ex-patients who employed various individual management strategies to deal with what they perceived to be their *own* problems—personal failings—ex-patient activists saw their problems not as personal failings or potentially stigmatizing attributes but as *societal* problems. To the extent that ex-patients viewed their situations in this manner, it allowed them to develop more positive self-images. Speaking of this process as one of "stigma conversion," Humphreys (1972:142) writes:

In converting his stigma, the oppressed person does not merely exchange his social marginality for political marginality. . . . Rather, he emerges from a stigmatized cocoon as a transformed creature, one characterized by the spreading of political wings. At some point in the process, the politicized "deviant" gains a new identity, an heroic self-image as crusader in a political cause.

Sally, a neophyte activist, placed the "blame" on society for her deviant self–image:

It's not any of our faults that we ended up the way we did. I felt guilty for a long time. . . . I crouched away feeling that I had something that made me "different" from everyone else, a pock on my life. . . . But I learned at the activist meetings that none of it was my fault. It was all society's fault—they're the one who can't deal with anything that is different. Now I realize that having mental illness is nothing to be ashamed of; it's nothing to hide. I'm now proud of who . . . . and what I am!

Just as political activism, as contrasted with other adaptive responses to stigma, sought, in repudiating the dominant value system, to provide ex-patients with positive, non-deviant statuses, so too, did it attempt to propagate this new positive, normal image of ex-psychiatric patient to others in society. Thus, through such activities as rallies, demonstrations, protest marches, attendance at conferences on human rights for patients, lobbyist activities directed toward politicians and the medical profession, and the production of newsletters, ex-patient activists sought to promote social change. Specifically, they sought to counter or remove the stigma associated with their "differentness" and present society with an image of former psychiatric patients as "human beings" capable of self-determination and political action. Abe, the president of the activist group, aptly summed up the aim of political activism during a speech to selected political figures, media personnel, and "upstanding" citizens:

Simply put, we're tired of being pushed around. We reject everything society says about us, because it's just not accurate. We reject the type of treatment we get . . . both in the hospital . . . and out. We don't like the meaning of the words [people] use to describe us—"mentals" and "nuts." We see ourselves differently, just as good and worthy as everybody out there. In our newsletter, we're trying to get across the idea that we're not the stereotypical mental patient you see in the movies. We're real people who want to be treated equally under the Charter of Rights. We're not sitting back now, we're fighting back!

In sum, then, through participation in political activist groups, many ex-patients internalized an ideology that repudiated societal values and conventional standards of normalcy, rejected their deviant identities and statuses, adopted more positive, non-deviant identities, and attempted to alter society's stereotypical perceptions about mental patients and mental illness in general. . . .

NOTES

1. The stigmatized status of individuals and the information management strategies they employ have been well-documented for other groups such as the retarded (Edgerton 1967), epileptics (Schneider and Conrad 1980), secret homosexuals (Humphreys 1975), involuntary childless women (Miall 1986), swingers (Bartell 1971), and lesbians (Ponse 1976), among others.

2. Chronicity, for the purposes of this study, was defined *not* in diagnostic terms, that is, "chronic schizophrenic"; rather, it was defined in terms of duration, continuity, and frequency of hospitalizations. Specifically, the term "non-chronic" refers to those individuals hospitalized for periods of less than two years, those institutionalized on a discontinuous basis, those hospitalized on fewer than five occasions, or those treated in psychiatric wards of general hospitals.

3. The term "chronic" ex-psychiatric patient refers to those institutionalized in psychiatric hospitals for periods of two years or more, institutionalized on a continual basis, or hospitalized on five or more occasions.

4. The decision to stratify the sample by chronicity was based upon my interests and prior fieldwork activities. My earlier study (Herman 1981) indicated that when we speak of "deinstitutionalized" or "discharged" patients, we cannot merely assume that they are one homogeneous grouping of individuals with like characteristics, and similar post-hospital social situations, experiences, and perceptions of reality. Rather, prior research led me to believe that there might be distinct subgroups of individuals with varying perceptions of reality and experiences (see Herman 1986).

5. In Estroff's (1981) ethnography on chronic ex-mental patients, she points out the catch-22 situation in which they find themselves. Ex-patients need to take their medications regularly in order to remain on the outside. Ironically, however, in an effort to become more like others, they take "meds" that make them "different." The various side effects reinforce their deviant identities.

6. See Herman (1986) for a detailed discussion of such other post-hospital problems.

7. Following Lyman's (1970) typology of deviant voluntary associations, ex-mental patient political activist groups represent an "instrumental-alienative" type of association. It is interesting to note that chronic ex-patients also employed one collective form of stigma management; specifically, they formed and participated in deviant subcultures (see Herman 1987).

REFERENCES

Adler, Patricia A. 1992. "The 'post' phase of deviant careers: Reintegrating drug traffickers." *Deviant Behavior* 13:103–126.

Adler, Patricia A., and Peter Adler. 1983. Shifts and oscillations in deviant careers: The case of upper-level drug dealers and smugglers. *Social Problems* 31:195–207.

Anspach, Renee. 1979. From stigma to identity politics: Political activism among the physically disabled and former mental patients. *Social Science and Medicine* 13A:765–773.

Bartell, Gilbert D. 1971. *Group Sex: A Scientist's Eyewitness Report on the American Way of Swinging*. New York: Wyden.

Bell, Alan, and Martin S. Weinberg. 1978. *Homosexualities: A Study of Diversity among Men and Women*. New York: Simon and Schuster.

Bogdan, Robert. 1974. *Being Different: The Autobiography of Jane Fry*. New York: Wiley.

Carrier, J. M. 1976. Family attitudes and Mexican male homosexuality. *Urban Life* 50:359–375.

Cheadle, A. J., H. Freeman, and J. Korer. Chronic schizophrenic patients in the community. *British Journal of Psychiatry* 132:221–227.

Christensen, Harold T. 1953. Studies in child spacing: Premarital pregnancy as measured by the spacing of the first birth from marriage. *American Sociological Review* 18:53–59.

Cogswell, B. 1967. Rehabilitation of the paraplegic: Processes of socialization. *Sociological Inquiry* 37:11–26.

Davis, Fred. 1961. Deviance disavowal: The management of strained interaction by the visibly handicapped. *Social Problems* 9:120–132.

———. 1963. *Passage through Crisis: Polio Victims and Their Families*. Indianapolis: Bobbs-Merrill.

Delph, E. 1978. *The Silent Community: Public Homosexual Encounters*. Beverly Hills, CA: Sage.

Edgerton, Robert. 1967. *The Cloak of Competence: Stigma in the Lives of the Mentally Retarded*. Berkeley: University of California Press.

Estroff, Sue E. 1981. *Making It Crazy: An Ethnography of Psychiatric Clients in an American Community*. Berkeley: University of California Press.

Faupel, Charles E. 1991. *Shooting Dope: Career Patterns of Hard-Core Heroin Users*. Gainesville: University of Florida Press.

Frazier, Charles. 1976. *Theoretical Approaches to Deviance*. Columbus, OH: Merrill.

Glassner, Barry, Margret Ksander, Bruce Berg, and Bruce D. Johnson. 1983. A note on the deterrent effect of juvenile vs. adult jurisdiction. *Social Problems* 31:219–221.

Goffman, Erving. 1961. *Asylums*. New York: Doubleday.

———. 1963. *Stigma*. Englewood Cliffs, NJ: Prentice-Hall.

Harris, Mervyn. 1973. *The Dilly Boys*. Rockville, MD: New Perspectives.

Herman, Nancy J. 1981. *The Making of a Mental Patient: An Ethnographic Study of the Processes and Consequences of Institutionalization upon Self-Images and Identities.* Unpublished master's thesis, McMaster University, Hamilton, Ontario.

————. 1986. *Crazies in the Community: An Ethnographic Study of Ex-Psychiatric Clients in Canadian Society—Stigma, Management Strategies and Identity Transformation.* Unpublished Ph.D. dissertation, McMaster University, Hamilton, Ontario.

————. 1987. "Mixed nutters" and "looney tuners": The emergence, development, nature, and functions of two informal, deviant subcultures of chronic, ex-psychiatric patients. *Deviant Behavior* 8:235–258.

Hewitt, J., and R. Stokes. 1975. Disclaimers. *American Sociological Review* 40:1–11.

Humphreys, Laud. 1972. *Out of the Closets: The Sociology of Homosexual Liberation.* Englewood Cliffs, NJ: Prentice-Hall.

————. 1975. *Tearoom Trade.* New York: Aldine de Gruyter.

Inciardi, James. 1975. *Careers in Crime.* Chicago: Rand McNally.

Irwin, John. 1970. *The Felon.* Englewood Cliffs, NJ: Prentice-Hall.

Kando, T. 1973. *Sex Change: The Achievement of Gender Identity among Feminized Transsexuals.* Springfield, IL: Charles C. Thomas.

Kitsuse, John. 1980. Presidential address. *Society for the Study of Social Problems* 9:1–13.

Lamb, J., and V. Goertzel. 1977. The long-term patient in the era of community treatment. *Archives of General Psychiatry* 34:679–682.

Levitin, T. 1975. Deviants as active participants in the labelling process: The case of the visibly handicapped. *Social Problems* 22:548–557.

Luckenbill, David F., and Joel Best. 1981. Careers in deviance and respectability: The analogy's limitations. *Social Problems* 29:197–206.

Lyman, Stanford M. 1970. *The Asian in the West.* Reno and Las Vegas, NV: Western Studies Center, Desert Research Institute.

McCaghy, Charles H. 1968. Drinking and deviance disavowal: The case of child molesters. *Social Problems* 16:43–49.

Meisenhelder, Thomas. 1977. An exploratory study of exiting from criminal careers. *Criminology* 15:319–334.

Miall, Charlene E. 1986. The stigma of involuntary childlessness. *Social Problems* 33(4):268–282.

Millman, Marcia. 1980. *Such a Pretty Face.* New York: Norton.

Pfuhl, Erdwin H., Jr. 1978. The unwed father: A "non-deviant" rule breaker. *Sociological Quarterly* 19(Winter):113–128.

Ponse, Barbara. 1976. Secrecy in the lesbian world. *Urban Life* 5:313–338.

Ray, Marsh. 1961. The cycle of abstinence and relapse among heroin addicts. *Social Problems* 9:132–140.

Reynolds, David K., and Norman Farberow. 1977. *Endangered Hope: Experiences in Psychiatric Aftercare Facilities*. Berkeley: University of California Press.

Schneider, J., and P. Conrad. 1980. In the closet with illness: Epilepsy, stigma potential and information control. *Social Problems* 28(1):32–44.

Scott, Robert. 1969. The socialization of blind children. In *Handbook of Socialization Theory and Research,* edited by D. Goslin. Chicago: Rand McNally.

Veevers, Jean. 1980. *Childless by Choice*. Toronto: Butterworths.

23

Confronting Deadly Disease

The Drama of Identity Construction

*among Gay Men with AIDS**

KENT L. SANDSTROM

The phenomenon of AIDS (acquired immunodeficiency syndrome) has been attracting increased attention from sociologists. A number of observers have examined the social meanings of the illness (Conrad 1986; Sontag 1989; Palmer 1989), the social influences and behavior involved in its onset and progression (Kaplan et al. 1987) and the larger social consequences of the AIDS epidemic (Ergas 1987). Others have studied the "psychosocial" issues faced by individuals who are either diagnosed with the illness (Nichols 1985; Baumgartner 1986: Weitz 1989) or closely involved with

*Thanks are given to Dennis Brissett, Vicki Kessler, Jeylan Mortimer, Joseph Kotarba, Robert Fulton, Gary Alan Fine, Barbara Laslett, John Clark, and the anonymous reviewers for their helpful comments on earlier drafts of this article. Thanks also to Dan Martin for his valuable assistance in interviewing some of the participants in this study. This research was supported by a grant from the Conflict and Change Project, University of Minnesota.

From: "Confronting Deadly Disease: The Drama of Identity Construction among Gay Men with AIDS," Kent L. Sandstrom, *Journal of Contemporary Ethnography,* Vol. 19, No. 3, October 1990. Reprinted by permission of Sage Publications, Inc.

someone who has been diagnosed (Salisbury 1986; Geiss, Fuller and Rush 1986; Macklin 1988).

Despite this growing interest in the social and psychosocial dimensions of AIDS, little attention has been directed toward the processes of social and self-interaction (Denzin 1983) by which individuals acquire and personalize an AIDS-related identity. Further, given the stigmatizing implications of AIDS, there has been a surprising lack of research regarding the strategies of stigma management and identity construction utilized by persons with this illness.

This article presents an effort to address these issues. It examines the dynamics of identity construction and management which characterize the everyday lives of persons with AIDS (PWAs). In doing so, it highlights the socially ambiguous status of PWAs and considers (a) the processes through which they personalize the illness, (b) the dilemmas they encounter in their interpersonal relations, (c) the strategies they employ to avoid or minimize potentially discrediting social attributions, and (d) the subcultural networks and ideologies which they draw upon as they construct, avow and embrace AIDS-related identities. Finally, these themes are situated within the unfolding career and lived experience of people with AIDS.

METHOD AND DATA

The following analysis is based on data gathered in 56 in-depth interviews with 19 men who had been diagnosed with HIV (human immunodeficiency virus) infections. On the average, each individual was interviewed on three separate occasions and each of these sessions lasted from 1 to 3 hours. The interviews, conducted between July 1987 and February 1988, were guided by 60 open-ended questions and were audiotaped. Most interviews took place in the participants' homes. However, a few participants were interviewed in a private university office because their living quarters were not conducive to a confidential conversation.

Participants were initially recruited through two local physicians who treat AIDS patients and through a local self-help organization that provides support groups and services for people with HIV infections. Those individuals who agreed to be interviewed early in the study spoke with friends or acquaintances and encouraged them to become involved. The majority of interviews were thus obtained through "snowball" or chain-referral sampling.

By employing a snowball sampling procedure, we were able to gain fairly rapid access to persons with AIDS-related diagnoses and to discuss sensitive issues with them. However, due to its reliance on self-selection processes and relatively small social networks, this method is not likely to reflect the range of variation which exists in the population of persons with advanced AIDS-related infections. This study is thus best regarded as exploratory.

All respondents were gay males who lived in a metropolitan area in the Midwest. They varied in age, income, and the stage of their illness. In age, they ranged from 19 to 46 years, with the majority in the 28 to 40 age bracket. Six persons were currently employed in professional or white-collar occupations. The remaining 13 were living marginally on Social Security or disability benefits. Several members of this latter group had previously been employed in either blue-collar or service occupations. Seven individuals were diagnosed with AIDS, 10 were diagnosed with ARC (AIDS-related complex), and 2 were diagnosed as HIV positive but both had more serious HIV-related health complications (e.g., tuberculosis).

ON BECOMING A PWA:
THE REALIZATION OF AN AIDS IDENTITY

For many of these men, the transformation of physical symptoms into the personal and social reality of AIDS took place most dramatically when they received a validating diagnosis from a physician. The following account reveals the impact of being officially diagnosed:

> She [the doctor] said, "Your biopsy did come out for Kaposi's sarcoma. I want you to go to the hospital tomorrow and to plan to spend most of the day there." While she is telling me this, the whole world is buzzing in my head because this is the first confirmation coming from outside as opposed to my own internal suspicions. I started to cry—it (AIDS) became very real . . . *very real.* . . .
>
> Anyway, everything started to roller coaster inside me and I was crying in the office there. The doctor said "You knew this was the way it was going to come out, didn't you?" She seemed kind of shocked about why I was crying so much, not realizing that no matter how much you are internally aware of something, to hear it from someone else is what makes it real. For instance, the first time I really accepted being gay was when other people said "You are gay!" . . . It's a social thing—you're not real until you're real to someone else.

This quote illustrates the salience of social processes for the validation and realization of an identity—in this case, an AIDS-related identity. Becoming a PWA is not simply a matter of viral infection, it is contingent on interpersonal interaction and definitions. As depicted in the quote, a rather momentous medical announcement facilitates a process of identity construction which, in turn, entails both interpersonal and subjective transformation. Within the interpersonal realm, the newly diagnosed "AIDS patient" is resituated as a social object and placed in a marginal or liminal status. He is thereby separated from many of his prior social moorings. On the subjective level, this separation produces a crisis, or a disruption of the PWA's routine activities and self-understanding. The diagnosed individual is prompted to "make sense" of the meaning of his newly acquired status and to feel its implications for future conceptions and enactments of self. . . .

INTERPERSONAL DILEMMAS
ENCOUNTERED BY PWAS

Stigmatization

Stigmatization is one of the most significant difficulties faced by people with AIDS as they attempt to fashion a personal and social meaning for their illness. The vast majority of our informants had already experienced some kind of stigma because of their gay identities. When they were diagnosed with AIDS, they usually encountered even stronger homophobic reactions and discreditation efforts. An especially painful form of stigmatization occurred when PWAs were rejected by friends and family members after revealing their diagnosis. Many respondents shared very emotional accounts of how they were ostracized by parents, siblings, or colleagues. Several noted that their parents and family members had even asked them to no longer return home for visits. However, rejections were not always so explicit. In many cases, intimate relationships were gradually and ambiguously phased out rather than abruptly or clearly ended.

A few PWAs shared stories of being stigmatized by gay friends or acquaintances. They described how some acquaintances subtly reprimanded them when seeing them at gay bars or repeatedly reminded them to "be careful" regarding any sexual involvements. Further, they mentioned that certain gay friends avoided associating with them after learning of their AIDS-related diagnoses. These PWAs thus experienced the problem of being "doubly stigmatized" (Kowalewski 1988), that is, they were devalued within an already stigmatized group, the gay community.

PWAs also felt the effects of stigmatization in other, more subtle ways. For example, curious and even sympathetic responses on the parts of others, especially strangers, could lead PWAs to feel discredited. One PWA, reflecting on his interactions with hospital staff, observed:

> When they become aware [of my diagnosis], it seemed like people kept looking at me . . . like they were looking for something. What it felt like was being analyzed, both physically and emotionally. It also felt like being a subject or guinea pig . . . like "here's another one." They gave me that certain kind of look. Kind of that look like pity or that said "what a poor wretch," not a judgmental look but rather a pitying one.

An experience of this nature can precipitate a crisis of identity for a person with AIDS. He finds himself being publicly stigmatized and identified as a victim. Such an identifying moment can seriously challenge prior conceptions of self and serve as a turning point from which new self-images or identities are constructed (Charmaz 1980). That is, it can lead a PWA to internalize stigmatizing social attributions or it can incite him to search for involvements and ideologies which might enable him to construct a more desirable AIDS identity.

Counterfeit Nurturance

Given the physical and social implications of their illness, PWAs typically desire some kind of special nurturing from friends, partners, family members, or health practitioners. Yet displays of unusual concern or sympathy on the part of others can be threatening to self. People with AIDS may view such expressions of nurturance as counterfeit or harmful because they highlight their condition and hence confirm their sense of difference and vulnerability.

The following observation illustrates the sensitivity of PWAs to this problem:

> One thing that makes you feel kind of odd is when people come across supportive and want to be supportive but it doesn't really feel like they are supportive. There is another side to them that's like, well, they are being nice to you because they feel sorry for you, or because it makes them feel good about themselves to help someone with AIDS, not because they really care about you.

PWAs often find themselves caught in a paradox regarding gestures of exceptional help or support. They want special consideration at times, but if they accept support or concern which is primarily focused on their condition, they are likely to feel that a "victim identity" is being imposed on them. This exacerbates some of the negative self-feelings that have already been triggered by the illness. It also leads PWAs to be more wary of the motivations underlying others' expressions of nurturance.

Given these dynamics, PWAs may reach out to each other in an effort to find relationships that are more mutually or genuinely nurturing. This strategy is problematic, though, because even PWAs offer one another support which emphasizes their condition. They are also likely to remind each other of the anomalous status they share and the "spoiled" features (Goffman 1963) of their identities qua PWAs.

Ultimately, suspicions of counterfeit nurturance can lead those diagnosed with AIDS to feel mistreated by almost everyone, particularly by caregivers who are most directly involved in helping them. Correctly or incorrectly, PWAs tend to share some feelings of ambivalence and resentment towards friends, lovers, family members, and medical personnel.

Fears of Contagion and Death Anxiety

Fears of contagion present another serious dilemma for PWAs in their efforts to negotiate a functional social identity. These fears are generated not only by the fact that people with AIDS are the carriers of an epidemic illness but also because, like others with a death taint, they are symbolically associated with mass death and the contagion of the dead (Lifton 1967). The situation may even be further complicated by the contagion anxiety which homosexuality triggers for some people.

In general, others are tempted to withdraw from an individual with AIDS because of their fears of contracting the virus. Even close friends of a PWA are apt to feel more fearful or distant toward him, especially when first becoming aware of the diagnosis. They may feel anxious about the possibility of becoming infected with the virus through interactions routinely shared with him in the past (e.g., hugging and kissing). They may also wish to avoid the perils of being stigmatized themselves by friends or associates who fear that those close to a PWA are a potential source of contagion.

Another dimension of contagion anxiety is reflected in the tendency of significant others to avoid discussing issues with a PWA that might lead them to a deeper apprehension of the death-related implications of his diagnosis. As Lifton (1967) suggested, the essence of contagion anxiety is embodied in the fear that "if I come too close to a death tainted person, I will experience his death and his annihilation" (p. 518).

This death-related contagion anxiety often results in increased strain and distance in a PWA's interactions with friends or family members. It can also inhibit the level of openness and intimacy shared among fellow PWAs when they gather together to address issues provoked by their diagnoses. Responses of grief, denial, and anxiety in the face of death make in-depth discussions of the illness experience keenly problematic. According to one respondent:

> Usually no one's ever able to talk about it [their illness and dying] without going to pieces. They might start but it only takes about two minutes to break into tears. They might say something like "I don't know what to do! I might not even be here next week!" Then you can just see the ripple effect it has on the others sitting back and listening. You have every possible expression from anger to denial to sadness and all these different emotions on people's faces. And mostly this feeling of "what can we do? Well . . . Nothing!"

Problems of Normalization

Like others who possess a stigmatizing attribute, people with AIDS come to regard many social situations with alarm (Goffman 1971) and uncertainty (Davis 1974). They soon discover that their medical condition is a salient aspect of all but their most fleeting social encounters. They also quickly learn that their diagnosis, once known to others, can acquire the character of a *master status* (Hughes 1945; Becker 1963) and thus become the focal point of interaction. It carries with it, "the potential for inundating the expressive boundaries of a situation" (Davis 1974, 166) and hence for creating significant strains or rupture in the ongoing flow of social intercourse.

In light of this, one might expect PWAs to prefer interaction contexts characterized by "closed" awareness (Glaser and Strauss 1968). Their health status would be unknown to others and they would presumably encounter fewer problems when interacting. However, when in these situations, they must remain keenly attuned to controlling information and concealing attributes relevant to their diagnosis. Ironically, this requirement to be dramaturgically "on" may give rise to even more feelings of anxiety and resentment.

The efforts of persons with AIDS to establish and maintain relationships within more "open" contexts are also fraught with complications. One of the major dilemmas they encounter is how to move interactions beyond an atmosphere of fictional acceptance (Davis 1974). A context of fictional acceptance is typified by responses on the part of others which deny, avoid, or minimize the reality of an individual's diagnosis. In attempting to grapple with the management of a spoiled identity, PWAs may seek to "break through" (Davis 1974) relations of this nature. In doing so, they often try to broaden the scope of interactional involvement and to normalize problematic elements of their social identity. That is, they attempt to project "images, attitudes and concepts of self which encourage the normal to identify with [them] (i.e., 'take [their] role') in terms other than those associated with imputations of deviance" (Davis 1974, 168).

Yet even if a PWA attains success in "breaking through," it does not necessarily diminish his interactional difficulties. Instead, he can become caught in an ambiguous dilemma with respect to the requisites of awareness and normalization. Simply put, if others begin to disregard his diagnosis and treat him in a normal way, then he faces the problem of having to remind them of the limitations to normalcy imposed by this condition. The person with AIDS is thus required to perform an intricate balancing act between encouraging the normalization of his relationships and ensuring that others remain sensitized to the constraining effects of such a serious illness. These dynamics promote the construction of relationships which, at best, have a qualified sense of normalcy. They also heighten the PWA's sense that he is located in an ambiguous or liminal position.

AVOIDING OR MINIMIZING DILEMMAS

In an attempt to avoid or defuse the problematic feelings, attributions, and ambiguities which arise in their ongoing interactions, PWAs engage in various forms of identity management. In doing so, they often use strategies which allow them to minimize the social visibility of their diagnoses and to carefully control interactions with others. These strategies include *passing, covering, isolation,* and *insulation.*

The particular strategies employed vary according to the progression of their illness, the personal meanings they attach to it, the audiences serving as primary referents for self-presentations, and the dynamics of their immediate social situation.

Passing and Covering

As Goffman (1963) noted in his classic work on stigma, those with a spoiled identity may seek to pass as normal by carefully suppressing information and thereby precluding others' awareness of devalued personal attributes. The

PWAs we interviewed mentioned that "passing" was a maneuver they had used regularly. It was easily employed in the early stages of the illness when more telltale physical signs had not yet become apparent and awareness of an individual's diagnosis was [confined] to a small social circle.

However, as the illness progresses, concealing the visibility of an AIDS-related diagnosis becomes more difficult. When a person with AIDS begins to miss work frequently, to lose weight noticeably, and to reduce his general level of activity, others become more curious or suspicious about what ailment is provoking such major changes. In the face of related questions, some PWAs elected to devise a "cover" for their diagnosis which disguised troubling symptoms as products of a less discrediting illness.

One informant decided to cover his AIDS diagnosis by telling co-workers that he was suffering from leukemia:

> There was coming a point, I wasn't feeling so hot. I was tired and the quality of my life was decreasing tremendously because all of my free time was spent resting or sleeping. I was still keeping up with work but I thought I'd better tell them something before I had to take more days off here and there to even out the quality of my life. I had already had this little plan to tell them I had leukemia . . . but I thought how am I going to tell them, what am I going to tell them, how am I going to convince them? What am I going to do if someone says, "You don't have leukemia, you have AIDS!"? This was all stuff clicking around in my mind. I thought, how could they possibly know? They only know as much as I tell them.

This quote reveals the heightened concern with information control that accompanies decisions to conceal one's condition. Regardless of the psychic costs, though, a number of our informants opted for this remedial strategy. A commonly used technique consisted of informing friends, parents, or co-workers that one had cancer or tuberculosis without mentioning that these were the presenting symptoms of one's AIDS diagnosis. Covering attempts of this kind were most often employed by PWAs when relating to others who were not aware of their gay identity. These relationships were less apt to be characterized by the suspicions or challenges offered by those who knew that an individual was both gay and seriously ill.

Isolation and Insulation

For those whose diagnosis was not readily visible, dramaturgical skills, such as passing and covering, could be quite useful. These techniques were not so feasible when physical cues, such as a pale complexion, emaciated appearance, or facial lesion made the nature of a PWA's condition more apparent. Under these circumstances, negotiations with others were more alarming and they were more likely to include conflicts engendered by fear, ambiguity, and expressions of social devaluation.

In turn, some PWAs came to view physical and social isolation as the best means available to them for escaping from both these interpersonal difficulties and their own feelings of ambivalence. By withdrawing from virtually all interaction, they sought to be spared the social struggles and psychic strains that could be triggered by others' recognition of their condition. Nonetheless, this strategy was typically an unsuccessful one. Isolation and withdrawal often exacerbated the feelings of alienation that PWAs were striving to minimize in their social relationships. Moreover, their desire to be removed from the interactional matrix was frequently overcome by their need for extensive medical care and interpersonal support as they coped with the progressive effects of the illness.

Given the drawbacks of extreme isolation, a number of PWAs used a more selective withdrawal strategy. It consisted of efforts to disengage from many but not all social involvements and to interact regularly with only a handful of trusted associates (e.g., partners, friends, or family members). Emphasis was placed on minimizing contacts with those outside of this circle because they were likely to be less tolerant or predictable. PWAs engaging in this type of selective interaction tried to develop a small network of intimate others who could insulate them from potentially threatening interactions. Ideally, they were able to form a reliable social circle within which they felt little need to conceal their diagnosis. They could thereby experience some relief from the burden of stigma management and information control.

BUILDING AND EMBRACING
AN AIDS IDENTITY

Strategies such as passing, covering, isolation, and insulation are used by PWAs, especially in the earlier stages of their illness, to shield themselves from the stigma and uncertainty associated with AIDS. However, these strategies typically require a high level of personal vigilance, they evoke concerns about information control, and they are essentially defensive in nature. They do not provide PWAs with a way to reformulate the personal meaning of their diagnosis and to integrate it with valued definitions of self.

In light of this, most PWAs engage in more active types of *identity work* which allow them to "create, present and sustain personal identities which are congruent with and supportive of the(ir) self-concept[s]" (Snow and Anderson 1987, 1348). Certain types of identity work are especially appealing because they help PWAs to gain a greater sense of mastery over their condition and to make better use of the behavioral possibilities arising from their liminal condition.

The most prominent type of identity work engaged in by the PWAs we interviewed was embracement. As Snow and Anderson (1987) argued, embracement refers to "verbal and expressive confirmation of one's acceptance of and attachment to the social identity associated with a general or specific role, a set of social relationships, or a particular ideology" (p. 1354). Among the PWAs involved in this study, embracement was promoted and reinforced through participation in local AIDS-related support groups.

Support Groups and Associational Embracement

People facing an existential crisis often make use of new memberships and social forms in their efforts to construct a more viable sense of self (Kotarba 1984). The vast majority of respondents in this study became involved in PWA support groups in order to better address the crisis elicited by their illness and to find new forms of self-expression. They typically joined these groups within a few months of receiving their diagnosis and continued to attend meetings on a fairly regular basis.

By and large, support groups became the central focus of identity work and repair for PWAs. These groups were regarded as a valuable source of education and emotional support that helped individuals to cope better with the daily exigencies of their illness. At support group meetings, PWAs could exchange useful information, share feelings and troubles, and relate to others who could see beyond the negative connotations of AIDS.

Support groups also facilitated the formation of social ties and feelings of collective identification among PWAs. Within these circles, individuals learned to better nurture and support one another and to emphasize the shared nature of their problems. Feelings of guilt and isolation were transformed into a sense of group identification. This kind of *associational embracement* (Snow and Anderson 1987) was conveyed in the comments of one person who proclaimed:

> I spend almost all of my time with other PWAs. They're my best friends now and they're the people I feel most comfortable with. We support one another and we know that we can talk to each other any time, day or night.

For some PWAs, especially those with a troubled or marginal past, support group relationships provided an instant "buddy system" that was used to bolster feelings of security and self-worth. Recently formed support group friendships even took on primary importance in their daily lives. Perhaps because of the instability and isolation which characterized their life outside of support groups, a few of these PWAs tended to exaggerate the level of intimacy which existed in their newly found friendships. By stressing a romanticized version of these relationships, they were able to preserve a sense of being cared for even in the absence of more enduring social connections.

Identity Embracement and Affirmation

Most of the PWAs we interviewed had come to gradually affirm and embrace an AIDS-related identity. Participation in a support group exposed them to alternative definitions of the reality of AIDS and an ongoing system of identity construction. Hence, rather than accepting public imputations which cast them as "AIDS victims," PWAs learned to distance themselves from such designations and to avow more favorable AIDS-associated identities. In turn, the process of *identity embracement* was realized when individuals proudly announced that they were PWAs who were "living and thriving with the illness."

Continued associations with other PWAs could also promote deepening involvement in activities organized around the identity of being a person with AIDS. A case in point is provided by a man who recounted his progression as a PWA:

> After awhile, I aligned myself with other people with AIDS who shared my beliefs about taking the active role. I began writing and speaking about AIDS and I became involved in various projects. I helped to create and promote a workshop for people with AIDS. . . . I also got involved in organizing a support group for family members of PWAs.

As involvement in AIDS-related activities increases, embracement of an AIDS-centered identity is likely to become more encompassing. In some cases, diagnosed individuals found themselves organizing workshops on AIDS, coordinating a newsletter for PWAs, and delivering speeches regularly at schools and churches. Virtually all aspects of their lives became associated with their diagnosis. Being a PWA thus became both a master status and a valued career. This process was described by a person who had been diagnosed with ARC for two years:

> One interesting thing is that when you have AIDS or ARC and you're not working anymore, you tend to become a veteran professional on AIDS issues. You get calls regularly from people who want information or who want you to get involved in a project, etc. You find yourself getting drawn to that kind of involvement. It becomes almost a second career!

This kind of identity embracement was particularly appealing for a few individuals involved in this study. Prior to contracting an AIDS-related infection, they had felt rejected or unrecognized in many of their social relationships (e.g., family, work, and friendships). Ironically, their stigmatized AIDS diagnosis provided them with an opportunity for social affirmation. It offered them a sense of uniqueness and expertise that was positively evaluated in certain social and community circles (e.g., public education and church forums). It could even serve as a springboard for a new and more meaningful biography.

Ideological Embracement: AIDS as a Transforming Experience

Support groups and related self-help networks are frequently bases for the production and transmission of subcultural perspectives which controvert mainstream social definitions of a stigma. As Becker (1963) argued, when people who share a deviant attribute have the opportunity to interact with one another, they are likely to develop a system of shared meanings emphasizing the differences between their definitions of who they are and the definitions held by other members of the society. "They develop perspectives on themselves and their deviant [attributes] and on their relations with other members of the society" (p. 81). These perspectives guide the stigmatized as they engage in processes of identity construction and embracement.

Subcultural perspectives contain ideologies which assure individuals that what they do on a continuing basis has moral validity (Lofland 1969). Among PWAs, these ideologies were grounded in metaphors of transformation which included an emphasis on *special mission* and *empowerment*.

One of the most prominent subcultural interpretations of AIDS highlighted the spiritual meaning of the illness. For PWAs embracing this viewpoint, AIDS was symbolically and experientially inverted from a "curse" to a "blessing" which promoted a liberating rather than a constricting form of identity transformation. The following remarks illustrate this perspective:

> I now view AIDS as both a gift and a blessing. That sounds strange, I suppose, in a limited context. It sounds strange because we [most people] think it's so awful, but yet there are such radical changes that take place in your life from having this illness that's defined as terminal. You go through this amazing kind of *transformation*. You look at things for the first time, in a powerful new way that you've never looked at them before in your whole life.

A number of PWAs similarly stressed the beneficial personal and spiritual transitions experienced as a result of their diagnosis. They even regarded their illness as a motivating force that led them to grapple with important existential questions and to experience personal growth and change that otherwise would not have occurred.

For many PWAs, *ideological embracement* (Snow and Anderson 1987) entailed identity constructions based on a quasi-religious sense of "special mission." These individuals placed a premium on disseminating information about AIDS and promoting a level of public awareness which might inhibit the further transmission of this illness. Some felt that their diagnosis had provided them with a unique opportunity to help and educate others. They subsequently displayed a high level of personal sacrifice and commitment while seeking to spread the news about AIDS and to nurture those directly affected by this illness. Most crucially, their diagnosis provided them with a heightened sense of power and purpose:

> Basically I feel that as a person with ARC I can do more for humanity in general than I could ever do before. I never before in my life felt like I belonged here. For the most part, I felt like I was stranded on a hostile planet—I didn't know why. But now with the disease and what I've learned in my life, I feel like I really have something by which I can help other people. It gives me a special sense of purpose.
>
> I feel like I've got a mission now and that's what this whole thing is about. AIDS is challenging me with a question and the question it asks is: If I'm not doing something to help others regarding this illness, then why continue to use up energy here on this earth?

The idea of a "special mission" is often a revitalizing formulation for those who carry a death taint (Lifton 1967). It helps to provide PWAs with a sense of mastery and self-worth by giving their condition a more positive or redemptive meaning. This notion also gives form and resolution to painful feelings of loss, grief, guilt, and death anxiety. It enables individuals to make use of these emotions, while at the same time transcending them. Moreover, the idea of special mission provides PWAs with a framework through which they can moralize their activities and continuing lives.

Beliefs stressing the empowering aspects of AIDS also served as an important focus of identity affirmation. These beliefs were frequently rooted in the sense of transformation provoked by the illness. Many of those interviewed viewed their diagnosis as empowering because it led them to have a concentrated experience of life, a stronger sense of purpose, a better understanding of their personal resources and a clearer notion of how to prioritize their daily concerns. They correspondingly felt less constrained by mundane aspects of the AIDS experience and related symptoms.

A sense of empowerment could additionally be derived from others' objectification of PWAs as sources of danger, pollution, or death. This was illustrated in the remarks of an informant who had Kaposi's sarcoma:

> People hand power to me on a silver platter because they are afraid. It's not fear of catching the virus or anything, I think it is just fear of identification with someone who is dying.

The interactional implications of such attributions of power were also recognized by this same informant:

> Because I have AIDS, people leave me alone in my life in some respects if I want them to. I never used to be able to get people to back off and now I can. I'm not the one who is doing this, so to speak. They are giving me the power to do so.

Most PWAs realized their condition offered them an opportunity to experience both psychological and social power. They subsequently accentuated the empowering dimensions of their lived experience of AIDS and linked these to an encompassing metaphor of transformation.

SUMMARY AND CONCLUSIONS

People with AIDS face many obstacles in their efforts to construct and sustain a desirable social identity. In the early stages of their career, after receiving a validating diagnosis, they are confronted by painful self-feelings such as grief, guilt, and death anxiety. These feelings often diminish their desire and ability to participate in interactions which would allow them to sustain favorable images of self.

PWAs encounter additional difficulties as a result of being situated (at least initially) as liminal persons. That is, their liminal situation can heighten negative self feelings and evoke a sense of confusion and uncertainty about the social implications of their illness. At the same time, however, it releases them from conventional roles, meanings, or expectations and provides them with a measure of power and maneuverability in the processes of identity construction.

In turn, as they construct and negotiate the meaning of an AIDS-related identity, PWAs must grapple with the effects of social reactions such as stigmatization, counterfeit nurturance, fears of contagion, and death anxiety. These reactions both elicit and reinforce a number of interactional ambiguities, dilemmas, and threats to self.

In responding to these challenges, PWAs engage in various types of identity management and construction. On one hand, they may seek to disguise their diagnoses or to restrict their social and interactional involvements. PWAs are most likely to use such strategies in the earlier phases of the illness. The disadvantage of these strategies is that they are primarily defensive. They provide PWAs with a way to avoid or adjust to the effects of problematic social reactions, but they do not offer a means for affirming more desirable AIDS-related identities.

On the other hand, as their illness progresses and they become more enmeshed in subcultural networks, most PWAs are prompted to engage in forms of identity embracement which enable them to actively reconstruct the meaning of their illness and to integrate it with valued conceptions of self. In essence, through their interactions with other PWAs, they learn to embrace affiliations and ideologies which accentuate the transformative and empowering possibilities arising from their condition. They also acquire the social and symbolic resources necessary to fashion revitalizing identities and to sustain a sense of dignity and self-worth.

Ultimately, through their ongoing participation in support networks, PWAs are able to build identities which are linked to their lived experience of AIDS. They are also encouraged to actively confront and transform the stigmatizing conceptions associated with this medical condition. Hence, rather than resigning themselves to the darker implications of AIDS, they learn to affirm themselves as "people with AIDS" who are "living and thriving with the illness."

REFERENCES

Baumgartner, G. 1986. *AIDS: Psychosocial factors in the acquired immune deficiency syndrome.* Springfield, IL: Charles C. Thomas.

Becker, H. S. 1963. *Outsiders.* New York: Free Press.

Charmaz, K. 1980. The social construction of pity in the chronically ill. *Studies in Symbolic Interaction* 3:123–45.

Conrad, P. 1986. The social meaning of AIDS. *Social Policy* 17:51–56.

Davis, F. 1974. Deviance disavowal and the visibly handicapped. In *Deviance and liberty,* edited by L. Rainwater, 163–72. Chicago: Aldine.

Denzin, N. 1983. A note on emotionality, self and interaction. *American Journal of Sociology* 89:402–9.

Ergas, Y. 1987. The social consequences of the AIDS epidemic. *Social Science Research Council/Items* 41:33–39.

Geiss, S., R. Fuller, and J. Rush. 1986. Lovers of AIDS victims: Psychosocial stresses and counseling needs. *Death Studies* 10:43–53.

Goffman, E. 1963. *Stigma.* Englewood Cliffs, NJ: Prentice-Hall.

———. 1971. *Relations in public.* New York: Harper & Row.

Glaser, B. S., and A. L Strauss. 1968. *Awareness of dying.* Chicago: Aldine.

Hughes, E. C. 1945. Dilemmas and contradictions of status. *American Journal of Sociology* 50:353–59.

Kaplan, H., R. Johnson, C. Bailey, and W. Simon. 1987. The sociological study of AIDS: A critical review of the literature and suggested research agenda. *Journal of Health and Social Behavior* 28:140–57.

Kotarba, J. 1984. A synthesis: The existential self in society. In *The existential self in society,* edited by J. Kotarba and A. Fontana, 222–33. Chicago: Aldine.

Kowalewski, M. 1988. Double stigma and boundary maintenance: How gay men deal with AIDS. *Journal of Contemporary Ethnography* 7:211–28.

Lifton, R. J. 1967. *Death in life.* New York: Random House.

Lofland, J. 1969. *Deviance and identity.* Englewood Cliffs, NJ: Prentice Hall.

Macklin, E. 1988. AIDS: Implications for families. *Family Relations* 37:141–49.

Nichols, S. 1985. Psychosocial reactions of persons with AIDS. *Annals of Internal Medicine* 103:13–16.

Palmer, S. 1989. AIDS as metaphor. *Society* 26:45–51.

Salisbury, D. 1986. AIDS: Psychosocial implications. *Journal of Psychosocial Nursing* 24 (12): 13–16.

Snow, D., and L. Anderson. 1987. Identity work among the homeless: The verbal construction and avowal of personal identities. *American Journal of Sociology* 1336–71.

Sontag, S. 1989. *AIDS and its metaphors.* New York: Farrar, Straus, & Giroux.

Weitz, R. 1989. Uncertainty and the lives of persons with AIDS. *Journal of Health and Social Behavior* 30:270–81 .

24

Illness Ambiguity and the Search for Meaning

A Case Study of a Self-Help Group for Affective Disorders

DAVID A. KARP

This article reports on a group of people suffering from depression or manic depression who met weekly in "sharing and caring" groups to talk about the nature of their "illness" and the various life contingencies associated with it. It shortly became apparent that one central function of this support group was to provide a forum for making more coherent an elusive and uncertain condition. Much of people's discussion in these groups centered on the very difficult problems that depression and mania created in their lives. Most of those attending the meetings had experienced multiple hospitalizations, and many described severely dysfunctional family and work lives. The material discussed in the meetings was often poignant, and one could not help but be struck by the courage these people displayed as they daily grappled with extraordinary and debilitating pain.

The group talk also illustrated the uncertainty, ambiguity, and lack of clarity attached to the experience of having an "affective disorder." Much of the discussion centered on the confusions surrounding the nature and causes of depression, the labels that most accurately described their condition, the problems in the use of psychotropic drugs, the utility of psychotherapy, the biochemical basis for their condition, and the extent to which they were victims of a disease beyond their control. The frequency with which these and related issues came up in group discussion revealed their problematic character. Group members could not easily achieve consensus on these matters, and they were often the basis for debate concerning the meaning of their common experience. How they collectively created explanations, understandings, agreements, and common "illness ideologies" in order to impose some order onto a hazy and ill-understood life condition is the subject matter for this article. . . .

Although group discussion covered a range of topics, certain questions and themes kept recurring, and it is around these issues that participants tried to

From: "Illness Ambiguity and the Search for Meaning: A Case Study of a Self-Help Group for Affective Disorders," David A. Karp, *Journal of Contemporary Ethnography,* Vol. 21, No. 2, 1992. Reprinted by permission of Sage Publications, Inc.

work toward consensus about the meaning of their shared condition. Each of these questions refers to an interpretive dilemma that had to be resolved by individuals as they tried to comprehend their illness. From the array of issues discussed, four interconnected questions constantly reasserted themselves:

1. What is involved in getting a proper diagnosis, and what are the appropriate labels to attach to our common difficulty?
2. To what extent are we responsible for causing the problem, and what is our personal responsibility for remedying it?
3. How much ought we, or must we, rely on medical experts to provide relief for our situation?
4. Are drugs the ultimate salvation for those with affective disorders?

. . . In the following section, I describe in greater detail the setting of this research and the problems faced in gathering the data. I then discuss, in turn, how the subjects thought about the diagnoses of depression and manic depression, assessed cause and responsibility for their condition, evaluated their relationship with therapeutic experts, and interpreted the experience of taking psychotropic medications. Taken together, these features of group talk illuminate how persons collaboratively constructed a *version* of what affective disorders are all about and what one ought to do about them.

SETTING AND METHOD

The organization studied had been in existence for 5 years. I became a participant and observer in August 1988 and attended meetings regularly through November 1990. When I first attended group meetings, they were being held at a local mental hospital that had allocated to the organization three small rooms in an out-of-the-way building on its grounds. At the time, approximately 35 to 50 people attended the Wednesday night meetings. By the end of the observation period, about 120 people attended the meetings each week, and the hospital had made available to the organization a number of rooms in its relatively new cafeteria building. On a typical evening, after the president of the group greeted those present and various announcements of general interest were made, the groups for that evening were indicated. Several "open" groups were available, but along with them were groups on such specific topics as "coping with mania," "coping with depression," "spiritual healing," and "Why do I live?" In addition there were groups each week for newcomers, for family and friends, and for young people.

Significantly, as the group increased in size and, by that criterion, became more successful, the running of the meetings became somewhat more formalized. When I first attended, decisions about which members would "run" the groups were made very informally. Normally, two or three of the group's officers volunteered to do this, but, as often, people who had been coming for a couple of months or more were approached on an ad hoc basis to run a group. Within the last 6 months of my participation, the decision was made to require potential group leaders to take a 12-week leadership training course given by a consultant on small group dynamics. Consequently, only those persons who had successfully completed the training could be group leaders. As a result, a relatively small cadre of individuals repeatedly served as group leaders.

At the outset of each group session, the leader read the specific rules that were expected to govern discussion. Then, the 15 to 20 people in each group introduced themselves, described the central dimensions of their problem, and indicated what they hoped to get out of the discussion. Particular attention was given to individuals who described themselves as having an immediate crisis. If no one had a special problem to raise, group leaders suggested possible foci for the discussion. Usually, conversation happened quite naturally and group leaders needed to do little more than ensure orderliness to the discussion.

The membership in the group was fluid. Some persons attended only once or twice without returning. Others came periodically, attending only when they were in the middle of an "episode" and stopped when they felt better. Others came regularly, even when they were feeling well. Of the approximately 120 persons who attended each week, I would estimate that between 20 and 30 constituted a core of "regulars" whom I knew by name. Because the population of persons attending shifted from week to week, it was difficult to establish with confidence the demographic characteristics of those who came. Observation, however, allows the generalizations that the group was composed about equally of men and women; that, beginning with the late teen years, all age categories were represented; and that the members were overwhelmingly White. Based on casual conversation with regulars and on individuals' self-identification during meetings, attendees were drawn from points all along the class structure and from a range of religious groups. An estimate, based on group discussion, is that about 80% of the members had been hospitalized at one time or another and that about 50% were unemployed for reasons largely related to their mental health.

Because this research involved entering into a group of people that met expressly to discuss personal identity information about a mental condition that is often deeply stigmatizing, questions about my research identity and matters of confidentiality were critical considerations from the outset. Although there is no consensus among social scientists about the appropriateness of disclosing one's research role, especially when it may compromise the study itself (see Douglas 1976; Humphreys 1970; Punch 1985), my own stance is closest to the view that researchers should not deliberately misrepresent their identities in order to gain access to a group (Bulmer 1982; Erikson 1970). My decision from the beginning was to present myself as an individual who has grappled periodically with depression but also as a social scientist who might eventually "be interested in writing about how people live with depression in their daily lives." This article, therefore, is truly an instance of a researcher attempting, as Mills (1959) put it, to "transform private troubles into public issues." . . .

It would have been illegitimate for someone to attend the meetings of this group without having experienced an affective disorder. This was partially confirmed at a later meeting when I approached Alex (pseudonymns used throughout), another of the organization's officers, who routinely tape-recorded lectures delivered to the group. I explained that I would like to get a copy of an earlier talk as I was considering the possibility of writing about depression. At this point, he said, "Well, you *are* someone who experiences this illness, right?" When I indicated that this was so, Alex seemed more than willing to provide me a copy of the tape. Moreover, at one time or another, nearly everyone participated in group discussion. People described their episodes of depression or mania, talked extensively about their experience with medications, their encounters with therapists of all sorts, and the negative impact that their problem has had on family and work life. To be a member in good standing, one needed an experiential basis for discussing such issues. It would have been hard to participate without having a "story" to tell. Of course, a researcher could invent stories, but to do so would be, in my view, an unacceptably cynical misrepresentation of identity. . . .

What follows is a rendering of the repeating themes in the talk of group members and an analysis of how that talk was directed toward the creation of a coherent reality about the meaning of having an affective disorder.

THE SOCIAL CONSTRUCTION OF AFFECTIVE DISORDERS

The data collected in the groups suggest that the persons studied faced a number of interpretive tasks as they endeavored to make sense out of their situation. Without exception persons had to define what was wrong with them and to attach a label to their experience. The process of acquiring an acceptable diagnosis was often arduous. In an ongoing way, everyone sought to understand the causes of their condition. Because no one definitively knows the causes of affective disorders, efforts to understand whether nature, nurture, or some combination of the two explained the "illness" were always incomplete. Although people eventually arrived at "theories" about causation, an inevitable dilemma was the inherent untestability of such theories. . . . All of the people in this group, once diagnosed, began an extensive "therapeutic career" involving relations with psychiatric experts who prescribed a variety of psychotropic drugs. Everyone, therefore, necessarily evaluated his or her relationship with these experts and the effects on them of mind–altering medications.

Talking About Diagnosis

Just as people introduce themselves at AA meetings, group sessions often began with personal introductions that took the form "My name is Dick and I'm a [manic] depressive." Such an introduction often included as well the year in which they were first diagnosed. Indeed, members of the group dated them-selves from the time of first diagnosis. The point of diagnosis was an important symbolic benchmark in a difficult illness career. The importance of having a diagnosis was also evidenced by the frequency with which people making an initial acquaintance asked, "And when were you diagnosed?" For purposes of self-identification in the context of this group, description of one's diagnosis (or diagnoses) was a key piece of information. Self-identification as having a unipolar or bipolar disorder, as being obsessive-compulsive, as suffering from agoraphobia, and so on often set the pattern of subsequent conversation.

Acquiring a clear conception of what one has and having a label to attach to confounding feelings and behaviors was significant because most individuals in the group, particularly older people, often went years without being able to give a name to their situation. People often remarked on the increase in knowledge about depression in the past 15 or 20 years and then told stories of years of absolute bewilderment during which psychiatric help was not widely available.

Although some individuals continued to doubt the veracity of the diagno-sis given them, most persons were thankful for being able to name what they experienced. To name something objectified it and gave it a tangible reality. The creation of an illness reality was, then, a first and critical step in the effort to make sense of things. On several occasions, persons, while complaining about their own current hardships, were reminded by group members, "At least you're diagnosed." Such a comment referred to the millions of people who cannot begin to take steps to help themselves because they do not know what is wrong with them. Illustrating the idea that people begin to understand what troubles them through a series of successive approximations, one man, probably in his 60s, reported,

> At first I thought I needed more sleep than other people. Then I realized that I had mood swings. Then I learned that I had depressive periods. Then I learned that I had bipolar depression. Then I learned from the doctors that I inherited this from my grandmother. This was a learning process that took several years.

Just as social scientists (Brown 1987; Rosenberg 1984) have been critical of the meaning of psychiatric diagnoses and the process through which they are established, group members were often critical of doctors who misdiag-nosed them for years. . . . The sense of consternation and anger at doctors, whom individuals believe gave incorrect diagnoses, is evident in the com-ments of a middle-aged woman who offered this history, starting with a major episode experienced while living abroad:

> I flew home. I don't know how I made it. Just all my symptoms came back. I didn't know what was happening to me, and my parents are saying it's my fault and I don't know what's up. I think to myself, "I've been totally fine, so why is this happening to me?". . . I tried to pull myself together. I still didn't have a diagnosis. I even decided to get remarried to

someone who was willing to put up with me, but I have to say that I didn't have a strong base of confidence. . . . I didn't think I knew any answers. . . . I was 26 at the time and in the middle of what I now see as my second depressive episode in 10 years. . . . So I walk into my old therapist's office and say, "I've got this problem. I can't sleep and I feel like I'm walking on water." And he said, "Oh, what I'd like you to do, I'd like you to go back on the medication (taken years earlier)." I said, "I'd like to get another opinion." . . . He never heard me say that before. I never thought to say that. I said, "What is my diagnosis?" He told me, "You are psycho-bizarre" but I couldn't find that anywhere in the lexicon. . . . I finally did see a consultant at [names hospital]. Well, in a 1-hour appointment . . . he said, "I think you're a manic depressive." Nowhere had I been told that. "And you have a disease." Nowhere had I been told that. I was not pleased, but he gave me some literature . . . and recommended a specialist.

. . . Group members, however, felt ambivalent about the labels associated with their diagnosis. For many, receiving a diagnosis was a double-edged thing. On one hand, knowing that you "have" something that doctors see as a specific illness imposes definitional boundaries onto an array of behaviors and feelings that previously had no name. To be diagnosed suggests the possibility that the condition can be treated and that one's suffering can be diminished. However, many individuals were unsettled by the "mental illness" label. At one meeting, for example, the question of labels came up as a topic of discussion. People around the room expressed their preferences, with some wishing to avoid the word "illness" altogether. They spoke about having a "chemical imbalance" or "emotional disorder." Some found the words "disease" and "illness" appropriate but tried to avoid the word "mental," referring instead to their "emotional" illness or disease. The discussion indicated that these people recognized and had thought a great deal about the power of words in influencing personal identity. Several agreed with one person's sentiment that "you do not want to think that you have a mental illness." . . .

Evaluating Cause and Responsibility

Among the stated goals of this organization, as described in one of its publications, was "to educate the patient and the public on the biochemical nature of mania and depression." Although other documents explicitly stated that the organization "has not and will not form any opinions or comments on politics, religion or other controversial matters or on the merits of various forms of medicine or therapy in the treatment of illness," it had plainly adopted the position that the root cause of affective disorders was biochemical. Only on the matter of cause had the organization taken an "official" position, one that also reflected the beliefs of most of the members of the group. Although there was no unanimity on the point, belief in the biochemical nature of the illness was widely shared in the group. . . .

Often, people voiced a fatalism surrounding the onset of episodes, suggest-ing the view that they were deterministically and involuntarily buffeted about by faulty brain chemistry. One woman, for example, said, "I need the doctor to tell me it's not just my attitude, that there is something chemical going on. You can't just snap out of a chemical depression." A related and common theme was that people fall into states of depression or mania even when there is no precipitating event. People often said things like "You just never know when it's [a period of depression] going to hit" or "I've had a few good days recently, but you never know when it's going to fall apart." One person described working as manageable "until disaster comes." Another described her life as "iffy and tentative" and elicited head nods from others in the groups when she said, "It just comes on. You can't explain it sometimes. It seems to happen sort of randomly." When they voiced the language of biochemical cause, direct analogies were often made to organic illnesses, as in the case of the woman who said, "The illness comes upon me, sort of like a bout of pneumonia out of the blue." . . .

To manage their identities in a way that preserved a positive sense of self, people effectively split themselves in two. It was quite common to hear people talk about the fact that when they were overly aggressive, acting out, irritable, and so on, "it was the illness talking," not them. One woman whose husband was a manic depressive said that sometimes "it is unclear whether the illness is talking or whether [he] is talking." In a different meeting, a woman, who described a history of beginning to take college courses only to drop out, offered the following analysis of her behavior: "I would always start courses and then drop out. I used to think I was easily bored. Now I don't blame myself or the class. It's not me or the class, it's the illness." Yet another woman who was undecided about returning to work as a preschool teacher wondered out loud: "Do I not want to go back to it, or is it because of the illness that I don't want to return?" These and similar comments were meant to reaffirm the view that people suffering from depression are actually intelligent, proper people whose interactional competence is sometimes compromised by a disease. . . .

Looking for Dr. Right

As several cultural observers (Derber, Schwartz, and Magrass 1990; Gross 1978; Lasch 1980) have noted, the behavior of persons in today's society is dominated by "experts." Experts advise us on virtually every aspect of our lives. Today, experts follow us through the life course. They are there when we are born and follow us each step along the way, eventually to our graves. Many have come to feel reliant on experts to tell them how to maintain their health, how to become educated, and how to raise their children. Freidson (1970) indicated the enormously expanded role of professionals in our daily lives with his com-ment that "the relation of the expert to modern society seems in fact to be one of the central problems of our time, for at its heart lie the issues of democracy and freedom and the degree to which ordinary men can shape the character of their own lives" (p. 336). Most critical to the present inquiry, however, is the fact that medical experts now tell us when our bodies and our "selves" need repair and the proper procedures for doing it. The medical profession has played a key role in the rise of what might be termed a "therapeutic culture" (Rieff 1966) in which ever greater numbers of behaviors fall under the aegis of a "medical model" (Conrad and Schneider 1980). According to some critics, especially of psychiatry (Goffman 1961; Laing 1967; Szasz 1974), the medical model is used to support an essentially political reality.

Like social scientists, those who had long histories as patients were uncertain about the proper role of psychiatrists in defining reality and the adequacy of their conceptual knowledge for curing the soul. The typical person in the group had years of experience with psychiatrists, and much conversation concerned their relationships with particular psychiatrists, their sense of dependency on them, and questions about the value of psychotherapy in combating affective disorders. In a word, the underlying feeling about psychiatry was "ambivalence." On one hand, these individuals saw psychiatrists as the professional experts who held out the possibility of helping them. However, for every person who talked of having found a warm, sensitive, and sensible psychiatrist, there were those who told psychiatric horror stories involving, as they saw it, insensitivity and incompetence. In one group meeting, when asked about possible topics for discussion, one person jokingly suggested "therapists I have known." Several people described themselves as "veterans of the system" and told of years searching for the "right" therapist. In a newsletter, one group member used an interesting analogy in talking about illness treatment and the search for good psychiatric help:

> The solution to the problem? Preventive maintenance. Any mechanic or fleet operator will tell you that it's far cheaper and more effective to change the oil regularly than it is to do major overhauls. And when trying to solve a particularly confounding problem with your car, it may require going to a couple of different mechanics until you find someone who knows how to do the job right. If you aren't satisfied with your current mechanic, don't go back. The same is true of affective disorders. Regular check-ups and medication maintenance add immeasurably to our ability to lead normal lives. When you find a competent therapist, you'll know it. Being in the care of someone you trust is a wonderfully reassuring feeling.

The phrase "When you find a competent therapist, you'll know it" is noteworthy because it parallels comments often heard in everyday conversation and popular music about falling in love. People's relations with a series of therapists might be seen as similar to a pattern of "serial monogamy"; finding someone you believe is right for you, becoming dependent on that person for meeting certain needs, making a commitment, eventually realizing that you may have made the wrong choice, leaving the relationship, and searching once more for the person who is *really* right for you. Stories interchangeable with the following were frequently heard:

My father died in March of 1977 and again I had another bad depression. My second therapist was having some of his own problems and deep down inside I realized that he could never help me. . . . I hired a new consultant and we both agreed that a new therapist was necessary at this time. This therapist, who is now still my therapist . . . is very caring. He is very insightful and he made me feel that he wanted to work with me, and he saw a lot of hope for me that I had not for myself. Despite the fact that I had no hope and was totally discouraged, he saw that I wanted to get better, and he did everything possible to show how much he believed in me. That was something I never had in my life, a truly caring person.

Although some persons claimed to have found the right psychiatrist, the more typical pattern appeared to be disillusionment about the efficacy of psychiatry. In the course of a discussion where two people were talking about their respective psychiatrists, one indicated that he was switching from Dr. X to Dr. Y. The second responded by saying, "They send you to Y when they don't know what to do with you." Others were heard to comment on different occasions: "Psychiatrists don't really understand, and often they're not helping the situation." "There are some good, some bad, but mostly bad." "I can't stand therapy. It never seems relevant to my disease." "I'd like to see a shrink with the disorder." If some were disillusioned with psychiatrists, others were downright angry:

> For all the research that's gone on since 1940 about drug research, psychiatrists still know jack-squat about it. This is a really frustrating thing for me. They are rip-off artists, but I need them to dispense my drugs.

Even those who spoke favorably about their psychiatrists ("My present therapist and therapy is very positive"; "With the work and encouragement of my current therapist, I am able to avoid regression and even step ahead") came to recognize that therapy could not relieve their uncertainty. One frequently heard the sentiment that after years of dealing with depression, the patient knows more about it than the doctor ("We really know ourselves inside out and they don't"). . . . Despite their physicians' best efforts, these patients eventually realized that their therapists could not clear away their confusion about depression. . . . What emerged from the group discussion was the definition of affective disorders as troubles that must ultimately be remedied by the individuals who suffered from them. Implicit in such a definition is an antipsychiatry ideology that demands, at the very least, a greater democracy between doctor and patient in efforts to treat the problem.

Interpreting the Drug Experience

Every person I met in this group had taken psychotropic medications at one point or another. Although some people periodically stopped taking drugs, the vast majority were currently taking medication(s). Many in the group saw two doctors: psychotherapists with whom they talked and psychopharmacologists who prescribed and monitored their medications. Over the years, most people had taken multiple drugs in what seemed to be a trial and error process. Of all the topics raised in the groups, shared experience about drug use was the one that came up at nearly all the meetings. Because drugs literally altered people's feeling and perceptions, their use required especially constant evaluation and interpretation.

Group discussion about the effectiveness of psychotropic medications affirmed the view that compliance or noncompliance with physicians' directives about drug use must be understood in terms of the daily lives of patients themselves. Rather than being passive recipients of medical advice, group members created "theories" about drug use in terms of their practical experience with illness. In this regard, persons taking psychotropic medications responded in much the same way as the epilepsy patients described by Conrad (1985). For both groups, medications rarely solved their problem, and in the face of uncertainty about the value of medications, patients constructed their own drug management strategies. Conrad's argument that "regulating medication represents an attempt to assert some degree of control over a condition that appears at times to be beyond control" (p. 36) fits well with the perspectives voiced and constructed in the group.

For many people in the group, there was a great deal of confusion surrounding the use of drugs. Conversation revealed that the vast majority had been taking medications for years, and many were on multiple medications simultaneously. The number of permutations and combinations of drugs and dosage (lithium, tricyclic antidepressants, monoamine oxidase [MAO] inhibitors, and tranquilizers) taken by individuals in the group was bewilderingly large, and arriving at the "right" combination of drugs was a puzzling and frustrating affair. Persons talked regularly about stopping one drug, beginning another, changing doses of their current medications, and sometimes making the decision to stop drugs altogether. Just as individuals became expert about different forms of psychotherapy, the majority of persons in the group had taken a wide variety of drugs over the years and had definite opinions about them.

Discussion in the group often centered on an individual who had just begun a new drug and did not quite know what to expect. Such a person could count on there being others in the group who were either currently taking the same drug or had taken it in the past. It was common for persons to trade information about dosage levels, side effects, and interactions among medications. The problem of making sense of the drug experience was compounded by the fact that the same medication that helped one individual could precipitate a terrible experience for another. Nevertheless, group members relied on others to determine whether a medication was working for them, when they could expect it to "kick in," and when the side effects were sufficiently noxious that they should either get off the drug or alter its dosage. It appears that the information supplied by this subculture of fellow drug users was at least as important as the advice offered by psychopharmacologists in determining a drug's effectiveness.

Although some individuals told of finding a drug that "saved my life," the more common response was the recognition that the drugs, like psychotherapy, constituted, at best, a partial answer to their problem. Just as there was a disillusionment that often set in with their therapy, many expressed disillusionment about medications. Stories suggested that individuals began with high hopes that a new medication would finally be the one to help them. Sometimes, it did dramatically change things for the better, but more often, it either modified their symptoms only slightly or exacerbated them. Sometimes, individuals reported feeling well for periods of time on certain medications, only to then have a depressive episode that obviously called into question the drug's effectiveness. Several times, I heard individuals describe having "broken through" their lithium (experiencing mania or depression in spite of the drug), and casual conversations often went like this:

"How are you?"

"I've been feeling really terrible for the last couple of weeks."

"But I thought you were one of Prosac's miracles [said somewhat tongue in cheek to imply that there are no miracles]."

. . . Despite confusion about the efficacy of drugs, the consensus was that one should not try "to go it alone" (without drugs). At one meeting, for example, a young man announced his plan to stop taking his lithium because, as he put it, "I don't like feeling drugged. I'm like a whirling dervish and that's just the way I am." In response, one individual offered the idea that wanting to get off medications "is a stage through which all of us go." Another concurred with the observation that "we are all struggling with what to do with medications." Not everyone, however, agreed that the best course is to keep taking medications. I periodically heard opinions equivalent to the following: "There is no reason to take pills all the time when your episodes are infrequent" or "Pills are not the absolute God." Such contrasts in view indicate that group discussion could only provide the broad interpretive frame within which individuals tried to understand the value of medications in helping them to maximize the quality of their lives. It was, however, a frame reasserting the view that patients are the ultimate experts on what it means to have an affective disorder and, consequently, that group members should, whenever possible, avoid giving doctors autonomy over their lives. In this regard, the group provided a nascent ideology posing a potential threat to medical dominance in the treatment of affective disorders.

CONCLUSION

This article proceeded from a cornerstone of sociological thought. A key idea of the symbolic interaction perspective (Blumer 1969; Hewitt 1986; Karp and Yoels 1986) is that persons' feelings and behaviors arise from their "definition of the situation": that to appreciate people's thoughts, feelings, and actions, we need to inquire into how they arrive at definitions of the situations in their lives. However, we need to acknowledge that some life situations require more extensive definitional efforts than others. This article has centered on the role of conversation in a support group to aid individuals in better comprehending a life situation—having an affective disorder—that was intrinsically ambiguous and problematic. Although groups of the sort described in this article could not absolutely resolve questions about the correctness of particular diagnoses, the extent to which persons ought to feel responsible for their condition, or how to deal with psychiatrists and the medications they prescribe, there was something very powerful for individuals in learning that others shared their confusions. Beyond this, however, participation in the group was compelling because members' talk provided a "vocabulary" through which to see what affective disorders were all about. . . .

REFERENCES

Blumer, H. 1969. *Symbolic interaction: Perspective and method.* Englewood Cliffs, NJ: Prentice-Hall.

Brown, P. 1987. Diagnostic conflict and contradiction in psychiatry. *Journal of Health and Social Behavior* 28:37–50.

Bulmer, M. 1982. *Social research ethics.* London: Macmillan.

Conrad, P. 1985. The meaning of medications: Another look at compliance. *Social Science and Medicine* 20:29–37.

Conrad, P., and J. Schneider. 1980. *Deviance and medicalization.* St. Louis: Mosby.

Derber, C., W. Schwartz, and Y. Magrass. 1990. *Power in the highest degree: Professionals and the rise of a new mandarin class.* New York: Oxford University Press.

Douglas, J. 1976. *Investigative social research.* Beverly Hills, CA: Sage.

Erikson, K. 1970. A comment on disguised observation in sociology. In *Qualitative sociology,* edited by W. Filstead, 252–60. Chicago: Markham.

Freidson, E. 1970. *Profession of medicine.* New York: Harper & Row.

Goffman, E. 1961. *Asylums.* New York: Anchor Books.

———. 1963. *Behavior in public places.* New York: Free Press.

Gross, M. 1978. *The psychological society.* New York: Random House.

Hewitt, J. 1986. *Self and society.* Boston: Allyn & Bacon.

Humphreys, L. 1970. *Tearoom trade: Impersonal sex in public places.* Chicago: Aldine.

Karp, D., and W. Yoels. 1986. *Sociology and everyday life.* Itasca, IL: Peacock.

Laing, R. 1967. *The politics of experience.* New York: Ballantine.

Lasch, C. 1980. Life in the therapeutic state. *New York Review of Books,* 12 June, 24–31.

Mills, C. W. 1959. *The sociological imagination.* New York: Oxford University Press.

Punch, M. 1985. *The politics and ethics of fieldwork.* Beverly Hills, CA: Sage.

Rieff, P. 1966. *Triumph of the therapeutic.* New York: Harper & Row.

Rosenberg, M. 1984. A symbolic interactionist view of psychosis. *Journal of Health and Social Behavior* 25:289–302.

Szasz, T. 1974. *The myth of mental illness.* New York: Harper & Row.

25

From Stigma to Identity Politics:

Political Activism among the

Physically Disabled and

Former Mental Patients

RENEE R. ANSPACH

Among other legacies of the 1960s was what might be termed the "politicization of life." As political organizers extended their efforts to the most disenfranchised groups—welfare recipients, poor tenants, and prisoners—power, once seen as elusive, was now considered attainable. And as "deviant" groups such as hippies, Hell's Angels, and Gay Liberationists began to articulate their demands in quasi-political terms, the once hard-and-fast distinction between social deviance and political marginality became blurred [1]. Most significantly, the sixties witnessed a widening definition of politics to embrace all aspects of the person. An ever-increasing array of personal habits, from long hair to the discarding of brassieres to vegetarianism, came to be equated with conscious rebellion against the confines of the normative order. The language of political protest and persuasion, no longer a

From: "From Stigma to Identity Politics: Political Activism among the Physically Disabled and Former Mental Patients," Renee R. Anspach, *Social Science and Medicine,* Vol. 13A, 1979. Reprinted by permission of Pergamon Press Ltd., Oxford, England.

specialized vocabulary, became the available idiom for the expression of discontent. Sociologically, these developments pointed to the existence of a political as well as a social construction of reality.

This politically-charged climate set the stage for the subject of this paper: a nascent political activism among former mental patients and the physically disabled. On April 6, 1977, the politicization of the handicapped was dramatically and vividly displayed to the American public. About 300 blind, deaf, and physically disabled activists assembled in front of Washington's Department of Health, Education, and Welfare, carrying placards, shouting slogans, and chanting songs from the civil rights movement. The issue was H.E.W. Secretary Califano's failure to sign Section 504 of the Rehabilitation Act, which forbids architectural and economic discrimination on the basis of disability. The 30-hr sit-in coincided with demonstrations in other major cities. In San Francisco, the sit-in lasted almost a month, until final victory was achieved with the signing of the bill [2, 3]. Former mental patients, too, have recently joined the ranks of the politically active. In 1975, for example, 250 persons from several states and Canada and representing about twenty organizations attended a conference on "Human Rights and Psychiatric Oppression." The names of the most militant organizations—Mental Patients' Liberation Front, Network Against Psychiatric Assault, and Committee Against Psychiatric Oppression—give testimony to their concern with such issues as the quality of mental hospitals, aftercare, civil rights, and forced medication and shock treatments. How many of the nation's disabled and former mental patients have been swept up by this tide of politicization, and how many others are "fellow travellers," passive onlookers, apathetic, or opposed? No one knows. While the scope of the new activism is elusive, it is clear from these events and from the spate of publications such as *Mainstream* and *Madness Network News* that a social movement has been born.

While it is difficult to delineate the contours of a social movement which is only now taking shape, it is clear that this growing politicization differs qualitatively from two traditional organizational modalities of the disabled, the voluntary organization and the self-help group. Since various forms of disability have historically captured the sympathetic imagination of the American public, charitable voluntary associations, such as the Muscular Dystrophy Association and the National Foundation for Infantile Paralysis, have proliferated. The Mental Hygiene Movement, founded at the turn of the century by Clifford Beers, established the tradition of advocacy and social brokerage characteristic of contemporary Mental Health Associations [4]. However, unlike these voluntary associations and lobbies, the new activist groups are composed of the disabled themselves, seeking social change through their own efforts, rather than through others acting as surrogates. The new activist groups also differ from self-help groups, such as colostomy clubs and Recovery, Incorporated. While the emerging activist groups borrow from the self-help movements the emphasis on indigenous organization and self-reliance, they are political, rather than therapeutic, in orientation. They seek not to modify their own behavior in conformance to a pre-existing normative mold, but rather to influence the behavior of groups, organizations, and institutions. For, unlike its predecessors, the new activism, with its spate of organizations and its outpouring of publications, is self-consciously polemical in tone, characterized by its open agitation for legislative change and architectural reform, its militant opposition to job discrimination, and its frequent reliance on demonstrations and the tactics of social protest.

This newly emergent political activism among the disabled and former mental patients is the subject of this paper. And, since this essay is exploratory and speculative in scope, its conclusions await the test of empirical research. Such speculative work carries an obvious disadvantage: it runs the risk of distorting, to a certain extent, the phenomenon under study. The meaning of events may be misinterpreted and motives incorrectly attributed to participants. However, speculation and theorizing may serve as a basis for future empirical study, can provide novel and imaginative ways of viewing subject matter, and are, therefore, an integral part of the sociological enterprise.

The following pages will explore one facet of the politicization of the handicapped: the implications of such activism for the identity of its participants. Political activity will be viewed as a symbolic arena wherein selves are continually created, dramatized, and enacted.

The interface between politics and the self can, in my view, add a new dimension to the study of social movements. Political social movements have characteristically been classified according to the goals of the actors and/or the theorist's conception of their social function. In *Symbolic Crusade*, for example, Gusfield creates a taxonomy of political action. The categories he delineates are not mutually exclusive, for actual political social movements contain an admixture of elements. Nevertheless, political movements can be differentiated according to their salient foci. *Instrumental* politics involve actual conflicts of interest among social groups. By contrast, in *expressive* politics, political participation is "not a vehicle of conflict but of catharsis." Midway between the instrumental and expressive realms stand *symbolic* politics, typified by the Temperance Movement. In symbolic politics, participants are not concerned with actual changing of behavior, but rather seek institutional affirmation for their values, life styles, and normative standards of conduct. Symbolic politics, according to Gusfield, often fall within the realm of collective morality [5].

But the political activism of the handicapped and former mental patients seems to represent a style of politics not entirely captured by the labels "instrumental," "expressive," or "symbolic." This fourth political modality, while equally symbolic, primarily concerns not status, life style, or morality, but rather *identity* or being. This type of politics is characteristic not only of the disabled but of many recent social movements, such as the radical feminist movement, the black power movement, or gay liberation. While such social movements may have strong instrumental components, insofar as they seek to effect changes in public policy, they consciously endeavor to alter both the self-concepts and societal conceptions of their participants. Political goals and strategies often become a vehicle for the symbolic manipulation of persona and the public presentation of self. Hence many disabled activists are preoccupied with style and tactics, and imagery often takes precedence over substance. They eschew the telethon's "politics of pity" and abhor the "poster child" image [6], demanding instead to be regarded (by themselves and others) as self-determining adults, capable of militant political action. In the words of one articulate muscular dystrophy victim:

> Thriving on a climate of increasing public tolerance and kindness, we are becoming presumptuous. Now we reject any view of ourselves as being lucky to be allowed to live. We reject too all the myths and superstitions that have surrounded us in the past. We are challenging society to take account of us, to listen to what we have to say, to acknowledge us as an integral part of society itself. We do not want ourselves, or anyone else treated as second class citizens and put away out of sight and out of mind [7].

Political activism among the handicapped and former mental patients, then, exemplifies a type of politics which we will term *identity politics*. Among its goals are forging an image or conception of self and propagating this self to attentive publics.

Not only is the fashioning of collective identity an explicitly articulated *goal* of the politicized disabled, but the very act of political participation in itself induces others to impute certain characteristics to the activist. For once (s)he has entered the realm of polity, the political actor assumes a certain persona. We attribute certain qualities to political action which sharply contrast to personal deviation, which is assumed to be individualistic and non-rational, seemingly without purpose. Political action, on the other hand, is viewed as prima facie evidence of rational, goal-directed, voluntaristic, and change-oriented behavior. The agitation of the disabled and former mental patients demonstrates, in word and deed, that they are capable of purposive political action.

POLITICAL ACTIVISM OF
THE DISABLED: AN ANOMALY

. . . Most of the prevailing thinking on deviance conceives of its phenomenon in a way that does not provide *sufficiently* for the possibility of politicized deviants, collectively engaged in attempts to reweave the fabric of identity. According to Horowitz and Liebowitz, theories which conceptualize deviance as a social, rather than as a political, problem tend to obscure its political dimensions. They fault prevailing theories for their reliance on a "social welfare model," which tends to evaluate deviance in largely therapeutic terms. Presumably, Horowitz and Liebowitz are referring to those structural-functional and sub-cultural perspectives which utilize notions of "improper socialization" or socially-induced individual pathology. Such theories, argue the authors, fail to recognize that deviance is a "conflict between superordinates who make the rules and subordinates whose behavior violates them" [1].

Instead, Horowitz and Liebowitz exhort sociologists to adopt a "conflict model" of deviance which turns out to be none other than labelling theory. In calling attention to the role of the responses of others and agents of social control in forging deviant identity, labelling theorists were among the first to take cognizance of issues of power and control. According to Becker:

> The question of the purpose or function of a group . . . is very often a political question. Factions within the group disagree and maneuver to have their definitions of the group's function accepted. . . . If this is true, then it is likewise true that the question of what rules are to be enforced, what behavior regarded as deviant, and which people labeled as outsiders must also be regarded as political. . . .
>
> Differences in the ability to make rules and apply them to other people are essentially power differentials [8].

But although labelling theorists may be credited with introducing a political dimension into the study of deviance, their notion of power ultimately proves to be limited and one-sided. Most studies of deviance generated by the labelling perspective portray the deviant as powerless, passive, and relatively uninvolved in the labelling process [9–11]. Part of this bias stems from the decision of many labelling theorists to focus on "total institutions" [12], where those labelled are perforce compelled to accept identities enforced by institutional dictates and where there is little latitude for negotiation with officials. But this portrait of the determined deviant goes much deeper, and is rooted in a *one-way* model of the role-taking process,[1] in which identities are simply *imposed* upon the deviant rather than negotiated. This problem is most apparent in the work of Scheff, who portrays the mental patient as a *tabula rasa*, who in one fell swoop casts off the trappings of an old self and steps into the strait jacket of an imposed, deviant role [13]. But the assumption of deviant passivity creeps into the work of almost all labelling theorists, reliant as they are on the notion of "victimization" and an ideological commitment to the "underdog." Labelling theory, then, is conceptually ill-equipped to deal with the identity politics of the "mentally ill" and the physically disabled. Demonstrations and social protest provide these politicized deviants with a forum where identity is negotiated on a grand scale with the American public.

Goffman's book-length essay, *Stigma* [14], is one of the most systematic attempts to deal with the problems of "spoiled" identity. It is an attempt which, in some senses, seems to avoid many of the pitfalls of labelling theory.

Stigmatized persons emerge not as powerless victims, but as strategists and con artists, engaged in the work of "information control" and "tension management." Yet even in this more sophisticated framework we find the ever-present theme of passivity and imposed identity. The strategies of the stigmatized person are essentially *defensive maneuvers*. For the stigmatized person ultimately subscribes to the definitions of "normals" and derives his/her identity reactively, in response to the imputations of the "wider society." (S)he is fated to remain at the mercy of the Invisible Hand of the Generalized Other. Yet the political activism of the disabled and former mental patients is not an endorsement of but rather an *assault* upon the generalized other. Seeking the power to define their situation for themselves, disabled activists repudiate societal imputations. In the words of one politically active woman:

> As a kid I was labelled handicapped and I pretty much fit that role. . . . And then one day I thought, "Hey, *I don't want to be defined by others.* I want to be defined for and by myself." Because when you start accepting society's definition of the handicapped, then you are really different.
>
> When I was once asked by a reporter how old I was, I answered, "Does it really matter?" . . . I think it's time that we stop judging people by their age, height, and weight. We've got to start looking at them as human beings, first and foremost . . . (quoted in [6]).

The politicization of the disabled represents an attempt to wrest definitional control of identity from "normals." Goffman makes only passing reference to political activism of the stigmatized (for it had not yet come into being as a full-blown social movement), and he takes a somewhat fatalistic view of this phenomenon, emphasizing that, paradoxically, militant activity is ineluctably bound up with the world of "normals":

> The problems associated with militancy are well known. When the ultimate political objective is to remove the stigma from the differentness, the individual may find that his very efforts can politicize his own life, rendering it even more different from the normal life initially denied him. Further in *drawing attention to the situation of his own kind* (italics added) he is in some respects consolidating a public image of his own differentness as a real thing, as constituting a real group. On the other hand, if he seeks some kind of separateness, not assimilation, he may find that he is presenting his militant efforts in the language and style of his enemies. Moreover, the plea he presents, the plight he reveals, the strategies he advocates are all part of *an idiom* of expression and feeling that belongs to the whole society [14].

In Goffman's view, then, political activism is doomed to failure because assimilationist tactics only serve to reify their differentness and because separatists are compelled to rely on the pre-cast political idiom and imagery of "normals." Yet such a depiction does not, in my view, quite do justice to contemporary identity politics. The new activists do not seek to assimilate themselves; nor does their reliance on the political strategies of "normals" necessarily spell defeat. While the disabled activists do not operate in a political vacuum, their actions do testify to their ontological status as independent, politically active beings. Whether "normals" eventually accept or internalize the image they sell is, of course, a different matter. . . .

I have been leading up to the following: sociological theories have, perhaps unwittingly, subscribed to the mythology of the helplessness of the handicapped and the "mentally ill." In one way or another, the social welfare model, labelling theory, and the dramaturgical model of the actor are tacitly infused with commonsense assumptions of the deviant actor as individualistic, passive, and powerless. The sociology of deviance must revise its conceptions to account for active attempts on the part of those labelled as deviant to mold their own identities. . . .

STRATAGEMS FOR IDENTITY MANAGEMENT

So far I have spoken as though identity politics were the only, or the prevailing, response to handicap or stigma. Political activism is, however, only one of an array of possible stratagems for the management of a somewhat problematic identity. It is theoretically worthwhile to contrast political activism with these other responses in order to delineate its essential features.

These stratagems can, I believe, be contrasted with respect to two central dimensions: the individual's conception of self and his/her relationship to prevailing societal values. To begin with, the handicapped and those labelled as "mentally ill" must develop some posture *vis-à-vis* certain salient cultural values which portray them in a less than favorable light. As Davis points out in his study of polio children and their families, the handicapped person must contend with that fact that

> he is at a disadvantage with respect to several important values emphasized in our society: e.g., physical attractiveness; wholeness and symmetry of body parts; and various physiognomic attributes felt to be prerequisite for a pleasant and engaging personality [15].

The former mental patient, while usually free from the obtrusiveness of an immediately visible stigma, is also said to violate some of the sacred principles of our society: the canons of rationality, self-determination, and full responsibility for the consequences of one's actions. No matter what the particulars of the stigma, the individual must ultimately, consciously or unwittingly, accept or reject the societal values which his/her very ontology contradicts. Moreover, the stigmatized person invariably adopts a stratagem which carries profound implications for his/her identity or conception of self. By combining these two dimensions—stance toward societal values and self-concept—it is possible to develop a typology of four[2] modal responses to stigma, represented schematically below:

**Stratagems for Stigma
Management**
Societal values

		Accepts	Rejects
Self-concept	Positive	Normalization	Political Activism
	Negative	Disassociation	Retreatism

The terms "normalization" and "disassociation" were drawn from Davis' study of polio children and their families. In the first stratagem, *normalization*, the individual is firmly committed to cultural conceptions of normalcy and endorses commonly-held assumptions about the "ideal" person. But while accepting these societal values and cognizant that (s)he fails to measure up to them, the individual makes a concerted effort to minimize, rationalize, explain away, and downplay the stigma attached to his/her differentness. The typical existential stance of the normalizer is that (s)he is "superficially different but basically the same as everyone else," or that ultimately "differences don't really matter." Generally, the normalizer attempts, insofar as possible, to participate in the round of activities available to "normals" and to aspire to "normal" attainments. As Davis indicates, the societal "idealization of the normal, healthy, and physically attractive," and the "democratic fiction" that differences wither away in the face of a fundamental equality of human beings, make normalization the favored response [15]. There is another societal factor which often impels the disabled toward normalization: the ever-present American myth of success. Our culture abounds with symbolic inducements to overcome even the weightiest of obstacles. F.D.R. and Helen Keller may be said to be the Horatio Algers of the handicapped, reminders that, given hard work, diligence, and individual effort, the courageous disabled can reach the loftiest pinnacles of achievement.

Normalization carries a certain undeniable advantage: the individual is able to maintain a relatively sanguine and confident attitude toward the self. Yet the normalizer purchases this positive self-conception at a certain price. First, interactions with "normals" are necessarily strained, for an obvious disability inevitably tends to obtrude upon the social situation. In their gambits to "disavow their deviance," normalizers must continually manage and contend with the inescapable tension of an interaction which is fragile, problematic, and easily subject to "slips" and disruptions [17].

Secondly, normalization is premised on a number of contradictory beliefs, and hence the normalizer is fated to experience a certain degree of "cognitive dissonance." Most obvious of these contradictions is the gap between the "democratic fiction"—the "You-can't-tell-a-book-by-its-cover" ideology—and the actual obsession with ascriptive attributes which plagues our culture. Then there is the discrepancy between the canons of civility—the norms of polite society which proffer a superficial acceptance to the handicapped—and the actual emotional displays conveyed non-verbally by "normals" in social intercourse. This gap between what Goffman [16] terms the "expressions given" and the "expressions given off" renders the existential state of the normalizer one of profound *distrust* and suspicion of a rejection that may lurk beneath the façade of civility. Finally, there is the economic gap between the myth of success and individual achievement propagated by rehabilitative ideology and the harsh realities of economic discrimination against the handicapped and former mental patients. The list of gaps between myth and reality could be multiplied indefinitely, but the point is that the normalizer's valiant attempts to preserve a positive self image are undoubtedly punctuated by moments of anguish, despair, and internal turmoil.

The next two responses—disassociation and retreatism—require little explanation and entail for the disabled person a less than felicitous conception of self. In *disassociation*, the individual remains attached to the values of the wider society and aware that (s)he is disqualified according to them. Unable to accept the harsh fact of their disqualification and unwilling to aspire to acceptance by "normals," those who disassociate live with an identity that is tainted and tarnished. This negative self-conception leads them to avoid contacts with "normals," perhaps in an effort to spare themselves the pain of impending rejection. Self-exclusion, resentment, and anger toward "normals"—these are the emotional concomitants of the disassociative response [15].

In the third response, *retreatism* (a term borrowed from Merton), the self-image of the individual is profoundly negative, and (s)he neither accepts societal values nor aspires to "normal" attainments. The profound despair of retreatists lead them to withdraw from the world of "normals" to a world of private hopes and fears. These are the people who have "given up"—found in mental hospitals, flophouses, welfare hotels, and other margins of "civilization."

Political activism has recently come to represent a viable collective alternative to the previous individualistic responses.[3] Like the normalizer, the activist seeks to attain a favorable conception of self, often asserting a claim to superiority over "normals." But unlike the normalizer, the activist relinquishes any claim to an acceptance which (s)he views as artificial and consciously repudiates prevailing societal values. The new activist demands institutional equality rather than friendship. Because of this separatist stance on the part of the activist, (s)he may be apt to experience less internal turmoil than is the normalizer. Although some emotional conflicts undoubtedly ensue whenever the allure of "acceptance" and the strain of militancy beckon the activist to return to a normalizing response, political activism is a response which is less conflictual and less frought with dissonance than is normalization. Paul Hunt, a disabled writer, provides perhaps the most eloquent statement of the activist's repudiation of societal values and exaltation of self:

> What I *am* rejecting is society's tendency to set up rigid standards of what is right and proper and to force the individual into a mould. . . .
>
> For the disabled person with a fair intelligence or other gifts, perhaps the greatest temptation is to use them just to escape from his disabledness,

to buy a place in the sun, a share in the illusory normal world where all is light and pleasure and happiness. Naturally we want to get away from and forget the sickness, depression, pain, loneliness, and poverty of which we probably see more than our share. But if we deny our special relation to the dark in this way, we shall have ceased to recognize the most important asset of disabled people in our society—the uncomfortable subversive position from which we act as a living reproach to any scale of values that puts attributes or possessions before the person [7]. . . .

THE STRATAGEMS OF IDENTITY POLITICS

Political activism, as contrasted with other responses to disability, seeks, in repudiating societal values, to elevate the self-concept of its participants. In this section I will explore in greater detail just how these two aspects of identity social movements—repudiation and self-elevation—are accomplished. In so doing, I am relying largely on the published statements of articulate disabled activists, especially those generated by the 1975 Conference on Human Rights and Psychiatric Oppression. While such pronouncements cannot be construed as "representative" of the breadth and scope of identity politics, and while they are biased toward the "official," such official ideologies do reveal basic themes of identity politics. And ideology, after all, is one medium through which selves are created.

Most activists endeavor to dispel stigma by viewing their deviation in non-medical terms. The politicized mental patient groups adhere to some notion of *societal etiology*. The cause of mental disorder is to be sought not in individual pathology, but rather in perceived political, economic, or social repression. The Mental Patients' Liberation Front, a Boston-based activist group, provides perhaps the most militant illustration of this tenet, when it states its goal to be "to transform the classist, racist, and sexist society, with its oppressive power relations, that caused our pain and incarceration." Project Release, an organization based in New York City, is concerned with the problem of recidivism, or the "revolving door," in which patients are released, only to return quickly to the confines of the mental hospital. But Project Release explains an otherwise conventional problem for the psychiatric profession by reference to a socio-economic model: its leaders state that their "ideology is that recidivism is primarily caused by *socio-economic conditions* and not by 'mental illness'" [18].

Another group attending the conference runs a halfway house based on Laingian principles, which emphasizes the creative aspects of madness and expressly condemns the use of psychotropic drugs. Its leaders expressed another variant of this theme:

The "system" is set up to keep many people struggling for a living, for a place in society, for a feeling of self worth. People in emotional crises are experiencing their pain for legitimate reasons. The pressures of living in the world today are many and can severely tax anyone's emotional stability. It is a myth that "emotional breakdown" is due to an individual's weakness or inadequacies, a myth perpetrated by existing "mental health" facilities and other social and economic power structures [18].

In this framework, those who suffer "crises" (and this term is substituted for "mental illness") are neither weak nor emotionally unstable. Rather, their crises are viewed as legitimate and reasonable responses to the travails of an oppressive civilization. In providing a rationale for suffering, the Laingian ideology legitimates mental patients' experience and, in so doing, allows them to sustain viable conceptions of self. This same group adds an additional nuance, echoed by other activist groups:

> In traditional psychiatric settings, people's expressions of emotional distress are usually labelled as "mental illness," and are invalidated rather than listened to with respect. The patients are cut off from their feelings by mind-numbing psychotropic drugs, and by their environment. . . . We will help people to find their own solutions to problems, rather than having solutions imposed by psychiatric authorities [18].

This blame of and disdain for "psychiatric authorities" and professionals is a salient theme found in the ideological pronouncements of most patient activist groups. Psychoactive drugs and shock treatments in particular, and enforced medical interventions in general, are viewed as infringements upon patients' integrity and dignity. The Network Against Psychiatric Assault, an organization in San Francisco claiming more than 100 members, has successfully halted enforced shock treatments at Langley-Porter Neuropsychiatric Institute, and has agitated, with some degree of success, for broader legislative restrictions on enforced treatments. Similar lawsuits and demonstrations are being carried out in other states. This struggle against institutions and professionals on the part of activist groups indicates that the rehabilitative model has been supplanted by an ideology of overt conflict.

Physically disabled activists, unlike mental patients, cannot fault "society" for their disability. Yet even here we find a variant on the theme of societal culpability: disabled activists blame society for the stigmatization of a mere difference. Paul Hunt, a muscular dystrophy victim, faults contemporary society for its mindless obsession with achievement, its fetishism of commodities and conformity, and its use of ascriptive attributes and material possessions as the salient yardstick of personal worth [7]. In creating such a portrait of the world around him, Hunt safeguards his own sense of integrity and the validity of the experience of disability.

Hunt's writings, with their religious overtones, begin to define the parameters of a "pedagogy of the oppressed" [19]. He not only legitimates the experience of disability but equates disability with revolution against the constraints of rigid normative standards. The handicapped, by their very *being*, reproach materialism and petty conformity, standing as a reminder to all of their own tragedy and finitude. The disabled, then, play a symbolic missionary role. Former patients who adhere to Laingian ideology also celebrate their own condition and equate madness with revolutionary experience:

> Our basic philosophy is that "breakdown" or extreme emotional crisis, with suitable conditions of emotional and physical support, can be a constructive and growth-producing experience; "breakdown" is potentially "breakthrough" [18].

In identifying handicap with subversion and madness with revolution, the activists conceive themselves as playing an *active* role in transforming the social order. Hence the tendency of the disabled and former patients to "elevate" their status by identifying themselves with more politicized and active minority groups and revolutionaries:

> Our constant experience of this pressure towards unthinking conformity in some way relates us to other obvious deviants and outcasts like the Jew in a gentile world, the Negro in a white world, homosexuals, the mentally handicapped; and also the more voluntary rebels in every sphere—artists, philosophers, prophets, who are essentially subversive elements in society. This is an area where disabled people can play an important role [7].

One can only marvel at the power of these stratagems of self-affirmation. Madness emerges not as affliction, but as creative rebellion, and the disabled emerge not as passive victims, but as prophets, visionaries, and revolutionaries.

Language, too, is of paramount importance. The older self-help groups and many professionals sought to replace the usual names for disabilities with more delicate labels, such as "hard of hearing" [18]. Some disabled activists have attempted, usually unsuccessfully, to find substitute terms which remove the pejorative connotations of the words "handicapped" and "disabled." (And this is testimony to the power of language in structuring experience.) But the most militant proudly and self-consciously flaunt the most degrading terms: "cripple," "inmate," "madness," and "proud paranoids" are found in their published statements. This stratagem—*de-euphemization*—demonstrates a subtle mastery of the language game of "normals." Like Richard Wright before them, who created a linguistic revolution when he coined the term "black power," these activists use the very terms which directly assault the "liberal toleration" of "normal" outsiders. Their words, then, are at once an affront to "normals," a signal of conflict and battle, a repudiation of the illusory acceptance implicit in the euphemism, and a testimony to the intrinsic viability of their own experience.

Societal etiology, identification of deviation with revolution, and de-euphemization—these are the stratagems by which the politicized disabled seek to reformulate their condition and to redefine their situation. Identity politics, then, is a sort of phenomenological warfare, a struggle over the social meanings attached to attributes rather than an attempt to assimilate these attributes to the dominant meaning structure.

CONCLUDING REMARKS

Throughout, I have used the concept of identity politics to describe the incipient political activism of the disabled and former mental patients. The new political activists attempt to create an identity for themselves and to propagate this newly-created sense of self to "normals." This is accomplished in two ways. First, the *actions* of the disabled, their militancy and their reliance on social protest, demonstrate that they are independent, rational beings, capable of self-determination and political action. These actions *symbolically* assault the prevailing commonsense (and sociological) imagery of passivity and victimization. Secondly, unlike other responses to stigma and disability, political activism creates an *ideology* which repudiates societal values and normative standards and, in so doing, creates a viable self-conception for participants.

The political activism of the handicapped and former mental patients has far-reaching significance for the professionals who work with them and the sociologists who study them. Many rehabilitative tenets, appropriate to an era when normalization was the favored and modal response to disability, no longer seem applicable to those whose aspirations transcend an often illusory social acceptance by "normals." The identity politics of the disabled and "mentally ill" also challenges the sociologies of deviance, stigma, and disability to create new conceptions of the deviant actor, which embrace concerted political action which resists, rather than subscribes to, societal imputations.

As other identity social movements, such as the Radical Feminist Movement and the Gray Panthers, create new images of sex roles and aging, we can anticipate that the sociologists who study them will re-examine their own assumptions in the light of new historical circumstances.

NOTES

1. By this I mean that the labelling theorists have actually altered the negotiated model of interaction first proposed in earlier symbolic interactionist formulations.

2. Davis and Goffman discuss a fifth stratagem, passing, in which the individual accepts the normal standard yet attempts to conceal his/her condition. This response, while available to the former mental patient, is, of course, impossible for the person with a visible handicap.

3. Political activism may, however, be seen as the collective form of a more diffuse strategy of redefinition, reconstitution, and value transcendence. It is possible to respond to stigma in this fashion without belonging to a larger social movement. However, only with the development of larger social movements is such a response viewed as a viable alternative for a significant number of persons.

REFERENCES

1. Horowitz I. and Liebowitz M. Social deviance and political marginality. *Soc. Probl.* 281–296, 280, 282. 1968.

2. *Washington Post.* April 7, 1977.

3. *Los Angeles Times.* April 6, 1977.

4. Deutsch A. *The Mentally Ill in America.* Doubleday. New York. 1937.

5. Gusfield J. R. *Symbolic Crusade.* p. 179. University of Illinois Press. Urbana. 1963.

6. Ritter B. The rise of the handicapped. *New West* 29, 55, 54. 1976.

7. Hunt P. A critical condition. In *Stigma: The Experience of Disability* (Edited by Hunt P.). pp. 129–130, 151, 159, 151, 144–199. Chapman. London. 1966.

8. Becker H. S. *Outsiders: Studies in the Sociology of Deviance.* p. 7, 17. Free Press. New York. 1963.

9. Davis N. *Sociological Constructions of Deviance Perspectives: Issues in the Field.* Brown. New York. 1975.

10. Mankoff M. Societal reaction and career deviance: a critical analysis. *Sociol. Q.* 12, 205–217. 1971.

11. Akers R. Problems in the sociology of deviance: social definitions and behavior. *Soc. Forces* 455–465. 1976.

12. Goffman E. *Asylums: Essays on the Social Situation of Patients and Other Inmates.* Doubleday. Garden City. 1961.

13. Scheff T. *Being Mentally Ill. A Sociological Theory.* Aldine. Chicago. 1968.

14. Goffman E. *Stigma: Notes on the Management of Spoiled Identity.* pp. 116–24. 1963.

15. Davis F. The family of the polio child: some problems of identity. In *Illness, Interaction and the Self* (Edited by Davis F.). p. 105, 107. Wadsworth. Belmont. 1972.

16. Goffman E. *The Preservation of Self in Everyday Life.* p. 2. Doubleday. Garden City. 1959.

17. Davis F. Deviance disavowal: the management of strained interaction by the visibly handicapped. In *Illness, Interaction and the Self* (Edited by Davis F.). pp. 133–149. Wadsworth. Belmont. 1972.

18. Conference on Human Rights and Psychiatric Oppression. *Newsletter* 1974.

19. Freire P. *The Pedagogy of the Oppressed.* Herder & Herder. New York. 1970.

PART VI

Organizing Deviants and Deviance

We now turn to a closer examination of the lives and activities of deviants. Once they get past dealing with outsiders, they must deal with other members of their deviant communities and with the specifics of managing their deviance. There are several ways of looking at how deviants organize their lives. We will do it here by looking at their relationships, their acts, and their careers.

RELATIONS AMONG DEVIANTS

Relationships among deviants can take many forms. They vary along a dimension of sophistication, involving complexity, coordination, and purposiveness. Deviant associations vary in the task specialization among members, the stratification within the group, and the amount of authority concentrated in the hands of a leader or leaders. Some groups of deviants are loose and flexible, with members entering or leaving at their own will, uncounted or monitored by anybody. Others maintain more rigid boundaries, with access granted only by the consent of one or more insiders. Membership rituals may vary from none to highly specific acts that must be performed by prospective inductees, thereby granting them not only membership but also a place in the pecking order once they are inside. In some ways, the rigidity inside a deviant group is related to its insulation from conventional society: the more its members withdraw into a social and economic world of their own, the more they will develop norms and rules to guide them, replacing those of the outside order.

Groups of deviants also vary in their organizational sophistication, with the organized groups capable of more complex activities. Such organized groups provide greater resources and services to their members; they pass on the norms, values, and lore of their deviant subculture; they teach novices specific skills and techniques where necessary; and they help one another out when they get in trouble. As a result, individuals who join more tightly knit deviant scenes tend to be better protected from the efforts of agents of social control and more deeply committed to a deviant identity.

The three readings in this area look at *subcultures, gangs,* and *rings* of deviants. These readings are organized along a continuum that progressively rises in organizational sophistication. Kathryn Fox's article, "Real Punks and Pretenders: The Social Organization of a Counterculture," examines the stratification within a midwestern city's punk scene, or subculture. Starting at the center, she looks at the hard-core punks and then makes her way outward to the soft-core and preppie punks. For each group she examines their immersion in the punk lifestyle, and commitment to punk ideals, activities, style of dress, and mode of survival. Membership in the hard-core inner group required a more serious dedication than did participation in the transitory outer fringe.

Columbus Hopper and Johnny Moore's "Women in Outlaw Motorcycle Gangs" offers a journey into a more dangerous and isolated deviant world. Gang members are more serious and involved with one another than members of the nihilistic punk scene, with strong barriers to achieving entry and status. Stratification marks insiders as well, as women hold a decidedly inferior position within this society. Hopper and Moore's essay is at once intriguing and chilling, as it offers a glimpse into a hidden subculture that inspires intense loyalty in some, fear in others. This portrait reveals the kinds of bonds that exist among members of gangs.

If scene participants have loose relationships characterized by socializing together and if gang participants have tighter relationships based on living, partying, and fighting together, members of rings share the bond of working together. Members of deviant rings, otherwise known as crews, operate intricate deviant capers designed to provide their primary, often sole, source of income. These groups include auto and bicycle theft rings, smuggling crews, and, as Robert Prus and C. R. D. Sharper describe, "Con Men: Professional Crews of Card and Dice Hustlers." Prus and Sharper's subjects, traveling crews of deviants who live and work together on the road for most of the year, are tightly bound. They have a highly specialized set of skills, equipment, roles, and rules, with clear codes that must not be violated if everyone's safety is to be ensured. Authority and responsibility are clearly assigned to specific individuals within the group who make the travel arrangements, arrange for entry into high-rolling games, replace retiring members, and handle legal problems when they arise. In a tightly knit operation like this, relations are better when they are clearly defined than when they are unstated or casual. These relations would become formalized if we ventured further up the relational ladder into groups of deviants in cartels or brotherhoods.

DEVIANT ACTS

Although the structure of deviant associations is revealing, we can also learn a lot by analyzing the characteristics of the acts of deviance themselves. These acts vary widely in character from those enduring over a period of months to the more fleeting encounters that last only a few minutes, from those conducted alone to those requiring the participation of several or many people, and from those where the participants are face to face to those where they are physically separated. At the same time, acts of deviance can be looked at in terms of what they have in common. Like the relations among deviants, deviant acts fall along a continuum of sophistication and organizational complexity. We arrange them here according to the minimum number of their participants and the intricacy of the relations among these participants.

Some deviant acts can be accomplished by a lone *individual,* without recourse to the assistance or presence of other people. This does not mean that others cannot accompany the deviant, either before or during the deviant act, or even that two deviants cannot commit acts of individual deviance together. Rather, the defining characteristic of individual deviance is that it can be committed by one person, to that person, on that person, alone. A teenager's suicide, a drug addiction, a skid row bum's alcoholism, and a self-induced abortion are all examples of individual behavioral deviance. Nonbehavioral forms of individual deviance include obesity, minority group status, a physical handicap, and a deviant belief system (such as alternative religious or political beliefs).

For this section we include a reading by James Myers on "Nonmainstream Body Modification: Genital Piercing, Branding, Burning, and Cutting." These decorations, forms of deviant body adornment growing in popularity on the outskirts of conventional living, are sported by people circulating in a variety of deviant subcultures. Myers takes us on a journey through a series of workshops offered by expert body modifiers who discuss and illustrate the intricacies of these practices. He then discusses some of the motivations underlying people's engaging in this deviant act, grounding his explanation within a cross-cultural and historical perspective.

A second type of deviant act involves the *cooperation* of at least two voluntary participants. This cooperation usually involves the transfer of illicit goods such as pornography, arms, or drugs or the provision of deviant services such as those in the sexual or medical realm. Cooperative deviant acts may involve the exchange of money. Where this is absent, participants usually trade reciprocal acts. They both come to the interaction wanting to give and get something. In deviant sales, one participant supplies an illicit good or service in exchange for money. One or more of the participants in such acts may be earning a living through this means.

Laud Humphreys's "Tearoom Trade: Homosexual Behavior in Public Restrooms" offers an excellent illustration of a type of cooperative deviant act that does not involve the transfer of money. He studied impersonal homosexual behavior that occurred in public restrooms between anonymous consenting adults. In a fascinating description of this hidden subculture, he lays out the progression of the deviant act, tracing the way participants located and identified each other, how they signaled their interest in engaging in the deviant act and their particular role, and how they culminated the act. Individuals engaging in this behavior often lived their lives as covert homosexuals, passing as straights to avoid the stigma of deviance.

A second example of cooperative deviance can be found in "Turn-Ons for Money: Interactional Strategies of the Table Dancer," by Carol Rambo Ronai and Carolyn Ellis. They describe a relatively recent form of strip-dancing practiced in some nightclubs. Dancers not only (partially) remove their clothing on the stage, but they also contract for semiprivate performances in specially designed pits where greater contact with the customer is possible. This article discusses how these women "work" the customers to sell their table dances, from contact through culmination, how individuals among them construct certain presentational styles to appeal to specific types of clients, and how they cultivate relationships with "regular" customers who form the basis for their steady income.

The final type of deviant act is one of *conflict* between the involved parties. A perpetrator forces the interaction on the unwilling other, or an act entered into through cooperation turns out with one party "setting up" the other. In either case, the core relationship between the interactants is one of hostility, with one person getting the more favorable outcome. Conflictual acts may be carried out through secrecy, trickery, or physical force, but they end up with one person giving up goods or services to the other, involuntarily or without adequate compensation. Conflictful acts may be highly volatile in character, as victims may complain to the authorities or enlist the aid of outside parties if they have the chance. To be successful, therefore, perpetrators must control not only their victims' activities but also their perception of what is going on. Such acts can range from kidnapping and blackmail to theft, fraud, arson, and assault.

Our first reading in this area, by Patricia Martin and Robert Hummer, considers the situation of "Fraternities and Rape on Campus." This compelling article examines the culture of masculinity that flourishes within such an environment and how men engage in bonding and status building by degrading women and making them objects of sexual conquest. It also specifically describes the strategies that fraternity men use to accomplish their rapes, from using alcohol as a weapon to employing violence. It concludes that basic features of both society and fraternities foster the continuation of such rampant behavior.

Second, Diane Vaughan looks at elite deviance in "Transactions Systems and Unlawful Organizational Behavior." She considers crime in the higher echelon suites, showing that the line between aggressive business behavior and deviance is often very thin. She analyzes how the structures through which financial organizations manage and transmit information facilitate the abuse of trust. In particular, she considers the case of Revco Drug Stores, Inc., a discount chain, and its involvement in Medicaid fraud. In dealing with a government bureaucracy, officials at Revco became frustrated at the red tape and constant contradictions in how their claims for reimbursement were being handled by the government. Rather than starting from scratch and resubmitting new claims with the government properly, the company made the cost-effective decision to circumvent the difficulty by fraudulently altering the old claims and refiling them. Vaughan notes that such white collar crime was facilitated by the computer link upon which all transmission of information was based. This conflictful deviant act was based on the secrecy and guile of the company's agents and the pressure they felt from their supervisors to promote a quick and efficient solution to the problem.

DEVIANT CAREERS

One of the fascinating things about people's involvement in deviance is that it evolves. Doing something for the first time is very different from doing it for the two hundreth time. Sociologists have documented various stages of people's participation in such things as drug use, drug dealing, fencing, carrying out a professional hit, engaging in prostitution, and shoplifting. Although these activities are very different in character, they have structural similarities in the way people experience them according to the stage of their involvement. In fact, the career analogy has been applied fruitfully to the study of deviance because people go through many of the same cycles of entry, upward mobility, achieving career peaks, aging in the career, burning out, and getting out of deviance as they do in work. The career phases most commonly analyzed are those at the beginning and the end, when participants are involved in the transition between deviance and the conventional world, but we will also look at some features of intermediate career deviance.

Sociologists have been most fascinated by the process through which people *enter deviance*. They have looked at "push" factors, "pull" factors, and subcultural factors. Although some people venture into deviance on their own, the vast majority do it with the encouragement and assistance of others, often joining cooperative deviant enterprises. The turning points that mark significant phases in their transitions have been explored, as well as their changing self-identities. Most commonly, people who become involved in deviance do so through a process of shifting their circle of friends. They drift into new peer groups as they are drifting into deviance, or their whole peer group drifts into deviance together as the members enter a new phase of the life cycle.

The selections in this section highlight two different ways of entering deviance: alone or with a group. In "Joining a Gang," Martin Sanchez-Jankowski shows the many complex factors that induce individuals to enter a highly cohesive and illicit group. These range from the physical to the economic and social. But joining a gang is not a one-way decision; not every individual is acceptable as a member. Gangs have organizational needs that must be met through the continuous recruitment of new people, people who will mesh with existing members and be reliable in uncertain or dangerous situations. They may choose to entice new members with their attractiveness, to pressure them into joining through a sense of duty, or to coerce them through threats. Sanchez-Jankowski also addresses the counterposed question of why some individuals do not join gangs.

Clinton Sanders considers why people get tattoos in "Marks of Mischief: Becoming and Being Tattooed." Although this is largely an individual decision, potential tattooees, influenced by others they know who are already similarly decorated, see a way of affiliating themselves with a larger, somewhat nebulous, deviant community. Having a tattoo puts one out of the mainstream of American middle-class society, and people wanting to show their disaffiliation with this normative position and lifestyle are often attracted to the rebellious symbolism of tattoos. Sanders shows with irony that although tattoos last permanently, they usually result from an impulse decision that people have been mulling over only loosely.

Being deviant holds different challenges. Participants must manage their deviance, their relationships within deviant communities, and their safety from agents of social control and must evolve a personal style for their deviance. They must balance their deviance with the nondeviant aspects of their lives, such as their relationships with family members, people in the community, and those on whom they rely to meet their legitimate needs.

Alan Klein discusses some of the ambivalence faced by a group of weight lifters in "Managing Deviance: Hustling, Homophobia, and the Bodybuilding Subculture." Dedicated to perfecting their bodies so they can advance in their sport, many male weight lifters have experienced a conflict between their bodybuilding needs and the time demands of conventional jobs. In order to support themselves, they may find themselves dating homosexuals willing to support them, even though they are adamantly homophobic. The conflicting demands of their double lives thus cause them problems, as they struggle to manage both straight friends and girlfriends as well as their bodybuilding and source of financial support.

One of the largest underground domestic cash crops, marijuana, grows hardily throughout much of the Midwest and the coastal portions of the United States. In "The Intangible Rewards from Crime: The Case of Domestic Marijuana Cultivation," Ralph Weisheit offers insights into the farmers who risk stigmatization and arrest by planting this crop. He shows a variety of reasons why people cultivate marijuana, and he identifies several types of marijuana growers. Participants get many diverse rewards from this practice, despite the risks.

Although people tend to think that getting into deviance represents the more difficult end of the career span, sociologists have found that there can be greater problems associated with *exiting deviance*. People change during their involvement with deviance and the easily obtained money they often find there; returning to a more restricted base of funds is not always easy. They also become accustomed to the freewheeling lifestyle and open value system associated with a deviant community, where conventional norms are disdained. Reentering the straight world with its morality may chafe. And what are they to do to earn a living if they have been involved in occupational deviance, where they were making money through illicit means? Who will hire them? What can they put on a résumé? How will they find a job?

Yet most people do not want to spend their whole lives engaged in deviance. We discuss the process by which people burn out of deviance in "Shifts and Oscillations in Deviant Careers: The Case of Upper-level Drug Dealers and Smugglers." After spending several years in the upper echelons of the drug trade, many marijuana smugglers find that the drawbacks of the lifestyle exceed the rewards. The initial excitement and thrills turn to paranoia, people whom they know get busted all around them, and their risk of arrest grows. Years of consuming drugs to excess takes its toll on them physically, and they come to regard the straight life they formerly rejected as boring in a new light. Yet they cannot easily quit dealing; they have developed a high-spending lifestyle that they are loath to abandon. When they try to retire from trafficking, they quickly use up all their money and are drawn back into the business. Their patterns of exiting, thus, often resemble a series of quittings and restartings, as they move out of deviance with great difficulty.

David Brown highlights an alternate route out of deviance that has recently become more common in society: capitalizing on one's former deviance. In "The Professional Ex-: An Alternative for Exiting the Deviant Career," Brown uses the case of former alcoholics who are helped to sobriety through counseling and then go on to become counselors themselves. They are drawn to this practice for two main reasons: remaining involved helps them maintain their sobriety, and their status as former deviants gives them an experiential credential that has become respected in society. Like ex-alcoholics, we commonly see ex-cult members, ex-eating disorder counselors, ex-addicts, and a host of other professional "ex"s trading on their past to reenter society.

26

Real Punks and Pretenders

The Social Organization
of a Counterculture

KATHRYN JOAN FOX

odern Western society has been characterized by a variety of anti-establishment style countercultures following in succession (i.e., the Teddy Boys of 1953–1957, the Mods and Rockers of 1964–1966, the Skinheads of 1967–1970, and the Punk Rockers of the late 1970s; Taylor, 1982). The punk culture is but the latest in this series. Since most studies of youth- and style-oriented groups are British (Frith, 1982), very little has been written from a sociological perspective about punks in the United States. This study is an attempt to fill that void.

Whether or not the punk scene in the United States could be legitimately classified as a social movement is debatable. Most writers on this contemporary phenomenon agree that American punks have a more amorphous, less articulated ideological agenda than punks elsewhere (Brake, 1985; Street, 1986). While the punk scene in England responded to youth unemployment and working-class problems, the phenomenon in the United States was more closely connected to style than to politics. Street (1986: 175) notes that even for English punks, "politics was part of the style." I would argue that this was even more the case for American punks. The consciousness of the youth in the United States did not parallel the identification with the plight of youth found in Europe. Nonetheless, the "style" code for punks in the United States contained an insistent element of conflict with the dominant value system. The consensual values among the punks, as ambiguous as they were, could best be understood by their contradictory quality with reference to mainstream society. In this respect, punk in America fit the definition of a "counterculture" offered by Yinger (1982: 22-23). According to this definition, the salient feature of a counterculture is its contrariness. Further, as opposed to individual deviant behavior, punks constituted a counterculture in that they shared a specific normative system. Certain behaviors were considered punk, while others were not. Indeed, style was the message and the means of expression. Observation of behaviors that were consistent with punk sensibilities

From: "Real Punks and Pretenders: The Social Organization of a Counterculture," Kathryn J. Fox, *Journal of Contemporary Ethnography,* Vol. 16, No. 3, 1987. Reprinted by permission of Sage Publications, Inc.

were viewed as indicative of punk "beliefs." These behaviors, along with verbal pronouncements, verified commitment. Within the groups of punks I studied, the degree of commitment to the counterculture lifestyle was the variable that determined placement within the hierarchy of the local scene.

Previous portrayals of youthful, antiestablishment style cultures have discussed their norms and values (Berger, 1967; Davis, 1970; Hebdige, 1981; Yablonsky, 1968), their relationship to conventional society (Cohen, 1972; Douglas, 1970; Flacks, 1967), their focal concerns and ideology (Flacks, 1967; Miller, 1958), and their relationship to social class (Brake, 1980; Hall and Jefferson, 1975; Mungham and Pearson, 1976). With the exception of Davis and Munoz (1968), Kinsey (1982), and Yablonsky (1959), few of the studies of antiestablishment, countercultural groups discuss their implicit stratification. In this essay I will describe and analyze the various categories of membership in the punk scene and show how members of these strata differ with regard to their ideology, appearance, taste, lifestyle, and commitment.

I begin by discussing how I became interested in the topic and the methods I employed to gain access to the group and to gather data. I then offer a description of the setting and the people who frequented this scene. Next I offer a structural portrayal of the social organization of this punk scene, showing how the layers of membership form. I then examine each of the three membership categories (hardcore punks, softcore punks, and preppie punks), as well as the spectator category, focusing on the differences in their attitudes, behavior, and involvement with this antiestablishment style culture. I conclude by outlining the contributions each of these types of members makes to the continuing existence of the punk movement and, more broadly, by describing the relation between the punk counterculture and conventional society.

METHODS

My interest in the punk movement dates back to 1978. At that time, I attended a local punk bar fairly regularly and wore my hair and clothing in "punk" style, albeit not the radical version. I also visited a major northeastern city at about that same time, when the punk scene was in full flower, and spent several nights visiting what are now famous punk hangouts. My early interest and involvement in this scene laid the groundwork for this study, as it permitted me to gain knowledge of the punk vernacular, styles, and motives. The research continued, with active, weekly participation, through the middle of 1986. I have continued to keep a close watch on national trends and developments in punk culture. Further, I continually frequent local punk bars in an effort to deepen my understanding and to observe the decline of this counterculture. However, I conducted the bulk of my interviews in the fall of 1983 over a period of about two months. During that time, I attended "punk night" at a local bar once a week. In addition, I was invited to other punk functions, such as parties, midnight jam sessions, and public property destruction events. I thereby observed approximately 30 members of this movement with some degree of regularity. I used mainly observational techniques, along with some participation. Although I frequented numerous punk gatherings, my participatory role was constrained by the limited time I spent there. I also had to tread a line between covert and overt roles. While some people knew I was researching this setting, I could not reveal this to others because they might have denied me further access to the group. This created a problem, much the same as that experienced by Henslin (1972) and Adler (1985), in that I had to be careful about what I said and to whom I confided my research interests. This "tightrope effect" severely limited my active participation in the scene.

Nevertheless, by following the investigative research techniques advocated by Douglas (1976), I was able to gain the trust of some key members. I tried to establish friendly relations by running errands for them, buying them drinks and food, and driving them to pick up their welfare checks and food stamps. After several weeks, when I began to be recognized, I was able to broach the topic of doing interviews with several people. I formally interviewed nine people at locations outside the bar. These tape-recorded interviews were unstructured and open-ended. Additionally, I conducted 15 informal interviews at the bar. Finally, I had countless conversations with members, nonmembers, interested bystanders, and social scientists who had an interest in the punk counterculture. In all, my somewhat punk appearance, similar age, regularity at the scene, and apparent acceptance of their lifestyle allowed me to move within the scene freely and easily.

SETTING

The research took place in a small cowboy bar, "The Glass Gun," which was transformed into a punk bar one night a week. The bar was situated in a southwestern city with a population of about 500,000. The city itself is located in the "Bible Belt," characterized by conservative religious and political views. The bar was a small, dark, and dilapidated place. There was a stage area where the bands played, surrounded by a wooden rail. Wobbly tables and torn chairs formed a U-shape around the stage. There was a pool table in the corner, which the punks rarely used. Most of the patrons of the club stood at or near the bar.

For the punks in this city, the Glass Gun was the only place to congregate regularly at that time. On these designated nights, local punk bands played to an audience of about 20 people; some were punks, others were not. The typical audience ranged in age from about 16 to 30, although a few were younger or older. Basically, the punk counterculture was a youth phenomenon. It seemed to attract young, single, mobile people. Snow et al. (1980) have suggested that these characteristics make a person more "structurally available" for movement recruitment. Within the punk scene the number of men and women was fairly equal.

The punk style codes were somewhat diverse. Different styles existed for different kinds of punk. Pfohl (1985: 381) has referred to Hebdige's description of punk style as "the outrageous disfigurement of commonsensical images of aesthetics and beauty and the abrasive, destructive codes of punk style. These are aesthetic inversions of the normal, or consensus-producing, rituals of the dominant culture's style." The basic identifiable element was a subculturally accepted punk hairstyle. These ranged from a very short, uneven haircut, sticking straight up in front, to an American Indian mohawk style, to a shaven head. Along with the haircut, a punk fashion prevailed. The two were inextricably associated. The fashion ranged from torn, faded jeans, T-shirts, and army boots to expensive leather outfits.

The punk dress code was also fairly androgynous. There was no real distinction between male and female fashions. Both men and women wore faded jeans, although leather pants and miniskirts were also quite common among the women. The middle-class punk women, who tended to be students, dressed in a more traditionally feminine manner, glorifying and exaggerating the "glamour girl" image reminiscent of the sixties. This included tight skirts, teased hair, and dark, heavy makeup. The other punk women identified with a more masculine, working-class image, deemphasizing their feminine attributes. Both sexes also wore and admired leather jackets. It was also quite common to see both men and women with multiple pierced earrings all the way around the outside of their ears. Men sometimes wore eye makeup as well. (One man wore miniskirts, makeup, and rhinestones. However, this type of behavior occurred infrequently.) Basically, punk style ran counter to what the dominant culture would deem aesthetically pleasing. One major reason punks dressed as they did was to set themselves apart and to make themselves recognizable. The image consisted of dark, drab clothing, short, spiky, "homemade" haircuts, and blank, bored, expressionless faces reminiscent of those of concentration camp prisoners.[1] The punks created a new aesthetic that revealed their lack of hope, cynicism, and rejection of societal norms.

THE SOCIAL ORGANIZATION
OF THE PUNK SCENE

Like the youth gangs that Yablonsky (1962) studied, members of this local punk scene constituted a "near-group." The membership was impermanent and shifting, members' expectations were not always clearly defined, consensus within the group was problematic, and the leadership was vague. Yet out of this uncertainty surfaced an apparent consensus about the stratification of the local community and the roles of the three types of members and peripheral hangers-on who participated in this scene. These four typologies can be hierarchically arranged by the presence (or absence) and intensity of their commitment to the punk counterculture and their consequent display of the punk affectations and belief system. They thus formed a series of outwardly expanding concentric circles, with the most committed members occupying the core, inner roles, and the least involved participants falling around the periphery.

Starting from the center, the number of members occupying each stratum progressively increased as the commitment level of the participants diminished.

The categories to which I refer come from the terms used by the participants themselves.[2] The *hardcore punks* were the most involved in the scene, and derived the greatest amount of prestige from their association with it. They set the trends and standards for the rest of the members. The *softcore punks* were less dedicated to the antiestablishment lifestyle and to a permanent association with this counterculture, yet their degree of involvement was still high. They were greater in number and, while highly respected by the less committed participants, did not occupy the same social status within the group as the hardcores. Their roles were, in a sense, dictated by the hardcores, whom they admired, and who defined the acceptable norms and values. The *preppie punks* were only minimally committed, constituting the largest portion of the actual membership. They were held in low esteem by the two core groups, following their lead but lacking the inner conviction and degree of participation necessary to be considered socially desirable within the scene. Finally, the *spectators* made up the largest part of the crowd at any public setting where a punk event transpired. They were not truly members of the group, and therefore did not necessarily revere the actions and dedication of the hardcores as did the two intermediary groups. They did not attempt to follow the standards of those committed to this near-group. They were merely outsiders with an interest in the punk scene.

These four groups constituted the range of participants who attended and were involved with, to varying degrees, the punk counterculture. I will now examine in greater detail their styles, beliefs, practices, intentions, and roles in the scene.

PUNKS AND COMMITMENT

At the time of this study, the group of punks was small and disorganized. The number of people at any given punk event had steadily declined since my first encounters with the scene in 1978. Punk was no longer a new phenomenon, and this particularly conservative community did not provide a very conducive atmosphere for a large countercultural group to flourish. Every member of the group expressed dissatisfaction and boredom with the events (or rather, lack of events) within the scene. Even within the limits necessitated by the relatively small size of the group, there was a great deal of variation in terms of punk roles and characteristics. The qualities attributed to the different roles were based upon commitment to the scene. The punks' perceptions of levels of commitment were based principally on their evaluation of physical appearances and lifestyles. The punks categorized members of the scene on these bases and invented terms to describe them. The four types of participants, described above, varied according to their level of commitment to the scene.

Hardcore Punks

Hardcore punks made up the smallest portion of the scene's membership. In the eyes of the other punks, though, they were the essence of the local movement. The hardcores expressed the greatest loyalty to the punk scene as a whole. Although the hardcores embodied punk fashion and lifestyle codes to the highest degree, their commitment to the counterculture went much deeper than that. As one hardcore punk said:

> There's been so much pure bullshit written about punks. Everyone is shown with a safety pin in their ear or blue hair. The public image is too locked into the fashion. That has nothin' to do with punk, really. . . . For me, it is just my way of life.

The feature that distinguished hardcore punks from other punks was their belief in, and concern for, the punk counterculture. In this sense, the hardcore punks had gone beyond commitment; they had undergone the process of conversion (Snow et al., 1986). In other words, not only did they have membership status, but they believed in and espoused the virtues and ideology of the counterculture. Although many hardcores differed on what the counterculture's core values were, they all expressed some concern with punk ideology. These values were ambiguous at best, but included a distinctly antiestablishment, anarchistic sentiment. Street (1986: 175) has described punk as celebrating chaos and "a life lived only for the moment." The associated value system of punk was understood by the incorporation of cynicism and a distrust of authority. In keeping with other subcultures that intentionally distinguish themselves from the dominant culture, the punk aesthetic, lifestyle, and worldview directly confronted those of the larger society and its traditions. While the other types of punks made no reference to group beliefs or values, the hardcores revered the counterculture. For them, being punk had a profound effect on all aspects of their lives. As one hardcore said:

> There are a lot of punks around, even real punks, who don't mean it. At least, not all the way, like I do. Sometimes I feel so good about punk that I cry. And when I see people getting into some band with real punk lyrics, it's like a religious experience.

It was precisely this belief in "punk" as an external reality, like a higher good, that set the hardcores apart. Similar to Sykes and Matza's (1957) "appeal to higher loyalties," hardcore punks based their rejection of conventional society on their commitment to their antiestablishment lifestyles and beliefs. This imbued their self-identity with a sense of seriousness and purpose. Unlike other punks, they did not view their punk identity as a temporary role or a transitory fashion, but as a permanent way of life. As one hardcore member said:

> Punk didn't influence me to be the way I am much. I was always this way inside. When I came into punk, it was what I needed all my life. I could finally be myself.

Without exception, the hardcores reported having always held the values or qualities associated with the punk counterculture. The local scene, in fact, was just a convenient way of expressing these ideas collectively. Perhaps the most essential value professed by the punks was a genuine disdain for the conventional system. Their use of the term *system* here referred to a general concept of the way the material world works: bureaucracies, power structures, and competition for scarce goods. This "system" further referred to the ethic of deferred gratification, conventional hard work for profit, and the concept of private property. While this bears some similarity to Flacks's (1967) discussion of the student movement and Davis's (1970) portrayal of hippies, hardcore punks generally had a disdain for these earlier youth subcultures. There was a general attitude among punks of the need to create and maintain their own distinctive style. Kinsey found this same feature in the antibourgeois "killum and eatum" subculture. According to Kinsey (1982: 316), "K and E offered an attractive setting as its ideology presented an excellent vehicle for expressing hostility toward conventional society." This contempt for authority and the conventional culture was, in fact, such an essential value for the punks that if one expressed prosystem sentiments or support for the present administration, one could not be considered a committed member, no matter how well one looked the part. Overt behavioral and physical attributes, though, were major ways hardcore punks showed disdain for the system. Particular characteristics were essential for consideration as hardcores. Most fundamentally, a verbal commitment to punk values and the punk scene, in general, was required. For example, John, the epitome of a hardcore punk, claimed to hate the system. He talked about the inequality of the system quite often. In John's words:

> Punk set me free. It let me out of the system. I can walk the streets now and do what I want and not live by the demands of the system. When I walk the streets, I am a punk, not a bum.

However, this verbal pronouncement had to be backed by a certain lifestyle that further indicated commitment to the group. This lifestyle consisted of escaping the system in some way. Almost all of the hardcores were unemployed and lived in old, abandoned houses or moved into the homes of friends for periods of time. Some survived from the charity of sycophantic, less committed punks. Others worked in jobs that they considered to be outside the system, such as musicians in rock bands or artists.

Another central feature of the lifestyle was the hardcores' use of dangerous drugs. Many hardcores indulged heavily in sniffing glue. Glue was inexpensive and readily available to the punks. Its use also symbolized the self-destructive, nihilistic attitude of hardcores and their desire to live outside of society's norms. As one member said:

> It is kind of like a competition, a show-off thing. . . . See who has the most guts by seeing who can burn his brain up first. It is like a total lack of care about anything, really.

This closely corresponds to Davis and Munoz's (1968) description of "freaks." Both punks and freaks were "in search of drug kicks as such, especially if [their] craving carries [them] to the point of drug abuse where [their] health, sanity and relations with intimates are jeopardized" (Davis and Munoz, 1968: 306). Again, here we see a rejection of anything the larger society sees as "sensible."

However, the most salient feature of the hardcore lifestyle was the radical physical appearance. In every case, people who were labeled as hardcore had drastically altered some aspect of their bodies. For example, in addition to the hairstyles discussed earlier, they often had tattoos, such as swastikas, on their arms or faces. Brake (1985: 78) has referred to the use of the swastika as a symbol for punks that was actually devoid of any political significance. Rather, the swastika was a "symbol of contempt" employed as a means of offending the traditional culture. The hardcore punks did their best to alienate them-selves from the larger society.

According to Kanter (1972), the first requisite in the principle of a gestalt sociology is that a group forms maximum commitment to this higher ideal by sharply differentiating itself from the larger society. The hardcore members did this by going through the initiation rite of passage: semipermanently altering their appearance. As one hardcore said:

> Did you see Russell's mohawk? I'm so glad for him. He finally decided to go for it. Now he is a punk everywhere . . . no way he can hide it now.

This was similar to certain religious cults, such as the Hare Krishnas, where a drastic change in appearance was required for consideration as a total convert (Rochford, 1985). The punk counterculture informally imposed the same pre-requisite. By doing something so out of the ordinary to their appearance, the punks voluntarily deprived themselves of some of the larger society's coveted goods. For example, many of the hardcores were desperately poor. They said that they knew all they would have to do to obtain a job would be to grow their hair into a conventional style; yet they refused. This kind of action based on commitment was what Becker (1960) has called "side bets," where com-mitted people act in such ways that affect their other interests separate and apart from their commitment interests. By making specific choices, people who are committed sacrifice the possible benefits of their other roles. An important characteristic of Becker's notion of side bets is that people are fully aware of the potential ramifications of their actions. This point was illustrated by one punk:

> Some of my friends that aren't punk say, "Why don't you get a job? All you'd have to do is grow your hair out or get a wig and you could get a job." I mean, I know I could. Don't they think I knew that when I did it? It was a big step when I finally cut my hair in a mohawk.

For the hardcore punk, being punk was worth the sacrifices; it was per-ceived as an inherently good quality. In this respect, the hardcores differed from other types of punks. They held the larger punk scene in esteem. As one loyal member said:

> It really pisses me off when people act like ours is the only punk group in the world. They don't even care what bigger and better groups exist. If

this whole thing ended tomorrow, they wouldn't care what happened to the whole punk scene.

The hardcores continually expressed their disgust with the local scene. Much like the hippies studied by Davis (1970), "the scene," in itself, was the message. While not enough people joined the group to satisfy the punks, they were, nonetheless, grateful that they had any kind of group environment to which they could attach themselves. The Glass Gun, with its regularly scheduled "punk night," nonhostile attitude toward them, and coterie of interested bystanders, at least gave them a place to express their values collectively. It was essential in maintaining the group's solidarity and social organization.

Softcore Punks

The softcore punks made up a larger portion of the local scene than the hardcores. There were around fifteen softcore punks. There was one fundamental difference between the hardcore and softcore punks. For the hardcores, it was not sufficient just to be antiestablishment or to wear one's hair in a certain way. Rather, one had to embody the punk lifestyle and ideology in all possible ways. As one hardcore punk put it:

> Everybody thinks she is hardcore because she looks so hardcore. I mean, yeah, she has a mohawk, and she won't get a job and she says she's for anarchy, but she doesn't care that much about being punk. She likes all these different kinds of music and stuff. She seems sometimes like she is just in it for fun. She even says she'll be whatever's in when punk goes out!

The softcore punks lived similar lifestyles to the hardcores. However, the element of "seriousness" about the scene, so pervasive among the hardcores, was absent among the softcores. Visually, the two types were basically indistinguishable. They were different only in their level of commitment. The commitment for softcores was to the lifestyle and the image only, not to "punk" as an ideology or an intrinsically valuable good. The softcores made no pretense of concern for either the larger counterculture or the feeling of permanence about their punk roles. As one softcore, Beth, said:

> Everyone thinks I am so serious about it because I have a mohawk. Some people just can't get past it. Sometimes I get tired of it. Other times, I like to play jokes on people; like another friend of mine who has a mohawk, we'll walk down the street and point at someone with regular hair and say, "Wow! Look at him, he's weird, he doesn't have a mohawk." The fact is, if everyone did have one, I'd do something different to my hair.

The softcores identified with the punk image only temporarily. This distinguished their level of commitment from the *conversion* of the hardcores. The softcores' interest in the scene had only to do with what it could offer them at the present time. While participating, they did what was considered a good job of being "punk." However, if a new cultural trend surfaced, it would be just as likely that they would use their energy effectively to create that particular image. As one softcore punk said:

> I've spent time identifying myself as a hippie, then as a women's libber, then an ecologist, and now as a punk. I'm punk now, but I am in the process of changing into something else. I don't know what. I'm getting bored with this scene. But for now, if I'm gonna do it, I'll do it right.

The softcore punks were somewhat committed in that they participated in some of the more drastic elements of punk lifestyle. For example, softcores had their hair cut in severe ways, just like the hardcores. They were, at least temporarily, committed to being punk (or playing punk) in that they "cut" themselves off from some of society's goods as well. However, the softcores did not share the self-destructive bent of the hardcores. The drugs that they consumed, such as marijuana, alcohol, and amphetamines, were not so potentially dangerous. Yet, because of their apparent visual commitment, and because of the lip service they gave to punk values, softcores were viewed as members in good standing. The hardcores liked and respected the softcores; the two groups associated freely. Some hardcores considered softcores to be simply members in transition. Lofland and Stark (1965) have suggested that movements themselves play a role in promoting the ideology in the new members, rather than the members coming to the movement because its ideology coincides with their own established beliefs. This was the case for the softcore punks. They did not claim to have held punk values before becoming punk. As Anne, a softcore, recalled:

> It was scary to me at first. The hype from the magazines and stuff—all this weird shit, y'know. Then I went there and just hung out. The reason it was frightening is that a lot of people had different ideas about life than me. And I had to change myself to be with them. I had to be more intense, be an outcast. It was exciting because there seemed to be an element of danger in it—like living on the edge.

A process of simply happening onto the scene was typical of softcore members. Many recounted the feelings of purposelessness that preceded the drift into the punk scene. This drift is similar to the drifts that occur in other deviant lifestyles (Matza, 1964). What Matza called the "mood of desperation" often caused people to drift into delinquency or deviant lifestyles. As Joanie said:

> I was really doing nothing with my life and I just kinda accidentally came into the punk scene. I gradually got involved in it that way. The music, and the people to an extent, really raised my consciousness about the system.

Softcores' verbal recognition of punk attitudes, such as awareness of the system, helped to validate their punk performances. Hardcores felt that verbal commitment was an essential first step to further commitment. For this reason, the hardcores accepted the softcores and considered them to be genuine and authentic in their punk identity. Such identification with punk values, along with a typical punk lifestyle, made the distinction between hardcores and soft-cores difficult. Again, the distinction became clear only with regard to the level of commitment, or seriousness, of the two types. One softcore made this qualification more apparent:

> They get mad at me and think I'm insincere or whatever 'cause I like to have fun. I take my politics serious, too, but I feel if you are here, you might as well enjoy it. They think being punk is so serious, they are depressed or stoned all the time.

This statement indicates that the hardcores defined the situation for the local scene. The hardcores decided what differentiated real punks (or commit-ted punks) from pretenders. The hardcores considered only themselves and the softcores to be real. The "realness" of a punk was based on the level of commitment. The level was judged on the basis of willingness to sacrifice other identities for the punk identity. To prove this, a member would have to make permanent his or her punk image. What the punk identity offered was status within its own subculture for those who could not or would not achieve it in conventional society (Cohen, 1955). However, commitment to the deviant identity did not stem from a forced label. On the contrary, com-mitment to the punk identity was a "self-enhancing attachment" (Goffman, in Stebbins, 1971). The punks' self-esteem was enhanced by the approval they received. It would follow, then, that the more consistent one's behavior was with the superficial signs of commitment, the more prestige one would be able to obtain. Doug, a softcore, commented on this aspect of subcultural prestige among the punks:

> In their own way, they're elitist. It's kind of like because they're not part of the general run of things, because they've actually *chosen* to be rejected in a lot of cases, they've kind of set up their own little social order. It seems to me like it's based on, like a contest, who can be more cool than who. With the really hardcore punks, it's who can self-destruct first; in the name of punk, I guess.

The hardcores and the softcores used the same criteria to judge commit-ment. Both types agreed that the difference between them was their levels of commitment. Both types fully realized that the softcores did not share the same loyalty to, and identification with, the punk counterculture as a whole. Although both types expressed some commitment to the punk identity and lifestyle, they both realized that the hardcores viewed their own identities as permanent and the softcores' as temporary.

Preppie Punks

The preppie punks made up an even larger portion of the crowd at punk events. The preppies frequented the scene, but approached it similarly to a costume party. They were concerned with the novelty and the fashion. The preppies bore some resemblance to Yablonsky's (1968) "plastic hippies" in that they were drawn to the excitement of the scene. Whereas the core members acted nonchalant and natural about being punk, the preppies could not hide their enthusiasm about being part of the scene. This feature contributed to the core members' perceptions of preppies as "not real" punks. As one core member said:

> It really kills me when these preppie girls come up to me and say "Oh, wow, you're so punk; you're so new wave," like I'm really trying or something.

Preppie punks did not lead the lifestyle of the core members. The preppie punks tended to be from middle-class families, whereas the core punks were generally from lower- or working-class backgrounds.[3] Preppie punks often lived with their parents; they tended to be younger, and were often in school or in respectable, system-sanctioned jobs. This quasi-commitment meant that preppies had to be able to turn the punk image on and off at will. For example, a preppie punk hairstyle, although short, was styled in such a versatile way that it could be manipulated to look punk sometimes and conventional at other times. Preppie fashion was much the same way. Mary, a typical preppie punk, put her regular clothes together in a way she thought would look punk. She ripped up her sorority T-shirt. She bought outfits that were advertised as having the "punk look." Her traditional bangs transformed into "punk" bangs, standing straight up using hair spray or setting gel. The distinguishing feature of preppie punks was the manufactured quality of their punk look. This obvious ability to change roles kept the preppies from being considered real or committed. The preppies were not willing to give anything up for a punk identity. As one softcore said:

> They come in with their little punk outfits from Ms. Jordan's [an exclusive clothing store] and it's written all over 'em: money. They think they can have their nice little jobs and their semipunk hairdo and live with mom and dad and be a real punk, too. Well, they can't.

Another said of preppies:

> It's a little hard to take when you have nothing and they try to have everything. Having all that goes against punk. They gotta choose to not have it. Otherwise, they're just playing a game.

The preppies liked to disavow their punk association in situations that would sanction them negatively for such associations. This state of "dual commitments," in which they never had to reject the conventional world in order to be marginally a part of the group, was characteristic of preppie punks (Cohen, 1955). Kanter (1972) has described a process of conversion and commitment that is commonly found in communes. The first step in the process was the renunciation of previous identities. According to this model, the preppie punks would not be considered committed at all. Thus they could not have been categorized as punks in any meaningful sense. As one core member said, "Being 'punk' to them is like playing cowboys and Indians."

Criticizing and joking about the preppies made up a large portion of core members' conversations. Some truly disliked the preppies and others were flattered by their feeble attempts at imitation of core behavior. For example, when a preppie punk approached one core member, he rolled his eyes and said, "Here comes my fan club," with a half-embarrassed smile and a distinct look of pleasure on his face.

Also, the financial function that the preppies served to core members made them more tolerable. Preppies almost always had jobs or survived by their parents' support. Many of the core members subsisted on the continued generosity of their devout fans. The preppies were more than willing to help the other punks. Preppies sometimes offered hardcores financial help in the form of buying them groceries, driving them places, and providing them with cigarettes, alcohol, and other drugs. Because of this, many punks felt that they could not afford to reject outwardly those who were less committed. As Anne said, "One of these days, this kindness is going to dry up."

Yet the joking and poking fun at preppies was a constant activity. It served to separate, for the committed punks, "us" from "them." It reinforced their sense of being the only real punks. Again the distinction made by core members was grounded in the preppies' attempts to play numerous roles. Haircut and clothing were the decisive clues. The real punks could spot a preppie from a distance; they never had to say a word. As one core member said,

Oh look, she's punked out her hair. Yes, we're impressed. Tomorrow she'll look just like a Barbie doll again.

Perhaps the most definitive statement separating the real punks from the preppies referred to lifestyle:

All I know is that I live this seven days a week, and they just do it on weekends.

The preppies, though, while definitely removed from core members, still played an important role in the scene.

Spectators

The category of spectators referred to everyone who observed the scene fairly regularly, but were not punks themselves. This type consisted of, literally, "everyone else." They made up the largest portion of the crowd at the Glass Gun on any given night. They were different from the preppie punks in that they did not try to look punk. They made no pretense of commitment to the scene at all. They did not identify themselves as punks; they had no stake in the scene. Spectators consisted of all different types of people and varied in their occupations, clothes, and reasons for being there. The only common denominator this group shared was the desire to stand back and watch, rather than to participate actively in punk activities. One spectator said of his involvement in this scene:

> People on the fringe are usually voyeurs of a sort. They like to be on the receiver's end of what's happening. Maybe punk is really their alter-ego. And maybe that need is satisfied just by watching and pretending. That's how it is with me, anyway.

The spectators liked to observe the fashion, to listen to the music, and to be "in the know" about the scene. They were, in other words, punk appreciators.

For the most part, spectators on the fringe were ignored by the core members. They never received the attention that preppies did because they made no attempts to "play punk." However, if a spectator appeared on the scene looking completely antithetical to punk, core members would simply laugh or say something derogatory about them and drop the subject. For example, one time a hippie-looking character came in and one punk said, "Oh my God, I think we're in a time warp," to which another punk responded, "Maybe we should tell him that Woodstock's over and that it is 1983."

Following such statements, the punks would watch the spectator's reaction to the scene. For the most part, except as a diversion, the punks were uninterested in the spectators. They did not generally associate with them or talk about them much. Presumably this was the case because of the tremendous turnover in spectator membership.

Most spectators either slowly began to identify with the group (most core members started out as spectators) or stopped frequenting the punk events. There were, however, some loyal spectators. They would frequent the club. They knew most of the punks at least slightly. The punks generally liked this sort of spectator because they provided the punks with an audience. Every type of punk thrived on an audience. The punks needed people to shock. The spectator served that function. The attitude that the members had toward the spectators was one of tolerance and indifference. As one core member said of them:

> They're into it for the novelty. It's like going to the circus for them, to be a part of something new and exciting. But that's okay. I like going to the circus, too; I just like being in it better.

Thus though spectators were only peripheral to the scene they provided an alternative set of norms that functioned to delineate the social boundaries of the counterculture. . . .

NOTES

1. I am indebted to David Matza and John Torpey for this analogy.
2. With the exception of the term *softcore,* all of the distinctions between categories came directly from the participants. The members did make a

distinction between hardcore and what I am calling softcore punks. However, the softcores were referred to simply as "punks" by the hard–cores, in an effort to distinguish the "hardcore" quality they attributed to themselves. I chose to refrain from using the term *punk* to apply to one specific category so that I can use the term more freely and generally, and to avoid confusion.

3. Very little information is provided in this text about the class, race, and ethnicity of these participants. The community from which these data come is relatively homogeneous. The few references to class are more impressionistic; that is, based upon knowledge of family occupations, school districts, and so on. However, the dearth of this kind of data stems from the fact that I was more interested in the features the members had in common than in the distinctions between them, with the exception of their differing levels of commitment and their styles.

REFERENCES

Adler, P. A. (1985) *Wheeling and Dealing*. New York: Columbia Univ. Press.

Becker, H. S. (1960) "Notes on the concept of commitment." *Amer. J. of Sociology* 66: 32–40.

Berger B. (1967) "Hippie morality—more old than new." *Transaction* 5: 19–27.

Brake, M. (1980) *The Sociology of Youth Culture and Youth Subcultures*. London: Routledge.

———. (1985) *Comparative Youth Culture: The Sociology of Youth Culture and Subcultures in America, Britain, and Canada*. London: Routledge.

Cohen, A. (1955) *Delinquent Boys*. Glencoe, IL: Free Press.

Cohen, S. (1972) *Folk Devils and Moral Panics*. New York: St. Martin's.

Davis, F. (1970) "Focus on the flower children: why all of us may be hippies some day," pp. 327–340 in J. Douglas (ed.) *Observations of Deviance*. New York: Random House.

Davis, F., and L. Munoz (1968) "Heads and freaks: patterns and meanings of drug use among hippies." *J. of Health and Social Behavior* 9: 156–164.

Douglas, J. (1970) *Youth in Turmoil*. Washington, DC: National Institute of Mental Health.

———. (1976) *Investigative Social Research*. Newbury Park, CA: Sage.

Flacks, R. (1967) "The liberated generation: an exploration of the roots of student protest." *J. of Social Issues* 23: 52–75.

———. (1971) *Youth and Social Change*. Chicago: Markham.

Frith, S. (1982) *Sound Effects*. New York: Pantheon.

Hall, S., and T. Jefferson [eds.] (1975) *Resistance through Rituals*. London: Hutchinson.

Hebdige, D. (1981) *Subcultures: The Meaning of Style.* New York: Methuen.

Henslin, J. (1972) "Studying deviance in four settings: research experiences with cabbies, suicides, drug users, and abortionees," pp. 35–70 in J. Douglas (ed.) *Research on Deviance.* New York: Random House.

Kanter, R. M. (1972) *Commitment and Community: Communes and Utopia in Sociological Perspective.* Cambridge, MA: Harvard Univ. Press.

Kinsey, B. A. (1982) "Killum and eatum: identity consolidation in a middle class poly-drug abuse subculture." *Symbolic Interaction* 5: 311–324.

Lofland, J., and R. Stark (1965) "Becoming a world saver: a theory of conversion to a deviant perspective." *Amer. Soc. Rev.* 30: 862–875.

Matza, D. (1964) *Delinquency and Drift.* New York: John Wiley.

Miller, W. (1958) "Lower class culture as a generating milieu of gang delinquency." *J. of Social Issues* 14: 5–19.

Mungham, G., and G. Pearson [eds.] (1976) *Working Class Youth Culture.* London: Routledge.

Pfohl, S. (1985) *Images of Deviance and Social Control.* New York: McGraw-Hill.

Rochford, E. B., Jr., (1985) *Hare Krishnas in America.* New Brunswick, NJ: Rutgers Univ. Press.

Snow, D. A., E. B. Rochford, Jr., S. K. Worden, and R. D. Benford (1986) "Frame alignment and mobilization." *Amer. Soc. Rev.* 51: 464–481.

Snow, D. A., L. Zurcher, Jr., and S. Ekland-Olson (1980) "Social networks and social movements: a micro-structural approach to differential recruitment." *Amer. Soc. Rev.* 45: 787–801.

Stebbins, R. A. (1971) *Commitment to Deviance.* Westport, CT: Greenwood.

Street, J. (1986) *Rebel Rock: The Politics of Popular Music.* Oxford: Basil Blackwell.

Sykes, G., and D. Matza (1957) "Techniques of neutralization." *Amer. Soc. Rev.* 22: 664–670.

Taylor, I. (1982) "Moral enterprise, moral panic, and law-and-order campaigns," pp. 123–149 in M. M. Rosenberg et al. (eds.) *A Sociology of Deviance.* New York: St. Martin's.

Yablonsky, L. (1959) "The delinquent gang as a near-group." *Social Problems* 7: 108–117.

———. (1962) *The Violent Gang.* New York: Macmillan.

———. (1968) *The Hippie Trip.* New York: Pegasus.

Yinger, J. M. (1982) *Countercultures: The Promise and Peril of a World Turned Upside Down.* New York: Free Press.

Young, J. (1973) "The hippie solution: an essay in the politics of leisure," pp. 182–208 in I. Taylor and L. Taylor (eds.) *Politics and Deviance.* Harmondsworth, England: Penguin.

27

Women in Outlaw Motorcycle Gangs

COLUMBUS B. HOPPER

AND JOHNNY MOORE

T his article is about the place of women in gangs in general and in outlaw motorcycle gangs in particular. Street gangs have been observed
in New York dating back as early as 1825 (Asbury, 1928). The earli-
est gangs originated in the Five Points district of lower Manhattan and were
composed mostly of Irishmen. Even then, there is evidence that girls or young
women participated in the organizations as arms and ammunition bearers during gang fights. . . .

Studies have shown that girls have participated in street gangs as auxil-
iaries, as independent groups, and as members in mixed-gender organizations.
While gangs have varied in age and ethnicity, girls have had little success in
gaining status in the gang world. As reported by Bowker (1978, Bowker and
Klein, 1983), however, female street gang activities were increasing in most
respects; he thought that independent gangs and mixed groups were increasing
more than were female auxiliary units.

Unlike street gangs that go back for many years, motorcycle gangs are relatively new. They first came to public attention in 1947 when the Booze
Fighters, Galloping Gooses, and other groups raided Hollister, California
(Morgan, 1978). This incident, often mistakenly attributed to the Hell's
Angels, made headlines across the country and established the motorcycle
gangs' image. It also inspired *The Wild Ones,* the first of the biker movies
released in 1953 , starring Marlon Brando and Lee Marvin.

Everything written on outlaw motorcycle gangs has focused on the men in
the groups. Many of the major accounts (Eisen, 1970; Harris, 1985;
Montegomery, 1976; Reynolds, 1967; Saxon, 1972; Thompson, 1967;
Watson, 1980; Wilde, 1977; Willis, 1978; Wolfe, 1968) included a few tan-
talizing tidbits of information about women in biker culture but in none were
there more than a few paragraphs, which underscored the masculine style of
motorcycle gangs and their chauvinistic attitudes toward women.

Although the published works on outlaw cyclists revealed the fact that
gang members enjoyed active sex lives and had wild parties with women, the
women have been faceless; they have not been given specific attention as

From: "Women in Outlaw Motorcycle Gangs," Columbus B. Hopper and Johnny
Moore, *Journal of Contemporary Ethnography,* Vol. 18, No. 4, 1990. Reprinted by permission of Sage Publications, Inc.

functional participants in outlaw culture. Indeed, the studies have been so one-sided that it has been difficult to think of biker organizations in anything other than a masculine light. We have learned that the men were accompanied by women but we have not been told anything about the women's backgrounds, their motivations for getting into the groups or their interpretations of their experiences as biker women.

From the standpoint of the extant literature, biker women have simply existed; they have not had personalities or voices. They have been described only in the contemptuous terms of male bikers as "cunts," "sluts," "whores," and "bitches." Readers have been given the impression that women were necessary nuisances for outlaw motorcyclists. A biker Watson (1980: 118) quoted, for example, summed up his attitude toward women as follows: "Hell," he said, "if I could find a man with a pussy, I wouldn't fuck with women. I don't like 'em. They're nothing but trouble."

In this article, we do four things. First, we provide more details on the place of women in arcane biker subculture, we describe the rituals they engage in, and we illustrate their roles as money-makers. Second, we give examples of the motivations and backgrounds of women affiliated with outlaws. Third, . . . we show how the place of biker women has changed over the years of our study and we suggest a reason for the change. We conclude by noting the impact of sex role socialization on biker women.

METHODS

The data we present were gathered through participant observation and interviews with outlaw bikers and their female associates over the course of 17 years. Although most of the research was done in Mississippi, Tennessee, Louisiana, and Arkansas, we have occasionally interviewed bikers throughout the nation, including Hawaii.[1] The trends and patterns we present, however, came from our study in the four states listed.

During the course of our research, we have attended biker parties, weddings, funerals, and other functions in which outlaw clubs were involved. In addition, we have visited in gang clubhouses, gone on "runs" and enjoyed cookouts with several outlaw organizations.

It is difficult to enumerate the total amount of time or the number of respondents we have studied because of the necessity of informal research procedures. Bikers would not fill out questionnaires or allow ordinary research methods such as tape recorders or note taking. The total number of outlaw motorcyclists we studied over the years was certainly several hundred. In addition to motorcycle gangs in open society, we also interviewed and corresponded with male and female bikers in state and federal prisons.

The main reason we were able to make contacts with bikers was the background of Johnny Moore, who was once a biker himself. During the 1960s, "Big John" was president of Satan's Dead, an outlaw club on the Mississippi Gulf Coast. He participated in the rituals we describe, and his own experience and observations provided the details of initiation ceremonies that we relate. As a former club president, Moore was able to get permission for us to visit biker clubhouses, a rare privilege for outsiders.[2]

Most of our research was done on weekends because of our work schedules and because the gangs were more active at this time. The bikers usually had a large party one weekend a month, or more often when the weather was nice, and we were invited to many of these.

At some parties, such as the "Big Blowout" each spring in Gulfport, there were a variety of nonmembers present to observe the motorcycle shows and "old lady" contests as well as to enjoy the party atmosphere. These occasions were especially helpful in our study because bikers were "loose" and easier to approach while partying. We spent more time with three particular "clubs," as outlaw gangs refer to themselves, because of their proximity.

In addition to studying outlaw bikers themselves, we obtained police reports, copies of Congressional hearings that deal with motorcycle gangs, and indictments that were brought against prominent outlaw cyclists. Our attempt was to study biker women and men in as many ways as possible. We were honest in explaining the purpose of our research to our respondents. They were told that our goal was only to learn more about outlaw motorcycle clubs as social organizations. . . .

Problems in Studying Biker Women

Although it was difficult to do research on outlaw motorcycle gangs generally, it was even harder to study the women in them. In many gangs, the women were reluctant to speak to outsiders when the men were present. We did not hear male bikers tell the women to refrain from talking to us. Rather, we often had a man point to a woman and say, "Ask her," when we posed a question that concerned female associates. Usually, the woman's answer was, "I don't know." Consequently, it took longer to establish rapport with female bikers than it did with the men.

Surprisingly, male bikers did not object to our being alone with the women. Occasionally, we talked to a female biker by ourselves and this is when we were able to get most of our information and quotations from them. In one interview with a biker and his woman in their home, the woman would not express an opinion about anything. When her man left to help a fellow biker whose motorcycle had broken down on the road, the woman turned into an articulate and intelligent individual. Upon the return of the man, however, she resumed the role of a person without opinions.

THE PLACE OF WOMEN IN
OUTLAW MOTORCYCLE GANGS

Although national,[3] outlaw motorcycle clubs of the 1980s had restricted their membership to adult males (Quinn, 1983), women were important in the outlaw life-style we observed. We rarely saw a gang without female associates sporting colors similar to those the men wore.

To the casual observer, all motorcycle gang women might have appeared the same. There were, however, two important categories of women in the biker world: "mamas" and "old ladies." A mama belonged to the entire gang. She had to be available for sex with any member and she was subject to the authority of any brother. Mamas wore jackets that showed they were the "property" of the club as a whole.

An old lady belonged to an individual man; the jacket she wore indicated whose woman she was. Her colors said, for example, "Property of Frog." Such a woman was commonly referred to as a "patched old lady." In general terms, old ladies were regarded as wives. Some were in fact married to the members whose patches they wore. In most instances, a male biker and his old lady were married only in the eyes of the club. Consequently, a man could terminate his relationship with an old lady at any time he chose, and some men had more than one old lady.

A man could require his old lady to prostitute herself for him. He could also order her to have sex with anyone he designated. Under no circumstances, however, could an old lady have sex with anyone else unless she had her old man's permission.

If he wished to, a biker could sell his old lady to the highest bidder, and we saw this happen. When a woman was auctioned off, it was usually because a biker needed money in a hurry, such as when he wanted a part for his motorcycle or because his old lady had disappointed him. The buyer in such transactions was usually another outlaw.

RITUALS INVOLVING WOMEN

Outlaw motorcycle gangs, as we perceived them, formed a subculture that involved rituals and symbols. Although each group varied in its specific ceremonies, all of the clubs we studied had several. There were rites among bikers that had nothing to do with women and sex but a surprising number involved both.

The first ritual many outlaws were exposed to, and one they understandably never forgot, was the initiation into a club. Along with other requirements, in some gangs, the initiate had to bring a "sheep" when he was presented for membership. A sheep was a woman who had sex with each member of the gang during an initiation. In effect, the sheep was the new man's gift to the old members.

Group sex, known as "pulling a train," also occurred at other times. Although some mamas or other biker groupies (sometimes called "sweetbutts") occasionally volunteered to pull a train, most instances of train pulling were punitive in nature. Typically, women were being penalized for some breach of biker conduct when they pulled a train.

An old lady could be forced to pull a train if she did not do something her old man told her to do, or if she embarrassed him by talking back to him in front of another member. We never observed anyone pulling a train but we were shown clubhouse rooms that were designated "train rooms." And two women told us they had been punished in this manner.

One of the old ladies who admitted having pulled a train said her offense was failing to keep her man's motorcycle clean. The other had not noticed that her biker was holding an empty bottle at a party. (A good old lady watched her man as he drank beer and got him another one when he needed it without having to be told to do so.) We learned that trains were pulled in vaginal, oral, or anal sex. The last was considered to be the harshest punishment.

Another biker ritual involving women was the earning of "wings," a patch similar to the emblem a pilot wears. There were different types of wings that showed that the wearer had performed oral sex on a woman in front of his club. Although the practice did not exist widely, several members of some groups we studied wore wings.

A biker's wings demonstrated unlimited commitment to his club. One man told us he earned his wings by having oral sex with a woman immediately after she had pulled a train; he indicated that the brothers were impressed with his abandon and indifference to hygiene. Bikers honored a member who laughed at danger by doing shocking things.[4]

The sex rituals were important in many biker groups because they served at least one function other than status striving among members. The acts ensured that it was difficult for law enforcement officials, male or female, to infiltrate a gang.

BIKER WOMEN AS MONEY-MAKERS

Among most of the groups we studied, biker women were expected to be engaged in economic pursuits for their individual men and sometimes for the entire club. Many of the old ladies and mamas were employed in nightclubs as topless and nude dancers. Although we were not able to get exact figures on the proportion of "table dancers" who were biker women, in two or three cities almost all of them were working for outlaw clubs.

A lot of the dancers were proud of their bodies and their dancing abilities. We saw them perform their routines in bars and at parties. At the "Big Blowout" in Gulfport, which is held in an open field outside of the city, in 1987 and 1988 there was a stage with a sound system set up for the dancers. The great majority of the 2,000 people in attendance were bikers from around the country so the performances were free.

Motorcycle women who danced in the nightclubs we observed remained under the close scrutiny of the biker men. The men watched over them for two reasons. First, they wanted to make sure that the women were not keeping money on the side; second, the cyclists did not want their women to be exploited by the bar owners. Some bikers in one gang we knew beat up a nightclub owner because they thought he was "ripping off" the dancers. The man was beaten so severely with axe handles that he had to be hospitalized for several months.

While some of the biker women limited their nightclub activities to dancing, a number of them also let the customers whose tables they danced on know they were available for "personal" sessions in a private place. As long as they were making good money regularly, the bikers let the old ladies choose their own level of nightclub participation. Thus some women danced nude only on stage; others performed on stage and did table dances as well. A smaller number did both types of dances and also served as prostitutes.

Not all of the money-making biker women we encountered were employed in such "sleazy" occupations. A few had "square" jobs as secretaries, factory workers, and sales persons. One biker woman had a job in a bank. A friend and fellow biker lady described her as follows: "Karen is a chameleon. When she goes to work, she is a fashion plate; when she is at home, she looks like a whore. She is every man's dream!" Like the others employed in less prestigious labor, however, Karen turned her salary over to her old man on payday.

A few individuals toiled only intermittently when their bikers wanted a new motorcycle or something else that required more money than they usually needed. The majority of motorcycle women we studied, however, were regularly engaged in work of some sort.

MOTIVATIONS AND
BACKGROUNDS OF BIKER WOMEN

In view of the ill treatment the women received from outlaws, it was surprising that so many women wanted to be with them. Bikers told us there was never a shortage of women who wanted to join them and we observed this to be true. Although it was unwise for men to draw conclusions about the reasons mamas and old ladies chose their life-styles, we surmised three interrelated factors from conversations with them.

First, some women, like the male bikers, truly loved and were excited by motorcycles. Cathy was an old lady who exhibited this trait. "Motorcycles have always turned me on," she said. "There's nothing like feeling the wind on your titties. Nothing's as exciting as riding a motorcycle. You feel as free as the wind."

Cathy did not love motorcycles indiscriminately, however. She was imbued with the outlaw's love for the Harley Davidson. "If you don't ride a Hog," she stated, "you don't ride nothing. I wouldn't be seen dead on a rice burner" (Japanese model). Actually, she loved only a customized bike or "chopper." Anything else she called a "garbage wagon."

When we asked her why she wanted to be part of a gang if she simply loved motorcycles, Cathy answered:

There's always someone there. You don't agree with society so you find
someone you like who agrees with you. The true meaning for me is to
express my individuality as part of a group.

Cathy started "putting" (riding a motorcycle) when she was 15 years old
and she dropped out of school shortly thereafter. Even with a limited education, she gave the impression that she was a person who thought seriously.
She had a butterfly tattoo that she said was an emblem of the freedom she felt
on a bike. When we talked to her, she was 26 and had a daughter. She had
ridden with several gangs but she was proud that she had always been an old
lady rather than a mama.

The love for motorcycles had not dimmed for Cathy over the years. She still
found excitement in riding and even in polishing a chopper. "I don't feel like
I'm being used. I'm having fun," she insisted. She told us that she would like to
change some things if she had her life to live over, but not biking. "I feel sorry
for other people; I'm doing exactly what I want to do," she concluded.

A mama named Pamela said motorcycles thrilled her more than anything
else she had encountered in life. Although she had been involved with four
biker clubs in different sections of the country, she was originally from
Mississippi and she was with a Mississippi gang when we talked to her. Pamela
said she graduated from high school only because the teachers wanted to get rid
of her. "I tried not to give any trouble, but my mind just wasn't on school."

She was 24 when we saw her. Her family background was a lot like most
of the women we knew. "I got beat a lot," she remarked. "My daddy and my
mom both drank and ran around on each other. They split up for good my
last year in school. I ain't seen either of them for a long time."

Cathy described her feelings about motorcycles as follows:

I can't remember when I first saw one. It seems like I dreamed about
them even when I was a kid. It's hard to describe why I like bikes. But I
know this for sure. The sound a motorcycle makes is really exciting—it
turns me on, no joke. I mean really! I feel great when I'm on one. There's
no past, no future, no trouble. I wish I could ride one and never get off.

The second thing we thought drew women to motorcycle gangs was a
preference for macho men. "All real men ride Harleys," a mama explained
to us. Generally, biker women had contempt for men who wore suits and
ties. We believed it was the disarming boldness of bikers that attracted
many women.

Barbara, who was a biker woman for several years, was employed as a secretary in a university when we talked to her in 1988. Although Barbara gradually withdrew from biker life because she had a daughter she wanted reared in
a more conventional way, she thought the university men she associated with
were wimps. She said:

Compared to bikers, the guys around here (her university) have no balls at
all. They hem and haw, they whine and complain. They try to impress

you with their intelligence and sensitivity. They are game players. Bikers come at you head on. If they want to fuck you, they just say so. They don't care what you think of them. I'm attracted to strong men who know what they want. Bikers are authentic. With them, what you see is what you get.

Barbara was an unusual biker lady who came from an affluent family. She was the daughter of a highly successful man who owned a manufacturing and distributing company. Barbara was 39 when we interviewed her. She had gotten into a motorcycle gang at the age of 23. She described her early years to us:

I was rebellious as long as I can remember. It's not that I hated my folks. Maybe it was the times (1960s) or something. But I just never could be the way I was expected to be. I dated "greasers," I made bad grades; I never applied myself. I've always liked my men rough. I don't mean I like to be beat up, but a real man. Bikers are like cowboys; I classify them together. Freedom and strength I guess are what it takes for me.

Barbara did not have anything bad to say about bikers. She still kept in touch with a few of her friends in her old club. "It was like a family to me," she said. "You could always depend on somebody if anything happened. I still trust bikers more than any other people I know." She also had become somewhat reconciled with her parents, largely because of her daughter. "I don't want anything my parents have personally, but my daughter is another person. I don't want to make her be just like me if she doesn't want to," she concluded.

A third factor that we thought made women associate with biker gangs was low self-esteem. Many we studied believed they deserved to be treated as people of little worth. Their family backgrounds had prepared them for subservience. Jeanette, an Arkansas biker woman, related her experience as follows:

My mother spanked me frequently. My father beat me. There was no sexual abuse but a lot of violence. My parents were both alcoholics. They really hated me. I never got a kind word from either of them. They told me a thousand times I was nothing but a pain in the ass.

Jeanette began hanging out with bikers when she left home at the age of 15. She was 25 when we talked to her in 1985. Although he was dominating and abusive, her old man represented security and stability for Jeanette. She said he had broken her jaw with a punch. "He straightened me out that time," she said. "I started to talk back to him but I didn't get three words out of my mouth." Her old man's name was tattooed over her heart.

In Jeanette's opinion, she had a duty to obey and honor her man. They had been married by another biker who was a Universal Life minister. "The Bible tells me to be obedient to my husband," she seriously remarked to us. Jeanette also told us she hated lesbians. "I go in lesbian bars and kick ass," she said. She admitted she had performed lesbian acts but she said she did so only when her old man made her do them. The time her man broke her jaw was when she objected to being ordered to sleep with a woman who was dirty.

Jeanette believed her biker had really grown to love her. "I can express my opinion once and then he decides what I am going to do," she concluded.

In the opinions of the women we talked to, a strong man kept a woman in line. Most old ladies had the lowly task of cleaning and polishing a motorcycle every day. They did so without thanks and they did not expect or want any praise. To them, consideration for others was a sign of weakness in a man. They wanted a man to let them know who was boss. . . .

THE CHANGING ROLE OF BIKER WOMEN

During the 17 years of our study, we noticed a change in the position of women in motorcycle gangs. In the groups we observed in the 1960s, the female participants were more spontaneous in their sexual encounters and they interacted more completely in club activities of all kinds. To be sure, female associates of outlaw motorcycle gangs have never been on a par with the men. Biker women have worn "property" jackets for a long time, but in the outlaw scene of 1989, the label had almost literally become fact.

Bikers have traditionally been notoriously active sexually with the women in the clubs. When we began hanging out with bikers, however, the men and the women were more nearly equal in their search for gratification. Sex was initiated as much by the women as it was by the men. By the end of our study, the men had taken total control of sexual behavior, as far as we could observe, at parties and outings. As the male bikers gained control of sex, it became more ceremonial.

While the biker men we studied in the late 1980s did not have much understanding of sex rituals, their erotic activities seemed to be a means to an end rather than an end in themselves, as they were in the early years of our study. That is to say, biker sex became more concerned with achieving status and brotherhood than with "fun" and physical gratification. We used to hear biker women telling jokes about sex but even this had stopped.

The shift in the position of biker women was not only due to the increasing ritualism in sex; it was also a consequence of the changes in the organizational goals of motorcycle gangs as evidenced by their evolving activities. As we have noted, many motorcycle gangs developed an interest in money; in doing so, they became complex organizations with both legal and illegal sources of income (McGuire, 1986).

When bikers became more involved in illegal behavior, they followed the principles of sex segregation and sex typing in the underworld generally. The low place of women has been well documented in the studies of criminal organizations (Steffensmeier, 1983). The bikers did not have much choice in the matter. When they got involved in financial dealings with other groups in the rackets, motorcycle gangs had to adopt a code that had prevailed for many years; they had to keep women out of "the business."

Early motorcycle gangs were organized for excitement and adventure; money-making was not important. Their illegal experiences were limited to individual members rather than to the gang as a whole. In the original gangs, most male participants had regular jobs, and the gang was a part-time organization that met about once a week. At the weekly gatherings, the emphasis was on swilling beer, soaking each other in suds, and having sex with the willing female associates who were enthusiastic revelers themselves. The only money the old bikers wanted was just enough to keep the beer flowing. They did not regard biker women as sources of income; they thought of them simply as fellow hedonists.

Most of the gangs we studied in the 1980s required practically all of the members' time. They were led by intelligent presidents who had organizational ability. One gang president had been a military officer for several years. He worked out in a gym regularly and did not smoke or drink excessively. In his presence, we got the impression that he was in control, that he led a disciplined life. In contrast, when we began our study, the bikers, including the leaders, always seemed on the verge of personal disaster.

A few motorcycle gangs we encountered were prosperous. They owned land and businesses that had to be managed. In the biker transition from hedonistic to economic interests, women became defined as money-makers rather than companions. Whereas bikers used to like for their women to be tattooed, many we met in 1988 and 1989 did not want their old ladies to have tattoos because they reduced their market value as nude dancers and prostitutes. We also heard a lot of talk about biker women not being allowed to use drugs for the same reason. Even for the men, some said drug usage was not good because a person hooked on drugs would be loyal to the drug, not to the gang.

When we asked bikers if women had lost status in the clubs over the years, their answers were usually negative. "How can you lose something you never had?" a Florida biker replied when we queried him. The fact is, however, that most bikers in 1989 did not know much about the gangs of 20 years earlier. Furthermore, the change was not so much in treatment as it was in power. It was a sociological change rather than a physical one. In some respects, women were treated better physically after the transition than they were in the old days. The new breed did not want to damage the "merchandise."

An old lady's status in a gang of the 1960s was an individual thing, depending on her relationship with her man. If her old man wanted to, he could share his position to a limited extent with his woman. Thus the place of women within a gang was variable. While all women were considered inferior to all men, individual females often gained access to some power, or at least they knew details of what was happening.

By 1989, the position of women had solidified. A woman's position was no longer influenced by idiosyncratic factors. Women had been formally defined as inferior. In many biker club weddings, for example, the following became part of the ceremony:

> You are an inferior woman being married to a superior man. Neither you nor any of your female children can ever hold membership in this club or own any of its property.

Although the bikers would not admit that their attitudes toward women had shifted over the years, we noticed the change. Biker women were completely dominated and controlled as our study moved into the late 1980s. When we were talking to a biker after a club funeral in North Carolina in 1988, he turned to his woman and said, "Bitch, if you don't take my dick out, I'm going to piss in my pants." Without hesitation, the woman unzipped his trousers and helped him relieve himself. To us, this symbolized the lowly place of women in the modern motorcycle gang.

CONCLUSION

Biker women seemed to represent another version of what Romenesko and Miller (1989) have referred to as a "double jeopardy" among female street hustlers. Like the street prostitutes, most biker women came from backgrounds in which they had limited opportunities in the licit or conventional world, and they faced even more exploitation and subjugation in the illicit or deviant settings they had entered in search of freedom.

It is ironic that biker women considered themselves free while they were under the domination of biker men. They had the illusion of freedom because they lived with men who were bold and unrestrained. Unlike truly liberated women, however, the old ladies and mamas did not compete with men; instead, they emulated and glorified male bikers. Biker women thus illustrated the pervasive power of socialization and the difficulty of changing deeply ingrained views of the relations between the sexes inculcated in their family life. They believed that they should be submissive to men because they were taught that males were dominant. While they adamantly stated that they were living the life they chose, it was evident that their choices were guided by values that they had acquired in childhood. Although they had rebelled against the strictures of straight society, their orientation in gender roles made them align with outlaw bikers, the epitome of macho men.

NOTES

1. We briefly observed the Alii ("Chiefs"), a native Hawaiian gang, located on the island of Hawaii, the "Big Island." Motorcycle gangs on the island of Oahu may have developed before the California clubs. Lord (1978) described a club that volunteered its services during the attack on Pearl Harbor in 1941. The bikers, wearing their colors, carried messengers and officers on motorcycles from one place to another because automobiles could not move efficiently during the traffic jams caused by the battle. This display of patriotism was not the only instance. Sonny Barger, the president of the Oakland chapter, sent President Lyndon Johnson a letter offering to send the Hell's Angels to Vietnam. Some outlaws were active in "Toys for Tots" drives and blood drives. These activities were seldom noted that [sic] led bikers to the slogan "When we do right, nobody remembers; when we do wrong, nobody forgets."

2. Bikers presented "courtesy cards" to people that they believed deserved the privilege of biker acceptance. "Big John," as Moore was called by outlaws, had cards from outlaw motorcycle clubs throughout the country.

3. A national club had chapters in different regions of the country. For more details on gang vocabulary and argot and the distribution of specific gangs, see Hopper and Moore (1983, 1984).

4. Although the initiation ceremony was the culmination of a biker's efforts to become a member of a club, it usually required a period of a year or more before a man would be made a member in full standing. During this time, a person was a "probate." He rode with the gang, and to the public he appeared to be a regular member, but a probate was not trusted until he proved himself worthy of complete membership. He did this by showing his courage and disregard for danger.

REFERENCES

Asbury, H. (1928) The Gangs of New York. New York: Alfred A. Knopf.

Bowker, L. (1978) Women, Crime, and the Criminal Justice System. Lexington, MA: D. C. Health.

Bowker, L., and M. Klein. (1983) "The etiology of female juvenile delinquency and gang membership: a test of psychological and social structural explanations." Adolescence 8: 731–751.

Eisen, J. (1970) Altamont. New York: Avon Books.

Harris, M. (1985) Bikers. London: Faber & Faber.

Hopper, C., and J. Moore (1983) "Hell on wheels: the outlaw motorcycle gangs." J. of Amer. Culture 6:58–64.

Hopper, C., and J. Moore (1984) "Gang slang." Harpers 261:34.

Lord, W. (1978) Day of Infamy. New York: Bantam.

McGuire, P. (1986) "Outlaw motorcycle gangs: organized crime on wheels." National Sheriff 38:68–75.

Montegomery, R. (1976) "The outlaw motorcycle subculture." Canadian J. of Criminology and Corrections 18:332–342.

Morgan, R. (1978) The Angels Do Not Forget. San Diego: Law and Justice.

Quinn, J. (1983) Outlaw Motorcycle Clubs: A Sociological Analysis. M.A. thesis: University of Miami.

Reynolds, F. (1967) Freewheeling Frank. New York: Grove Press.

Rice, R. (1963) "A reporter at large: the Persian queens." New Yorker 39:153.

Romenesko, K., and E. Miller (1989) "The second step in double jeopardy: appropriating the labor of female street hustlers." Crime and Delinquency 35:109–135.

Saxon, K. (1972) Wheels of Rage. (privately published)

Steffensmeier, D. (1983) "Organization properties and sex-segregation in the underworld: building a sociology theory of sex differences in crime." Social Forces 61:1010–1032.

Thompson, H. (1967) Hell's Angels. New York: Random House.

Watson, J. (1980) "Outlaw motorcyclists as an outgrowth of lower class values." Deviant Behavior 4:31–48.

Wilde, S. (1977) Barbarians on Wheels. Secaucus, NJ: Chartwell Books.

Willis, P. (1978) Profane Culture. London: Routledge & Kegan Paul.

Wolfe, T. (1968) The Electric Kool-Aid Acid Test. New York: Farrar, Straus & Giroux.

28

Con Men: Professional Crews

of Card and Dice Hustlers

ROBERT PRUS

AND C. R. D. SHARPER

PROFESSIONAL HUSTLING

While anyone might attempt to cheat at card and dice games on a situational and/or systematic basis, like Sutherland's professional thieves (1937: 3–26), professional card and dice hustlers have several qualities that, taken together, define them as highly proficient and distinguish them from other card and dice hustlers. First, although the professional may have a number of hustling options at any given point in time, he is oriented toward card and dice hustling as a full-time career activity; any other (side) involvements are viewed as "moonlighting" or as sustaining activities when one hits "slack times." To be a successful card and dice hustler requires extensive dedication; for the requisites of the job, if met adequately, demand that one's other life activities be structured around this identity.

Second, through association with other hustlers, professionals acquire a specialized set of skills; they achieve a level of verbal/physical dexterity that sets them apart from amateurs and those in a rough hustle. Their hustle is

From: Robert Prus and C. R. D. Sharper, *Road Hustler* (New York: Kaufman and Greenberg Publishers, 1991). Reprinted by permission of Robert Prus.

extremely social, and one in which victims are deceived and conned into parting with their money without any inference of force. Victims seldom realize that they have been "fleeced" and may play again, blaming their losses on game strategies or bad luck. Professional card and dice hustlers, are, in varying degrees, combinations of magicians and public relations men. It is also expected that they, as professionals, will be able to perform under considerable pressure.

Third, although moves and routines are worked out in advance (systematic cheating), games are highly variable and it is expected that a professional will be able to make adjustments to fit the prevailing action. They possess the ability to maximize their profits in a highly problematic situation.

Fourth, professionals are mobile and "well connected." While it is rather unlikely that professional card and dice hustlers will have any Mafia-type allegiances, they do know how to locate action and whom to see in order to obtain specific commodities. Through their association with hustlers and thieves they acquire a working knowledge of thief subculture. This "street knowledge," along with their more cosmopolitan contacts, complemented by their mobility and financial assets, enables them to become quite resourceful.

Although we may speak of differentials between professionals and other card and dice hustlers, it is important to recognize that there is no set point at which one automatically becomes a professional. While any capable amateur or rough hustler is likely to have some professional qualities, professional status is accorded as one is seen as approximating an ideal. Thus, a hustler will be accorded professional status to the extent that he: hustles regularly, exclusively, and successfully; possesses interpersonal and equipment manipulatory skills; has extensive hustling related contacts; is considered dedicated, cool, and sharp; and is acknowledged as a desirable co-worker.

Professional Involvements

There are two major routes by which individuals become involved with professional crews; both, however, require that one have contacts with persons already hustling at that level. The first and less common route one might term co-optation; in this sequence a professional group decides to involve a "nonhustler" in their operations. Although this routing is relatively infrequent, occasionally a crew will encounter someone who seems a natural and may decide to capitalize on his particular abilities. Although this person is defined as a nonhustler (cards and dice), it should not be assumed that he is naive to hustling in a general sense.

The second and more frequent route to professional affiliation is considerably less direct and involves a "drifting into professionalism." This is the route most professional hustlers experience. They are, thus, likely to have spent several years bouncing around in a rough hustle prior to contact and acceptance by a professional crew. While not necessarily the best of the rough hustlers at the time of contact, they have somehow or other come to the attention and favor of a professional crew.

THE CREW

Although one may find some professional card and dice hustlers who operate on a solitary basis, most work in groups, making an examination of these "crews" essential to understanding card and dice hustling. Not only is most professional hustling done on a crew basis, but the crews represent a critical entry point vis-à-vis becoming a professional hustler. . . . Operating from a business perspective complicated by legal and physical liabilities, professional crews are concerned with profit, competence, and safety. Accordingly, one finds that these concerns are integrally related to crew size, role specializations, ratings of crew members, crew compatibility, and the recruitment of replacements.

Crew Size

Most card and dice hustling crews consist of three to four members. Although one occasionally encounters crews as large as eight or ten persons, larger crews offer few advantages and have several disadvantages, as they are more likely to arouse suspicion at an event, are costlier to operate, offer little guarantee of greater success, and yield smaller percentage earnings for individual members:

> Crew size is affected by capabilities. Let's say you have three people who are capable—one or two mechanics and a good shoot-up man to win the money. Now, if the shoot-up man has some muscle, that's all you really need. Then, the money is split up three ways. When there are four of you, it's easier to work the game. That extra man can provide shade when the mechanic is making his moves or he might be another shoot-up man or mechanic and take some of the heat off his partners that way. You can go stronger with four, but now your winnings are split four ways rather than three. Still, if you have four guys who are capable, that would be an ideal situation. But, if one of the four is not doing his share, maybe he is unreliable or is losing his nerve, it's better to have three. . . .

> If a crew is larger, say six to eight members, you can spread the money around easier, but it ceases to be profitable, and it's a little riskier. Sure, you can go in and win the money, boom, boom, boom, but a bunch of strangers, winning all the money at once doesn't look right. You try to blend in with the locals and as you get larger, this is harder to do. And, then, even if you win more money, it's cut up so many ways and your expenses are so much greater that it just doesn't pay. Sure, you're much stronger if there's a hey rube, but if you're careful and take your time, that shouldn't happen very often. You try to carry only as many men as you need to do the job; any more, and it's money out of your pocket.

Role Specifications

While a given crew member may assume multiple roles, three game roles and two nongame roles define the structure of the crew. The game roles are those of the "mechanic," the "shoot-up" (public relations) man, and the "muscle" man. The major nongame crew roles are those of the "contacts" man and the "boss."

By and large, the game roles are centered on the mechanic; he manipulates the game equipment and it is his job to control the outcomes of the games. Mechanics may be capable of manipulating cards or dice or (ideally) both. While most crew members will be capable of making a few moves, the mechanic is expected to be able to operate under more demanding conditions, exhibiting greater diversity and finesse than other crew members. The mechanic's role is pivotal to game success; if he can't produce, the other crew game roles become irrelevant. As the game manipulator, the mechanic experiences the most pressure. He may have some equipment on him (for example, weighted dice) and it is he who is most vulnerable if accused:

> The mechanic has to have an acceptable appearance. He has to look conventional and solid. Also, he has to watch his mannerisms. He can't yell and scream or make any false or funny moves, he doesn't want to draw attention to himself. He has to look natural while doing the manipulations. A lot of people watch the dealer or the man handling the craps, so he has to appear casual. He's under a lot of pressure, but he has to be able to control the game and relate to the people casually. And you never know what might happen next, so he has to be on top of things. So, the mechanic has to be capable, he has to be cool and be able to relate to the people, and he has to be alert.

Next to the mechanic, the shoot-up man experiences the most pressure. While all crew members will do some public relations work in any game and while all are likely to win some of the money, one or two members of the crew may come to be known as shoot-up men. It is their job to establish personal credibility among the patrons attending the event and to entice them into playing. Once they have involved other players, they try to increase the size of the bets while cooling out the losers. As shoot-up men it is their responsibility to "look good winning a large percentage of the money." If a man has established himself as a respectable and likable fellow among the other players, he is less likely to arouse suspicion or the sort of resentment that a stranger who is winning is likely to encounter. Thus, shoot-up men engage in public relations work throughout the evening:

> It's a special skill. Sure, you can tell a guy to get in and mix with the suckers. But, if you have a man with the gift of gab, it is great. He's in there mixing with the people, relaxing with them, enticing them. That man is worth his weight in gold! You or I might go and win that money and they might hate us, but he wins and it's accepted. Sure, the crew takes its time, so that he wins and loses with the suckers, but over the evening, he

will win the big money. . . . He also has to look the part, to be the kind of person they would like to associate with; an older, nice looking, respectable appearing man. Sure, he's joking around with the people, but he's trying to sustain confidence in our game, so he has to look credible and acceptable. . . . If you have two good public relations men in a game, it's a real pleasure to work. The action is usually good and the people are relaxed.

The muscle man is optional. When he is used, however, it is to provide protection for the other members of the team rather than to intimidate the other players. The appearances/mannerisms of muscle men are also of concern to the crew. Endeavoring to be unobtrusive, crews define an obvious muscle man as a threat to their definition of the game as a casual and friendly one. Additionally, unless the muscle man is capable of a few manipulations or winning some money, he tends to be regarded by the other crew members as a costly resource:

If you have a shoot-up/muscle man combination, that's great. But a man who just provides muscle is just so much deadweight; they get in the way more than anything. They don't know what they're doing half the time and they think they have to fight to win the money! They always seem nervous, it's like they're looking for ghosts someplace. But, if you have a man that's good with people and has a little strength, you're in good shape. . . . If you're younger and smaller, they may try to take advantage of you, where they wouldn't try to bully an older fellow that looked good and appeared to have some strength. . . . But straight muscle men, they have problems realizing that there's no fighting, they think they have to fight to win. He doesn't have to beat up anybody, just give us a chance to leave unmolested. . . .

Take this one man who was with us. At first, he thought that he made the crew go. We had to cool him out and teach him how to mix with people. Now, I don't think he fought more than twice in seven years of crew hustling and even then he might have avoided that. Once he got carried away with a bunch of hockey players and they were going to get him. Little did they know this guy was a Joe Palooka. He whipped four of them! Then, he came to the car mad as hell and we took off. But we lost out because we couldn't go back there. Now, if he had just left without fighting with them, we would have been all right. But, he whipped all four of them, bam, boom, bam, every one of them! It was funny, we were sitting in the car watching him and this one fellow says to me, "Do you think we should help him?" I said, "Are you kidding?", and sure enough, as soon as I said that, we saw those guys go flying. Here's what happened, we had already been to one party, so we came in late and we were going to shoot the money up. Then, they said, "Lights out in twenty minutes!" So we tried to win all the money then and there. We knew it was a mistake, but you are there and you have to try. . . . Usually you try to talk your way out of situations. Physical violence is a last resort, and, even then, all you want is a chance to run out, someone to block or interfere for you while you get away. . . . If you can find a big man with some larceny sense, and teach him a few moves and how to shoot-up the money, that's a big edge, because it's nice to have that muscle around.

Two nongame roles, those of the "contacts man" and the "boss," are important for putting card and dice hustling in perspective. The contacts man locates spots for crew action; he is someone who knows a variety of promising events and who is able to locate current events. Although other members are encouraged to help in locating events for the crew, the contacts man (generally the boss) is the "workhorse" of the crew; for, in addition to spending a great deal of time locating events and planning schedules, he has his game roles to fulfill as well. If a crew is to work full time, it has to be able to locate a minimum of four or five promising, accessible, and nonexclusive parties a week with most of this burden falling on the contacts man. If he cannot consistently locate "good" spots for the team to work, he will have trouble holding the crew together. In a solid crew, the contacts man will often have the next month's events scheduled in advance, so that the crew can maximize their earnings relative to traveling time and expenses.

A good contacts man will not only have a listing of events but, more important, will also have information on the background of the party. This background information allows the crew to assess the nature of any gaming activities, the chances of a good "pot," the type of clientele the crew is likely to encounter, local sources of assistance, troublesome persons in the community, and so forth. While most hustlers realize that knowing events and persons to contact about events is an integral component of their trade, many are content to let others do the work of locating promising events. However, if a hustler is hardworking and diligent, he may, over a period of time, accumulate a listing of places and persons that can provide him with considerable independence. If he has good working relations with the present crew, he can enrich their opportunities by offering an additional source of events from which to choose. If he finds that the present crew is not to his liking, he will have contacts for joining other crews or possibly forming his own crew. While all hustlers will develop some contacts, a well-connected man is a highly valued crew member.

The boss may be best defined as the manager of the crew and is generally the most experienced member of the crew. He typically discusses plans with members of the crew, but it [is] thought of as "his crew." While the boss is most likely the contacts man for the crew, he is also likely to be a skilled mechanic and a good shoot-up man. Like other supervisors, he may have problems motivating and disciplining members of his crew. Although crew bosses make the major decisions, these are generally influenced by the interests and anticipated cooperation of his partners, particularly those with more experience, talent, and contacts. Since he has no binding formal contracts, a boss's power over his members rests on their recognition of his talent as a manager and game partner.

Member Ratings

Assuming that other things were equal, within-crew ratings would reflect one's contacts, one's manipulative skills, one's public relations/shoot-up skills, and muscle, respectively. Other things are seldom equal, however, and one finds that in-crew evaluations of partners are more strongly affected by overall notions or reliability than by role specializations.

Defining their activity in a profit context and attempting to reduce physical and legal risks, crew members highly esteem partners on whom they can depend. Thus, although a crew requires a mechanic to effectively work a game, they also want a man on whom they can rely. In this sense, reliability subsumes several concerns: Can a man handle his specialty? Is he multitalented? Is he a dedicated team member? Is he able to perform under pressure?

Although having a particular talent is a prerequisite to crew membership, one's in-crew rating is affected by his partners' perception of his contributions to their game successes and by comparing his capabilities with other hustlers with similar talents. Thus, while a contacts man may know a lot of spots or while a mechanic may be able to manipulate, their in-crew assessments will be affected by their partners' perception of how well they perform compared to other contacts men, mechanics, and others with whom they are familiar. If however, a hustler is defined as *multitalented,* his in-crew ratings are likely to increase, for he is seen as adding versatility and depth to the crew:

> Say your main shoot-up man is also capable of manipulating cards. Maybe your regular mechanic is getting a lot of heat in a game or has to leave the game or perhaps he's sick that evening. You wouldn't lose that much. Your shoot-up man becomes the mechanic in the game and the others win the money. Usually, you find that the professionals are capable of doing it all, so if you have three or four people like that, maybe with a little muscle for protection, you can rip them off left and right. . . . If you have a new man, you try to teach him how to help out all around. It's not that you want to do him a favor by teaching him anything, but if he's stronger, then the crew can work easier and more profitably. Anyone who has one good talent is valuable, but a man who has mastered two or three skills adds a lot of strength to the crew. . . . You see a similar thing happening when a good card mechanic becomes capable of moving craps. His esteem really goes up.

Being a reliable team member entails much more than being able to handle a specialty or being multitalented. It refers to rather unexceptional, but basic, concerns, such as persons being punctual for their meetings, being willing to share the nongame work load (for example, driving/phoning) and sustaining game-related appearances and activities. Thus, although individuals might be defined as very capable specialists, if they are considered to be lazy, drinking too much, indiscreet, or overly aggressive, they will lose in-crew credibility:

> To have a good crew rating, you have to be dependable on an everyday basis and in the game itself. If you pull your weight, do your share of the work, and keep your word to the crew, they will think more of you. Also, the way you handle yourself in the game is very important. For example, you wouldn't yell and scream, "Bet you fifty!" And, if there's a disagreement, you wouldn't yell at the guy and get mad at him. The rule is "Don't get mad at the sucker," because he's liable to have money, or it's liable to interfere with the game. You always let the sucker have the benefit of the doubt. The customer is always right! That's the attitude you have to take. If you take out time to argue with all the suckers, they are going to want to know who you are and what you're doing there. Now, you'll find a lot of people who, once they have taken advantage of you, will keep taking edges in the game, but you just try and hold them down where they won't go too far.

Maintaining one's temperament under game pressure is another component of being considered a reliable teammate. Some very skilled mechanics, for example, may tense up in the dramatic atmosphere of the game and either won't switch the dice or may make noticeably rigid movements. Such unreliability is unacceptable in a group of hustlers who may have traveled some distance to attend a particular event. When a mechanic "loses his nerve," crews are reluctant to use him in that capacity. He may, if the crew has another mechanic, become a shoot-up man or he may become involved in other hustles:

> We were going to work this spot with another crew and as we had some spare time, this one guy and I decided to play a little golf. While we were waiting to tee off, I was crackling these dice. I turned to this guy and said, "Hey - - -, show me what you can do." He was fantastic! So that evening, when we went to the party, I figured that he was going to get the tools, me being a relative novice. But, my partner gave me the dice and this guy got nothing. I said, "What are you doing, - - -'s a champ!" He said, "Are you kidding, - - - wouldn't switch dice if his life depended on it." Sure enough, this guy was afraid to win the money, never mind switching the dice. . . .

> But that goes to show you, if you don't keep on the ball, you are going to lose both money and credibility. Hustlers talk with one another saying "Gee, that - - -'s not doing his job lately. He's been drinking and carrying on." If you know that they have doubts about you, it makes it harder in the game situation. I've met many hustlers who can switch dice like it was nothing when they're sitting around drinking in a motel room, but they just can't produce at a game. . . . Now, if a man shows a lot of cool when it's a tough game, then his credibility really goes up amongst the other hustlers. You don't go banzai, you try to work the game using discretion and finesse.

Generally speaking, if a group member has contacts, mechanical and social skills, coolness, and is hardworking, he will have considerable bargaining power in a group. Group members will define him as a highly desirable partner, feeling that he would be able to fit into other groups quite readily and that he would be difficult to replace.

Crew Compatibility

If a group of people are to work together for any length of time as a hustling crew, it is important that they be compatible—that they be relatively congenial toward one another. In part, compatibility is promoted by team members who define one another as reliable workers, but unless crews achieve relative consensus on other issues, they are likely to split up.

One set of concerns that affects compatibility is contingent on definitions of success (that is, How much money should we be making?) and work loads (that is, How long/hard should we work?). If a crew is financially successful by all members' definitions, many disagreements will be tolerated. However, unless there is consensus on work assignments, distribution of the earnings, working styles, and a shared willingness to persevere, it is unlikely that a crew will be successful:

> If everyone feels that everyone is doing their job and putting in his share, the crew has a much better chance of staying together. Say one guy was getting all the contacts, and he was the mechanic, the boss, and the public relations man, he would get tired of doing that. He would say, "What do I need all these people for?" And say another guy is just loafing off, instead of getting on the phone a few hours, he's out drinking cocktails with some chick. If everyone puts in their effort as a unit, things just seem to work out better. . . . Another big thing is the division of money. Now you might shortcake a new man, but if you are going to work with a group of guys for some time, you have to be reasonably square with them. They're usually pretty sharp and have a good idea of what's going on, so if you try to shortchange them, it might work a few times, but pretty soon there's going to be a lot of bad feelings. . . .

> I'll give you an example with this one shoot-up/muscle man we had. After a party we pool our money, right. "How much do you have? I have so much." Now, this guy might say "two-fifty" and I figure he took in over four hundred. He was a good man at the game, but we couldn't trust him to turn over his winnings. You don't want to be fighting with your partners. We let him know that we knew what was going on, but he didn't change, so we let him go. I guess he had been in a rough hustle for a while and thought he could do the same thing with us. . . . In a solid crew, it's just standard procedure, you get up and you know what you're doing. You have everything planned out for the next week or two weeks. Everyone knows their job and everyone does it. Nobody slacks off.

> Otherwise, you waste a lot of time and energy fighting amongst each
> other. You have to run it like a business.

Even a carefully defined set of rules will not, however, eliminate another
important source of in-crew conflict, that of establishing operating procedures
in ongoing game situations. Coming from a variety of different backgrounds,
hustlers in a crew may have different notions of: lucrative parties; how/when
to take a party off; and risktaking in game situations:

> Okay, you might agree on something like "no rough stuff." But, when
> you're in the game you might have different ideas of when to take this
> game off. Do you start right in? Do you slowly grind it out? Do you try to
> keep even and beat them later? And exactly how are you going to beat
> them? These games are so hard to predict. What do you do if you're bet-
> ting twenty and some sucker wants to bet two hundred. Do you make a
> move and lock up that two hundred then and there or do you take your
> chances? It's these sorts of things that have to be worked out. Now, if a
> crew has been working together for a long time, you can pretty well tell
> what one another will do. But if you're in a new crew, these sorts of
> things can cause a lot of arguments. . . . This risk thing is also very impor-
> tant. You might work with a crew in which one man runs to the car as
> soon as anything happens or there's a little trouble. You get tired of that
> after a while. We had one guy who ran to the car three times in one
> night. He was a big guy, but scared to death. You would have to go to the
> car after him and tell him, "It's all right, they like you." Then he'd come
> back, but he'd say, "I'm sure that guy knows me."

Given the physical, legal, and financial implications of an incident, crew members may be more/less willing to continue playing in a game or they may
decide that more finesse is required than that displayed by a particular partner.
Thus, conflict may arise between members who desire greater safety at the
expense of financial gains and other members who desire to "milk" the game
as long as possible. Over time, one finds a "mutual-selection process" in operation, wherein persons find partners with compatible levels of coolness. When
crew members feel that a partner's "cold feet" caused them to miss out on
extra money, they tend to be quite resentful of this person who has, in effect,
"taken money out of their pockets." On the other hand, considerable in-crew
hostility may result when one crew member is seen to have acted recklessly or
irresponsibly. Although eventual game success may nullify some of this hostility, were it to result in a bad incident, it might mean the termination of a
member.

As the group is also defined in a business framework, members are select-
ed/retained on the basis of performance capabilities. Thus, although other
crew members may personally like a fellow whom they define as having
become troublesome, they will terminate anyone they feel is costing them
money and jeopardizing their operations. It is, however, most unlikely that a

regular crew member would be released at the first sign of personal difficulties. Before a regular member is dismissed, the boss, as well as other crew members, is likely to have tried to reform their "deviant" partner.

Terminations of established crew members are generally defined as a "falling-out" by the involved parties and, as far as possible, an attempt is made to maintain good parting relations. The crew may anticipate that the deviant's services may be useful at another time and/or they may be concerned that the departing partner might take others to their best parties. The person who leaves the crew may anticipate working with this crew in other settings and may share their assessment of the lack of compatibility between their interests and his own at this time.

FINDING REPLACEMENTS

If a crew is successful and the members find one another compatible, it will remain together indefinitely. Some of the crews in our sample have been together for over a decade. However, if a group loses or terminates a member, or if conflicts develop within a group and the crew splits up, vacancies are created, providing an opportunity for other persons to become part of the team. In discussing newcomers, it is important to note that many of the concerns operative in evaluating existing crew members also shape definitions of prospective members. Thus, any crew that considers terminating a "problem member" will first ask if they can find a suitable replacement. If not, they will ask if they can effectively operate without this partner until they can find a replacement. If the others are unhappy with a particular partner's habits or performance, but consider him an integral cog in their operations, they will try to work around his weaknesses until they can find a satisfactory solution to his problem or a suitable replacement.

In attempting to locate replacements, crews typically seek the services of established hustlers, those they know to be capable, experienced, and reliable. Desiring the best available, a crew may try luring an established hustler from another crew. If a seasoned veteran is not available, the crew will expand their field of recruitment, but they are likely to become increasingly concerned about the specific usefulness, general reliability, and compatibility of any prospective members. As most crew members can perform some manipulations, even a team that lacks a full-time mechanic can survive in the interim. The rule of thumb crews use in recruiting is not to involve anyone they don't trust. Accordingly, initial involvements with outsiders will reflect personal contacts, a perception of larceny sense, and tentative probationary involvements.

Although known hustlers, whose habits and styles are better known and more predictable, are seen as better risks, crews generally have a number of reservations in involving any new person in their operations. Will the newcomer disrupt general crew routines? Does he represent an information control threat? Will he jeopardize their game style? Can he take game pressure? Is he going to cost them money, either at games or as "excess baggage"? Is he a legal liability (that is, "wanted," or have a bad record)? If the prospect seems "hep" and trustworthy, the crew will put him through a trial period. During this time, they will endeavor to ascertain the reliability and compatibility of this person, relative to their working style. While the more experienced hustlers will be taken to somewhat better parties, novices are typically taken to the roughest and poorest-paying events. In general, the crew is concerned with maintaining exclusive knowledge of the better annuals and will test potential members in situations that place their pocketbooks in less jeopardy:

> Okay, let's say that you need another person and, let's say that none of the top hustlers are available. Most likely you know someone who is reasonably hep and who looks presentable. So you get in touch with him—does he want to go to a game and make a few bucks? You tell him as little as possible. You might tell him to play his regular game, that might be all he has to do. Afterwards, we split the pot. If we like his style and he seems solid and fits in with the crew, we might ask him to another party. You try the fellow out, but you don't commit yourself. Then, if you feel that he is an asset to the crew, you might work him in a little more. If he is willing to learn and digs up a few parties for us, he might become a regular. It depends on his dedication to the crew and how the crew stands at that time. He might work with the crew for quite a while before we tell him anything about our operations. We want to test him—can he work under pressure, is he reliable, is he honest. . . . Usually, we take the new guy to these ten-dollar stags. You take him to your poorest parties, and they're usually the roughest because the people there can't afford to lose a hundred dollars. Now, you seldom have to run out of a fifty- or hundred-dollar stag, but you don't want to take new people or even experienced hustlers that you're not sure of to your better parties. If the man can take the pressure of these rougher parties and you see that he can mix well with different people, maybe he's worth something to the crew. He has to have desire if he wants to be a regular, he's got to be willing to put his ass on the line.

From his perspective, the prospect finds that he is getting occasional invitations to join a couple of other fellows in what he may perceive as a casual activity. Unless the prospect is an established hustler, it is rather unlikely that he would be able to define these experiences in the context of professional crew hustling. Lacking a solid benchmark from which to define their activities, novices have little idea of their worth to the crew. While some prospects develop an exaggerated notion of their worth, others may be much more valuable than they realize:

> So you tell them as little as possible and some of them feel very comfortable in a poker game. They don't see anything and they're getting good hands. Sometimes they say, "Gee, did you guys do anything?" I had one

guy who had the nerve to say, "Well, since I won most of the money legitimately, I think I should get the biggest share." So you look at that guy and you figure, "Oh, my God! We let the guy win and now he thinks he did it all himself! Now he doesn't need us!!!" But, you can't tell them much about your manipulations at first, because you might frighten them or they might spend the evening watching you. Now instead of being involved in the game, they are watching to see if you make a move. If the other players notice them watching you they may get suspicious. So, you tell these guys as little as possible. Later, if the man proves himself, it might be useful to tell him a little more. . . .

Now, sometimes these guys don't realize how useful they are. Like this one guy, he was a businessman, but he had been in a few other hustles. He had good social contacts, mixed well with people, and had good size. An older, respectable looking man with muscle, who was hep. It didn't take us long to realize his potential as a public relations/shoot-up man. But, he didn't know what he had going for him. He figured that the manipulations was all there was to card and dice hustling. He didn't realize the strength of socializing. We worked with the guy fairly steadily for three or four weeks in his area. Then we said, "Look, we're going on the road, do you want to work steadily?" He said, "Sure." He was really interested and he worked hard. It was a fluke to find a man like that. He saw the strength of hustling and he worked out fine. But you just can't pick someone up and say, "Come with us!" You try to learn about the guy's background and see what he is capable of doing. You want a man that's dedicated and who will fit in. . . .

Now, if the man looks solid, you try to see that he does well. You try to show him that it's a business and that he can make good steady money. Sure, you test him, try him out, but if he can make money for the crew, you let him know that he can make money with the crew.

. . . In sum, knowledge of the structure and operating concerns of crews is essential to the understanding of professional card and dice hustling. These partnerships are clearly predicated on performance criteria with partners being recruited/retained on the basis of their dedication to their work and their ability to fit into existing crew routines. Having indicated the central organizational features of the crew, we now turn our attention to the socialization processes that take place within these crews.

REFERENCE

Sutherland, Edwin. 1937. *The Professional Thief*. Chicago: University of Chicago Press.

29

Nonmainstream Body Modification

Genital Piercing, Branding,

Burning, and Cutting

JAMES MYERS

The term *body modification* properly includes cosmetics, coiffure, ornamentation, adornment, tattooing, scarification, piercing, cutting, branding, and other procedures done mostly for aesthetic reasons. It is a phenomenon possibly as old as genus *Homo,* or at least as ancient as when an intelligent being looked down at some clay on the ground, daubed a patch of it on each cheek, and caught the pleasing reflection on the surface of a pond. Appropriate to my overall topic is Thevos's (1984) observation that a "self retouching impulse" distinguishes humans from other animals (p. 3).

At the outset, it is important to distinguish between the two main types of body modification: permanent (or irreversible) and temporary. Permanent modifications, such as tattooing, branding, scarification, and piercing result in indelible markings on the surface of the body. With the exception of branding, these marks involve the application of sharp instruments to the skin. Dental alterations, skull modeling, and modern plastic surgery are also forms of permanent body modification. Temporary modifications include body painting, cosmetics, hair styling, costume, ornamentation, and any other alteration that can be washed off, dusted away, or simply lifted off the body. This article focuses on permanent body modifications in contemporary United States, especially genital piercing, branding, and cutting.[1]

The literature of anthropology abounds with descriptive and analytical accounts of body modification among humans, but almost all of it emanates from people living or who had lived in the non-Western traditional societies of the world. From Mayan tongue piercing to Mandan flesh skewering, Ubangi lip stretching to Tiv scarification, there is a vast and incredibly varied body of literature that seeks to explain it all—anthropologically, psychologically, sociologically, and biologically. Curiously, very little research has been done on contemporary, nonmainstream American body modification. When one considers the huge amount of literature devoted to the subject among traditional non-Western peoples, this paucity of data becomes glaringly evident. For example, one of the best recent sources on body modification is Rubin's

From: "Nonmainstream Body Modification: Genital Piercing, Branding, Burning, and Cutting," James Myers, *Journal of Contemporary Ethnography,* Vol. 21, No. 3, 1992. Reprinted by permission of Sage Publications, Inc.

(1988) *Marks of Civilization,* but even in this excellent publication, most of the articles deal with tattoos and cicatrization and none are devoted to such contemporary Euro-American practices as multiple piercing, scarification, cutting, and branding.

That a tattoo renaissance has been occurring in the United States since the late 1960s is now quite evident in the popular media and to a growing extent in scholarly publications and papers presented at professional conferences (see especially Govenar 1977; Rubin 1988; Sanders 1986, 1988a, 1988b, 1989; St. Clair and Govenar 1981).[2] This void in the literature is probably due more to a simple lack of awareness of the practice than it is a lack of interest, as the population of people involved in multiple piercing, scarification, branding, and cutting is minuscule compared to tattooing. In addition, because the modification and/or jewelry involved typically creates even greater revulsion in the general public's eye than tattoos, much of the work is kept secret among recipients and their intimates.

My observations and conclusions regarding nonmainstream body modification run counter to the general public's assessment that people so involved are psychological misfits bent on disfigurement and self-mutilation. None of the people I interviewed, however deep and varied their involvement in body alteration, fit the standard medical models for "self-mutilation." . . .

METHOD AND POPULATION

My original plan was to concentrate my research efforts on tattooing, but 4 months into the 24-month study period, I shifted my focus almost entirely to piercing, cutting, burning, and branding. The change was brought about by my increasing awareness of the growing popularity of nonmainstream modification other than tattoos and the realization that research on the subject was scant. I was also intrigued by the deep feelings of revulsion and resentment held by mainstream American society against these forms of body modification.[3]

Using participant observation and interviews as primary data-gathering techniques, I involved myself in six workshops organized especially for the San Francisco SM (sadomasochist) community by Powerhouse (fictitious name), a San Francisco Bay Area SM organization.[4] Tattoo and piercing studios were also a rich source of data, as was the 5th Annual Living in Leather Convention held in Portland, Oregon in October 1990. I gathered additional data from a small but dedicated group of nonmainstream body modifiers at my university and the city in which it is located. Interviews with several medical specialists and an examination of pertinent medical literature provided an important perspective, as did solicited and unsolicited commentary from hundreds of mainstream society individuals who viewed my body modification slides and/or heard me lecture on the topic. Finally, chance encounters with devotees served to broaden my awareness and understanding of the various forms of nonmainstream body modification.

Entree to the workshops was of paramount importance to the study; thus early in the fieldwork, I contacted the primary organizer of Powerhouse and introduced myself as a straight, male anthropologist interested in attending the workshops in order to gather data on nonmainstream body modification for use in my university classroom and publication in a scholarly journal. Her response was immediate:

> Good God, yes! You're welcome to come. We need people to see that just because we're kinky doesn't mean we're crazy, too. You'll see people here with all kinds of sexual interests. We learn from each other and have a heckuva lot of fun while we're at it.

As is true for most ethnographic participant observation situations, the largest amount of my data from the workshops were gathered from observation. I participated in the true sense of the word on two occasions, once during a play piercing demonstration and again during a playing with fire demonstration.[5] The rest of my participation involved such typical "interested involvement" as mingling, asking questions as an audience member, introducing myself around, helping arrange chairs, setting up demonstration paraphernalia, taking photographs, conducting interviews, and generally lending a hand whenever possible. At the Living in Leather Convention in Portland, I was able to expand my involvement by showing my body modification slides to several people, attending parties, and helping out at the host organization's hospitality suite.

The population of body modifiers in my study included males and females, heterosexuals and homosexuals (lesbians and gays), bisexuals, and SMers. It is important to note that the single largest group was composed of SM homosexuals and bisexuals. Although this skewing likely resulted from my extended contact with the Powerhouse workshops and several SM body modifiers whom I interviewed at the Living in Leather Convention in Portland, it is supported by a 1985 piercing profile of subscribers to *Piercing Fans International Quarterly* (Nichols 1985). The survey determined that 37% of the group was gay, 15% bisexual, and 57% involved in dominant-submissive play, a keystone of SM activity. Also of interest from the *PFIQ* profile, 83% had attended college, 24% had college degrees, and 33% had undertaken postgraduate study. Caucasians represented 93% of the survey.

Like any fieldwork, this research had its pleasant and difficult aspects. On the positive side was the subject matter itself. Body modification is inherently fascinating to human beings. In addition, there was the relative ease with which I was able to gather empirical data on the topic. The people I interviewed and observed were for the most part barely subdued exhibitionists who took joy in displaying and discussing their body and its alterations. This was especially true when a group was together and a sense of trust pervaded the room. On such occasions, an exuberant "show and tell" was the order of the day. To field-workers accustomed to tight-lipped, monosyllabic responses and other forms of "informant lockjaw" from people they are studying, and who have been advised on occasion what they could do with their camera, it should be understandable why it was a pleasure to work with this uninhibited, communicative population.[6]

Such rapport presupposes that an element of trust has been achieved between the field-worker and the individuals or group being studied. Many people whom I interviewed were keenly aware that because mainstream society regarded them as deviants, there was a high probability that harm to themselves or their life-style was never far away. Thus interaction between the field-worker and the individuals being studied must occur early to establish the trust necessary to conduct a worthwhile study. . . .

BODY MODIFICATION WORKSHOPS

Most of the ethnographic data in my study were derived from the body modification SM workshops I observed and the contacts I made while in attendance. These workshops were part of a series of continuing programs sponsored by Powerhouse and were designed to "enhance the SM experience."[7] Taught by individuals who were regarded as professional practitioners of various nonmainstream body modifications, the workshops were limited to a top enrollment of 50 people. The six workshops I attended were on male piercing, female piercing, branding and burning, cutting, play piercing, and playing with fire.[8] The audience at each workshop was markedly homogeneous. With the exception of myself and perhaps a half-dozen others, each session was typically attended by SM-oriented lesbians, gays, and bisexuals. Participants ranged in age from their late teens to their late 50s, with most attendees in their mid-20s and 30s. Leather was predominant—jackets, trousers, skirts, chaps, trucker's caps, gloves, boots, arm bands, wrist bands, and gauntlets. Heavily laden key rings, hunting knives in leather scabbards, slave collars, and T-shirts with sexual preference messages were also omnipresent. Tattoos, lip and nasal septum piercings, and multiple pierced ears were quickly visible, whereas more intimate piercings, such as nipple, navel, and genital would become evident as the workshops proceeded. It is fair to say that the groups would have attracted some attention were they to have gathered in a suburban shopping mall.

The four workshops described here were held on Saturday afternoons in an upstairs room of a liberal church in San Francisco.[9] Two other workshops I attended but do not describe in this article were conducted in a small room above a popular San Francisco gay bar.

Male Piercing

The first workshop in the series was on male piercing. Jim Ward, the teacher, was the president of Gauntlet, Inc., one of the few firms in the world that manufactures nonmainstream piercewear. Recognized as a "master piercer," Ward has been piercing since the mid-1970s and has estimated that he has done 15,000 piercings in the 14 years between 1975 and 1989. He is also the editor and publisher of *Piercing Fans International Quarterly (PFIQ)*, a successful glossy publication devoted exclusively to the subject of piercing. . . .

Ward arrived at the workshop early to set up his piercing equipment and a massage table that would serve as a piercing couch. He was wearing Levi's, a studded belt, black boots, and a black T-shirt that had the logo "Modern Primitives" (see Vale and Juno 1989) printed above 12 white-bordered rectangles, each of which contained a graphic drawing of one of the most popular genital piercings. Ward's lover and assistant set out several jewelry display cases and arranged chairs for the audience. He had multiple ear piercings, a bonelike tusk in his nasal septum, and a Gauntlet button on his T-shirt that proclaimed "We've got what it takes to fill your hole."

Ward's popularity and fame were evident as several arrivees paid their respect by shaking his hand or hugging and kissing him. Even though the workshop was on male piercing, one third of the audience was women, a crossing-over evident at each of the workshops regardless of the gender-specific body modification being highlighted. Ward welcomed the group, confirmed that his prearranged volunteers were present, and began his discussion of male piercing. It was evident that he had been through the routine many times, which he had, both before live audiences and in his continuing series in *PFIQ*, "Piercing With a Pro." His presentation was divided into halves, the first of which was a general discussion of the topic, or as he said, "the ins and outs of piercings," and the second consisting of actual demonstrations. As was true of each of the Powerhouse workshops, there was much emphasis on safety, cleanliness, sterilization, and proper hygiene after the procedure. Assuming that most of his audience was already involved in or at least aware of piercing, Ward dispensed without definition such esoteric piercing terminology as ampallang, dydoe, frenum, Prince Albert, guiche, and so on. Questions were asked about autoclave temperatures, rubber gloves, anesthetics, antiseptics, play piercing versus permanent piercing, the dangers of AIDS and hepatitis, body rejection, jewelry selection, and the like.

During the break, I asked Ward about his own piercings:

> Well, you can see them in my ear lobe and tragus, but I also have a Prince Albert in my cock and a nipple ring on each tit. Oh, I've got a guiche with a piece of cord in it, too. I'm wearing all that stuff right now. I've been piercing myself for 20 years, but I don't wear jewelry in most of the holes. I travel all over the country, and I can tell you it's a real mind-fuck to get on an airplane and sit next to some hunk knowing you've got all this sexy stuff on.

I also talked with an audience member who was not interested in getting pierced but wanted to see what the attraction was for his pierced friends:

> I don't feel the need to get pierced. Actually, I'm deathly afraid of needles. I don't think I have to look like a pin cushion in order to look sexy. When I'm out cruising I might stuff my balls through a coupla cock rings.

Gives me a great feeling and enough basket to turn a few heads. Best of all, no artificial holes in my body to get infected.

Ward's first volunteer after the break was a leather-clad male who wanted his left nipple repierced. He sat shirtless on the table as his companion offered him a reassuring hug. Ward examined the nipple and told the group the scar tissue from the previous piercing would make this one more difficult. He also took advantage of the audience's concern to note the difference between pain and sensation in piercing and that he preferred the latter term to best describe the feeling. The volunteer's facial expression gave the impression that he had some doubts about Ward's evaluation. Before starting the piercing, Ward summoned his second volunteer and explained to the group that before he did the nipple job he needed to prep Number 2 for his forthcoming Prince Albert, a procedure that requires the application of a local anesthetic because the needle pierces the urethra, a particularly sensitive area. Number 2 dropped his leather trousers to his ankles, and Ward casually tamped a xylacane-coated cotton swab into the urethra about 1 inch. A male in the audience teased, "I bet he wishes there wasn't any anesthetic on that Q-tip®." Laughter. Ward directed the volunteer to step over to one side of the room and wait for the anesthetic to numb the area. The volunteer, leather trousers still at his ankles and undershorts dropped below his knees, hopped over to the wall where he waited patiently, with the Q-tip® jauntily protruding from the tip of his penis.

Ward returned to the first volunteer and spent several moments discussing different types of male nipples and the particular piercing technique warranted by each. Then he scrubbed the volunteer's nipple with Hibiclens and Betadine, marked each side of the nipple with a dot to guide the needle path, clamped a Pennington forceps on the nipple to keep it from retracting and to afford better manageability, and expertly pushed a needle through the guide dots. An audible sharp gasp and a rigid tensing of the volunteer's body confirmed Ward's earlier comment about the likelihood of tougher tissue in repiercings. There was an immediate sigh of relief from the group accompanied by applause and congratulatory whoops. One end of the jewelry was used to push the needle the rest of the way through the nipple, thus resulting in the needle being expulsed and the jewelry attached in one continuous movement. The entire procedure had taken less than 3 minutes.

The second volunteer was invited back to the table, Q-tip® still in place as he waddled across the room. Ward had him sit on the table, then decided that it would be better if he stood on the table. There was some concern in the group about this stance, as the volunteer was visibly trembling, a circumstance that was all the more worrisome because the table itself began to shake. It was not clear whether the bare-legged volunteer was simply cold or whether he was suffering from pre-op jitters. Nevertheless, Number 2 balanced precariously atop the uncertain table while Ward, who had now gained an eye-level view of his work site, examined the about-to-be-pierced penis with his eyes and his fingers. As he worked, Ward maintained a running commentary on the history of the Prince Albert, noting that "it was originally designed to tether the penis to either the right or left pant leg for a neater looking appearance, but now it's strictly erotic." After completing the usual prepping around the piercing area, he deftly pushed the needle into the underside of the penis just behind the head, into the urethra, up toward the tip of the penis and the still lodged Q-tip®. As he pushed, the Q-tip® suddenly popped out of the urethra—"a sure sign I'm on course"— followed by the tip of the gleaming needle. The volunteer gazed warily down at the sight while being steadied by a friend. Applause and cheers. The jewelry was attached and Number 2 was eased down from the table. Still shaking, he pulled up his shorts and trousers. Ward peeled off his rubber gloves and disposed of them while discussing his thoughts on abstinence during the healing process.

The third volunteer was to receive a dydoe, a piercing that would pass through both sides of the upper edge of the glands. This volunteer, in his late 40s, removed his trousers and undershorts and stretched out calmly on the table. With more than 2 hours of discussion and demonstrations behind him, Ward was now much quieter. The usual preliminaries were undertaken while the volunteer chattered about his piercing history. The group was only mildly interested in the disclosures, but full attention resumed when Ward began the actual piercing. As with the first two volunteers, this one emitted a controlled but audible gasp, then relaxed. The jewelry was attached and the volunteer hopped off the table and dressed while the group applauded. Later, this person expressed his feelings about piercing to me:

> I like the jewelry very much, but the real turn-on comes from having my body penetrated. Everytime I see that sharp, shiny needle heading towards my flesh I know I'm going to get either a dick orgasm or a head orgasm or maybe both.[10]

Several of the audience congratulated the new piercees and expressed their appreciation to Jim Ward. The first workshop was over.

Branding and Burning

The second workshop of the series was devoted to branding and burning and was taught by Fakir Musafer. Musafer, who pierced his own penis at age 13, was 58 years old at the time of the workshop. Recognized by many as the doyen of "modern primitivism" in the United States, there is little in the practice of body modification and "body play" that he has not experienced on his own body. He has been tattooed, burned, cut, pierced, skewered, and electrically shocked. He has fasted, deprived himself of sleep, rolled on beds of thorns, and constricted and compressed various parts of his body with belts and corsets. He has also gilded, flagellated, punctured, and manacled himself. In addition to frequently suspending himself with fleshhooks à la the Sun Dance, he has reclined on a bed of nails or blades, "negated" his scrotum and penis by sealing them in plaster, and conversely "enhanced" the same by such practices as scrotum enlargement and penis elongation (a procedure accomplished by regular stretch workouts with 3-pound weights, or, what he refers to as "rock on a cock" exercises). His fame continues to spread in the United States and abroad through his continuing involvement with Jim Ward's *PFIQ* and the release of "Dances Sacred and Profane," a widely distributed videotape that highlights his Sun Dance ritual. Musafer undoubtedly represents the distant end on any scale of contemporary nonmainstream body modifiers, the great percentage of whom are content with their tattoos and/or multiple piercings.

Musafer arrived 30 minutes before the start of the workshop. He was wearing a black T-shirt and baggy khaki cotton trousers with large flapped mid-leg pockets. Puffing on a cigarette, his hair dyed sable brown, and bereft of any visible piercing jewelry, he looked like any other middle-aged ad executive enjoying his weekend. I volunteered to help him unload his van. The contents of the box I carried up the stairs vaguely hinted of his workshop's topic—acetylene torch, metal snips, needle-nose pliers, wire, matches, incense sticks, candles, several strips of copper and tin, mirror, two potatoes, and various other oddities.

Like Jim Ward at the previous workshop, Musafer was hugged and kissed by many arrivees. And, as in all the workshops, a spirit of *bonhomie* prevailed as arriving couples and singles of both sexes hugged and kissed acquaintances, chatted amiably with others, and shared their latest body art. A few moments before the starting time, Musafer shed his T-shirt, revealing a small ring in each nipple. He removed the rings and deftly replaced them with hollow metal tubes, each, according to his admission, 7/8 of an inch in diameter by 1 inch in length. Then he quickly inserted white teflon tubes through the holes on each side of his chest that had been created several years earlier for his Sun Dance hooks. Each tube was about the size of a king-sized cigarette and ran vertically through the flesh behind each nipple. Finally he reached into a pocket and pulled out a large nasal septum ring, which, with the aid of an audience member, was installed in his nose. He pulled his T-shirt back on and commented, "A-h-h, that's more like it!" Now, with open arms he welcomed everyone.

"The Fakir" as he frequently refers to himself, is an old pro at this sort of presentation and glibly but knowledgeably discussed his first topic of the day: branding. With the aid of an easel and predrawn charts, we learned about technique ("Don't go too deep, you're not doing a 'Mighty Dog' brand"), patterns ("The simpler the better"), important reminders ("Remember, each mark in the final scar will be two to four times thicker than the original imprint"), desirable locations ("The flatter the surface the better. Try to stick with the chest, back, tummy, thighs, butt, leg, upper arm"), and tools and materials and where to get them.

A prearranged volunteer indicated a preference for a 2-inch skull to be branded on the calf of her left leg. She reclined on the table with her skirt pulled up over her knees and her left leg extended toward Musafer's work site. A previously drawn pattern was transferred onto her left calf. As we watched, he fashioned the skull shape out of a strip of metal cut from a coffee can and showed us how to heat and apply the brand. Holding the brand in his needle-nosed pliers, he heated it in the acetylene torch until it was red hot, then quickly applied it to a piece of cardboard to check the design's appearance. Musafer also cut a potato in half and applied the reheated brand to the cut surface, noting that this was a good way for beginners to practice depth control before doing the real thing on human skin. The volunteer, a professional piercer and cutter in her early 30s, laughed with the audience as the potato hissed and smoked from Musafer's strike. "The Fakir" was now ready and alerted his client. The brand was heated, precisely positioned over the desired part of the pattern, and struck. There was an immediate hiss and a crackling sound, followed by a wisp of smoke and the odor of scorched flesh. The volunteer scarcely twitched. Musafer examined his first strike and proceeded to do six more to finish the skull, complete with stylized eyes, mouth, and teeth. The only time the volunteer reacted to the hot brand, and intensively so, was when Musafer inadvertently brushed the edge of her left foot with a "cooled" brand as he returned it to the torch for renewed heating. Musafer rubbed some Vaseline on the brand and the foot burn, and the volunteer sat up and put her low-cut boot back on. The audience applauded.

Musafer's next demonstration was on burning.[11] Displaying a row of seven or eight self-imposed circular burns on the front of his right thigh, each about the size of a penny and resembling inoculation scars, he discussed different types of burnings ("Cigarette burns are nasty but nice"), and techniques to cause them ("I prefer incense sticks because they work real well and smell so delightful during the burn"). There was no prearranged volunteer for burning, but a woman in her early 20s volunteered from the audience. Her friends cheered her as she removed her Levi's and sat on the table. Musafer touched her gently on the legs and softly said, "You are giving your flesh to the gods." He also instructed her on the importance of deep breathing and visualization as he glued a 3-inch length of incense stick to her left thigh and lit it. Within 4 or 5 minutes, the stick had burned down to skin level and extinguished itself. Although the stick was slightly less than a pencil in diameter, the circular burn mark quickly expanded to the penny sized marks I had viewed earlier on Musafer's thigh. Throughout the burning, the volunteer kept her eyes tightly closed and followed Musafer's instructions on deep breathing. She now opened her eyes and the audience applauded. Several of the audience members came up to where she was sitting and examined the burn, while one of her friends asked Musafer for some extra incense sticks to take home. Musafer doled out some sticks and jokingly admonished, "Watch it, this stuff is catching!" Many chuckles. Musafer responded to several last-minute questions while packing up his paraphernalia. This workshop had ended.

Female Piercing

The female piercing workshop was taught by Raelynn Gallina, a woman recognized throughout the Bay Area as a professional "total body" piercer. In addition to her popularity as a piercer and cutter (some of her clients glowingly refer to has as "Queen of the Blood Sports"), Raelynn is a successful designer and manufacturer of jewelry. Although she pierces males ("above the navel only"), the greatest proportion of her clientele is female, most of whom are, like herself, lesbian. She had been piercing for 7 years at the time of the workshop. Two thirds of the workshop's audience of 45 people were women, all but a few of whom were lesbians involved in sadomasochism. Most of the men present were gay SMers. Raelynn announced that she had four volunteers for the afternoon and would pierce a nipple, a clitoris hood, an inner labia, and a nasal septum.

The first half of the workshop was devoted to a "do it yourself" clinic on technique, tools, antiseptics, and various do's and don'ts and ended with the admonition that it was safer and better to be pierced by a pro:

> I get a lot of people who want me to fix up their bungled piercings. Usually turns out they were heavy into a torrid scene when someone says, "Heh! Wouldn't it be hot if we pierced each other!" All I'm saying is if you do get carried away, make sure you know what you're doing.

After the break, Raelynn's spiritual-psychological bent revealed itself as she emphasized the importance of centering, grounding, visualization, client-practitioner compatibility, and the relationship between individual personality and type of piercing jewelry to be worn. She later commented to me:

> Piercing is really a rite of passage. Maybe a woman is an incest victim and wants to reclaim her body. Maybe she just wants to validate some important time in her life. That's why I like to have a ceremony to go along with my piercing and why I do it in a temple—my home. Most of my clientele are bright, sensitive women. It's not as if a bunch of "diesel dykes" are busting into my place to prove how tough they are by getting their boobs punched through with needles.

Raelynn's first volunteer was an achondroplastic dwarf in her 30s. Dressed in leather trousers and field boots, it was obvious that she was extremely popular with many in the group. At least three different women lifted her up and danced merrily around with her in their arms during the break, while numerous others bent down to kiss her on the cheeks or lips. As Raelynn described the nipple piercing she was about to perform, the volunteer peeled off her blouse and climbed up on a long-legged director's chair. Clearly visible across her left breast in what appeared to be a recent cutting were the words "The bottom from Hell." There was much laughter, expressions of encouragement, and joking about anticipated pain, needles, second doubts, and the like. Raelynn scrubbed the volunteer's left breast and nipple with Hibiclens, applied a good coating of Betadine, and clamped a Pennington forcep on the nipple. The exact penetration and exit points for the needle were marked with a pen, and Raelynn quickly and expertly forced a needle through the nipple and into a small cork at the exit point. The forceps were removed, and a gold ring was inserted in the place of the needle. The only sign of pain from the volunteer was a short gasp as the needle pierced the nipple. As in the previous piercing procedures, there was a collective sigh of relief from the group, followed by applause and various congratulatory remarks. The volunteer climbed down from her perch and hugged Raelynn around the hips. Raelynn scooped her up, returned the hug, and kissed her.

By the time the first volunteer had put her blouse back on, the second volunteer had already removed her jeans and underpants and was sitting on the chair. In her mid 20s, this volunteer would receive a clitoris hood piercing and jewelry. During most of the procedure, she held a hand mirror over her pubic area to better monitor the procedure. Raelynn advised the client to close her eyes and visualize the process. The piercing and jewelry attachment was completed within 3 minutes, again with the client showing minimal reaction to the actual piercing. As the volunteer pulled her tight jeans on, Raelynn reminded the group that it was important to wear loose-fitting clothes when getting a genital piercing.

The third volunteer, in her mid-20s, had spent the first half of the workshop curled up in the laps of three different women. She removed her cotton skirt and hopped onto the chair. Pantyless and clean-shaven, two labia rings and a clitoris hood ring were easily visible. Raelynn announced that this client would be getting a third labia ring today and with a theatrical leer, added, "And she has asked me to do it real-l-l slow and with a twist." The audience responded with mock moaning and various teasing expressions. While Raelynn talked about genital piercing in general, the client sat spread-legged and observed with the hand mirror her present labia piercings and jewelry. Raelynn noted that a clamp was not usually needed for labia piercings and started the procedure. An audible intake of air and a slight tensing of the body were the only signs that the needle had pierced the flesh. The jewelry was quickly attached and the usual applause delivered.

I was unable to remain for the last piercing of the session, a nasal septum procedure through the nose of the first volunteer.

Cutting

Raelynn was also the teacher for the workshop on cutting. She had arrived in the room an hour before the scheduled starting time to set up her equipment and a videocamera. As people entered the room, she greeted them, occasionally examining a piercing or cutting that she had apparently performed at a previous occasion. By the starting time, 43 people had arrived and there was the usual happy buzz of chatter. Three fourths of the group were women, most of whom were wearing leather. As in the other workshops, the majority of the group members were gay and lesbian. Couples held hands, snuggled, kissed, and engaged in animated conversations. Raelynn sat on a table with knees crossed and officially greeted everyone. Although she welcomed the group by saying "Hello fellow blood sluts" and there was a button attached to her equipment case that read "I'm hungry for your blood," she quickly stated she does her cutting for aesthetic reasons and not just for the joy of blood. Some of the audience responded in unison, "O-h-h, su-u-re!" She told the group she had been cutting for approximately 8 years and that she got her start while "caught up in some heavy SM scenes."

During the 2 1/2 hour workshop, she discussed where cuttings should be done on the body ("Fleshy areas like the butt, thighs, back—not on the neck, joints, places where there are veins"), cleanliness ("Cutting is a clean procedure, not a sterile one"), use of rubber gloves, and concern about AIDS and hepatitis, depth of cut, design, tools, and various other bits of information regarding her subject. She also distinguished between her style of cutting and the types of scarification and cicatrization done in several preliterate populations of the world.

After a short break, Rosie, a prearranged volunteer in her 40s, stripped to the waist to receive her cutting. She and Raelynn had decided earlier on a design that consisted of a pattern of stylized animal scales in a triangular shape. The design was large and would be cut into the upper left area of the back near the shoulder blade. Raelynn scrubbed Rosie's back with the usual antiseptics, dried it off, and covered the area with stick deodorant to facilitate the transfer of the design. Using a No. 15 disposable scalpel ("Toss it after it's been used"), she started her first incision. As she cut, she explained that cutting was not really a painful procedure because of the sharp scalpel and the shallow cuts. Someone in the group wondered aloud, "Is it bleeding yet?" to which someone else reported, "I certainly hope so!" Raelynn also urged would-be cutters to remember to start cutting at the bottom of the design so that the dripping blood would not wash away the uncut design. Both the cutter and the client were obviously moved by the procedure. Daubing away some blood, Raelynn told the group, "Once you start cutting someone, you get a very high, heady experience." Rosie, her eyes closed and mouth sensuously open, emitted several soft sighs during the 10-minute procedure. At one point, Rosie squeezed her companion's hand and whispered, "This is intense, wonderfully intense."

When the cutting was completed, Raelynn blotted the design several times to soak up the still bleeding incisions. Rosie was alerted to brace herself for the alcohol rinsings, which were done several times over the cutting. A towel around Rosie's waist kept the alcohol from dribbling further down her body. Raelynn then ignited a fresh rinsing of alcohol with her cigarette lighter. A loud poof was heard, and a bluish flame danced across the entire left side of Rosie's back. The flame was quickly doused as Raelynn announced, "Rosie asked me to do that because she's into fire." Referred to as "slash and burn" by Raelynn, the fire event was repeated two more times as the audience oohed and aahed. Finally, Raelynn rubbed black ink over the entire design and the wound was covered with a protective surgical wrap. In a few days, the excess ink would be scrubbed away, leaving the lines of the cutting colored black.

Kay, the second prearranged volunteer, wanted the fish cutting already on her back touched up. The original cutting had been done by Raelynn 9 months earlier, and although the design was still easily discernible, Kay liked the idea of having it redone. Raelynn prepared the area, unpackaged a new scalpel, and recut the design in less than 10 minutes. No ink was rubbed into the wound as Kay preferred the natural look.

Raelynn ended the workshop with a discussion of different body reactions to cutting, explaining that some people scarred nicely, well enough that there was no evidence of any cutting, whereas some keloided into large amorphous bumps.

While people were socializing after the session, I talked with a woman I recognized from earlier workshops. She told me Raelynn had pierced her labia, navel, and both nipples. She also compared Fakir Musafer and Jim Ward unfavorably with Raelynn:

> Those guys are out on the edge! They're an embarrassment. I mean, bones through the noses, the branding, the fleshhooks, the pain. You heard them—"If a client is in pain, you just keep on pushing and jabbing. It'll be over before you know it." Raelynn is great because she is gentle and looks for any special aspects of your personality that will help her do a better piercing or cutting. . . .

CONCLUSIONS

Taken as a whole, the responses from my informants portray a group of individuals who for a variety of reasons enthusiastically involve themselves in nonmainstream body modification. They readily admit that their body modification interests are statistically outside the average range, but none transfer this conclusion to a statement regarding a deficiency in mental health. The medical literature on the topic presents a picture of deeply disturbed individuals engaging in self-mutilation for various psychopathological reasons (for an understanding of the medical interpretation of self-mutilation, see American Psychiatric Association 1987; Eckert 1977; Greilsheimer and Grover 1979; Pao 1969; Phillips and Muzaffer 1961; Tsunenari et al. 1981). This view is supported by the general nonparticipating public. My empirical observations lead me to disagree with the latter assessment. The overwhelming number of people in my study appear to be remarkably conventional sane individuals. Informed, educated, and employed in good jobs, they are functional and successful by social standards. . . .

The number of contemporary Americans who have become involved with nonmainstream body modification is presently small. However, it is important to remember that the practices discussed in this article are a relatively new phenomenon in this culture. Each year, American society is bombarded with new body alterations, many of which are quickly assessed as unacceptable for one reason or another and fail to enter mainstream society. However, recent history also shows that some initially rejected alterations may take hold in a subculture and eventually catapult their way into the larger society. For example, ear piercing in America moved from nonmainstream to mainstream society in less than a decade and multiple ear piercing among both males and females is now relatively common. Lip and nose piercing is increasingly tolerated, but whether nipple piercing will follow suit remains to be seen.

A growing number of people in American culture believe that the penis and the clitoris are just as deserving of gilding as are earlobes. These individuals, like the style setters in earlier times who defied American society's strictures on body alteration by experimenting with such daring embellishments as lipstick, rouge, painted nails, eye makeup, and radical hairstyles, join human beings around the world in using their bodies to express a symbolic language that reveals their sentiments, dispositions, and desired alliances. Through adornment, the naked skin moves one from the biological world to the cultural world. As Claude Lévi-Strauss observed in Vale and Juno's (1989) book,

> The unmarked body is a raw, inarticulate, mute body. It is only when the body acquires the "Marks of Civilization" that it begins to communicate and becomes an active part of the social body. (p.158)

NOTES

1. Of course, most so-called "irreversible" body modifications are not truly so. Tattoos fade, scars may flatten, cutting may heal without scars, and piercings, if not tended, will close.

2. A *Newsweek* magazine article (January 7, 1991) recognized the current popularity of tattooing in the United States and noted, "It's the most painful trend since whalebone corsets: tattooing, the art of the primitive and the outlaw, has been moving steadily into the fashion mainstream." This same article also observed that in the past 20 years, the number of professional tattoo studios had jumped from 300 to 4,000.

3. Interestingly, today many tattooed people regard piercing, branding, and scarification as repugnant. For example, the following warning was displayed prominently on the wall of a Northern California tattoo studio (it was apparently part of a registration form for a 1982 national tattoo convention): "This convention is for Tattoo Artists and Fans who care about the Tattoo Profession. Anyone breaking the following rules will be asked to leave with no refunds. Facial tattoos other than cosmetic (eyebrows, lines, etc.) not permitted. Piercing of the private parts of the anatomy not permitted to be shown at any time. Any facial piercing with bones, chains, etc. must be removed during entire convention."

4. I was accompanied at each workshop by Craig Moro, a Berkeley resident who had served several months as a volunteer for the San Francisco Sexual Information Switchboard (SIS), a call-in telephone service for people seeking sexual information.

5. The first occasion was a workshop on play piercing, a procedure that involves brief piercing with hypodermic or sewing needles, fishhooks,

staples, and the like for fun and enjoyment and, unlike "permanent" piercing, is not done with the intention of installing some type of jewelry in the hole. After a lengthy introduction to the techniques, hygiene, and materials involved, the workshop leader divided audience volunteers into piercers and piercees. As a piercer, I selected an experienced partner to pierce, donned my rubber gloves, popped a hypodermic needle from its protective capsule using the recommended technique, and was within inches of making the jab when it suddenly occurred to me that I had some serious doubts about what I was doing. Even though the teacher was exquisitely clear on the need for care and safety during the "scene," my congenital clumsiness and concern about the blood being splattered here and there caused me to back out as gracefully as possible at that late instant. Another volunteer happily replaced me, and I don't believe my rapport was damaged by the event.

The second participation occurred during the workshop on playing with fire. Here it was simply a matter of overcoming my innate fear of being burned and joining the group of eager volunteers brushing each other with a lighted small torch soaked in a 70% solution of isopropyl alcohol. To my surprise, the activity was enjoyable and caused some of my fellow volunteers to wonder if I was considering changing my sexual proclivities.

6. The intellect of many SMers whom I interviewed and observed during my fieldwork was confirmed to me at one San Francisco workshop entitled "Playing with Fire." During the workshop, the instructor's knowledgeable discourse on the ignition points of various isopropyl alcohols was interrupted by several audience members who had extemporaneously launched into an animated conversation on such matters as flammability versus combustionability, chemical structures of alcohol and gasoline, and the medical definitions and implications of the various types of skin burns. After listening to the display a few moments, the Instructor broke in by expressing her astonishment at the esoteric outpouring, only to be sharply reminded by one audience member, "There are no dumb SMers!"

7. Because SM practices are so varied, it is difficult to provide a satisfactory single definition of SM behavior. Charles Moser, a psychotherapist who specializes in SM clients and wrote his Ph.D. dissertation on sadomasochism, uses five criteria to identify people involved in SM (in Truscott 1989): (1) appearance of dominance and submission, (2) role-playing, (3) consensuality, (4) sexual content, and (5) mutual definition (i.e., people involved recognize that what they are doing is different from the "norm"). Townsend (1983) suggested "a short list of characteristics" that he believed are present in most scenes that he would classify as SM: a dominant-submissive relationship, a giving and receiving of pain that is pleasurable to both parties, fantasy and role-playing, humiliation, fetish

involvement, and the acting out of one or more ritualized interactions (bondage, flagellation, etc.).

SM does not have to involve pain. There are many people who prefer a gentler approach to what otherwise would be considered SM; thus one sees the letters "D and S" for "dominance and submission," or "B and D" for "bondage and discipline." For an extended discussion of the pain issue, see Baumeister (1988), Gebhard (1969), Reik (1957), and Weinberg (1987).

For a scholarly presentation of SM behavior, see Weinberg's (1987) review of recent sociological literature on sadomasochism in the United States. For a nonacademic but informative introduction to SM behavior, I recommend two popular books by insiders: *Urban Aboriginals* (Mains 1984) and *The Leatherman's Handbook 2* (Townsend 1983). Because these sources are male oriented (very little academic work has appeared regarding female SM), the reader interested in female SM will find helpful the *Sandmutopia Guardian and Dungeon Journal* or *Dungeonmaster* magazine. Pat Califia has also written knowledgeably of the lesbian SM community (see Califia 1987).

Not surprising, the SM people I interviewed were infuriated with the standard psychological characterization of SM as aberrant behavior. For example, *The DSM–III–R,*the official diagnostic manual for the American Psychiatric Association, lists both sadism and masochism as psychosexual disorders.

8. Other Powerhouse workshops scheduled during my fieldwork period were Creative Bondage, Electrical Toys, Clothespins and Staples, Male Tit and Genitorture, Tit Play, Cock and Ball Torture, Whipping and Caning, Mummification, and Equestrian Restraints.

9. The church's liberal reputation was confirmed to me one sunny Saturday afternoon as our group huddled over a spread-legged woman undergoing a clitoris hood piercing to the accompaniment of an a cappella choir rehearsing Handel's "Hallelujah Chorus" in a downstairs room—a surrealistic scene that would not have escaped Van Gennep (1960).

10. It is not unusual to hear SM people remark on whether a particular body modification or play technique would produce a genital orgasm or an equal thrill in the mind ("head orgasm") or both.

11. A distinction may be made between two different types of burning that I witnessed during my fieldwork with the SM body modifiers. One type, as described in the Musafer burning demonstration, was intended to leave a mark. Another type, "play burning," capitalizes on the classic SM goals of trust and the *threat* of pain and/or injury, but does so without leaving intentional burn marks. The ritual use of fire in SM scenes is not uncommon. The cross-cultural use of fire as a means of "cooking" a person symbolically was discussed by Lévi Strauss (1970). See also Tonkinson (1978) on the use of fire on Mardudjara initiates and Warner (1964) on fire jumping among the Ngoni.

REFERENCES

American Psychiatric Association. 1987. *Diagnostic and statistical manual of mental disorders.* Washington, DC: American Psychiatric Association.

Baumeister, R. 1988. Masochism as escape from self. *Journal of Sex Research* 25:29.

Califia, P. 1987. A personal view of the history of the lesbian community and movement in San Francisco. In *Samois: Coming to power,* 243–87. Boston, MA: Alyson.

Eckert, G. 1977. The pathology of self-mutilation and destructive acts: A forensic study and review. *Journal of Forensic Sciences* 22:54.

Gebhard, P. 1969. Fetishism and sadomasochism. In *Dynamics of deviant sexuality,* edited by J. H. Masserman, 71–80. New York: Grune & Stratton.

Govenar, A. 1977. The acquisition of tattooing competence: An introduction. *Folklore Annual of the University Folklore Association* 7 and 8.

Greilsheimer, H., and J. Grover. 1979. Male genital self-mutilation. *Archives of General Psychiatry* 36:441.

Lévi-Strauss, C. 1970. *The raw and the cooked.* New York: Harper.

Mains, G. 1984. *Urban aboriginals.* San Francisco: Gay Sunshine Press.

Nichols, M. 1985. The piercing profile evaluated. *Piercing Fans International Quarterly* 24:14–15.

Pao, P. 1969. The syndrome of delicate self-cutting. *British Journal of Medical Psychiatry* 42:195.

Phillips, R., and A. Muzaffer. 1961. Aspects of self-mutilation in the population of a large psychiatric hospital. *Psychiatric Quarterly* 35:421.

Reik, T. 1957. *Masochism in modern man.* New York: Grove.

Rubin, A. 1988. *Marks of civilization.* Los Angeles: Museum of Cultural History.

Sanders, C. 1986. Tattooing as fine art and client work: The art work of Carl (Shotsie) Gorman. *Appearances* 12:12–13.

———. 1998a. Drill and fill: Client choice, client typologies and interactional control in commercial tattoo settings. In *Marks of civilization,* edited by A. Rubin, 219–31. Los Angeles: Museum of Cultural History.

———. 1988b. Marks of mischief: Becoming and being tattooed. *Journal of Contemporary Ethnography* 16:395–432.

———. 1989. *Customizing the body: The art and culture of tattooing.* Philadelphia: Temple University Press.

St. Clair, L., and A. Govenar. 1981. *Stoney knows how: Life as a tattoo artist.* Lexington: University Press of Kentucky.

Thevos, M. 1984. *The painted body.* New York: Rizzoli.

Tonkinson, R. 1978. *The Mardudjara aborigines: Living the dream in Australia's desert*. New York: Holt, Rinehart & Winston.

Townsend, L. 1983. *The leatherman's handbook 2*. New York: Modernismo.

Truscott, C. 1989. Interview with a sexologist: Dr. Charles Moser. *Sandmutopia Guardian and Dungeon Journal* 4:23–24.

Tsunenari, S., et al. 1981. Self-mutilation: Plastic spherules in penile skin in *yakuza,* Japan's racketeers. *American Journal of Forensic Medical Pathology* 2:203.

Vale, V., and A. Juno. 1989. *Modern primitives*. San Francisco: Re/Search.

Van Gennep, A. 1960. *The rites of passage*. Chicago: University of Chicago Press.

Warner, W. L. 1964 *A black civilization: A study of an Australian tribe*. New York: Harper.

Weinberg, T. 1987. Sadomasochism in the United States: A review of recent sociological literature. *Journal of Sex Research* 23:50–69.

30

Tearoom Trade: Homosexual Behavior in Public Restrooms

LAUD HUMPHREYS

O.K., here goes—no self-respecting homosexual in his right mind should condone sex in public places, but let's face it, it's fun. . . . The danger adds to the adventure. The hunt, the cruise, the rendezvous, a great little game. Then more likely than not, "instant sex." That's it.[1]

The nature of sexual activity presents two severe problems for those who desire impersonal one-night-stands. In the first place, except for masturbation, sex necessitates collective action; and all collective action requires communication. Mutually understood signals must be conveyed, intentions expressed, and the action sustained by reciprocal encouragement. Under normal circumstances, such communication is ritualized in those patterns of word and movement we call courtship and love-making. Verbal agreements are reached and intentions conveyed. Even when deception is involved in such exchanges, as it often is, self-revelation and commitment are likely by-products of courtship rituals. In the search for impersonal, anonymous sex, however, these ordinary patterns of collective action must be avoided.

A second problem arises from the cultural conditioning of Western man. For him, sex is invested with personal meanings: interpersonal relationship, romantic love, and an endless catalogue of sentiments. Sex without "love" meets with such general condemnation that the essential ritual of courtship is almost obscured in rococo accretions that assure those involved that a respectable level of romantic intent has been reached. Normal preludes to sexual action thus encourage the very commitment and exposure that the tearoom participant wishes to avoid. Since ordinary ways reveal and involve, special ritual is needed for the impersonal sex of public restrooms.

Both the appeal and the danger of ephemeral sex are increased because the partners are usually strangers to one another. The propositioning of strangers for either heterosexual or homosexual acts is dangerous and exciting—so much so that it is made possible only by concerted action, which progresses in stages of increasing mutuality. The special ritual of tearooms, then, must be both noncoercive and noncommital.

APPROACHING

The steps, phases, or general moves I have observed in tearoom games all involve somatic motion. As silence is one of the rules of these encounters, the strategies of the players require some sort of physical movement: a gesture with the hands, motions of the eyes, manipulation and erection of the penis, a movement of the head, a change in stance, or a transfer from one place to another.

The approach to the place of encounter, although not a step within the game, resembles moves of the latter sort. Although occurring outside the interaction membrane, the approach may affect the action inside. An automobile may circle the area a time or two, finally stopping in front of the facility. In what I estimate to be about a third of the cases, the driver will park a moderate distance away from the facility—sometimes as far as 200 feet to the side or in back, to avoid having his car associated with the tearoom.

Unless hurried (or interested in some particular person entering, or already inside, the facility), the man will usually wait in his auto for five minutes or longer. While waiting, he looks the situation over: Are there police cars near? Does he recognize any of the other autos? Does another person waiting look like a desirable partner? He may read a newspaper and listen to the radio, or even get out and wipe his windshield, invariably looking up when another car approaches. The purpose here is to look as natural as possible in this setting, while taking the opportunity to "cruise" other prospective players as they drive slowly by.

Sometimes he will go into the restroom on the heels of a person he has been watching. Should he find the occupant of another auto interesting, he may decide to enter as a signal for the other man to follow. If no one else approaches or leaves, he may enter to see what is going on inside. Some will wait in their autos for as long as an hour, until they see a desirable prospect approaching or sense that the time is right for entry.

From the viewpoint of those already in the restroom, the action of the man outside may communicate a great deal about his availability for the game. Straights do not wait; they stop, enter, urinate, and leave. A man who remains in his car while a number of others come and go—then starts for the facility as soon as a relatively handsome, young fellow approaches—may be revealing both his preferences and his unwillingness to engage in action with anyone "substandard."

Whatever his behavior outside, any man who approaches an occupied tearoom should know that he is being carefully appraised as he strides up the path. While some are evaluating him from the windows, others may be engaged in "zipping the fly."

POSITIONING

Once inside the interaction membrane, the participant has his opportunity to cruise those already there. He will have only the brief time of his passage across the room for sizing-up the situation. Once he has positioned himself at the urinal or in a stall, he has already begun his first move of the game. Even the decision as to which urinal he will use is a tactical consideration. If either of the end fixtures is occupied, which is often the case, an entering party who takes his position at the center of the three urinals is "coming on too strong." This is apt to be the "forward" sort of player who wants both possible views. Should both ends be occupied, it is never considered fair for a new arrival to take the middle. He might interrupt someone else's play. For reasons other than courtesy, however, the skilled player will occupy one of the end urinals because it leaves him more room to maneuver in the forthcoming plays.

If the new participant stands close to the fixture, so that his front side may not easily be seen, and gazes downward, it is assumed by the players that he is straight. By not allowing his penis to be seen by others, he has precluded his involvement in action at the urinals. This strategy, followed by an early departure from the premises, is all that those who wish to "play it straight" need to know about the tearoom game. If he makes the positioning move in that manner, no man should ever be concerned about being propositioned, molested, or otherwise involved in the action. (For defecation, one should seek a facility with doors on the stalls.)

A man who knows the rules and wishes to play, however, will stand comfortably back from the urinal, allowing his gaze to shift from side to side or to the ceiling. At this point, he may notice a man in the nearest stall peer over the edge at him. The next step is for the man in the stall (or someone else in the room) to move to the urinal at the opposite end, being careful to leave a "safe" distance between himself and the other player.

My data indicate that those who occupy a stall upon entering (or who move into a stall after a brief stop at the urinal) are playing what might be called the Passive-Insertee System. By making such an opening bid, they indicate to other participants their intention to serve as fellator. In the systematic observation of fifty encounters ending in fifty-three acts of fellatio, twenty-seven of the insertees opened in this manner (twenty-five sitting on stools, two standing). Only two insertors opened by sitting on stools and four by standing in stalls.

Positioning is a far more "fateful" move for those who wish to be insertees than for others. In sixteen of the observed encounters, the fellator made no further move until the payoff stage of the game. The strategy of these men was to sit and wait, playing a distinctly passive role. Those who conclude as insertors, however, are twice as apt to begin at the urinals as are the insertees. A few of each just stood around or went to window during the positioning phase of the game.

SIGNALING

The major thesis of Scott's work on horse racing is that "the proper study of social organization is the study of the organization of information."[2] To what extent this holds for all organizations is not within my realm of knowledge. For gaming encounters, however, this is undoubtedly true, with "skill" inhering in the player's ability to convey, interpret, assimilate, and act upon the basis of information given and received. Every move in the gaming encounter is not only a means of bettering one's physical position in relation to other participants but also a means of communication.

Whereas, for most insertees, positioning is vital for informing others of their intentions, about half of the eventual insertors convey such information in the signaling phase. The primary strategy employed by the latter is playing with one's penis in what may be called "casual masturbation."

Respondent: The thing he [the potential insertee] is watching for is "handling," to see whether or not the guy is going to play with himself. He's going to pretend like he is masturbating, and this is the signal right there. . . .

Interviewer: So the sign of willingness to play is playing with oneself or masturbation?

Respondent: Pseudomasturbation.

The willing player (especially if he intends to be an insertor) steps back a few inches from the urinal, so that his penis may be viewed easily. He then begins to stroke it or play with the head of the organ. As soon as another man at the urinals observes this signal, he will also begin autoerotic manipulation. Usually, erection may be observed after less than a minute of such stimulation.

The eyes now come into play. The prospective partner will look intently at the other's organ, occasionally breaking his stare only to fix directly upon the eyes of the other. "This mutual glance between persons, in distinction from the simple sight or observation of the other, signifies a wholly new and unique union between them."[3] A few of the players have been observed to move directly from positioning to eye contact, but this seems to happen in only about 5 per cent of the cases.

Through all of this, it is important to remember that showing an erection is, for the insertor, the one essential and invariable means of indicating a willingness to play. No one will be "groped" or otherwise involved in the directly sexual play of the tearooms unless he displays this sign. This touches on the rule of not forcing one's intentions on another, and I have observed no exceptions to its use. On the basis of extensive and systematic observation, I doubt the veracity of any person (detective or otherwise) who claims to have been "molested" in such a setting without first having "given his consent" by showing an erection. Conversely, anyone familiar with this strategy may become involved in the action merely by following it. He need not be otherwise skilled to play the game.

Most of those who intend to be insertors will engage in casual masturbation at a urinal. Others will do so openly while standing or sitting in a stall. Rarely, a man will begin masturbation while standing elsewhere in the room and then only because all other facilities are occupied.

In about 10 per cent of the cases, a man will convey his willingness to serve as insertee by beckoning with his hand or motioning with his head for another in the room to enter the stall where he is seated. There are a few other signals used by men on the stools to attract attention to their interests. If there are doors on the stalls, foot-tapping or note-passing may be employed. If there is a "glory hole" (a small hole, approximately three inches in diameter, which has been carefully carved, at about average "penis height," in the partition of the stall), it may be used as a means of signaling from the stall. This has been observed occurring in three manners: by the appearance of an eye on the stool-side of the partition (a very strong indication that the seated man is watching you), by wiggling fingers through the hole, or by the projection of a tongue through the glory hole.

Occasionally, there is no need for the parties to exchange signals at this stage of the game. Others in the room may signal for a waiting person to enter the stall of an insertee. There may have been conversation outside the facility—or acquaintance with a player—which precludes the necessity of any such communication inside the interaction membrane. This was the case in about one-sixth of the acts I witnessed.

MANEUVERING

The third move of the game is optional. It conveys little information to other players and, for this reason, may be skipped. As the Systematic Observation

Sheets show, twenty-eight of the eventual insertees and thirty-five of the insertors (out of fifty-three sexual acts) made no move during this phase of the interaction. This is a time of maneuvering, of changing one's position in relation to other persons and structures in the room. It is important at this point in the action, first, because it indicates the crucial nature of the next move (the contract) and, second, because it is an early means of discerning which men wish to serve as insertees.

Twenty of the thirty-three players observed in motion during this stage of the encounter later became insertees. Two-thirds of these used the strategy of moving closer to someone at the urinal:

> X entered shortly and went to third urinal. Y entered in about a minute and went to first urinal. . . . O stood and watched X and Y. Y was masturbating, as was X. X kept looking over shoulder at me. I smiled and moved over against far wall, lit cigarette. *X moved to second urinal* and took hold of Y's penis and began manipulating it. I moved to door to observe park policeman (in plain clothes with badge), who was seated on park bench. Then I went back to position by wall. By this time, X was on knees in front of urinals, fellating Y. I went back to door, saw A approaching, and coughed loudly.

Others may use this stage to move closer to someone elsewhere in the room or to move from the urinal to an unoccupied stall. All of these strategies are implemental but nonessential to the basic action patterns. The restroom's floor plan, I have found, is the strongest determinant of what happens during this phase of the game. If there are only two urinals in the facility, the aggressor's maneuver might be no more than to take a half-step toward the prospective partner.

CONTRACTING

Positions having been taken and the signals called, the players now engage in a crucial exchange. Due to the noncoercive nature of tearoom encounters, the contract phase of the game cannot be evaded. Initially, participants have given little consent to sexual interaction. By means of bodily movements, in particular the exposure of an erect penis, they have signaled such consent. Now *a contract must be agreed upon*, setting both the terms of the forthcoming sexual exchange and the expression of mutual consent.

Eighty-eight per cent of the contracts observed are initiated in one of two ways, depending upon the intended role of the initiator. One who wishes to be an insertee makes this move by taking hold of his partner's exposed and erect penis. One who wants to be an insertor under the terms of the contract steps into the stall where the prospective insertee is seated. If neither of these moves is rejected, the contract is sealed.

Manipulation of the other's organ is reciprocated in about half of the cases. Some respondents have indicated that they appreciate this gesture of

TABLE 4.1 *Major Strategy Systems in Tearoom Encounters.*
(Source: systematic observations of 48 encounters ending in fellatio.)

	Insertee	Insertor
Active	position: urinal signal: casual 　masturbation contract: manipulates 　partner's penis 　(27%)★	position: urinal signal: casual 　masturbation contract: steps into 　partner's stall 　(41%)★★
Passive	position: sits in stall signal: masturbation 　(sometimes beckoning) contract: accepts partner's 　entry (50%)★	position: urinal signal: masturbation contract: accepts partner's 　manipulation (27%)★★
Totals	N = 37 (77%)★	N = 33 (68%)★★

★Percentage of total insertees.
★★Percentage of total insertors.
Note: Eleven insertees (23%) and fifteen insertors (32%) followed a variety of minor strategy systems, combining elements of the major systems with idiosyncratic moves.

mutuality, but it is not at all essential to the agreement reached. The lack of negative response from the recipient of the action is enough to seal the contract. It is interesting, in this connection, to note that such motions are seldom met with rejection (only one of my systematic observations records such a break in the action). By moving through gradual stages, the actors have achieved enough silent communication to guarantee mutuality.

In the positioning and signaling phases of the game, the players have already indicated their intentions. This stage, then, merely formalizes the agreement and sets the terms of the payoff. One should note, in this connection, that a party's relative aggressiveness or passivity in this phase of the game does not, in itself, indicate the role to be acted out at the climax of the interaction. In connection with the positioning of the first move, however, it does provide an indication of future role identification: the man who is seated in the stall *and* is the passive party to the contract will generally end up as an insertee; the man who stands at the urinal *and* is passive in the contract stage, however, will usually be an insertor at the payoff. The more active insertees play from the urinal and initiate the contract by groping, but active insertors play from the urinal (or elsewhere in the room) and initiate the contract by entering the stall of a passive insertee (see Table 4.1).

The systems of strategy illustrated in Table 4.1 account for 77 per cent of the patterns of play for the insertees observed and for 68 per cent of the insertors' moves. (Again, the insertee role seems to be most stable in the encounters.) The Active-Insertee and Passive-Insertor Systems have already been illustrated and discussed in detail. An illustration from the systematic observation reports of the other two systems follows:

> [This was the fourth encounter observed in this tearoom within an hour. The man here identified as X had been the fellator in the second of these actions and had remained seated on the stool throughout the third. I estimated his age at about fifty. He was thin, had grey hair, and wore glasses. Y was about thirty-five, wearing white jeans and a green sport shirt, and was described as "neat." He was well-tanned, had black hair, was balding, and drove up in a new, luxury-class automobile.]
>
> Saw Y approaching tearoom from bridge, so I got into room just before him. I stood at first urinal for a minute, until he began to masturbate at other one—then I moved to the window and looked out on the road. I could see X peering through glory hole at Y when I was at the urinal. Y moved to opposite window and looked out. He then turned to look at X and me. He was stroking his penis through his pants, maintaining erection. I nodded to him to go ahead. He moved into first stall where X began to fellate him. This took less than five minutes. He wiped penis on tissue and left. X got up, zipped pants and left.

"X" played the Passive-Insertee System throughout these encounters. "Y" followed the Active-Insertor System, with some reassurance from the observer. The total time of this encounter—from entry of "Y" until his departure—was ten minutes. Note that the glory hole has three functions (the first two of which were employed in this encounter): as a peephole for observation, as a signaling device, and as a place of entry for the penis into the insertee's stall. The latter is very rare in the tearooms observed, and I have only twice seen these openings used in such a manner.

There may be forms of contracting other than the two I have described. I once observed a contract effected by the insertor's unzipping his pants directly in front of a prospective partner in the middle of the restroom. Another time, I noticed an active insertee grope a man whose erection was showing while his pants were still zipped. The move was not rejected, and the object of this strategy then played the insertor role. In a very few instances, my observations indicate that the insertee entered a stall where the eventual insertor was seated or that the insertor took hold of the partner's exposed penis at the urinals. These exceptions, while rare, make it necessary to withhold judgment as to what roles are being played until the payoff phase itself.

FOREPLAY

Although optional and quite variable, sexual foreplay may be seen as constituting a fifth phase of the tearoom encounters. Like maneuvering, it has very little communicative function and is not essential to production of the payoff. From positioning to payoff, nearly all players—and some of the waiters—

engage in automanipulation. There is little need, therefore, to prepare the insertee for fellatio by any other means of stimulation.

Unlike coitus, oral-genital sex does not require rigidity of the organ for adequate penetration of the orifice. Whereas an erection is a necessary signal in the early phases of the game, interruptions and repositioning between the contract and payoff stages occasionally result in the loss of an erection by the prospective insertor. The observer has noted that it is not uncommon for a fellator to take the other man's penis in his mouth even in its flaccid state. The male sex organ is a versatile instrument. With the proper psychosocial circumstances (varying with the individual's prior conditioning), it can reach the orgasmic phase in less than a minute. The authors of *Human Sexual Response* briefly discuss the many factors that intersect in determining the length of the "sexual response cycle":

> The first or excitement phase of the human cycle of sexual response develops from any source of somatogenic or psychogenic stimulation. The stimulative factor is of major import in establishing sufficient increment of sexual tensions to extend the cycle. If the stimulation remains adequate to individual demand, the intensity of response usually increases rapidly. In this manner the excitement phase is accelerated or shortened. If the stimulative approach is physically or psychologically objectionable, or is interrupted, the excitement phase may be prolonged greatly or even aborted. This first segment and the final segment (resolution phase) consume most of the time expended in the complete cycle of human sexual response.[4]

Foreplay may help in maintaining the level of stimulation required for advancing the response cycle. Such strategies as mutual masturbation and oral contact in the pubic area may not only add appreciably to the sensual pleasure of the players but may help to precipitate orgasm when the participants are operating under the pressure of time and threatened intrusion:

> It was now raining hard. O remained standing at window, saw X leave car and enter. Y drove up as X was walking toward tearoom, waited for about three minutes in car and run through rain to tearoom. X went to urinal nearest window. Y went to other urinal, urinated and began to play with his penis, stroking head slowly. Couldn't see what X was doing. X then moved over by Y (they had both been looking at one another) and took hold of his penis. Y did not reciprocate or withdraw. Y then moved over to far stall, still masturbating. X went over and stood by him, taking hold of his penis again. X's pants were zipped and I could not see evidence of an erection. He unbuttoned Y's pants and slipped them down to his knees, as he did with his shorts. Playing with Y's testicles and stroking his legs, he began to fellate him. . . .

It should be noted that, due to the danger of interruption, participants in this gaming encounter seldom lower their pants to the floor or unbutton any other clothing. They generally remain ready to engage in covering action at a moment's notice. Perhaps the rain gave these men . . . a . . . sense of invulnerability.

THE PAYOFF

The action now moves into its culminating stage. As is illustrated by continuing the above narrative, intrusions may temporarily detach the payoff phase from the action that leads up to it, providing moments of incongruous suspense:

> Two kids, B and C, came running toward facility with fishing poles. I coughed. Y sat down on stool and X moved over to window. B and C entered, talking loudly and laughing at the rain. They rearranged some fishing gear in a box, then ran back outside, the rain having let up slightly.
>
> D, an older boy around fourteen, came riding up on a bicycle from the bridge. I did not see him coming, but X and Y were still separated. D entered, went to urinal, urinated, looked out window by me and said, "Sure is raining out!" I remarked that it was letting up. "Guess I'll make a dash for it," he said. He left and rode off toward street. . . .
>
> X and Y resumed activity with X working his head back and forth and rubbing his hands up under shirt of Y, who was again standing. I saw A coming up walk and coughed. X and Y broke—Y sat down—X moved back to window. A entered and I recognized him. I nodded to X and Y, who resumed fellating position. A peered around edge of stall to watch, then stood up and looked over partition to get a better view, masturbating as he stood. Y moaned at orgasm and pressed on back of X's head. X stood up and continued masturbating Y even after orgasm. Y withdrew in a minute and pulled up pants. X moved back to window. Y looked for paper in other stall but it had been used up. He tucked his penis in pants, zipped up and left. X came over to window by me and looked me over. I smiled and left. A remained on stool.

Among other things illustrated here, one may notice the importance of hand play in the sexual act itself. The observations indicate that body and hand movements carry the action through stages that, lacking conversation, might otherwise be awkward. Primarily by means of the hands, the structure of the encounter is well maintained, in spite of the absence of verbal encouragement. Caresses, friendly pats, relaxed salutes, support with the hands, and thrusting motions are all to be observed throughout the action. Normally, the man who takes the insertor role will sustain the action of the fellator by clasping the back of his head or neck or by placing his hands on the partner's shoulders. As a frequent insertee points out:

> When you are having sex, it's not just that the sexual organ is being activated. The whole body comes into it. And you want to use your

whole body, and your hands are a very important part in sex. Next to the organs themselves, I think the hands are the most important part, even more important than the mouth for kissing. I really think the hands are more important—because you can do fantastic things, if you know how to do them. You can do fantastic things with your hands to another person's body.

Without the use of scientific instruments other than the human eye, it is impossible to say what proportion of the hand play during the sexual act is voluntary or involuntary. During the orgasmic phase, undoubtedly, there is a great deal of involuntary, spasmodic movement of the extremities, such as that described by Masters and Johnson:

> This involuntary spasm of the striated musculature of the hands and feet is an indication of high levels of sexual tension. Carpopedal spasm has been observed more frequently during male masturbatory episodes than during intercourse, regardless of body positioning.[5]

For physiological reasons, such spasmodic clutching of the hands is engaged in only by the insertor and confined mostly to the period surrounding orgasm.

The insertee, however, may have certain functional reasons for handling his partner. Some respondents have spoken of clutching the base of the penis with a hand, in order to ward off a thrust that may cause them to gag or choke:

> If the man has a very large piece of meat—I know from experience—I will not have somebody ram that thing down my throat. I'm sorry, but that hurts! It can cause a person to vomit. [Like if you put your fingers down your throat?] Exactly, and this can be very embarrassing. So, ordinarily, I will try to hold on. I know just about how much I can take. Then I am going to put my hand in a certain place, and I know it can't go any further than my hand. . . .

The same participant continues by describing another functional use of the hands during fellatio:

> Then I use my hands on the balls, too—on the scrotum. This can do wild things! [You said something about the hips. Or did I imagine that?] The hips or the backs of the legs. Now, there is one value in this which some people don't realize: these muscles contract first at the point of orgasm. This is one of the first signs of orgasm. When these muscles back in here begin to contract (the legs stiffen, these muscles contract or flex, or whatever—they get hard), you can tell at this point the orgasm is about to be reached. It is very helpful to know these things, especially if you are doing somebody. Because you can tell to go faster—or keep doing what you are doing. You can at least get ready and know not to pull away all of a sudden.

I suspect that, if one were to concentrate on observing peripheral matters in a study of heterosexual intercourse, he would find the same pattern of hand involvement: exploration of the partner's body, support of the head or pressure on the back, stimulation of the erotic zones, numerous caresses. At least in the payoff phase, silence in sexual encounters is not confined to the tearooms. When body communicates directly with body, spoken language is no longer essential. Thus far in history, the action of sex is the only universal language—perhaps because the tongue is but one among many members to convey the message, and the larynx is less important than the lips.

As has been indicated, it is a lack of such physical involvement—along with the silence—that tends to make tearoom sex less personal. When hand play does occur, therefore, it tends to raise the involvement level of the sexual action. Perhaps for that reason, some people attempt to avoid it:

> I saw X's hand as he motioned Y over into his stall. Y entered, stood facing X and unzipped. X ran his hands all over Y's buttocks, the back of his legs and up under his shirt while sucking. Y stood rather still with hands held out just far enough from his sides to give X freedom of movement *without touching him.*

The primary physical connection between the partners is that of the mouth, lips, and tongue of the fellator with the penis of the insertor. The friction and sucking action in the meeting of these organs is what produces the orgasm upon which the encounter focuses. A number of my respondents claim that the physical sensation of oral-genital copulation, while not unlike that of coitus, is actually more stimulating—or "exciting," as they generally word it. While some of the married men among the cooperating participants say that they actually prefer the sensations of fellatio to those of coitus, most agree that this is true only when certain other variables are held constant, when both acts take place in bed, for example. Many tend to look on tearoom sex as only a substitute for "the real thing."

> It's different—and I like both. I guess you could say I'd rather have sex with my wife. Getting a blow job isn't like having the real thing, but it has its points, too. I just don't know. I hadn't thought about it that way. I guess you really can't compare the two. Let's just say I like them both.

Some insertees retain the seminal fluid and swallow it: others clear their throats and spit it out. In one-fifth of the encounters observed, I noted that the insertee spit following the ejaculation of his partner. One respondent claimed that he only spits it out "when it tastes bad":

> The variety of tastes is unbelievable! You can almost tell what a person's diet is by what it tastes like. A person with a good, well-balanced ordinary diet, the fluid has a very mild, tangy, salty flavor. A person who has been drinking heavily—even if they aren't drunk or suffering from a hang-over, if they drink a lot—the stuff tastes like alcohol. And I mean pure, rot-gut alcohol, the vilest taste in the world!

I was unable to find any medical references to the taste of ejaculatory fluid, so I have not been able to verify this connoisseur's judgment. Other respondents will say only that they do think "some men taste different than others."

Acts of fellatio generally take place within the stall. This puts the insertee in a more comfortable position than crouching on his haunches elsewhere in the room. It also has an advantage in case of an intrusion, in that only one party to the action needs to move. There are certain tactical advantages as well. If the man who prefers the insertee position takes a stall and remains there for any length of time, he legitimizes himself, indicates the role he wishes to play, and needs only to wait for a partner to arrive.

Another twenty-nine per cent of the observed acts took place in front of the urinals. From that position, both may turn to face the fixtures in case of an intrusion. The fellator is poorly braced for his action, however, and probably quite uncomfortable. Occasionally, the act will occur in front of a window. I am informed that this is especially true when no lookout is present, because it has the advantage of enabling the insertor to double as lookout. When the oral-genital contact takes place away from the stool, the insertee will generally squat or drop to his knees to make the necessary contact.

During oral copulation, other men may come from around the room into viewing position. Many will proceed to masturbate while watching, sometimes without opening their pants for the automanipulation. Seldom does the exchange that is the focal point of this attention last more than a few minutes. In looking over my data, I found indications that I had grossly overestimated the amount of time lapsed between insertion and orgasm. What seemed to me like "a long time" (sometimes recorded as five or ten minutes) was probably a reflection only of my nervousness during the payoff stage. Since I was attempting to pass as a voyeur-lookout (both aspects of the role requiring my closest attention during these moments), it was impossible for me to use my watch in timing. No true voyeur would glance at his watch in the middle of a sex act! Actually, I suspect that the oral penetration ranged from ten seconds to five minutes, not counting interruptions.

I have twice seen couples engaging in anal intercourse. This is a form of sexual activity rare in most tearooms, probably due to the great amount of time required and the drastic rearrangement of clothing involved, both of which tend to increase the danger of being apprehended in the act. Mutual masturbation is an occasional means of reaching orgasm, particularly by the urinals or elsewhere in a crowded room.

CLEARING THE FIELD

Once the sexual exchange is accomplished, most insertors step into a stall to use the toilet paper. After the penis is cleansed, clothes are rearranged and flies zipped. In those rare cases in the observed facilities where a workable wash basin is provided, the participants may wash their hands before leaving.

Nearly always in the observation records, when a man took the insertor role he left for his car immediately after cleansing. The insertee may leave, too, but he frequently waits in the tearoom for someone else to enter. Sometimes he becomes the insertor in a subsequent encounter, as in the following account:

> [X is about forty-five, wearing a green banlon shirt, light blue slacks, driving a red, late model sports coupe. Y is about thirty, driving a green Ford convertible. He wears a light blue shirt, dark blue slacks, and a conservative tie. He is described as being tan, masculine, well dressed. B is about forty, balding, thin, tanned, wearing horn-rimmed glasses and a grey sport shirt. It is 2:25 on a beautiful Thursday afternoon, and there are few people in the park.]
>
> B was seated on stool when O entered. O stood at urinal a minute, noticed B watching him through glory hole. Crossed to far window, looked out and lit cigarette. Y entered and went to first urinal. X came in soon after Y and went to third urinal. They stood there for about five minutes. X kept peering over edge of stall at B and also at me. I crossed to opposite window and looked out missing pane. X was masturbating. Y went to second stall, lowered pants and sat down. I went back to window on right. Y spread his legs and began masturbating. (He had removed his coat and hung it over the edge of stall.) He had slumped on the stool seat as if sitting in camp chair, legs stretched out almost straight in front of him but spread apart and was masturbating obviously. I went back to window overlooking street. X then moved to window by Y, stood there a minute, then leaned over and took hold of Y's penis and began stroking it. He then knelt on the floor to begin fellatio. B just sat in his stall and masturbated. Y moaned a bit at climax. He then wiped and X stood back by window and masturbated while watching Y. Y flushed toilet, put on his coat, zipped pants and left. X stepped in B's stall and was sucked by him. This didn't take more than a minute. X then went to urinal, cleared throat, spit and left. (I was able to see autos of both X and Y through window.)

This is what I have labeled a series encounter. Generally, in order to facilitate the eventual analysis, I have broken these up into "Encounter A," "Encounter B," etc. In the above instance and in a few others among my systematic observations, I was not able to do so because of the rapid succession of events. During the hunting season, series encounters are the most common variety. Once the action begins to "swing," a series may last throughout the day, each group of participants trading upon the legitimation process of the previous game.

Another type of action that tends to swell the volume of sexual acts in a given facility is the simultaneous encounter, in which more than one sexual act is in process at the same time. The payoff phases are seldom reached at exactly the same time in these encounters, but they are staggered as in a round.

[It is a warm, humid, Friday afternoon. A few youngsters are playing ball in the park and some heterosexual couples are parked nearby. X is about thirty-five, tough looking, tattooed, dirty working clothes, drives an old Chevrolet. Y is about forty-five, lean and tanned, wearing tan work clothes. A is about thirty-two, neatly dressed with sport shirt and tie. I describe him as "masculine looking but wore pinky ring." He drove a new, foreign economy car. B is about fifty, heavy set, grey hair, sports clothes, rather unkempt. C is about forty, with a pot belly, wearing white sports shirt, dark blue pants.]

When O entered, X and Y were seated on stools with A standing by far window facing into room. While O urinated, he noticed that X was watching him through glory hole. O lingered at urinal for about four minutes, during which time A moved into stall with X (X is a noisy sucker, much "slurping," so I could tell what was happening but could not see). O crossed to far window, lit cigarette and peered out. A left first stall and stepped into space between Y and O. He stood there, masturbating both himself and Y, who had stood up. Meanwhile, C, who was sitting on bridge watching tearoom when O entered, came into the room. O saw him approaching through window and coughed. Y and A broke contact for the moment but, recognizing C, returned to action. C stood at urinal less than a minute, halted for another minute opposite stalls, then went into stall with X, who proceeded to fellate him. Y then stood on the toilet seat, watching X and C, while A sucked Y (A had to crouch but continued to masturbate). It was getting crowded on that side of the room, so O moved to opposite window. From this position, he could only see part of A's backside, Y's face and shoulders, and the backside of C. X kneaded C's buttocks and ran his hands up and down the backs of C's legs. When C finished, he left without wiping. A finished with Y about this time, went to urinal number three and spit. Y wiped and left. A stepped over to window by O, peered out through broken pane. His pants were still unzipped and he proceeded to masturbate and to look at O suggestively. O, feeling uncomfortable, went back to far window. B entered and A zipped up pants. B looked around as he went to middle urinal but stayed there for a brief time. He then moved over to stall with X. No one seemed to be made uncomfortable by B and seemed to recognize him. O then left, followed closely by A, who engaged him in conversation by water fountain. . . . All of this took place in twenty-five minutes.

The reader should be able to sense, at this point, that what I have described as a rather simple, six-step game (only four of which are essential to the action) may be acted out with infinite variety and confusing modifications. Every encounter reduces, ultimately, to the basic steps of positioning, signaling, contracting, and payoff; but no two of them are quite alike.

A pat on the shoulder, a wave of the hand, an occasional whispered "thanks" concludes the action. The departure ritual is simple and brief. Once the field is cleared, some individuals go to their homes or jobs, others return to their cars and await the arrival of fresh players, and a few may venture to a different tearoom to take their positions in another encounter.

The length of these games was observed to range between five and forty minutes, with an average duration of about eighteen minutes. Tearoom encounters, then, require relatively little time—a quarter of an hour if one knows where to go and how to play the game. Many suburban housewives may think their husbands delayed by the traffic when, in reality, the spouses have paused for a tearoom encounter. . . .

NOTES

1. Letter in "Open Forum: Sex in Public Places," edited by Larry Carlson in *Vector*, Vol. 3, No. 6 (May, 1967), p. 15. In my opinion, *Vector* is the best of the homophile journals. It is published monthly by the Society for Individual Rights, 83 Sixth Street, San Francisco, California.

2. Marvin B. Scott, *The Racing Game* (Chicago: Aldine, 1968), p. 3.

3. From Georg Simmel, *Soziologie*, as quoted in Goffman, *Behavior in Public Places* (New York: The Free Press, 1963), p. 93. For a thorough discussion of the use of eye contact in face-to-face engagements, see pp. 91–96 of Goffman's book.

4. William H. Masters and Virginia E. Johnson, *Human Sexual Response* (Boston: Little, Brown, 1966), pp. 5–6.

5. *Ibid.*, p. 173. See also pp. 296–297.

31

Turn-Ons for Money

Interactional Strategies of the Table Dancer

CAROL RAMBO RONAI

AND CAROLYN ELLIS

She swayed from side to side above him, her hands on his shoulders, her knee brushing gently against the bulge in his pants. He looked up at the bottom of her breasts, close enough to touch, but subtly forbidden. His breath came in ever shorter gasps.

This is the world of the table dancer—a world where women exchange titillating dances for money. Our study looks at the dynamic processes of interaction that occur in the exchange. Previous studies (Carey et al., 1974; Gonos, 1976; McCaghy and Skipper, 1969, 1972; Salutin, 1971; Skipper and McCaghy, 1970, 1971) have concentrated on "burlesque" or "go-go" dancers, sometimes referring to them more generally as stripteasers. Dancers' interactions with customers were restricted, for the most part, to the stage setting where they danced and received money from customers. Because investigators in these studies occupied positions as researchers or researchers as customers, and relied to a large extent on survey and interview techniques, this work led to a static description of this occupation.

Boles and Garbin (1974) have looked at customer-stripper interaction in a setting where strippers sold drinks in addition to performing stage acts. Although they described interaction, they interpreted it in terms of norms, club motif, and customer goals. They found that the conflict between customer's goals and strippers' goals resulted in "counterfeit intimacy" (Foote, 1954), a situation in which an aura of intimacy masked mutually exploitative interactions.

Although counterfeit intimacy is a structural reality in such contexts, this description created another model of behavior that ignored the interactive, dynamic nature of the exchanges and set up in its place stiff caricatures behaving in an unbending, cardboard manner. As actors get caught up in dialogue, they exchange symbols, extract meanings, and modify expectations of what goals they can reasonably expect to reach. Interaction has a tentative quality (Blumer, 1969; Turner, 1962); goals are in a constant state of flux.

From: "Turn-Ons for Money: Interactional Strategies of the Table Dancer," Carol Rambo Ronai and Carolyn Ellis, *Journal of Contemporary Ethnography,* Vol. 18, No. 3, 1989. Reprinted by permission of Sage Publications, Inc.

The nature of selling and performing table dances that we describe yields more opportunity for interaction between customer and dancer than in previous studies. A table dancer must be a charming and sexy companion, keep the customer interested and turned on, make him feel special, and be a good reader of character and a successful salesperson; at the same time, she must deal with her own negative feelings about the customer or herself, negotiate limits, and then keep him under control to avoid getting fired by management.

Much of the early research literature has described stripping as a deviant occupation. Later, Prus and Irini (1980) looked at stripping as conforming to the norms of a bar subculture. Demystifying this "deviant" activity even further, we show that bargaining strategies in the bar actually mirror "respectable" negotiation in mainstream culture.

We begin by discussing the methods we used to elicit in-depth understanding of strategies used by table dancers. After describing the dance club setting, we turn to a description and analysis of particular tactics used on the stage, at the tables between stage acts, and then during the table dances in the pits. Our conclusion analyzes how this exchange reflects buying and selling in service occupations as well as the negotiation of gender relationships in mainstream society.

METHODS

Our study approaches stripping from the point of view of dancers and the dancer as researcher, the people with the most access to the thoughts, feelings, and strategies of exotic dancers. Dancers concentrate on manipulating men as they pursue money in exchange for a turn-on. In order for their strategies to work, they must understand and coordinate them with the games of men.

Our information comes primarily from the experiences of the first author who danced during 1984 and 1985 to pay her way through school. As a "complete-member-researcher" (Adler and Adler, 1987), she conducted opportunistic research (Riemer, 1977), that is, she studied a setting in which she was already a member. She interviewed dancers to find out how and why they began this occupation and kept a journal of events that happened while dancing. Later, she reconstructed, in chronological field notes, a retrospective account of her own dancing history, paying special attention to strategies, emotion work, and identity issues. She used "systematic sociological introspection" (Ellis, forthcoming) to put herself mentally and emotionally back into her experiences and record what she remembered (see Bulmer's, 1982, concept of "retrospective participant observation").

In May 1987, the first author danced in one strip bar for the explicit purpose of gathering data for a master's thesis, chaired by the second author. With approval of bar management, but without the knowledge of other dancers, she acted in the dual capacity of researcher and dancer. This time her primary identity was that of researcher, although as a complete-member-researcher she attempted to "become the phenomenon" (Adler and Adler, 1987; Jorgensen,

1989; Mehan and Wood, 1975). When she danced, she took on the identity of a dancer, suffered identity conflicts similar to those she had experienced during earlier dancing, and shared a common set of experiences and feelings with other dancers. She kept field notes of events, which were buttressed by "interactive introspection" (Ellis, 1988), whereas the second author talked her through her experiences, probing at and recording her feelings and thoughts. She conducted informal interviews in the dressing room with dancers and on the floor with customers. Sometimes she revealed her dual role to customers as a strategy to keep them interested in spending more money and to get them to introspect about their own motives for being in the bar.

Because this article is concerned with describing dancers' subtle manipulation strategies that occurred semiprivately, we pulled much of our material from episodes engaged in by the first author, in which process was most easily observed. Because we believe that sociologists should acknowledge the role of their own introspection in their research (Ellis, forthcoming), the first author reveals which of the experiences in the article are hers. Throughout this article, we refer to the first author by her dancer name, Sabrina.

We realize the bias inherent in using introspection primarily from one source. For example, Sabrina, more than most dancers, tended to attract customers interested in mental stimulation as well as physical turn-on. Yet we could not have gained an in-depth understanding of intimate exchange, for example during table dances, in any other way. To understand this bias, we compared Sabrina's strategies and experiences with those of other dancers we observed and other bar participants with whom we talked. Later in 1987, we conducted interviews with four strippers, eight customers, four managers, three bar owners, and a law officer. This article then uses a triangulated method (Denzin, 1978; Webb et al., 1965) to present typical responses from field work and in-depth ones from current and retrospective introspection.

SETTING

An exotic dance club located in the Tampa Bay area of Florida provided the setting for this study. Since liquor was served, full nudity was prohibited by state law. Appearing individually in full costume on stage, each stripper gradually removed her clothing during a dance routine. By the end of the act, the dancer wore pasties that concealed her nipples and panties that covered genitals, pubic hair, and the cheeks of her derriere. Men handed out tips to dancers during performances.

Between acts, dancers strolled around the floor, making themselves available to spend time with customers. They made money if customers bought them drinks. However, the main attraction and source of income in this bar was the table dance. A dancer "sold" dances in a complicated negotiation process through which she convinced the client that he was turned on to her and/or that she was turned on to him. At the same time, she controlled the situation so that she was not caught disobeying "house" rules, many of which corresponded to what county authorities considered illegal. For example, since "charging" for a table dance was considered soliciting, the dancer, using word games similar to those used by the masseuse studied by Rasmussen and Kuhn (1976), suggested that there was "generally a contribution of $5."

After a dancer successfully sold a dance, she led her customer to one of the two elevated corners of the bar, known generically as the "The Pit," and affectionately nicknamed by customers as "Horny Holler" and "The Passion Pit." Railings and dim lights offered an artificial boundary between this area and the rest of the bar. Clothed in a bralike top and full panties or other revealing costume, the dancer leaned over a seated patron, her legs inside his, and swayed suggestively in rhythm to the music playing in the bar. Theoretically, customers were allowed to touch only the hips, waist, back, and outside of a dancer's legs. Many men tried and some succeeded in doing more. Disobeying rules prohibiting direct sexual stimulation or touching meant more money for dancers, but it also meant risking that management might reprimand them or that a "customer" would turn out to be an undercover officer or a representative looking for infractions on behalf of club management.

ELEMENTS OF STRATEGY

On the Stage

A dancer used symbols that appealed to her audience. At the same time, these symbols distanced her from customers and denoted that the stage was a performance frame (Goffman, 1974; Mullen, 1985). Her appearance, eye contact, manner, and choice of music made up her main expressive equipment.

Having a "centerfold" figure was an obvious asset for dancers. But the best looking woman did not always make the most money. A dancer's presentation of self was also a crucial factor in a customer's decision to tip her. Similar to strippers described by Gonos (1976) and Robboy (1985), women often portrayed exaggerated stereotypes through their clothing style and movement. For instance, a "vamp style" dancer wore suggestive street clothing such as a leather micro-mini skirt, spike-heeled boots, and a halter-style top while strutting around the stage displaying overt sexual mannerisms such as "flushing" (opening her shirt to reveal her pasty-clad breasts). Others had a "gimmick." For example, one woman was an acrobat; another stood on her head while twirling her large breasts. In contrast, a more sensual dancer dressed in sexy bedroom clothing such as a corset and garters or a teddy, and displayed subtle sensual behavior such as slow undulation of the hips.

A dancer chose symbols that drew a certain type of customer to her. Dressing the part of the vamp, for example, reflected an extroverted attitude that attracted customers out to have a good time. Overtly sexual dancers were more likely to perform sexual favors in the bar or meet a man for sex outside the bar. The sensual presentation of self attracted customers who were interested in a "serious," private interaction. Customers interpreted each dancer's symbols as cues to what it might be like to interact with her or, specifically, to have sex with her. For example, Jim, a regular customer, discussed Samantha, a sensual dancer: "She is nothing to look at. God, she's only twenty-six, and we both know she looks like forty. But the way she moves, man! She promises the moon and stars in bed."

Most dancers used eye contact to "feel out" a patron. Managing frequent eye contact while dancing on stage usually meant a tip for the dancer and made a customer feel as if a dancer was specifically interested in him.

A dancer's first close contact with a customer often occurred while accepting a tip. During the exchange, the dancer formed impressions about how the customer was reacting to her, and the customer decided whether he was attracted to the woman. The customer stood at the side of the stage holding currency, which signaled the dancer that he wanted to tip her. The dancer greeted him while accepting the tip in her garter and said "thanks," perhaps giving him a "special" look.

At this point, a dancer might choose from several courses of action, such as "coming on" to a customer, doting on a customer, and using humor. When dancers "came on" to customers, they grinned, wiggled their breasts, spread their legs, struck their buttocks, suggestively sucked their fingers, talked dirty, or French kissed.

Others, such as the sensual dancer, doted on a customer for a few seconds. She caressed his arm, wrapped her arms around his neck, and smiled while he tipped her. If she felt confident of his interest, typical comments she might make were: "I would love a chance to get to know you," or "I look forward to sitting with you," which meant accompanying him to his table after her stage performance.

Humor was an effective and safe tool for generating a good impression while accepting a tip on stage. Customers generally construed a funny statement made by a dancer as friendly and spontaneous. Often it made a nervous client more at ease. Sabrina noted lines she used: "What's a nice guy like you doing in a dump like this?" or "I bet you'd look better up here than I do."

Familiar with the usual "acts" of dancers, such as coming on and showing phony interest, customers were pleased when they thought a woman had "dropped the routine." Often this meant only that she had staged a less frequently displayed one. A dancer had to be careful not to use the same line more than once on the same person, or let a customer overhear it being used on another man. No matter a customer's taste, he wanted a sincere performance.

Dick, a customer who was feeling jilted one evening, commented to Sabrina: "The thing with that chick, Dana, is that she makes a big deal out of you while she is onstage, but if you watch her real close, you notice she looks at everyone who tips her 'that way.'" Another customer reported he did not like a dancer in the bar named Tammy because she was insincere: "She

frenched me and told me to insert my dollar deeply (in her garter). Now I
ain't stupid. I know a come on like that is a fake."

A dancer's music affected how a customer viewed her. This was reflected
in Tim's comment about Jessica: "That girl has a great body, but every time I
hear her music [heavy metal] I get the creeps thinking about what she must be
like." While most women danced to top-40 music, some used other music to
attract a tip from a particular kind of client. Mae, an older dancer in her late
thirties, played country music and presented herself as a country woman.
Bikers and blue-collar workers were loyal to Mae, expressing sentiments like:
"She's the only *real* woman in the bar."

On the Floor

Offstage, interaction was even more complex. Between stage performances, a
dancer circulated among customers and offered her company. Body language,
expressions, and general appearance helped define each customer's interest in
her and the difficulty of being with him. Once a dancer located an interested
customer and introduced herself, or followed up on a contact made while performing on stage, she then had to convince him that he wanted to spend time
with her. Ordinarily, her eventual goal was to sell a table dance.

CHOOSING A CUSTOMER

The ideal customer had a pleasant disposition, was good looking, had time and
money to spend, and was sitting at one of the tables on the floor. Most customers did not meet all these criteria. Dancers weighed these features for each
customer and also compared them against the circumstances of the evening.
Sabrina often asked herself: "What do I want more right now? Money or
someone nonthreatening to sit with?" Her answer was different depending
upon time of night, how much money she had made already, and how she felt
at the moment. Other dancers made the same calculations. For example, three
hours before the bar closed one night, Naomi said, "I know this guy I'm sitting with doesn't have a lot of money, but I've made my hundred for the
night so I can afford to take it easy." Another time, Vicky said, "God! I know
I should be out there hustling instead of drinking with Jim, but I just can't get
into it. I guess I'll just get fucked-up and blow it off today." Darcy displayed a
more typical attitude, "It's twelve thirty already and I haven't made shit! This
guy I'm sitting with better cough it up or I'm taking off." Negotiations with
oneself and with the customer were always in process. Throughout the interaction, each participant tried to ascertain what she or he was willing to give
and how much could be acquired from the other.

Attractive customers appeared, at first, more appealing. They were pleasant
to look at and the dancer could pretend to be on a date while sitting with
them. But these men seemed to know they were more desirable than others in
the bar and were more likely to bargain with those resources. The end result

was that the dancer spent most of the interaction trying to convince the customer to spend money while he tried to persuade her to go out on a date.

When Sabrina was new to the profession, she decided one evening to sit with a good looking, blonde-haired man. She reported the following:

> I started talking to him and eventually led the conversation to the point where I asked, "Would you care for a table dance?"
>
> "Later," he replied.
>
> I continued to make small talk. "Do you come in here often?"
>
> "I stop in once every few months," he responded.
>
> For the next 15 minutes we covered various topics of conversation such as his job and my schooling. Then I asked him again, "Do you want a table dance?"
>
> "Are you going to go to 'le Bistro' with me tomorrow night?"
>
> "I'll think about it," I responded, in hopes of getting a table dance out of him before I turned him down. "Do you want that table dance?"
>
> "Will you go out with me?" he insisted.
>
> "I'm still considering it," I lied.
>
> We volleyed back and forth for 30 minutes. Finally, he told me, "I don't want a dance. I just want to know if you will go out with me."

This customer was aware that Sabrina would not stay with him unless she thought he might want a dance. Both used strategies and gambled time hoping one would give in to the other's goals. Each lost a bet.

Sometimes customers who were old, heavy, unattractive, or otherwise weak in social resources came into the bar. Many women avoided these men, while others, like Sabrina, realized unattractive men were eager for company and tended to treat a dancer better and spend more money than their more attractive competitors would. With the right strategies, dancers could control these men. For example, a dancer might corner a customer into treating her as he would his granddaughter by acting polite and addressing him as "sir." This insinuated that, of course, he would never act inappropriately. Some accepted the role to such an extent that they acted like grandfathers. One man told Scarlet that she was cute, tweaked her cheek, and compared her to his granddaughter.

When scanning the bar and deciding whom to approach first, a dancer tried to find the man who appeared to have the most money. Logically, the better a customer was dressed, the more likely he was to have money. However, he also had a higher probability of already being in the company of another dancer.

Making sure a customer was not spoken for by another dancer was important. It was considered dangerous (one could get into an argument) and rude to sit with another dancer's customer. Some regular customers, for instance, visited the bar to see particular dancers. These customers often turned down another dancer's offer of company by saying they were "waiting for someone." When a dancer entered the bar, she immediately scanned the room, paying particular attention to which women were seated with which customers. If she noticed later that a woman had left a table for a long period of time, she then asked her if it was okay to sit with that customer. This served the dual purpose of following tacit rules (i.e., being polite) and gave the dancer an opportunity to gather information about the customer in question.

Sabrina was warned about a customer in this manner. Upon asking Debbie if she was finished with "the old man in the corner wearing a hat," Debbie replied, "Sure, you can have him. That's 'Merv the perv.' He has lots of money, but he'll want to stick his finger up your asshole for twenty bucks a feel."

A dancer might ignore all other customers to sit with one of her "regulars." When two or more of her regulars were in the bar, she had to juggle them, first sitting with one and then the other. It was difficult to table dance for both of them and still portray "special attachment." Eventually, she had to offer an account (Scott and Lyman, 1968) to one of them. One excuse was to appeal to the principle of fairness: "I really want to be with you, but he came in first and now I have to be with him." Or she might appeal to higher loyalties (Sykes and Matza, 1957), insinuating that the decision was out of her control: "I have to go sit with another customer now. My bosses know I avoid him and they're watching me."

Time in the bar correlated with decreased spending. If a customer had been spending for a while, it was fair to assume that he would run out of money or would soon decide to leave, that is, unless he was intoxicated and freely using a credit card. Dancers in this situation risked having to deal with and control a problematic person who did not remember or pay for the correct number of dances purchased. On the other hand, a dancer might convince a drunk credit card customer to pay for more dances than he actually bought.

A customer's location in the bar indicated his attitude toward female company. In this club, sitting at the bar meant little interest in interacting with dancers. Patrons near the stage wanted to see the show. Being seated at one of the tables in the floor area was conducive to interaction with dancers and to inquiries about table dances.

AT THE TABLES

Once a customer accepted an offer of company, a dancer sat with him and introduced herself. Her overall goal remained fairly consistent—money with no hassle. Many women also enjoyed the attention they received and got an exhibitionist thrill out of being desired and told how beautiful they were. Others believed the compliments were just part of the game. Some liked the feeling of conquering and being in control. Others felt degraded and out of control.

The customer's manifest goal was impersonal, sexual turn-ons for money; a close examination showed other objectives that shadowboxed with and sometimes transcended this more obvious goal. Although most customers initially focused on the pursuit of sex in or outside the bar, they also came looking for a party, to feel good about themselves, to find a friend or companion, or to develop a relationship. A dancer's strategies varied depending on her personality and her perception of the customer.

Some women said nothing. A customer who wanted passive indifference from an attractive female willing to turn him on liked this approach. Sex, not conversation, was his goal. The dancer did not have to initiate activity nor get to know the customer. Her role was to respond as a sexual nonperson by allowing him to kiss and fondle her body. Verbal interaction potentially endangered the continuance of the exchange.

Most customers wanted a dancer to interact with them. Seduction rhetoric (Rasmussen and Kuhn, 1976) became part of the dancer's sexual foreplay before the table dance as well as a vehicle for the customer to persuade the dancer to see him outside the bar. By talking "dirty" and acting "like a whore"—for example, telling stories about kinky sex in her life outside the bar—a dancer could keep a customer "going," eager to buy the next dance, ready to believe the dancer might have sex with him later.

If a customer wanted a prostitute, he dropped hints such as, "Do you do work on the side?" or "Where does a guy go for a good time around here?" or "Do you date?" Sometimes he propositioned outright: "Will you go to bed with me for a hundred dollars?" The more blatant proposals told the dancer that the customer was not a police officer; all of the requests informed her he had money to spend and opened up the possibility of using strategies to extract it.

One strategy dancers used in this situation was to mislead a customer into thinking she might meet him later if he bought table dances from her now. From the first author:

> Ted bought dances from me two at a time. "After several of these, he asked, "Are you going to see me at the Holiday Inn tonight?"
> "Why should I?" I responded.
> "Because I am new in town and have lots of money."
> "I don't go out with strange men," I said.
> "Well, why don't you get to know me then," he said. He bought two more dances, then asked, "Do you know me now?" I smiled at him.
> He continued, "Why don't you meet me after you're done working. What time do you get off?"
> In an effort to shift the focus of the conversation, I said suggestively, "When do you get off?"
> "I get off on you baby!" He exclaimed. "I'm in room 207. Will you be there?"
> To keep him going while not committing myself, I said, "I don't know."
> We talked a while, and then he asked again. I replied, "I've never turned a trick in my life. I'm not sure I ever will."

"So we won't do it for money," he said. "Come see me tonight." He buys two more dances and we sit down again. I start the conversation first this time to keep him interested yet deter him from bringing up my meeting him. "Tell me, Ted, what is the kinkiest thing you have ever done in bed." This conversation kept us busy for a while, until, sixty dollars later, he asks, "Do I go to the bank machine or not?"

"What do you mean?" I ask.

"If you are going to see me tonight, I need to go to the teller. I'm out of money."

I had a big grin on my face and asked, "Will you be back here after the teller?"

"Probably not," he replied.

"Too bad," I said.

"Would you see me if I bought more dances?" he asked. I was tempted to say maybe, but I thought at this point I was being too obvious.

"Probably not," I said.

He stood to leave. "You show up tonight at room 207 if you want. It was fun."

Similar to the strippers discussed by Prus and Irini (1980), a few women used the bar setting as a place to make contacts for their prostitution careers, while many more had sex occasionally outside the bar to augment their incomes. Before accepting an offer, a woman usually asked other dancers about the customer or spent time getting to know him. Interacting with him then gave her an opportunity to make money table dancing. Most women claimed they had sex "only for the money." A few, such as Sasha, seemed to enjoy sexual contact in and out of the bar. Sasha's enthusiasm—"I'm so horny, I want a cock tonight"—was deemed deviant by the other dancers, who ostracized her—usually avoided her and talked behind her back—for her overt enjoyment.

The customer who wanted a date outside the bar could be handled in a similar manner to the customer looking for a prostitute. Often a dancer conveyed the impression, "if only I knew you were safe" by saying: "You could be Jack-the-Ripper," "You could be a cop," "It's not safe to date everyone you meet in here." Then she suggested interest by saying, "I need a chance to get to know you better." The logical way for a dancer to get to know the customer was for him to spend time and money buying drinks and table dances from her. Lured by the offer of expensive dinners or vacations, and sometimes attracted by a man she liked, most dancers occasionally accepted dates.

If customers were in the bar "to party" (to be entertained) in groups, such as bachelor parties, a dancer wasted no time on interaction. She asked immediately if they wanted a dance. These men interacted mostly with each other, requiring dancers to be lively and entertaining hostesses while treating them like sex objects. Often they commented on her body—her big tits, nice ass, or ugly face—as though she were not there. Party groups purchased dances with the same attitude and frequency as they bought rounds of drinks.

Most men who came to the bar seemed to want to find a friend or companion, or in some other way be treated as a special person. One of Sabrina's customers left the bar twice during an evening to change shirts, just to see if she recognized him when he returned. The best ploy in this situation was for the dancer to put on an honest front, altercasting (Weinstein and Deutschberger, 1963) her customer into the role of being special and "different" from other men.

Most successful dancers were able to hold conversations with these men. Asking his name, where he lived, occupation, and what he did with his spare time provided initial interaction. Finding common ground helped conversation run smoothly. Asking questions at a leisurely pace, making comments, and showing interest both verbally and nonverbally afforded a semblance of credibility to the conversational process. This dialogue helped the dancer to "check out" (Rasmussen and Kuhn, 1976) the customer to make sure he was not a police officer, determine how much money he had to spend and which of her interactional strategies might make him willing to part with it. Giving the customer an opportunity to talk about himself and to demonstrate whatever expertise he had made him feel good about himself. A customer pleased with his presentation of self was more apt to spend money. Sabrina told this story:

> In the field, I had a regular customer, Ray, who was a systems' analyst. I was shopping for a computer at the time, so I enlisted Ray's assistance. Ray had an opportunity to show off his expertise, and feel like he was helping. He turned-on to the contrast of seeing me as intellectual and a sex object.

The best way for a dancer to convince a customer that she found him appealing and unique was to find a likable characteristic about the customer and continually tell him how impressed she was with him and with that trait. For example, some men liked to be praised for their appearance, success, intelligence, sexual desirability, trustworthiness, or sensitivity. The dancer had to convey to him directly that she preferred his company to others in the bar, or indirectly through such statements as "You're not as vulgar as the rest of these guys in here"; "You're more intelligent than most men I meet in here"; "You're not just another one of these assholes," or "I appreciate your spending time with me. When I'm sitting with you I'm safe from those animals out there." The message was that because of his specialness, she could be "straight" with him, be who she really was, instead of putting on one of her usual acts.

This tactic worked best with customers the dancer liked and enjoyed talking to; otherwise, it was difficult to muster up and maintain the sincerity necessary for a believable performance. When this strategy worked, the dancer had close to total control of the interaction. Then the customer tried hard to meet the dancer's expectations, spending money and treating her like a date or friend to avoid disappointing her. If he stopped spending money, the dancer might say, and sometimes mean, "I'll see you later. Don't get angry with me. I know you understand that I have to make money, although I would rather spend time with you. If I don't find anything, I'll come back and visit." Sometimes the customer responded by spending more money to keep the dancer around. If not, he was forced to "understand" her leaving because he and the dancer had an honest relationship and she had been "straight" with him about the nature of her job. This strategy was an effective way to cultivate regular customers.

Sometimes a dancer did not have anything in common with a customer. Over time, most dancers worked up routine questions to keep conversation flowing. Sabrina frequently used lines such as: "What do you look for in a woman?" "Why do you visit strip bars?" "What is your opinion of that dancer over there?" "I try," she said, "to get the customer to share something personal with me. I like for him to feel like there is something more solid than a salesperson–customer relationship."

Some regular customers acted as if they were involved in a long-term, serious relationship with a dancer. They bought her expensive gifts such as diamonds, minks, cars, and flowers. These customers seemed to forget the businesslike nature of the bar setting.

Dancers in these interactions appeared involved with the customers. However, most did not take the relationship outside the bar, since this would have cut off a source of income. But they tried to convince the men of their desire to leave the bar scene and be saved by them, even though it was impossible now. Sabrina, for example, had many offers from men who wanted to rescue her from the bar. She developed a routine to solicit this desire from men—it usually meant more money for her in the bar—but that allowed her to reject their proposals without causing anger. She explained:

> I presented myself as attractive and intelligent, but helpless, trapped by circumstances. When they asked me to leave the bar, I told them I had to work to pay for school. When they suggested setting me up in a place of my own, I told them I was independent and wanted to do it on my own. This put them off, but kept them interested and earned their respect.

Mae, a dancer mentioned earlier, seemed to have a knack for cultivating these types of relations. Sabrina describes a discussion with Mae while sharing a ride home.

> Mae had been given a mink coat that night by a customer and she had given it back to him.
> Always intimidated by this woman, I took a moment to get up some nerve and finally asked, "Why did you give back the mink?" "I couldn't hock it for very much, and I won't use it here in Florida. I'd rather get money," she stated.
> "How are you going to get money?" I asked.
> "I'll get more money from him by being the type of person who gives this stuff back than if I keep it. I have lots of customers who give me nicer stuff than that mink."

She spoke to the driver, "Hey, do you remember that necklace Tom gave me?"

The driver replied, "It's true, Mae can really get them going. That necklace was a grand, easy."

"Did you keep the necklace?" I asked.

"Hell yes!" she responded.

Mae had a routine that could "really get them going." But she and other dancers, usually the older ones, who used this technique often, took some aspect of the relationship seriously. They saw these men as "options" or possibilities for a life change. On the other hand, they felt this was too good to be true, or were unsure about making the change because of other factors in their lives, such as a husband or children. Keeping the interaction going, yet not allowing it to take place outside the bar, meant they were able to have romance, feel appreciated, and, to some extent, have a relationship while they continued making money in their occupations. However, the occasional relationship that did work out in the bar kept everyone hoping. Sabrina, for example, met her husband there.

CLOSING THE SALE

A dancer rapidly closed a sale on a table dance to a man who wanted sexual favors in the bar. But since these men often violated rules regarding touching and sexual stimulation, some dancers did not feel that they were worth the trouble. For example, one night Annette came into the dressing room and announced, "I just left this old geezer who wanted me to rub him off with my knee. I'm not into it. If someone else wants to, go for it."

The same problems existed after a quick sale to men in the bar for a party. In this situation, a dancer had to concentrate on not acting offended long enough to perform table dances and collect her money. For some dancers, the money was not worth the degradation. As a result, they avoided the bachelor parties.

The customer who wanted to be treated as special took more time. Questioning allowed time for the dancer to convince him that he wanted a table dance from her. It was important that she not appear pushy, yet she needed to determine quickly whether she could make money from this person. Would he buy table dances? Did he want to spend time getting to know a dancer or go directly to a dance? Answers to such questions guided the dancer in constructing her behavior toward the customer.

If a customer purchased a drink for a dancer, she then knew that he was interested enough to spend some time with her. Some customers, however, bought drinks for dancers but refused to purchase table dances, claiming table dances got them "worked up for nothing." If a customer acknowledged that right away, a dancer then had to make a decision about staying or leaving based on the availability of other moneymaking opportunities in the bar. If the action in the club was slow, she might stay with him since she made $1 on every drink he bought for her. Regular customers were always good for a drink: "I'll go sit with Jim today," said Sharon. "At least I know he'll buy me a drink if nothing else." Often a dancer gave the waitress a secret signal indicating that no liquor should be put in her glass. The waitress brought the drink in a special glass, placed a dollar under the dancer's napkin and the drink on top of it.

Most women closed on a dance after the first drink had arrived and it was apparent that the customer liked her. If the customer said no, most dancers left fairly quickly. But in rare cases a customer paid $50–$100 for a dancer to sit with him for a while. This guaranteed the dancer money without trouble and bought the customer companionship. Customers who saw themselves in an involved relationship with a dancer generally rejected table dances in favor of company. When these customers bought dances they treated the dancer gently, barely touching her for fear of offending her.

Even when a dancer was not paid for her company, it was not always a good idea for her to leave immediately when a man refused a table dance. As a rare and novel routine, staying made the dancer appear sincere in her interest and less concerned about making money. Sabrina occasionally used this approach:

> "Why are you still sitting here?" the customer asked immediately after he had turned me down for a table dance.
>
> "I'm finishing my drink," I replied.
>
> "Then you are leaving?" he asked.
>
> "Oh, sir, I had no idea you wanted me to go. You must be waiting for someone. Forgive me for being so rude," I said tongue in cheek. I stood to leave.
>
> "Hold it, hold it. Sit back down. I don't necessarily want you to leave. The girls always leave after you say no to a dance. You must be new here. You really should leave when customers say no. You won't make any money this way." During this exchange he was clutching my arm. He loosened his grip. "Wouldn't that be rude to just up and walk off?" I asked incredulously. He stares at me a minute, and then smiles. "Lady," he says. "You are a card. I want a table dance." He bought four.

In the Pits

Once a customer agreed to a table dance, another set of complex exchanges took place. Although interaction varied with the particular dancer and customer, common routines offered promise of what was to come. Leading the customer to the pit, one acrobatic dancer followed a routine of bending from the waist and peering at her customer from between straight legs. Ascending the stairs to the pit, she performed various kicks and other gestures to demonstrate her flexibility. Another dancer sashayed gracefully in an elegant and poised, yet seductive, manner. Sabrina's style was to talk in a sexy way as she walked: "See that corner. That's my corner. I love to take my men there."

Once in the pit, a woman sat close to the man. Often she put her hand on his leg, draped an arm on his shoulder, or swung a leg over his lap. Some girls necked with their customers, French kissing with a frenzied passion. Other dancers allowed kisses only on the cheek.

If a customer tried to French kiss when a dancer did not want it, she had several "routines" to control him. Leveling a questioning look at the customer and then backing away from him was enough to stop most men. When a client voiced dissatisfaction over the limitation—"What did you do that for?" or "What's your problem? Why are you so cold?"—it usually indicated an aggressive and potentially problematic customer. Sabrina's response to this was, "Imagine if I kissed every guy in the bar like that before I kissed you. Would that be a turn-on for you?" Most customers backed off then with comments such as, "You're absolutely right. I never thought of that before." By their continuous attempts, however, it was apparent that some were being insincere, assuming, like the dancer, that if they moved more slowly, they would get more of what they wanted. But sometimes the restriction reflected positively on the customer's impression of the dancer. One customer stated to Sabrina after she used this routine: "You have a lot of respect for yourself. I like that."

While some women danced immediately, many waited one or two songs before actually starting a table dance. Sabrina noted that she rarely danced on the first available song because it gave off the impression that she was just interested in making money quickly. She preferred to sit with a customer for a while, talk, drink, and get to know him better. This created a sexual or intimate atmosphere and convinced him that she liked spending time with him. Often this cultivated customers who were likely to buy a greater number of dances, and return to visit her later.

At the beginning of a new song, a dancer might say: "Would you like that table dance now?" or "Let's go for it, baby," depending on the type of interaction in which they were involved. Sexually oriented behavior on the part of the customer called for aggressive behavior from the dancer; less sexually overt actions required more subtle requests.

TABLE DANCES

Strategy became important during a table dance; close quarters meant a dancer's presentation could be difficult to maintain and a customer hard to control. Normally, a dancer attempted to maintain eye contact with a patron, operating on the premise that it demonstrated interest and that if he had his eyes on her, he wouldn't have his hands on her as much. Sabrina hypothesized that a customer confronting a dancer's eyes was forced to acknowledge her "personhood," and that he then was less likely to violate it. Another impression given off (Goffman, 1959) by the dancer's body language was that the intimate exchange demonstrated by this eye contact might be impinged upon by the customer's groping at her body. Sometimes eye contact was difficult if a customer caused the dancer to laugh or feel disgusted (for example, if he was ugly or panting). In this situation, a dancer could turn away from him and make an impersonal shaking of her derriere part of her dance.

Sexual activity was illegal during table dances, but it sometimes occurred. Customers and dancers acknowledged that "hand jobs," oral sex, and intercourse happened, although infrequently. Once a customer requested that Sabrina wear a long skirt during a table dance so that intercourse could take place unobserved.

More common were body-to-penis friction and masturbation. The most frequent form consisted of the customer sliding down to the end of his seat, spreading his legs, and pulling the dancer in close to him where she could then use her knees discreetly to rub his genitals while she danced. Customers sometimes wore shorts without underwear to allow their genitals to hang out the side, or they unzipped their pants to bare their genitals, or masturbated themselves by hand while watching the dancer.

If a customer insisted on violating rules—putting his fingers inside the dancer's briefs or touching her breasts—a dancer might dance much faster than normal, or sway quickly side to side, to escape the wandering hands. If he was insistent, a dancer might grab his wrists teasingly, but firmly, and say, "No, no," addressing him as if he were a misbehaving child.

These attempts to control the customer could not be too aggressive at the outset, or the customer would be turned off. A subtle game was being played: The customer attempted to get the dancer to go as far as she would, and bend the rules, without antagonizing her so much that she stopped dancing; the dancer attempted to keep him in line, but in such a way that he still wanted to buy dances from her. A particularly good strategy at this point was for the dancer to make it look as if she was interested in what he wanted to do, but, because of management, was unable to oblige him: "Look, this would be fine, but I'm going to get in trouble with management. They're going to catch us if you keep acting like this." This disclaimer (Hewitt and Stokes, 1975) shifted the focus of the patron's annoyance to management and away from her and reasserted the idea that this was a respectable occupation with rules (see Hong et al., 1975).

If a man continued to act inappropriately, the dancer most likely lost her money and the negotiation process broke down. If the customer did not pay after the dance, the dancer had no recourse. Her only power was her seductiveness or ability to persuade the customer subtly that he "owed" it to her. Fights between customers and dancers started occasionally because a man did not want to pay a woman who "didn't give him a good dance." Management quickly squelched these and fired or fined dancers who were involved.

Most dances, however, were successful. After one of these, a dancer might give the customer a reward for "being good." Sabrina reported that she kissed the customer, closed mouthed, on the cheek or on the corner of his mouth. By gently resting her fingers on his chin, tilting up his head, and delivering a kiss, she left the impression, "I'm involved with you. I like you."

After a table dance had been completed, the next goal was to keep the interaction going so that the customer would buy more dances. If a customer continued to hold onto a dancer after the song ended, it usually signaled that he wanted her to dance through the next song. If he let her go, a dancer might look inquisitively at the customer and ask, "Is that all for now? Do you want to continue?" or "Will you want a dance later?" The questions asked depended on the dancer's impression of how involved the customer was with the dance. At the least, she encouraged him to look her up the next time he returned to the bar.

EXCHANGE FROM THE BOTTOM UP

Interaction in strip bars reflects negotiation in "respectable" society. What is being exchanged—economic resources for sexual titillation, ego gratification, and submission—is viewed in our society as honorable (Lasch, 1977; Lipman-Blumen, 1984; Safilios-Rothschild, 1977). The strategies dancers use to sell their product are similar to those used by sellers in reputable service occupations (Bigus, 1972; Browne, 1973; Davis, 1959; Henslin, 1968; Prus, 1987; Katovich and Diamond, 1986). Unlike many deviant sales (Luckenbill, 1984), dancers and customers normally are protected by a structured, bureaucratic setting with formal rules.

Interaction in a strip club represents negotiation in a buyer's market: sexual turn-on is available for the asking. Although men show some interest in being customers simply by walking through the door, they must be persuaded to "buy" from a certain dancer. To establish control, women use facilitating (Prus, 1987) or cultivating techniques (Bigus, 1972), much like those used by service workers trying to sell a product directly to a client. To acquire customers, a dancer must develop mutual trust. The most important weapon in the arsenal of interaction is to present oneself as sincere: be warm and imply realness, appear spontaneous, give out insider information to show loyalty, accentuate honesty, demonstrate that one is different from others in similar positions, or tell hard-luck stories. At the same time, a dancer must attempt to determine the trustworthiness of her customer: Will he pay for the dance, and will he hassle her later?

Once trust is established, the dancer must promote repeat patronage and customer loyalty (Prus, 1987). This is done by calling on the norm of reciprocity (Gouldner, 1960). The expectation is that the customer will repay friendship, special attention, and favors with money. Thus a hard sell often is not as productive as other more indirect techniques, such as taking personal interest in customers (Prus, 1987), nurturing pseudo friendships, or effecting obligation (Bigus, 1972). Much like any business relationship, the seller must gauge time spent in an encounter to pay-off potential.

Interaction in the bar also reflects power dynamics in mainstream society. As a subordinate group, women in general have responded to men's macro-manipulation of societal institutions by using micromanipulation—interpersonal behaviors and practices—to influence the power balance (Lipman-Blumen, 1984). Women in the bar play a game that they know well; in some form, they have been forced to play it for years. They are accustomed to anticipating male behavior, pleasing and charming men, appearing to be what they want, and following their rules. At the same time, dancers are skilled at manipulating to get their own needs met. The bar is a haven for them; they are old hands.

Women who dance for a living have fewer resources or opportunities to manipulate the macrostructure than do most women. Many come from broken homes where fathers often were absent. They frequently had distant relations with parents and left home at an early age. They had sexual experience earlier than other females had. Financial crisis often served as the impetus for starting this occupation. Few have sufficient training or education to make as much money in other occupations (Carey et al., 1974; Skipper and McCaghy, 1971).

Although dancers often have few resources, they are used to taking care of themselves. The occupation of stripping demands that they be tough. It provides them with a context of control. Being the purveyors and gatekeepers of sexuality has always provided powerful control for women (Safilios-Rothschild, 1977); it served this function even more for those women who make sexual turn-on into an occupation.

In male-female relationships, sex is "shrouded in romantic mystique" (Salutin, 1971). It has been okay for women to exchange sex for financial security (Salutin, 1971), as long as they confined the exchange to the context of love and marriage (Safilios-Rothschild, 1977). On this level, the activity in the bar is deviant. There this shroud is removed, revealing the rawness of the exchange, the unequal distribution of macropower, and the often cold, calculating nature of the microstrategies. There, sexuality is carried out in public between people who are often strangers. The dancers use sex as a direct currency of exchange: turn-ons for money. They are not likely to have illusions of love. For them, this is a job. When they are tempted to redefine the situation, their histories with men or the realities of their lives remind them otherwise.

For some dancers then, there is a feeling that they have won the ultimate game in American society, which continues to judge the value of women by their attractiveness and seductiveness (Chernin, 1982). Dancers get validation, attention, and money for displaying these characteristics and argue that they are doing nothing more than most women do, not as much as some.

Yet, this world is not a haven for women. If they could make the same money and have the same freedom in another occupation, most dancers would pursue an alternative to table dancing, but they cannot (Prus and Irini, 1980; Robboy, 1985). Most also have internalized "honorable" exchange, and, without the shroud of romance, outright trading of their bodies some-times breaks through as degrading (Prus and Irini, 1980; Salutin, 1971; Skipper and McCaghy, 1971). They suffer identity problems as they take on the negative attitudes of mainstream society toward their occupation (Rambo [Ronai], 1987; Skipper and McCaghy, 1970; 1971). Many are disillusioned with males to the point that they characterize their audience as degenerates (McCaghy and Skipper, 1969), yet these same degenerates decide their take-home pay.

The negotiation process we have described then is a case study of exchange between those differentially empowered. As in other occupations in which a person's job requires emotion management, stripping has high emo-tional costs (Hochschild, 1983). Stripping, as a service occupation, pays well, but costs dearly.

REFERENCES

Adler, P. A., and P. Adler (1987) Membership Roles in Field Research. Newbury Park, CA: Sage.

Bigus, O. (1972) "The milkman and his customer: a cultivated relationship." Urban Life and Culture 1:131–165.

Blumer, H. (1969) Symbolic Interactionism: Perspective and Method. Englewood Cliffs, NJ: Prentice-Hall.

Boles, J., and A. P. Garbin (1974) "The strip club and customer-stripper pat-terns of interaction." Sociology and Social Research 58:136–144.

Browne, J. (1973) The Used-Car Game: A Sociology of the Bargain. Lexington, MA: Lexington Books.

Bulmer, M. (1982) "When is disguise justified? alternatives to covert partici-pant observations." Qualitative Sociology 5:251–264.

Carey, S. H., R. A. Peterson, and L. K. Sharpe (1974) "A study of recruit-ment and socialization in two deviant female occupations." Soc. Symposium 11:11–24.

Chernin, K. (1982) The Obsession: Reflections on the Tyranny of Slenderness. New York: Harper Collophon.

Davis, F. (1959) "The cab driver and his fare: facets of a fleeting relationship." Amer. J. of Sociology 65:158–165.

Denzin, N. K. (1978) The Research Act. New York: McGraw-Hill.

Ellis, C. (1988) "Keeping emotions in the sociology of emotions." University of South Florida. (unpublished)

———. (forthcoming) "Sociological introspection and emotional experience." Symbolic Interaction 13.1.

Foote, N. N. (1954) "Sex as play." Social Problems 1:159–163.

Goffman, E. (1959) The Presentation of Self in Everyday Life. Garden City, NY: Doubleday.

———. (1974) Frame Analysis: An Essay on the Organization of Experience. Cambridge, MA: Harvard Univ. Press.

Gonos, G. (1976) "Go-Go dancing: a comparative frame analysis." Urban Life 9:189–219.

Gouldner, A. (1960) "The norm of reciprocity." Amer. Soc. Rev. 25:161–178.

Henslin, J. (1968) "Trust and the cab driver," pp. 138-155 in M. Truzzi (ed.) Sociology and Everyday Life. Englewood Cliffs, NJ: Prentice-Hall.

Hewitt, J., and R. Stokes (1975) "Disclaimers." Amer. Soc. Rev. 40:1–11.

Hochschild, A. (1983) The Managed Heart: Commercialization of Human Feeling. Berkeley: Univ. of California Press.

Hong, L. K., W. Darrough, and R. Duff (1975) "The sensuous rip-off: consumer fraud turns blue." Urban Life and Culture 3:464–470

Jorgensen, D. L. (1989) Participant Observation. Newbury Park, CA: Sage.

Katovich, M. A., and R. L. Diamond (1986) "Selling time: situated transactions in a noninstitutional environment." Soc. Q. 27:253–271.

Lasch, C. (1977) Haven in a Heartless World. New York: Basic Books.

Lipman-Blumen, J. (1984) Gender Roles and Power. Englewood Cliffs, NJ: Prentice-Hall .

Luckenbill, D. F. (1984) "Dynamics of the deviant sale." Deviant Behavior 5:337–353.

McCaghy, C. H., and J. K. Skipper (1969) "Lesbian behavior as an adaptation to the occupation of stripping." Social Problems 17:262–270.

———. (1972) "Stripping: anatomy of a deviant life style," pp. 362-373 in S. D. Feldman and G. W. Thielbar (eds.) Life Styles: Diversity in American Society. Boston: Little, Brown.

Mehan, H., and H. Wood (1975) The Reality of Ethnomethodology. New York: John Wiley.

Mullen, K. (1985) "The impure performance frame of the public house entertainer." Urban Life 14:181–203

Prus, R. (1987) "Developing loyalty: fostering purchasing relationships in the marketplace." Urban Life 15:331–366.

Prus, R., and S. Irini (1980) Hookers, Rounders, and Desk Clerks: The Social Organization of the Hotel Community. Salem, WI: Sheffield.

Rambo (Ronai), C. (1987) "Negotiation strategies and emotion work of the stripper." University of South Florida. (unpublished)

Rasmussen, P., and L. Kuhn (1976) "The new masseuse: play for pay." Urban Life 5:271–292.

Riemer, J. W. (1977) "Varieties of opportunistic research." Urban Life 5:467–477.

Robboy, H. (1985) "Emotional labor and sexual exploitation in an occupational role." Presented at the annual meetings of the Mid South Sociological Society, Little Rock, AK.

Safilios-Rothschild, C. (1977) Love, Sex, and Sex Roles. Englewood Cliffs, NJ: Prentice-Hall.

Salutin, M. (1971) "Stripper morality." Transaction 8:12–22.

Scott, M. B., and S. M. Lyman (1968) "Accounts." Amer. Soc. Rev. 33:46–62.

Skipper, J. K., and C. H. McCaghy (1970) "Stripteasers: the anatomy and career contingencies of a deviant occupation." Social Problems 17: 391–405.

———. (1971) "Stripteasing: a sex oriented occupation," pp. 275–296 in J. Henslin (ed.) The Sociology of Sex. New York: Appleton Century Crofts.

Sykes, G., and D. Matza (1957) "Techniques of neutralization: a theory of delinquency." Amer. Soc. Rev. 22:664–670.

Turner, R. (1962) "Role-taking: process versus conformity," pp. 20–40 in A. M. Rose (ed.) Human Behavior and Social Process. Boston: Houghton Mifflin.

Webb, E. J., D. T. Campbell, R. D. Schwartz, and L. Sechrest (1965) Unobtrusive Measures. Chicago: Rand McNally.

Weinstein, Eugene A., and Paul Deutschberger (1963) "Some dimensions of altercasting." Sociometry 26:454–466.

32

Fraternities and Rape on Campus

PATRICIA YANCEY MARTIN

AND ROBERT A. HUMMER

Rapes are perpetrated on dates, at parties, in chance encounters, and in specially planned circumstances. That group structure and processes, rather than individual values or characteristics, are the impetus for many rape episodes was documented by Blanchard (1959) 30 years ago (also see Geis 1971), yet sociologists have failed to pursue this theme (for an exception, see Chancer 1987). A recent review of research (Muehlenhard and Linton 1987) on sexual violence, or rape, devotes only a few pages to the situational contexts of rape events, and these are conceptualized as potential risk factors for individuals rather than qualities of rape-prone social contexts.

Many rapes, far more than come to the public's attention, occur in fraternity houses on college and university campuses, yet little research has analyzed fraternities at American colleges and universities as rape-prone contexts (cf. Ehrhart and Sandler 1985). Most of the research on fraternities reports on samples of individual fraternity men. One group of studies compares the values, attitudes, perceptions, family socioeconomic status, psychological traits (aggressiveness, dependence), and so on, of fraternity and nonfraternity men (Bohrnstedt 1969; Fox, Hodge, and Ward 1987; Kanin 1967; Lemire 1979; Miller 1973). A second group attempts to identify the effects of fraternity membership over time on the values, attitudes, beliefs, or moral precepts of members (Hughes and Winston 1987; Marlowe and Auvenshine 1982; Miller 1973; Wilder, Hoyt, Doren, Hauck, and Zettle 1978; Wilder, Hoyt, Surbeck, Wilder, and Carney 1986). With minor exceptions, little research addresses the group and organizational context of fraternities or the social construction of fraternity life (for exceptions, see Letchworth 1969; Longino and Kart 1973; Smith 1964).

Gary Tash, writing as an alumnus and trial attorney in his fraternity's magazine, claims that over 90 percent of all gang rapes on college campuses involve fraternity men (1988, p. 2). Tash provides no evidence to substantiate this claim, but students of violence against women have been concerned with fraternity men's frequently reported involvement in rape episodes (Adams and Abarbanel 1988). Ehrhart and Sandler (1985) identify over 50 cases of gang rapes on campus perpetrated by fraternity men, and their analysis points to many of the conditions that we discuss here. Their analysis is unique in focusing on conditions in fraternities that make gang rapes of women by fraternity

From: "Fraternities and Rape on Campus," Patricia Y. Martin and Robert A. Hummer, *Gender & Society*, Vol. 3, No. 4, 1989. Reprinted by permission of Sage Publications, Inc.

men both feasible and probable. They identify excessive alcohol use, isolation from external monitoring, treatment of women as prey, use of pornography, approval of violence, and excessive concern with competition as precipitating conditions to gang rape (also see Merton 1985; Roark 1987).

The study reported here confirmed and complemented these findings by focusing on both conditions and processes. We examined dynamics associated with the social construction of fraternity life, with a focus on processes that foster the use of coercion, including rape, in fraternity men's relations with women. Our examination of men's social fraternities on college and university campuses as groups and organizations led us to conclude that fraternities are a physical and sociocultural context that encourages the sexual coercion of women. We make no claims that all fraternities are "bad" or that all fraternity men are rapists. Our observations indicated, however, that rape is especially probable in fraternities because of the kinds of organizations they are, the kinds of members they have, the practices their members engage in, and a virtual absence of university or community oversight. Analyses that lay blame for rapes by fraternity men of "peer pressure" are, we feel, overly simplistic (cf. Burkhart 1989; Walsh 1989). We suggest, rather, that fraternities create a sociocultural context in which the use of coercion in sexual relations with women is normative and in which the mechanisms to keep this pattern of behavior in check are minimal at best and absent at worst. We conclude that unless fraternities change in fundamental ways, little improvement can be expected.

METHODOLOGY

Our goal was to analyze the group and organizational practices and conditions that create in fraternities an abusive social context for women. We developed a conceptual framework from an initial case study of an alleged gang rape at Florida State University that involved four fraternity men and an 18-year-old coed. The group rape took place on the third floor of a fraternity house and ended with the "dumping" of the woman in the hallway of a neighboring fraternity house. According to newspaper accounts, the victim's blood-alcohol concentration, when she was discovered, was .349 percent, more than three times the legal limit for automobile driving and an almost lethal amount. One law enforcement officer reported that sexual intercourse occurred during the time the victim was unconscious: "She was in a life-threatening situation" (*Tallahassee Democrat*, 1988b). When the victim was found, she was comatose and had suffered multiple scratches and abrasions. Crude words and a fraternity symbol had been written on her thighs (*Tampa Tribune*, 1988). When law enforcement officials tried to investigate the case, fraternity members refused to cooperate. This led, eventually, to a five-year ban of the fraternity from campus by the university and by the fraternity's national organization.

In trying to understand how such an event could have occurred, and how a group of over 150 members (exact figures are unknown because the fraternity refused to provide a membership roster) could hold rank, deny knowledge of the event, and allegedly lie to a grand jury, we analyzed newspaper articles about the case and conducted open-ended interviews with a variety of respondents about the case and about fraternities, rapes, alcohol use, gender relations, and sexual activities on campus. Our data included over 100 newspaper articles on the initial gang rape case; open-ended interviews with Greek (social fraternity and sorority) and non-Greek (independent) students (N = 20); university administrators (N = 8, five men, three women); and alumni advisers to Greek organizations (N = 6). Open-ended interviews were held also with judges, public and private defense attorneys, victim advocates, and state prosecutors regarding the processing of sexual assault cases. Data were analyzed using the grounded theory method (Glaser 1978; Martin and Turner 1986). In the following analysis, concepts generated from the data analysis are integrated with the literature on men's social fraternities, sexual coercion, and related issues.

FRATERNITIES AND THE SOCIAL CONSTRUCTION OF MEN AND MASCULINITY

Our research indicated that fraternities are vitally concerned—more than with anything else—with masculinity (cf. Kanin 1967). They work hard to create a macho image and context and try to avoid any suggestion of "wimpishness," effeminacy, and homosexuality. Valued members display, or are willing to go along with, a narrow conception of masculinity that stresses competition, athleticism, dominance, winning, conflict, wealth, material possessions, willingness to drink alcohol, and sexual prowess vis-à-vis women.

Valued Qualities of Members

When fraternity members talked about the kind of pledges they prefer, a litany of stereotypical and narrowly masculine attributes and behaviors was recited and feminine or woman-associated qualities and behaviors were expressly denounced (cf. Merton 1985). Fraternities seek men who are "athletic," "big guys," good in intramural competition, "who can talk college sports." Males "who are willing to drink alcohol," "who drink socially," or "who can hold their liquor" are sought. Alcohol and activities associated with the recreational use of alcohol are cornerstones of fraternity social life. Nondrinkers are viewed with skepticism and rarely selected for membership.[1]

Fraternities try to avoid "geeks," nerds, and men said to give the fraternity a "wimpy" or "gay" reputation. Art, music, and humanities majors, majors in traditional women's fields (nursing, home economics, social work, education), men with long hair, and those whose appearance or dress violate current norms are rejected. Clean-cut, handsome men who dress well (are clean, neat, conforming, fashionable) are preferred. One sorority woman commented that "the top ranking fraternities have the best looking guys."

One fraternity man, a senior, said his fraternity recruited "some big guys, very athletic" over a two-year period to help overcome its image of wimpiness. His fraternity had won the interfraternity competition for highest grade-point average several years running but was looked down on as "wimpy, dancy, even gay." With their bigger, more athletic recruits, "our reputation improved; we're a much more recognized fraternity now." Thus a fraternity's reputation and status depends on members' possession of stereotypically masculine qualities. Good grades, campus leadership, and community service are "nice" but masculinity dominance—for example, in athletic events, physical size of members, athleticism of members—counts most.

Certain social skills are valued. Men are sought who "have good personalities," are friendly, and "have the ability to relate to girls" (cf. Longino and Kart 1973). One fraternity man, a junior, said: "We watch a guy [a potential pledge] talk to women . . . we want guys who can relate to girls." Assessing a pledge's ability to talk to women is, in part, a preoccupation with homosexuality and a conscious avoidance of men who seem to have effeminate manners or qualities. If a member is suspected of being gay, he is ostracized and informally drummed out of the fraternity. A fraternity with a reputation as wimpy or tolerant of gays is ridiculed and shunned by other fraternities. Militant heterosexuality is frequently used by men as a strategy to keep each other in line (Kimmel 1987).

Financial affluence or wealth, a male-associated value in American culture, is highly valued by fraternities. In accounting for why the fraternity involved in the gang rape that precipitated our research project had been recognized recently as "the best fraternity chapter in the United States," a university official said: "They were good-looking, a big fraternity, had lots of BMWs [expensive, German-made automobiles]." After the rape, newspaper stories described the fraternity members' affluence, noting the high number of members who owned expensive cars (*St. Petersburg Times*, 1988).

The Status and Norms of Pledgeship

A pledge (sometimes called an associate member) is a new recruit who occupies a trial membership status for a specific period of time. The pledge period (typically ranging from 10 to 15 weeks) gives fraternity brothers an opportunity to assess and socialize new recruits. Pledges evaluate the fraternity also and decide if they want to become brothers. The socialization experience is structured partly through assignment of a Big Brother to each pledge. Big Brothers are expected to teach pledges how to become a brother and to support them as they progress through the trial membership period. Some pledges are repelled by the pledging experience, which can entail physical abuse; harsh discipline; and demands to be subordinate, follow orders, and engage in demeaning routines and activities, similar to those used by the military to "make men out of boys" during boot camp.

Characteristics of the pledge experience are rationalized by fraternity members as necessary to help pledges unite into a group, rely on each other, and join together against outsiders. The process is highly masculinist in execution as well as conception. A willingness to submit to authority, follow orders, and do as one is told is viewed as a sign of loyalty, togetherness, and unity. Fraternity pledges who find the pledge process offensive often drop out. Some do this by openly quitting, which can subject them to ridicule by brothers and other pledges, or they may deliberately fail to make the grades necessary for initiation or transfer schools and decline to reaffiliate with the fraternity on the new campus. One fraternity pledge who quit the fraternity he had pledged described an experience during pledgeship as follows:

> This one guy was always picking on me. No matter what I did, I was wrong. One night after dinner, he and two other guys called me and two other pledges into the chapter room. He said, "Here, X, hold this 25 pound bag of ice at arms' length 'til I tell you to stop." I did it even though my arms and hands were killing me. When I asked if I could stop, he grabbed me around the throat and lifted me off the floor. I thought he would choke me to death. He cussed me and called me all kinds of names. He took one of my fingers and twisted it until it nearly broke. . . . I stayed in the fraternity for a few more days, but then I decided to quit. I hated it. Those guys are sick. They like seeing you suffer.

Fraternities' emphasis on toughness, withstanding pain and humiliation, obedience to superiors, and using physical force to obtain compliance contributes to an interpersonal style that de-emphasizes caring and sensitivity but fosters intragroup trust and loyalty. If the least macho or most critical pledges drop out, those who remain may be more receptive to, and influenced by, masculinist values and practices that encourage the use of force in sexual relations with women and the covering up of such behavior (cf. Kanin 1967).

Norms and Dynamics of Brotherhood

Brother is the status occupied by fraternity men to indicate their relations to each other and their membership in a particular fraternity organization or group. Brother is a male-specific status; only males can become brothers, although women can become "Little Sisters," a form of pseudomembership. "Becoming a brother" is a rite of passage that follows the consistent and often lengthy display by pledges of appropriately masculine qualities and behaviors. Brothers have a quasi-familial relationship with each other, are normatively said to share bonds of closeness and support, and are sharply set off from non-members. Brotherhood is a loosely defined term used to represent the bonds that develop among fraternity members and the obligations and expectations incumbent upon them (cf. Marlowe and Auvenshine [1982] on fraternities' failure to encourage "moral development" in freshman pledges).

Some of our respondents talked about brotherhood in almost reverential terms, viewing it as the most valuable benefit of fraternity membership. One senior, a business-school major who had been affiliated with a fairly high-status fraternity throughout four years on campus, said:

> Brotherhood spurs friendship for life, which I consider its best aspect, although I didn't see it that way when I joined. Brotherhood bonds and unites. It instills values of caring about one another, caring about community, caring about ourselves. The values and bonds [of brotherhood] continually develop over the four years [in college] while normal friendships come and go.

Despite this idealization, most aspects of fraternity practice and conception are more mundane. Brotherhood often plays itself out as an overriding concern with masculinity and, by extension, femininity. As a consequence, fraternities comprise collectivities of highly masculinized men with attitudinal qualities and behavior norms that predispose them to sexual coercion of women (cf. Kanin 1967; Merton 1985; Rapaport and Burkhart 1984). The norms of masculinity are complemented by conceptions of women and femininity that are equally distorted and stereotyped and that may enhance the probability of women's exploitation (cf. Ehrhart and Sandler 1985; Sanday 1981, 1986).

Practices of Brotherhood

Practices associated with fraternity brotherhood that contribute to the sexual coercion of women include a preoccupation with loyalty, group protection and secrecy, use of alcohol as a weapon, involvement in violence and physical force, and an emphasis on competition and superiority.

Loyalty, Group Protection, and Secrecy Loyalty is a fraternity preoccupation. Members are reminded constantly to be loyal to the fraternity and to their brothers. Among other ways, loyalty is played out in the practices of group protection and secrecy. The fraternity must be shielded from criticism. Members are admonished to avoid getting the fraternity in trouble and to bring all problems "to the chapter" (local branch of a national social fraternity) rather than to outsiders. Fraternities try to protect themselves from close scrutiny and criticism by the Interfraternity Council (a quasi-governing body composed of representatives from all social fraternities on campus), their fraternity's national office, university officials, law enforcement, the media, and the public. Protection of the fraternity often takes precedence over what is procedurally, ethically, or legally correct. Numerous examples were related to us of fraternity brothers' lying to outsiders to "protect the fraternity."

Group protection was observed in the alleged gang rape case with which we began our study. Except for one brother, a rapist who turned state's evidence, the entire remaining fraternity membership was accused by university and criminal justice officials of lying to protect the fraternity. Members consistently failed to cooperate even though the alleged crimes were felonies, involved only four men (two of whom were not even members of the local chapter), and the victim of the crime nearly died. According to a grand jury's findings, fraternity officers repeatedly broke appointments with law enforcement officials, refused to provide police with a list of members, and refused to cooperate with police and prosecutors investigating the case (*Florida Flambeau*, 1988).

Secrecy is a priority value and practice in fraternities, partly because full-fledged membership is premised on it (for confirmation, see Ehrhart and Sandler 1985; Longino and Kart 1973; Roark 1987). Secrecy is also a boundary-maintaining mechanism, demarcating in-group from out-group, us from them. Secret rituals, handshakes, and mottoes are revealed to pledge brothers as they are initiated into full brotherhood. Since only brothers are supposed to know a fraternity's secrets, such knowledge affirms membership in the fraternity and separates a brother from others. Extending secrecy tactics from protection of private knowledge to protection of the fraternity from criticism is a predictable development. Our interviews indicated that individual members knew the difference between right and wrong, but fraternity norms that emphasize loyalty, group protection, and secrecy often overrode standards of ethical correctness.

Alcohol as Weapon   Alcohol use by fraternity men is normative. They use it on weekdays to relax after class and on weekends to "get drunk," "get crazy," and "get laid." The use of alcohol to obtain sex from women is pervasive—in other words, it is used as a weapon against sexual reluctance. According to several fraternity men whom we interviewed, alcohol is the major tool used to gain sexual mastery over women (cf. Adams and Abarbanel 1988; Ehrhart and Sandler 1985). One fraternity man, a 21-year-old senior, described alcohol use to gain sex as follows: "There are girls that you know will fuck, then some you have to put some effort into it. . . . You have to buy them drinks or find out if she's drunk enough."

A similar strategy is used collectively. A fraternity man said that at parties with Little Sisters: "We provide them with 'hunch punch' and things get wild. We get them drunk and most of the guys end up with one." "'Hunch punch,'" he said, "is a girls' drink made up of overproof alcohol and powdered Kool-Aid, no water or anything, just ice. It's very strong. Two cups will do a number on a female." He had plans in the next academic term to surreptitiously give hunch punch to women in a "prim and proper" sorority because "having sex with prim and proper sorority girls is definitely a goal." These women are a challenge because they "won't openly consume alcohol and won't get openly drunk as hell." Their sororities have "standards committees" that forbid heavy drinking and easy sex.

In the gang rape case, our sources said that many fraternity men on campus believed the victim had a drinking problem and was thus an "easy make." According to newspaper accounts, she had been drinking alcohol on the evening she was raped; the lead assailant is alleged to have given her a bottle of wine after she arrived at his fraternity house. Portions of the rape occurred in a shower, and the victim was reportedly so drunk that her assailants had difficulty holding her in a standing position (*Tallahassee Democrat*, 1988a). While raping her, her assailants repeatedly told her they were members of another fraternity under the apparent belief that she was too drunk to know the difference. Of course, if she was too drunk to know who they were, she was too drunk to consent to sex (cf. Allgeier 1986; Tash 1988).

One respondent told us that gang rapes are wrong and can get one expelled, but he seemed to see nothing wrong in sexual coercion one-on-one. He seemed unaware that the use of alcohol to obtain sex from a woman is grounds for a claim that a rape occurred (cf. Tash 1988). Few women on campus (who also may not know these grounds) report date rapes, however; so the odds of detection and punishment are slim for fraternity men who use alcohol for "seduction" purposes (cf. Byington and Keeter 1988; Merton 1985).

Violence and Physical Force Fraternity men have a history of violence (Ehrhart and Sandler 1985; Roark 1987). Their record of hazing, fighting, property destruction, and rape has caused them problems with insurance companies (Bradford 1986; Pressley 1987). Two university officials told us that fraternities "are the third riskiest property to insure behind toxic waste dumps and amusement parks." Fraternities are increasingly defendants in legal actions brought by pledges subjected to hazing (Meyer 1986; Pressley 1987) and by women who were raped by one or more members. In a recent alleged gang rape incident at another Florida university, prosecutors failed to file charges but the victim filed a civil suit against the fraternity nevertheless (*Tallahassee Democrat*, 1989).

Competition and Superiority Interfraternity rivalry fosters in-group identification and out-group hostility. Fraternities stress pride of membership and superiority over other fraternities as major goals. Interfraternity rivalries take many forms, including competition for desirable pledges, size of pledge class, size of membership, size and appearance of fraternity house, superiority in intramural sports, highest grade-point averages, giving the best parties, gaining the best or most campus leadership roles, and, of great importance, attracting and displaying "good looking women." Rivalry is particularly intense over members, intramural sports, and women (cf. Messner 1989).

FRATERNITIES'
COMMODIFICATION OF WOMEN

In claiming that women are treated by fraternities as commodities, we mean that fraternities knowingly, and intentionally, *use* women for their benefit. Fraternities use women as bait for new members, as servers of brothers' needs, and as sexual prey.

Women as Bait Fashionably attractive women help a fraternity attract new members. As one fraternity man, a junior, said, "They are good bait." Beautiful, sociable women are believed to impress the right kind of pledges and give the impression that the fraternity can deliver this type of woman to its members. Photographs of shapely, attractive coeds are printed in fraternity brochures and videotapes that are distributed and shown to potential pledges. The women pictured are often dressed in bikinis, at the beach, and are pictured hugging the brothers of the fraternity. One university official says such recruitment materials give the message: "Hey, they're here for you, you can have whatever you want," and, "We have the best looking women. Join us and you can have them too." Another commented: "Something's wrong when males join an all-male organization as the best place to meet women. It's so illogical."

Fraternities compete in promising access to beautiful women. One fraternity man, a senior, commented that "the attraction of girls [i.e., a fraternity's success in attracting women] is a big status symbol for fraternities." One university official commented that the use of women as a recruiting tool is so well entrenched that fraternities that might be willing to forgo it say they cannot afford to unless other fraternities do so as well. One fraternity man said, "Look, if we don't have Little Sisters, the fraternities that do will get all the good pledges." Another said, "We won't have as good a rush [the period during which new members are assessed and selected] if we don't have these women around."

In displaying good-looking, attractive, skimpily dressed, nubile women to potential members, fraternities implicitly, and sometimes explicitly, promise sexual access to women. One fraternity man commented that "part of what being in a fraternity is all about is the sex" and explained how his fraternity uses Little Sisters to recruit new members:

> We'll tell the sweetheart [the fraternity's term for Little Sister], "You're gorgeous; you can get him." We'll tell her to fake a scam and she'll go hang all over him during a rush party, kiss him, and he thinks he's done wonderful and wants to join. The girls think it's great too. It's flattering for them.

Women as Servers The use of women as servers is exemplified in the Little Sister program. Little Sisters are undergraduate women who are rushed and selected in a manner parallel to the recruitment of fraternity men. They are affiliated with the fraternity in a formal but unofficial way and are able, indeed required, to wear the fraternity's Greek letters. Little Sisters are not full-fledged fraternity members, however; and fraternity national offices and most universities do not register or regulate them. Each fraternity has an officer called Little Sister Chairman who oversees their organization and activities. The Little Sisters elect officers among themselves, pay monthly dues to the fraternity, and have well-defined roles. Their dues are used to pay for the fraternity's social events, and Little Sisters are expected to attend and hostess fraternity parties and hang around the house to make it a "nice place to be."

One fraternity man, a senior, described Little Sisters this way: "They are very social girls, willing to join in, be affiliated with the group, devoted to the fraternity." Another member, a sophomore, said: "Their sole purpose is social—attend parties, attract new members, and 'take care' of the guys."

Our observations and interviews suggested that women selected by fraternities as Little Sisters are physically attractive, possess good social skills, and are willing to devote time and energy to the fraternity and its members. One undergraduate woman gave the following job description for Little Sisters to a campus newspaper:

> It's not just making appearances at all the parties but entails many more responsibilities. You're going to be expected to go to all the intramural games to cheer the brothers on, support and encourage the pledges, and just be around to bring some extra life to the house. [As a Little Sister] you have to agree to take on a new responsibility other than studying to maintain your grades and managing to keep your checkbook from bouncing. You have to make time to be a part of the fraternity and support the brothers in all they do. (*The Tomahawk*, 1988)

The title of Little Sister reflects women's subordinate status; fraternity men in a parallel role are called Big Brothers. Big Brothers assist a sorority primarily with the physical work of sorority rushes, which, compared to fraternity rushes, are more formal, structured, and intensive. Sorority rushes take place in the daytime and fraternity rushes at night so fraternity men are free to help. According to one fraternity member, Little Sister status is a benefit to women because it gives them a social outlet and "the protection of the brothers." The gender-stereotypic conceptions and obligations of these Little Sister and Big Brother statuses indicate that fraternities and sororities promote a gender hierarchy on campus that fosters subordination and dependence in women, thus encouraging sexual exploitation and the belief that it is acceptable.

Women as Sexual Prey Little Sisters are a sexual utility. Many Little Sisters do not belong to sororities and lack peer support for refraining from unwanted sexual relations. One fraternity man (whose fraternity has 65 members and 85 Little Sisters) told us they had recruited "wholesale" in the prior year to "get lots of new women." The structural access to women that the Little Sister program provides and the absence of normative supports for refusing fraternity members' sexual advances may make women in this program particularly susceptible to coerced sexual encounters with fraternity men.

Access to women for sexual gratification is a presumed benefit of fraternity membership, promised in recruitment materials and strategies and through brothers' conversations with new recruits. One fraternity man said: "We always tell the guys that you get sex all the time, there's always new girls. . . . After I became a Greek, I found out I could be with females at will." A university official told us that, based on his observations, "no one [i.e., fraternity men] on this campus wants to have 'relationships.' They just want to have fun [i.e., sex]." Fraternity men plan and execute strategies aimed at obtaining sexual gratification, and this occurs at both individual and collective levels.

Individual strategies include getting a woman drunk and spending a great deal of money on her. As for collective strategies, most of our undergraduate interviewees agreed that fraternity parties often culminate in sex and that this outcome is planned. One fraternity man said fraternity parties often involve sex and nudity and can "turn into orgies." Orgies may be planned in advance, such as the Bowery Ball party held by one fraternity. A former fraternity member said of this party:

> The entire idea behind this is sex. Both men and women come to the party wearing little or nothing. There are pornographic pinups on the walls and usually porno movies playing on the TV. The music carries sexual overtones. . . . They just get schnockered [drunk] and, in most cases, they also get laid.

When asked about the women who come to such a party, he said: "Some Little Sisters just won't go. . . . The girls who do are looking for a good time, girls who don't know what it is, things like that."

Other respondents denied that fraternity parties are orgies but said that sex is always talked about among the brothers and they all know "who each other is doing it with." One member said that most of the time, guys have sex with their girlfriends "but with socials, girlfriends aren't allowed to come and it's their [members'] big chance [to have sex with other women]." The use of alcohol to help them get women into bed is a routine strategy at fraternity parties.

CONCLUSIONS

In general, our research indicated that the organization and membership of fraternities contribute heavily to coercive and often violent sex. Fraternity houses are occupied by same-sex (all men) and same-age (late teens, early twenties) peers whose maturity and judgment is often less than ideal. Yet fraternity houses are private dwellings that are mostly off-limits to, and away from scrutiny of, university and community representatives, with the result that fraternity house events seldom come to the attention of outsiders. Practices associated with the social construction of fraternity brotherhood emphasize a macho conception of men and masculinity, a narrow, stereotyped conception of women and femininity, and the treatment of women as commodities. Other practices contributing to coercive sexual relations and the cover-up of rapes include excessive alcohol use, competitiveness, and normative support for deviance and secrecy (cf. Bogal-Allbritten and Allbritten 1985; Kanin 1967).

Some fraternity practices exacerbate others. Brotherhood norms require "sticking together" regardless of right or wrong; thus rape episodes are unlikely to be stopped or reported to outsiders, even when witnesses disapprove.

The ability to use alcohol without scrutiny by authorities and alcohol's frequent association with violence, including sexual coercion, facilitates rape in fraternity houses. Fraternity norms that emphasize the value of maleness and masculinity over femaleness and femininity and that elevate the status of men and lower the status of women in members' eyes undermine perceptions and treatment of women as persons who deserve consideration and care (cf. Ehrhart and Sandler 1985; Merton 1985).

Androgynous men and men with a broad range of interests and attributes are lost to fraternities through their recruitment practices. Masculinity of a narrow and stereotypical type helps create attitudes, norms, and practices that predispose fraternity men to coerce women sexually, both individually and collectively (Allgeier 1986; Hood 1989; Sanday 1981, 1986). Male athletes on campus may be similarly disposed for the same reasons (Kirshenbaum 1989; Telander and Sullivan 1989).

Research into the social contexts in which rape crimes occur and the social constructions associated with these contexts illumine rape dynamics on campus. Blanchard (1959) found that group rapes almost always have a leader who pushes others into the crime. He also found that the leader's latent homosexuality, desire to show off to his peers, or fear of failing to prove himself a man are frequently an impetus. Fraternity norms and practices contribute to the approval and use of sexual coercion as an accepted tactic in relations with women. Alcohol-induced compliance is normative, whereas, presumably, use of a knife, gun, or threat of bodily harm would not be because the woman who "drinks too much" is viewed as "causing her own rape" (cf. Ehrhart and Sandler 1985).

Our research led us to conclude that fraternity norms and practices influence members to view the sexual coercion of women, which is a felony crime, as sport, a contest, or a game (cf. Sato 1988). This sport is played not between men and women but between men and men. Women are the pawns or prey in the interfraternity rivalry game; they prove that a fraternity is successful or prestigious. The use of women in this way encourages fraternity men to see women as objects and sexual coercion as sport. Today's societal norms support young women's right to engage in sex at their discretion, and coercion is unnecessary in a mutually desired encounter. However, nubile young women say they prefer to be "in a relationship" to have sex while young men say they prefer to "get laid" without a commitment (Muehlenhard and Linton 1987). These differences may reflect, in part, American puritanism and men's fears of sexual intimacy or perhaps intimacy of any kind. In a fraternity context, getting sex without giving emotionally demonstrates "cool" masculinity. More important, it poses no threat to the bonding and loyalty of the fraternity brotherhood (cf. Farr 1988). Drinking large quantities of alcohol before having sex suggests that "scoring" rather than intrinsic sexual pleasure is a primary concern of fraternity men.

Unless fraternities' composition, goals, structures, and practices change in fundamental ways, women on campus will continue to be sexual prey for fraternity men. As all-male enclaves dedicated to opposing faculty and administration and to cementing in-group ties, fraternity members eschew any hint of homosexuality. Their version of masculinity transforms women, and men with womanly characteristics, into the out-group. "Womanly men" are ostracized; feminine women are used to demonstrate members' masculinity. Encouraging renewed emphasis on their founding values (Longino and Kart 1973), service orientation and activities (Lemire 1979), or members' moral development (Marlowe and Auvenshine 1982) will have little effect on fraternities' treatment of women. A case for or against fraternities cannot be made by studying individual members. The fraternity qua group and organization is at issue. Located on campus along with many vulnerable women, embedded in a sexist society, and caught up in masculinist goals, practices, and values, fraternities' violation of women—including forcible rape—should come as no surprise.

NOTE

1. Recent bans by some universities on open-keg parties at fraternity houses have resulted in heavy drinking before coming to a party and an increase in drunkenness among those who attend. This may aggravate, rather than improve, the treatment of women by fraternity men at parties.

REFERENCES

Adams, Aileen, and Gail Abarbanel. 1988. *Sexual Assault on Campus: What Colleges Can Do.* Santa Monica, CA: Rape Treatment Center.

Allgeier, Elizabeth. 1986. "Coercive versus Consensual Sexual Interactions." G. Stanley Hall Lecture to American Psychological Association Annual Meeting, Washington, DC, August.

Blanchard, W. H. 1959. "The Group Process in Gang Rape." *Journal of Social Psychology* 49:259–66.

Bogal-Allbritten, Rosemarie B., and William L. Allbritten. 1985. "The Hidden Victims: Courtship Violence among College Students." *Journal of College Student Personnel* 43:201–4.

Bohrnstedt, George W. 1969. "Conservatism, Authoritarianism and Religiosity of Fraternity Pledges." *Journal of College Student Personnel* 27:36–43.

Bradford, Michael. 1986. "Tight Market Dries Up Nightlife at University." *Business Insurance* (March 2):2, 6.

Burkhart, Barry. 1989. Comments in Seminar on Acquaintance/Date Rape Prevention: A National Video Teleconference, February 2.

Burkhart, Barry R., and Annette L. Stanton. 1985. "Sexual Aggression in Acquaintance Relationships." Pp. 43–65 in *Violence in Intimate Relationships*, edited by G. Russell. Englewood Cliffs, NJ: Spectrum.

Byington, Diane B., and Karen W. Keeter. 1988. "Assessing Needs of Sexual Assault Victims on a University Campus." Pp. 23–31 in *Student Services: Responding to Issues and Challenges*. Chapel Hill: University of North Carolina Press.

Chancer, Lynn S. 1987. "New Bedford, Massachusetts, March 6, 1983–March 22, 1984: The 'Before and After' of a Group Rape. *Gender & Society* 1:239–60.

Ehrhart, Julie K., and Bernice R. Sandler. 1985. *Campus Gang Rape: Party Games?* Washington, DC: Association of American Colleges.

Farr, K. A. 1988. "Dominance Bonding through the Good Old Boys Sociability Network." *Sex Roles* 18:259–77.

Florida Flambeau. "Pike Members Indicted in Rape." (May 19):1, 5.

Fox, Elaine, Charles Hodge, and Walter Ward. 1987. "A Comparison of Attitudes Held by Black and White Fraternity Members." *Journal of Negro Education* 56:521–34.

Geis, Gilbert. 1971. "Group Sexual Assaults." *Medical Aspects of Human Sexuality* 5:101–13.

Glaser, Barney G. 1978. *Theoretical Sensitivity: Advances in the Methodology of Grounded Theory*. Mill Valley, CA: Sociology Press.

Hood, Jane. 1989. "Why Our Society Is Rape-Prone." *New York Times*, May 16.

Hughes, Michael J., and Roger B. Winston, Jr. 1987. "Effects of Fraternity Membership on Interpersonal Values." *Journal of College Student Personnel* 45:405–11.

Kanin, Eugene J. 1967. "Reference Groups and Sex Conduct Norm Violations." *The Sociological Quarterly* 8:495–504.

Kimmel, Michael, ed. 1987. *Changing Men: New Directions in Research on Men and Masculinity*. Newbury Park, CA: Sage.

Kirshenbaum, Jerry. 1989. "Special Report, an American Disgrace: A Violent and Unprecedented Lawlessness Has Arisen among College Athletes in all Parts of the County." *Sports Illustrated* (February 27):16–19.

Lemire, David. 1979. "One Investigation of the Stereotypes Associated with Fraternities and Sororities." *Journal of College Student Personnel* 37:54–57.

Letchworth, G. E. 1969. "Fraternities Now and in the Future." *Journal of College Student Personnel* 10:118–22.

Longino, Charles F., Jr., and Cary S. Kart. 1973. "The College Fraternity: An Assessment of Theory and Research." *Journal of College Student Personnel* 31:118–25.

Marlowe, Anne F., and Dwight C. Auvenshine. 1982. "Greek Membership: Its Impact on the Moral Development of College Freshmen." *Journal of College Student Personnel* 40:53–57.

Martin, Patricia Yancey, and Barry A. Turner. 1986. "Grounded Theory and Organizational Research." *Journal of Applied Behavioral Science* 22:141–57.

Merton, Andrew. 1985. "On Competition and Class: Return to Brotherhood." *Ms.* (September):60–65, 121–22.

Messner, Michael. 1989. "Masculinities and Athletic Careers." *Gender & Society* 3:71–88.

Meyer, T. J. 1986. "Fight against Hazing Rituals Rages on Campuses." *Chronicle of Higher Education* (March 12):34–36.

Miller, Leonard D. 1973. "Distinctive Characteristics of Fraternity Members." *Journal of College Student Personnel* 31:126–28.

Muehlenhard, Charlene L., and Melaney A. Linton. 1987. "Date Rape and Sexual Aggression in Dating Situations: Incidence and Risk Factors." *Journal of Counseling Psychology* 34:186–96.

Pressley, Sue Anne. 1987. "Fraternity Hell Night Still Endures." *Washington Post* (August 11):B1.

Rapport, Karen, and Barry R. Burkhart. 1983. "Personality and Attitudinal Characteristics of Sexually Coercive College Males." *Journal of Abnormal Psychology* 93:216–21.

Roark, Mary L. 1987. "Preventing Violence on College Campuses." *Journal of Counseling and Development* 65:367-70.

St. Petersburg Times. 1988. "A Greek Tragedy." (May 29):1F, 6F.

Sanday, Peggy Reeves. 1981. "The Socio-Cultural Context of Rape: A Cross-Cultural Study." *Journal of Social Issues* 37:5–27.

———. 1986. "Rape and the Silencing of the Feminine." Pp. 84–101 in *Rape*, edited by S. Tomaselli and R. Porter. Oxford: Basil Blackwell.

Sato, Ikuya. 1988. "Play Theory of Delinquency: Toward a General Theory of 'Action.'" *Symbolic Interaction* 11:191–212.

Smith, T. 1964. "Emergence and Maintenance of Fraternal Solidarity." *Pacific Sociological Review* 7:29–37.

Tallahassee Democrat. 1988a. "FSU Fraternity Brothers Charged" (April 27):1A, 12A.

———. 1988b. "FSU Interviewing Students about Alleged Rape" (April 24):1D.

———. 1989. "Woman Sues Stetson in Alleged Rape" (March 19):3B.

Tampa Tribune. 1988. "Fraternity Brothers Charged in Sexual Assault of FSU Coed." (April 27):6B.

Tash, Gary B. 1988. "Date Rape." *The Emerald of Sigma Pi Fraternity* 75(4):1–2.

Telander, Rick, and Robert Sullivan. 1989. "Special Report, You Reap What You Sow." *Sports Illustrated* (February 27):20–34.

The Tomahawk. 1988. "A Look Back at Rush, A Mixture of Hard Work and Fun" (April/May):3D.

Walsh, Claire. 1989. Comments in Seminar on Acquaintance/Date Rape Prevention: A National Video Teleconference, February 2.

Wilder, David H., Arlyne E. Hoyt, Dennis M. Doren, William E. Hauck, and Robert D. Zettle. 1978. "The Impact of Fraternity and Sorority Membership on Values and Attitudes." *Journal of College Student Personnel* 36:445–49.

Wilder, David H., Arlyne E. Hoyt, Beth Shuster Surbeck, Janet C. Wilder, and Patricia Imperatrice Carney. 1986. "Greek Affiliation and Attitude Change in College Students." *Journal of College Student Personnel* 44:510–19.

33

Transaction Systems and Unlawful Organizational Behavior

DIANE VAUGHAN

oleman (1974) points out that the rise and increase of complex organizations have changed the nature of the social structure over time. New sets of interaction patterns have resulted. In addition to individuals interacting with individuals through roles, individuals now interact with organizations, and organizations interact with other organizations. These new sets of relationships have natural consequences for unlawful behavior. Where once both victim and offender were individuals, increasingly both roles have been played by complex organizations (Vaughan, 1980; 1983). Though inter-organizational relations are receiving intense sociological scrutiny (Aldrich, 1979; Evan, 1978; Pfeffer and Salancik, 1978), unlawful conduct between organizations has seldom been analyzed from an inter-organizational perspective (Vaughan, 1980). To do so draws attention to a neglected but critical element of inter-organizational relations: the nature of transactions and the transaction systems of complex organizations. Transactions, transaction systems, and the way in which they facilitate and in some cases *generate* unlawful organizational behavior are the subject of this paper. . . .

From: "Transaction Systems and Unlawful Organizational Behavior," Diane Vaughan, *Social Problems*, Vol. 29, No. 4, 1982. © 1982 Society for the Study of Social Problems. Reprinted by permission of University of California Press Journals and the author.

THE SYSTEM INTERFACE PROBLEM AND
UNLAWFUL ORGANIZATIONAL BEHAVIOR

A system interface problem occurs when the language, rules, procedures, and recording and processing systems of two organizations fail to mesh, so that a transaction is inhibited rather than facilitated. Resource exchange may stall and become difficult to complete to the satisfaction of both parties. One or both of the organizations concerned may have to adjust their system. Negotiations may often flounder in a between-system lag induced by formalized communications. The problem may be short- or long-term, depending on the two organizations, the nature of their interdependence, their frequency of interaction, the task around which the specific exchange revolves, and the resources each can devote to correcting the difficulty. Should one of the organizations be unwilling or unable to devote resources to legitimate resolution, or require immediate completion of the exchange in order to gain resources, the transaction system itself may be the chosen mechanism for bypassing the system interface problem. The result may be unlawful behavior, as the Revco case illustrates (Vaughan, 1980; 1983).

In 1977, Revco Drug Stores, Inc., one of the largest discount drug chains in the United States, was found guilty of Medicaid-provider fraud: specifically, a computer-generated double-billing scheme that resulted in the loss of over half a million dollars in Medicaid funds to the Ohio Department of Public Welfare. Revco was engaged in exchange on a contractual basis with the welfare department, as a provider of pharmacy goods and services to Medicaid recipients. Prescriptions were given to recipients by Revco pharmacists, then submitted to the welfare department for reimbursement. That a system interface problem existed is indicated by the history of high rejection rates for Revco claims submitted to the welfare department for reimbursement. Documents and memos showed that Revco had experienced reimbursement problems since the welfare department first began processing and paying claims by computer in 1972.

According to representatives of both organizations, Revco's high rejection rates were a function of welfare department rules for allowable Medicaid recipient claims and procedures for claim-filing on the part of providers. The number of rules and procedures was overwhelming; they changed frequently, and providers received periodic notification of how computer claims submission procedures should be altered to suit new criteria. The modifications could be costly and time-consuming, and, in addition to the computer changes, required assimilation by pharmacists in each store. Errors were common, and computer modifications either lagged behind or were not made. Claims were rejected for all three reasons. As a consequence, Revco was not reimbursed for filling the prescriptions which had been dispensed to recipients in the belief that the reimbursement would be forthcoming. When a claim is rejected, reimbursement is withheld until the error is corrected, and the claim successfully resubmitted. In Revco's case, rather than correct the rejected claims for resubmission, rejected claims accumulated. Over 50,000 claims, rejected by welfare department computers and representing more than half a million dollars in accounts receivable, accumulated at Revco headquarters.

Two Revco officials initiated a plan to bring the company's outstanding accounts receivable back into balance. A temporary clerical staff of six was hired to alter the rejected claims to make them acceptable to the welfare department computer. According to Revco officials, the decision to falsify prescription claims was influenced by four factors. (1) They had faced this situation before. Revco had a history of stalled negotiations with the welfare department which impeded reimbursement for provider services. Revco executives believed the corporation had repeatedly been victimized in this same manner by the welfare department (Vaughan, 1980). (2) The two executives believed that the rejected claims represented resources legitimately owed to Revco. (3) They calculated that the cost of legitimate correction and resubmission would be more than the average amount of the claims. (4) They believed that the funds could be retrieved without being detected by the department's screening system. This belief was based on the skills possessed by the two executives. One was a licensed pharmacist and the other a computer specialist who knew the welfare department's computer system well. They thoroughly understood the intermesh between the two organizations' transaction systems. To take back resources they believed belonged to the corporation, the executives falsified prescriptions equal in number to those rejected, and submitted them through the transaction system.

The Revco case suggests that when a system interface problem ties up resources or inhibits resource delivery, unlawful attainment of resources may be a function of: (1) demand for the resources; (2) legitimate access blocked by cost and delay; (3) structured opportunity to secure the resources through the transaction system; (4) low probability of detection and sanctioning (Vaughan, 1983); and (5) redefinition of property rights concerning possession of organizational resources (Dynes and Quarantelli, 1974).

Because the Revco incident is a case study, no conclusions can be drawn about the extensiveness of the system interface problem. In exchange between Medicaid providers and the welfare department in Ohio, all providers routinely had claims rejected. This fact might indicate that in this particular welfare department, the system interface problem runs rampant. However, the department monitors the rate of provider rejection for two reasons: to work out system interface problems, and to detect fraud. If a provider's rejection rate is higher than the average for all providers, or from its own rejection history, either system interface difficulties or intentional fraud could be the explanation. The Revco case is an example of a system interface problem which led to fraud. System interface difficulties can exist without fraud, or as in the market signaling example, fraud may be the principle purpose of a transaction, rather than the solution to a bureaucratic snag.

Without examining rejection rates of other providers and identifying the facts of each case, generalizations cannot be made about system interface problems in the Ohio Medicaid system. In this case, a system interface problem did occur, and unlawful behavior was used to resolve it. System interface problems occasionally occur for nearly all organizations, demanding varying amounts of resources to complete transactions that stall. With transactions encumbered by formalization, complex processing methods and mechanisms, and general rather than specific monitoring practices, some organizations may resolve their difficulties unlawfully. For some organizations, system interface problems may be the rule, rather than the exception, increasing the likelihood of fraudulent resolution. When this is the case, the transaction system itself may be labelled criminogenic.

SUMMARY AND IMPLICATIONS

Transaction systems which develop to cope with the complex legitimate exchanges between complex organizations can also be used to secure gains unlawfully. Because transactions of complex organizations are characterized by high degrees of formalization, intricate processing and recording methods, exchange based on trust, and general monitoring procedures, illegal conduct can be carried out with little risk of detection. Not only are opportunities for unlawful behavior present in each of these four characteristics, but also in the ways they combine in a transaction system. Indeed, the transaction systems of some organizations may be criminogenic, repeatedly generating violations between organizations. Unlawful conduct may occur regularly when exchange between organizations relies upon market signals as the basis for decision-making. Signals can be falsified, and transaction systems with incomplete information gathering and broad monitoring procedures facilitate the fraudulent attainment of resources. Transaction systems may also systematically generate illegality by creating system interface problems. Unlawful conduct may occur because the transaction system itself blocks legitimate access to resources.

Opportunities for violations multiply as transaction system complexity increases. While not all organizations in a market signaling situation or confronting system interface problems will resort to unlawful behavior,[1] some organizations may be more likely to become offenders, depending upon the complexity of their transaction systems and those of the various organizations in their set. This variability across organizations is important, for the Medicaid system examined here is highly-complex: a criminogenic transaction system plagued by violations as a result of the four factors functioning individually to generate illegal behavior, as well as acting in combination to present opportunities for violations both through a market signaling situation and system interface problems.

Of the organizational characteristics believed to be associated with unlawful organizational behavior (Gross, 1980; Needleman and Needleman, 1979; Stone, 1975; Vaughan, 1980; 1983), the transaction systems of organizations have been least addressed by scholars and activists, and yet are perhaps the most vulnerable to manipulation. Because the complexity of inter-organizational exchange appears to have systematic consequences for illegality, reducing the complexity of transactions may also reduce rates of violations. To realize this possibility, organizations must assume a social responsibility to recognize the relationship between transaction complexity and unlawful conduct and, hence, to monitor and adjust their own transaction systems to minimize the possibility that they are the source of violations. Reducing transaction complexity between organizations may result in fewer violations by decreasing opportunities for illegality, improving the risk of detection, and/or decreasing the probability of system interface problems, thus eliminating a source of motivation to pursue resources unlawfully.

What could motivate organizations to reduce the complexity of their own transaction systems? System complexity not only increases the potential for an organization to engage in unlawful behavior, but also promotes the possibility that an organization will be victimized—by other organizations or by its own members. Changes that decrease the possibility that an organization will commit an offense also decrease the possibility that the organization will be victimized; resources lost to an organization in this manner can be conserved. Though revamping transaction systems may create a strain on the organization, in the long run savings from forestalled victimization can be applied to the legitimate pursuit of organizational goals: Thus, self-surveillance and modification of transaction systems is in an organization's best interest. Furthermore, reducing the complexity of inter-organizational exchange is socially responsible because it may reduce the costs of unlawful organizational behavior and its control, which ultimately fall upon the public.

NOTE

1. For an analysis of the variation in patterns of organizational behavior under circumstances conducive to illegality, see Vaughan, 1983.

REFERENCES

Aldrich, Howard E. 1979. Organizations and Environments. Englewood Cliffs, N.J.: Prentice-Hall.

Coleman, James S. 1974. Power and the Structure of Society. Philadelphia: University of Pennsylvania Press.

Dynes, Russell R., and E. L. Quarantelli. 1974. "Organizations as victims in mass civil disturbances." Pp. 67–77 in Israel Drapkin and Emilio Viano (eds.), Victimology: A New Focus. Lexington, Mass.: D. C. Heath.

Evan, William M. (ed.) 1978. Interorganizational Relations. Philadelphia: University of Pennsylvania Press.

Gross, Edward. 1980. "Organizational structure and organizational crime." Pp. 52–76 in Gilbert Geis and Ezra Stotland (eds.), White-Collar Crime: Theory and Research. Beverly Hills: Sage.

Needleman, Martin L., and Carolyn Needleman. 1979. "Organizational crime: Two models of criminogenesis." Sociological Quarterly 20:517–528.

Pfeffer, Jeffrey, and Gerald Salancik. 1978. The External Control of Organizations: A Resource Dependence Perspective. New York: Harper and Row.

Stone, Christopher D. 1975. Where the Law Ends: The Social Control of Corporate Behavior. New York: Harper and Row.

Vaughan, Diane. 1980. "Crime between organizations: Implications for victimology." Pp. 77–97 in Gilbert Geis and Ezra Stotland (eds.), White-Collar Crime: Theory and Research. Beverly Hills: Sage.

———. 1983. On the Social Control of Organizations. Chicago: The University of Chicago Press.

34

Joining a Gang

MARTÍN SÁNCHEZ-JANKOWSKI

> Now it is thought to be the mark of a man of practical wisdom to be able to
> deliberate well about what is good and expedient for himself, not in some
> particular respect, e.g. about what sorts of thing conduce to health or to
> strength, but about what sorts of thing conduce to the good life in general.
>
> Aristotle
> *The Nicomachean Ethics*

> We are looking for a few good men.
>
> U.S. Marine Corps Recruiting Poster

In chapter 1, I argued that one of the most important features of gang members was their defiant individualist character. I explained the development of defiant individualism by locating its origins in the material conditions—the competition and conflict over resource scarcity—of the low-income neighborhoods of most large American cities. These conditions exist for everyone who lives in such neighborhoods, yet not every young person joins a gang. Although I have found that nearly all those who belong to gangs do exhibit defiant individualist traits to some degree, not all those who possess such traits join gangs. This chapter explores who joins a gang and why in more detail.

Many studies offer an answer to why a person joins a gang, or why a group of individuals start a gang. These studies can be divided into four groupings. First, there are those that hold the "natural association" point of view. These studies argue that people join gangs as a result of the natural act of associating with each other.[1] Their contention is that a group of boys, interrelating with each other, decide to formalize their relationship in an attempt to reduce the fear and anxiety associated with their socially disorganized neighborhoods. The individual's impetus to join is the result of his desire to defend against conflict and create order out of the condition of social disorganization.

From: Martín Sánchez-Jankowski, *Islands in the Street: Gangs and American Urban Society*
(Berkeley: University of California Press, 1991). Copyright © 1991 The Regents of the
University of California. Reprinted by permission of the publisher.

The second group of studies explains gang formation in terms of "the subculture of blocked opportunities": gangs begin because young males experience persistent problems in gaining employment and/or status. As a result, members of poor communities who experience the strain of these blocked opportunities attempt to compensate for socioeconomic deprivation by joining a gang and establishing a subculture that can be kept separate from the culture of the wider society.[2]

The third group of studies focuses on "problems in identity construction." Within this broad group, some suggest that individuals join gangs as part of the developmental process of building a personal identity or as the result of a breakdown in the process.[3] Others argue that some individuals from low-income families have been blocked from achieving social status through conventional means and join gangs to gain status and self-worth, to rebuild a wounded identity.[4]

A recent work by Jack Katz has both creatively extended the status model and advanced the premise that sensuality is the central element leading to the commission of illegal acts. In Katz's "expressive" model, joining a gang, and being what he labels a "badass," involves a process whereby an individual manages (through transcendence) the gulf that exists between a sense of self located within the local world (the here) and a reality associated with the world outside (the there). Katz argues that the central elements in various forms of deviance, including becoming involved in a gang and gang violence, are the moral emotions of humiliation, righteousness, arrogance, ridicule, cynicism, defilement, and vengeance. "In each," he says, "the attraction that proves to be most fundamentally compelling is that of overcoming a personal challenge to moral—not material—existence."[5]

Most of these theories suffer from three flaws. First, they link joining a gang to delinquency, thereby combining two separate issues. Second, they use single-variable explanations. Third, and most important, they fail to treat joining a gang as the product of a rational decision to maximize self-interest, one in which both the individual and the organized gang play a role. This is especially true of Katz's approach, for two reasons. First, on the personal level, it underestimates the impact of material and status conditions in establishing the situations in which sensual needs/drives (emotions) present themselves, and overestimates/exaggerates the "seductive" impact of crime in satisfying these needs. Second, it does not consider the impact of organizational dynamics on the thought and action of gang members.

In contrast, the data presented here will indicate that gangs are composed of individuals who join for a variety of reasons. In addition, while the individual uses his own calculus to decide whether or not to join a gang, this is not the only deciding factor. The other deciding factor is whether the gang wants him in the organization. Like the individual's decision to join, the gang's decision to permit membership is based on a variety of factors. It is thus important to understand that who becomes a gang member depends on two decision-making processes: that of the individual and that of the gang.

THE INDIVIDUAL AND THE
DECISION TO BECOME A MEMBER

Before proceeding, it is important to dismiss a number of the propositions that have often been advanced. The first is that young boys join gangs because they are from broken homes where the father is not present and they seek gang membership in order to identify with other males—that is, they have had no male authority figures with whom to identify. In the ten years of this study, I found that there were as many gang members from homes where the nuclear family was intact as there were from families where the father was absent.[6]

The second proposition given for why individuals join gangs is related to the first: it suggests that broken homes and/or bad home environments force them to look to the gang as a substitute family. Those who offer this explanation often quote gang members' statements such as "We are like a family" or "We are just like brothers" as indications of this motive. However, I found as many members who claimed close relationships with their families as those who denied them.

The third reason offered is that individuals who drop out of school have fewer skills for getting jobs, leaving them with nothing to do but join a gang. While I did find a larger number of members who had dropped out of school, the number was only slightly higher than those who had finished school.

The fourth reason suggested, disconfirmed by my data, is a modern version of the "Pied Piper" effect: the claim that young kids join gangs because they are socialized by older kids to aspire to gang membership and, being young and impressionable, are easily persuaded. I found on the contrary that individuals were as likely to join when they were older (mid to late teens) as when they were younger (nine to fifteen). I also found significantly more who joined when they were young who did so for reasons other than being socialized to think it was "cool" to belong to a gang. In brief, I found no evidence for this proposition.

What I did find was that individuals who live in low-income neighborhoods join gangs for a variety of reasons, basing their decisions on a rational calculation of what is best for them at that particular time. Furthermore, I found that they use the same calculus (not necessarily the same reasons) in deciding whether to stay in the gang, or, if they happen to leave it, whether to rejoin.

Reasons for Deciding to Join a Gang

Most people in the low-income inner cities of America face a situation in which a gang already exists in their area. Therefore the most salient question facing them is not whether to start a gang or not, but rather whether to join an existing one. Many of the reasons for starting a new gang are related to issues having to do with organizational development and decline—that is, with the existing gang's ability to provide the expected services, which include those that individuals considered in deciding to join. Those issues will be treated in chapter 4, which deals with the internal dynamics of the organization. This section deals primarily, although not exclusively, with the question of what influences individuals to join an existing gang. However, many of these are the same influences that encourage individuals to start a new gang.

Material Incentives

Those who had joined a gang most often gave as their reason the belief that it would provide them with an environment that would increase their chances of securing money. Defiant individualists constantly calculate the costs and benefits associated with their efforts to improve their financial well-being (which is usually not good). Therefore, on the one hand, they believe that if they engage in economic ventures on their own, they will, if successful, earn more per venture than if they acted as part of a gang. However, there is also the belief that if one participates in economic ventures with a gang, it is likely that the amount earned will be more regular, although perhaps less per venture. The comments of Slump, a sixteen-year-old member of a gang in the Los Angeles area, represent this belief:

> Well, I really didn't want to join the gang when I was a little younger because I had this idea that I could make more money if I would do some gigs [various illegal economic ventures] on my own. Now I don't know, I mean, I wasn't wrong. I could make more money on my own, but there are more things happening with the gang, so it's a little more even in terms of when the money comes in. . . . Let's just say there is more possibilities for a more steady amount of income if you need it.

It was also believed that less individual effort would be required in the various economic ventures in a gang because more people would be involved. In addition, some thought that being in a gang would reduce the *risk* (of personal injury) associated with their business ventures. They were aware that if larger numbers of people had knowledge of a crime, this would increase the risk that if someone were caught, others, including themselves, would be implicated. However, they countered this consideration with the belief that they faced less risk of being physically harmed when they were part of a group action. The comments of Corner, a seventeen-year-old resident of a poor Manhattan neighborhood, represent this consideration. During the interview, he was twice approached about joining the local gang. He said:

> I think I am going to join the club [gang] this time. I don't know, man, I got some things to decide, but I think I will. . . . Before I didn't want to join because when I did a job, I didn't want to share it with the whole group—hell, I was never able to make that much to share. . . . I would never have got enough money, and with all those dudes [other members of the gang] knowing who did the job, you can bet the police would find out. . . . Well, now my thinking is changed a bit 'cause there's more people involved and that'll keep me safer. [He joined the gang two weeks later.]

Others decided to join the gang for financial security. They viewed the gang as an organization that could provide them or their families with money in times of emergency. It represented the combination of a bank and a social security system, the equivalent of what the political machine had been to many new immigrant groups in American cities.[7] To these individuals, it provided both psychological and financial security in an economic environment of scarcity and intense competition. This was particularly true of those who were fifteen and younger. Many in this age group often find themselves in a precarious position. They are in need of money, and although social services are available to help during times of economic hardship, they often lack legal means of access to these resources. For these individuals, the gang can provide an alternative source of aid. The comments of Street Dog and Tomahawk represent these views. Street Dog was a fifteen-year-old Puerto Rican who had been in a New York gang for two years:

> Hey, the club [the gang] has been there when I needed help. There were times when there just wasn't enough food for me to get filled up with. My family was hard up and they couldn't manage all of their bills and such, so there was some lean meals! Well, I just needed some money to help for awhile, till I got some money or my family was better off. They [the gang] was there to help. I could see that [they would help] before I joined, that's why I joined. They are there when you need them and they'll continue to be.

Tomahawk was a fifteen-year-old Irishman who had been in a gang for one year:

> Before I joined the gang, I could see that you could count on your boys to help in times of need and that meant a lot to me. And when I needed money, sure enough they gave it to me. Nobody else would have given it to me; my parents didn't have it, and there was no other place to go. The gang was just like they said they would be, and they'll continue to be there when I need them.

Finally, many view the gang as providing an opportunity for future gratification. They expect that through belonging to a gang, they will be able to make contact with individuals who may eventually help them financially. Some look to meet people who have contacts in organized crime in the hope of entering that field in the future. Some hope to meet businessmen involved in the illegal market who will provide them with money to start their own illegal businesses. Still others think that gang membership will enable them to meet individuals who will later do them favors (with financial implications) of the kind fraternity brothers or Masons sometimes do for each other. Irish gang members in New York and Boston especially tend to believe this.

Recreation

The gang provides individuals with entertainment, much as a fraternity does for college students or the Moose and Elk clubs do for their members. Many

individuals said they joined the gang because it was the primary social institution of their neighborhood—that is, it was where most (not necessarily the
biggest) social events occurred. Gangs usually, though not always, have some
type of clubhouse. The exact nature of the clubhouse varies according to
how much money the gang has to support it, but every clubhouse offers
some form of entertainment. In the case of some gangs with a good deal of
money, the clubhouse includes a bar, which sells its members drinks at cost.
In addition, some clubhouses have pinball machines, soccer-game machines,
pool tables, Ping-Pong tables, card tables, and in some cases a few slot
machines. The clubhouse acts as an incentive, much like the lodge houses of
other social clubs.[8]

The gang can also be a promoter of social events in the community, such
as a big party or dance. Often the gang, like a fraternity, is thought of as the
organization to join to maximize opportunities to have fun. Many who joined
said they did so because the gang provided them with a good opportunity to
meet women. Young women frequently form an auxiliary unit to the gang,
which usually adopts a version of the male gang's name (e.g., "Lady Jets").
The women who join this auxiliary do so for similar reasons—that is, opportunities to meet men and participate in social events.[9]

The gang is also a source of drugs and alcohol. Here, most gangs walk a
fine line. They provide some drugs for purposes of recreation, but because
they also ban addicts from the organization, they also attempt to monitor
members' use of some drugs.[10]

The comments of Fox and Happy highlight these views of the gang as a
source of recreation.[11] Fox was a twenty-three-year-old from New York and
had been in a gang for seven years:

> Like I been telling you, I joined originally because all the action was hap-
> pening with the Bats [gang's name]. I mean, all the foxy ladies were going
> to their parties and hanging with them. Plus their parties were great. They
> had good music and the herb [marijuana] was so smooth. . . . Man, it was
> a great source of dope and women. Hell, they were the kings of the com-
> munity so I wanted to get in on some of the action.

Happy was a twenty-eight-year-old from Los Angeles, who had been a gang
member for eight years:

> I joined because at the time, Jones Park [gang's name] had the best club-
> house. They had pool tables and pinball machines that you could use for
> free. Now they added a video game which you only have to pay like five
> cents for to play. You could do a lot in the club, so I thought it was a
> good thing to try it for awhile [join the gang], and it was a good thing.

A Place of Refuge and Camouflage

Some individuals join a gang because it provides them with a protective group
identity. They see the gang as offering them anonymity, which may relieve

the stresses associated with having to be personally accountable for all their actions in an intensely competitive environment. The statements of Junior J. and Black Top are representative of this belief. Junior J. was a seventeen-year-old who had been approached about becoming a gang member in one of New York's neighborhoods:

> I been thinking about joining the gang because the gang gives you a cover, you know what I mean? Like when me or anybody does a business deal and we're members of the gang, it's difficult to track us down 'cause people will say, oh, it was just one of those guys in the gang. You get my point? The gang is going to provide me with some cover.

Black Top was a seventeen-year-old member of a Jamaican gang in New York:

> Man, I been dealing me something awful. I been doing well, but I also attracted me some adversaries. And these adversaries have been getting close to me. So joining the brothers [the gang] lets me blend into the group. It lets me hide for awhile, it gives me refuge until the heat goes away.

Physical Protection

Individuals also join gangs because they believe the gang can provide them with personal protection from the predatory elements active in low-income neighborhoods. Nearly all the young men who join for this reason know what dangers exist for them in their low-income neighborhoods. These individuals are not the weakest of those who join the gang, for all have developed the savvy and skills to handle most threats. However, all are either tired of being on the alert or want to reduce the probability of danger to a level that allows them to devote more time to their effort to secure more money. Here are two representative comments of individuals who joined for this reason. Chico was a seventeen-year-old member of an Irish gang in New York:

> When I first started up with the Steel Flowers, I really didn't know much about them. But, to be honest, in the beginning I just joined because there were some people who were taking my school [lunch] money, and after I joined the gang, these guys laid off.

Cory was a sixteen-year-old member of a Los Angeles gang:

> Man I joined the Fultons because there are a lot of people out there who are trying to get you and if you don't got protection you in trouble sometimes. My homeboys gave me protection, so hey, they were the thing to do. . . . Now that I got some business things going I can concentrate on them and not worry so much. I don't always have to be looking over my shoulder.

A Time to Resist

Many older individuals (in their late teens or older) join gangs in an effort to resist living lives like their parents'. As Joan Moore, Ruth Horowitz, and others have pointed out, most gang members come from families whose parents are underemployed and/or employed in the secondary labor market in jobs that have little to recommend them.[12] These jobs are low-paying, have long hours, poor working conditions, and few opportunities for advancement; in brief, they are dead ends.[13] Most prospective gang members have lived through the pains of economic deprivation and the stresses that such an existence puts on a family. They desperately want to avoid following in their parents' path, which they believe is exactly what awaits them. For these individuals, the gang is a way to resist the jobs their parents held and, by extension, the life their parents led. Deciding to become a gang member is both a statement to society ("I'll not take these jobs passively") and an attempt to do whatever can be done to avoid such an outcome. At the very least, some of these individuals view being in a gang as a temporary reprieve from having to take such jobs, a postponement of the inevitable. The comments of Joey and D.D. are representative of this group. Joey was a nineteen-year-old member of an Irish gang in Boston:

> Hell, I joined because I really didn't see anything in the near future I wanted to do. I sure the hell didn't want to take that job my father got me. It was a shit job just like his. I said to myself, "Fuck this!" I'm only nineteen, I'm too young to start this shit. . . . I figured that the Black Rose [the gang] was into a lot of things and that maybe I could hit it big at something we're doing and get the hell out of this place.

D.D. was a twenty-year-old member of a Chicano gang in Los Angeles:

> I just joined the T-Men to kick back [relax, be carefree] for a while. My parents work real hard and they got little for it. I don't really want that kind of job, but that's what it looked like I would have to take. So I said, hey, I'll just kick back for a while and let that job wait for me. Hey, I just might make some money from our dealings and really be able to forget these jobs. . . . If I don't [make it, at least] I told the fuckers in Beverly Hills what I think of the jobs they left for us.

People who join as an act of resistance are often wrongly understood to have joined because they were having difficulty with their identity and the gang provided them with a new one. However, these individuals actually want a new identity less than they want better living conditions.

Commitment to Community

Some individuals join the gang because they see participation as a form of commitment to their community. These usually come from neighborhoods where gangs have existed for generations. Although the character of such gangs may have changed over the years, the fact remains that they have continued to exist. Many of these individuals have known people who have been in gangs, including family members—often a brother, but even, in considerable number of cases, a father and grandfather. The fact that their relatives have a history of gang involvement usually influences these individuals to see

the gang as a part of the tradition of the community. They feel that their families and their community expect them to join, because community members
see the gang as an aid to them and the individual who joins as meeting his
neighborhood obligation. These attitudes are similar to attitudes in the larger
society about one's obligation to serve in the armed forces. In a sense, this
type of involvement represents a unique form of local patriotism. While this
rationale for joining was present in a number of the gangs studied, it was most
prevalent among Chicano and Irish gangs. The comments of Dolan and Pepe
are representative of this line of thinking. Dolan was a sixteen-year-old member of an Irish gang in New York:

> I joined because the gang has been here for a long time and even though
> the name is different a lot of the fellas from the community have been
> involved in it over the years, including my dad. The gang has helped the
> community by protecting it against outsiders so people here have kind of
> depended on it. . . . I feel it's my obligation to the community to put in
> some time helping them out. This will help me to get help in the commu-
> nity if I need it some time.

Pepe was a seventeen-year-old member of a Chicano gang in the Los
Angeles area:

> The Royal Dons [gang's name] have been here for a real long time. A lot
> of people from the community have been in it. I had lots of family in it so
> I guess I'll just have to carry on the tradition. A lot of people from outside
> this community wouldn't understand, but we have helped the community
> whenever they've asked us. We've been around to help. I felt it's kind of
> my duty to join 'cause everybody expects it. . . . No, the community
> doesn't mind that we do things to make some money and raise a little hell
> because they don't expect you to put in your time for nothing. Just like
> nobody expects guys in the military to put in their time for nothing.

In closing this section on why individuals join gangs, it is important to
reemphasize that people choose to join for a variety of reasons, that these reasons are not exclusive of one another (some members have more than one),
that gangs are composed of individuals whose reasons for joining include all
those mentioned, that the decision to join is thought out, and that the individual believes this was best for his or her interests at the moment.

ORGANIZATIONAL RECRUITMENT

Deciding whether or not to join a gang is never an individual decision alone.
Because gangs are well established in most of these neighborhoods, they are
ultimately both the initiators of membership and the gatekeepers, deciding
who will join and who will not.

Every gang that was studied had some type of recruitment strategy. A gang
will frequently employ a number of strategies, depending on the circumstances

in which recruitment is occurring. However, most gangs use one particular style of recruitment for what they consider a "normal" period and adopt other styles as specific situations present themselves. The three most prevalent styles of recruitment encountered were what I call the fraternity type, the obligation type, and the coercive type.

The Fraternity Type of Recruitment

In the fraternity type of recruitment, the gang adopts the posture of an organization that is "cool," "hip," the social thing to be in. Here the gang makes an effort to recruit by advertising through word of mouth that it is looking for members. Then many of the gangs either give a party or circulate information throughout the neighborhood, indicating when their next meeting will be held and that those interested in becoming members are invited. At this initial meeting, prospective members hear a short speech about the gang and its rules. They are also told about the gang's exploits and/or its most positive perks, such as the dances and parties it gives, the availability of dope, the women who are available, the clubhouse, and the various recreational machinery (pool table, video games, bar, etc.). In addition, the gang sometimes discusses, in the most general terms, its plans for creating revenues that will be shared among the general membership. Once this pitch is made, the decision rests with the individual. When one decides to join the gang, there is a trial period before one is considered a solid member of the group. This trial period is similar, but not identical, to the pledge period for fraternities. There are a number of precautions taken during this period to check the individual's worthiness to be in the group. If the individual is not known by members of the gang, he will need to be evaluated to see if he is an informant for one of the various law enforcement agencies (police, firearms and alcohol, drug enforcement). In addition, the individual will need to be assessed in terms of his ability to fight, his courage, and his commitment to help others in the gang.

Having the *will* to fight and defend other gang members or the "interest" of the gang is considered important, but what is looked upon as being an even more important asset for a prospective gang member is the *ability* to fight and to carry out group decisions. Many researchers have often misinterpreted this preference by gangs for those who can fight as an indication that gang members, and thus gangs as collectives, are primarily interested in establishing reputations as fighters.[14] They interpret this preoccupation as being based on adolescent drives for identity and the release of a great deal of aggression. However, what is most often missed are the functional aspects of fighting and its significance to a gang. The prospective member's ability to fight well is not looked upon by the organization simply as an additional symbol of status. Members of gangs want to know if a potential member can fight because if any of them are caught in a situation where they are required to fight, they want to feel confident that everyone can carry his or her own responsibility. In addition, gang members want to know if the potential gang member is disciplined enough to avoid getting scared and running, leaving them vulnerable.

Often everyone's safety in a fight depends on the ability of every individual to fight efficiently. For example, on many occasions I observed a small group of one gang being attacked by an opposing gang. Gang fights are not like fights in the movies: there is no limit to the force anybody is prepared to use—it is, as one often hears, "for all the marbles." When gang members were attacked, they were often outnumbered and surrounded. The only way to protect themselves was to place themselves back to back and ward off the attackers until some type of help came (ironically, most often from the police). If someone cannot fight well and is overcome quickly, everyone's back will be exposed and everyone becomes vulnerable. Likewise, if someone decides to make a run for it, everyone's position is compromised. So assessing the potential member's ability to fight is not done simply to strengthen the gang's reputation as "the meanest fighters," but rather to strengthen the confidence of other gang members that the new member adds to the organization's general ability to protect and defend the collective's interests. The comments of Vase, an eighteen-year-old leader of a gang in New York, highlight this point:

> When I first started with the Silk Irons [gang's name], they checked me out to see if I could fight. After I passed their test, they told me that they didn't need anybody who would leave their butts uncovered. Now that I'm a leader I do the same thing. You see the guy over there? He wants to be in the Irons, but we don't know nothing about whether he can fight or if he got no heart [courage]. So we going to check out how good he is and whether he going to stand and fight. 'Cause if he ain't got good heart or skills [ability to fight], he could leave some of the brothers [gang members] real vulnerable and in a big mess. And if [he] do that, they going to get their asses messed up!

As mentioned earlier, in those cases where the gang has seen a prospective member fight enough to know he will be a valuable member, they simply admit him. However, if information is needed in order to decide whether the prospective gang member can fight, the gang leadership sets up a number of situations to test the individual. One favorite is to have one of the gang members pick a fight with the prospective member and observe the response. It is always assumed that the prospective member will fight; the question is, how well will he fight? The person selected to start the fight is usually one of the better fighters. This provides the group with comparative information by which to decide just how good the individual is in fighting.[15] Such fights are often so intense that there are numerous lacerations on the faces of both fighters. This test usually doubles as an initiation rite, although there are gangs who follow up this test phase with a separate initiation ritual where the individual is given a beating by all those gang members present. This beating is more often than not symbolic, in that the blows delivered to the new members are not done using full force. However, they still leave bruises.

Assessing whether a prospective gang member is trustworthy or not is likewise done by setting up a number of small tests. The gang members are concerned with whether the prospective member is an undercover agent for law enforcement. To help them establish this, they set up a number of criminal activities (usually of medium-level illegality) involving the individual(s); then they observe whether law enforcement proceeds to make arrests of the specific members involved. One gang set up a scam whereby it was scheduled to commit an armed robbery. When a number of the gang members were ready to make the robbery, the police came and arrested them—the consequence of a new member being a police informer. The person responsible was identified and punished. Testing the trustworthiness of new recruits proved to be an effective policy because later the gang was able to pursue a much more lucrative illegal venture without the fear of having a police informer in the organization.

Recruiting a certain number of new members who have already established reputations as good fighters does help the gang. The gang's ability to build and maintain a reputation for fighting reduces the number of times it will have to fight. If a gang has a reputation as a particularly tough group, it will not have as much trouble with rival gangs trying to assume control over its areas of interest. Thus, a reputation acts as an initial deterrent to rival groups. However, for the most part, the gang's concern with recruiting good fighters for the purpose of enhancing its reputation is secondary to its concern that members be able to fight well so that they can help each other.

Gangs who are selective about who they allow in also scrutinize whether the individual has any special talents that could be useful to the collective. Sometimes these special talents involve military skills, such as the ability to build incendiary bombs. Some New York gangs attempted to recruit people with carpentry and masonry skills so that they could help them renovate abandoned buildings.

Gangs that adopt a "fraternity recruiting style" are usually quite secure within their communities. They have a relatively large membership and have integrated themselves into the community well enough to have both legitimacy and status. In other words, the gang is an organization that is viewed by members of the community as legitimate. The comments of Mary, a 53-year-old garment worker who was a single parent in New York, indicate how some community residents feel about certain gangs:

> There are a lot of young people who want in the Bullets, but they don't let whoever wants to get in. Those guys are really selective about who they want. Those who do get in are very helpful to the whole community. There are many times that they have helped the community . . . and the community appreciates that they have been here for us.

Gangs that use fraternity style recruitment have often become relatively prosperous. Having built up the economic resources of the group to a level that has benefited the general membership, they are reluctant to admit too many new members, fearing that increased numbers will not be accompanied by increases in revenues, resulting in less for the general membership.

Hackman, a twenty-eight-year-old leader of a New York gang, represented this line of thought:

> Man, we don't let all the dudes who want to be let in. We can't do that, or I can't 'cause right now we're sitting good. We gots a good bank account and the whole gang is getting dividends. But if we let in a whole lot of other dudes, everybody will have to take a cut unless we come up with some more money, but that don't happen real fast. So you know the brothers ain't going to dig a cut, and if it happens, then they going to be on me and the rest of the leadership's ass and that ain't good for us.

The Obligation Type of Recruitment

The second recruiting technique used by gangs is what I call the "obligation type." In this form, the gang contacts as many young men from its community as it can and attempts to persuade them that it is their duty to join. These community pressures are real, and individuals need to calculate how to respond to them, because there are risks if one ignores them. In essence, the gang recruiter's pitch is that everyone who lives in this particular community has to give something back to it in order to indicate both appreciation of and solidarity with the community. In places where one particular gang has been in existence for a considerable amount of time (as long as a couple of generations), "upholding the tradition of the neighborhood" (not that of the gang) is the pitch used as the hook. The comments of Paul and Lorenzo are good examples. Paul was a nineteen-year-old member of an Irish gang in New York:

> Yeah, I joined this group of guys [the gang] because they have helped the community and a lot of us have taken some serious lumps [injuries] in doing that. . . . I think if a man has any sense of himself, he will help his community no matter what. Right now I'm talking to some guys about joining our gang and I tell them that they can make some money being in the gang, but the most important thing is they can help the community too. If any of them say that they don't want to get hurt or something like that, I tell'm that nobody wants to get hurt, but sometimes it happens. Then I tell them the bottom line, if you don't join and help the community, then outsiders will come and attack the people here and this community won't exist in a couple of years.

Lorenzo was a 22-year-old Chicano gang member from Los Angeles. Here he is talking to two prospective members:

> I don't need to talk to you dudes too much about this [joining a gang]. You know what the whole deal is, but I want you to know that your barrio [community] needs you just like they needed us and we delivered. We all get some battle scars [he shows them a scar from a bullet wound], but that's the price you pay to keep some honor for you and your barrio. We all have to give something back to our community.[16]

This recruiting pitch is primarily based on accountability to the community. It is most effective in communities where the residents have depended on the gang to help protect them from social predators. This is because gang recruiters can draw on the moral support that the gang receives from older residents.

Although the power of this recruiting pitch is accountability to the community, the recruiter can suggest other incentives as well. Three positive incentives generally are used. The first is that gang members are respected in the community. This means that the community will tolerate their illegal business dealings and help them whenever they are having difficulty with the police. As Cardboard, a sixteen-year-old member of a Dominican gang, commented:

> Hey, the dudes come by and they be putting all this shit about that I should do my part to protect the community, but I told them I'm not ready to join up. I tell you the truth, I did sometimes feel a little guilty, but I still didn't think it was for me. But now I tell you I been changing my mind a little. I thinking more about joining. . . . You see the dudes been telling me the community be helping you do your business, you understand? Hey, I been thinking, I got me a little business and if they right, this may be the final straw to get me, 'cause a little help from the community could be real helpful to me. [He joined the gang three weeks later.]

The second incentive is that some members of the community will help them find employment at a later time. (This happens more in Irish neighborhoods.) The comments of Andy, a seventeen-year-old Irish-American in Boston, illustrate this view:

> The community has been getting squeezed by some developers and there's been a lot of people who aren't from the community moving in, so that's why some of the Tigers [gang's name] have come by while we've been talking. They want to talk to me about joining. Just like they been saying, the community needs their help now and they need me. I really was torn because I thought there might be some kind violence used and I don't really want to get involved with that. But the other day when you weren't here, they talked to me and told me that I should remember that the community remembers when people help and they take care of their own. Well, they're right, the community does take care of its own. They help people get jobs all the time 'cause they got contacts at city hall and at the docks, so I been thinking I might join. [He joined three weeks later.]

The third incentive is access to women. Here the recruiter simply says that because the gang is a part of the community and is respected, women look up to gang members and want to be associated with them. So, the pitch continues, if an individual wants access to a lot of women, it will be available through the gang. The comments of Topper, a fifteen-year-old Chicano, illustrate the effectiveness of this pitch:

> Yeah, I was thinking of joining the Bangers [a gang]. These two home-boys [gang members] been coming to see me about joining for two months now. They've been telling me that my barrio really needs me and I should help my people. I really do want to help my barrio, but I never really made up my mind. But the other day they were telling me that the *mujeres* [women] really dig homeboys because they do help the communi-ty. So I was checking that out and you know what? They really do! So, I say, hey, I need to seriously check the Bangers out. [One week later he joined the gang.]

In addition to the three positive incentives used, there is a negative one. The gang recruiter can take the tack that if a prospective member decides not to join, he will not be respected as much in the community, or possibly even within his own family. This line of persuasion can be successful if other mem-bers of the prospective recruit's family have been in a gang and/or if there has been a high level of involvement in gangs throughout the community. The suggestion that people (including family) will be disappointed in him, or look down on his family, is an effective manipulative tool in the recruiting process in such cases. The comments of Texto, a fifteen-year-old Chicano, provide a good example:

> I didn't want to join the Pearls [gang's name] right now cause I didn't think it was best for me right now. Then a few of the Pearls came by to try to get me to join. They said all the stuff about helping your barrio, but I don't want to join now. I mean I do care about my barrio, but I just don't want to join now. But you heard them today ask me if my father wanted me to join. You know I got to think about this, I mean my dad was in this gang and I don't know. He says to me to do what you want, but I think he would be embarrassed with his friends if they heard I didn't want to join. I really don't want to embarrass my dad, I don't know what I'm going to do. [He joined the gang one month later.]

The "obligation method of recruitment" is similar to that employed by governments to secure recruits for their armed services, and it meets with only moderate results. Gangs using this method realized that while they would not be able to recruit all the individuals they made contact with, the obligation method (sometimes in combination with the coercive method) would enable them to recruit enough for the gang to continue operating.

This type of recruitment was found mostly, although not exclusively, in Irish and Chicano communities where the gang and community had been highly integrated. It is only effective in communities where a particular gang or a small number of gangs have been active for a considerable length of time.

The Coercive Type of Recruitment

A third type of recruitment involves various forms of coercion. Coercion is used as a recruitment method when gangs are confronted with the need to increase their membership quickly. There are a number of situations in which this occurs. One is when a gang has made a policy decision to expand its operations into another geographic area and needs troops to secure the area and keep it under control. The desire to build up membership is based on the gang's anticipation that there will be a struggle with a rival gang and that, if it is to be successful, it will be necessary to be numerically superior to the expected adversary.

Another situation involving gang expansion also encourages an intense recruitment effort that includes coercion. When a gang decides to expand into a geographic area that has not hitherto been controlled by another gang, and is not at the moment being fought for, it goes into the targeted area and vigorously recruits members in an effort to establish control. If individuals from this area are not receptive to the gang's efforts, then coercion is used to persuade some of them to join. The comment of Bolo, a seventeen-year-old leader of a New York gang, illustrates this position:

> Let me explain what just happened. Now you might be thinking, what are these dudes doing beating up on somebody they want to be in their gang? The answer is that we need people now, we can't be waiting till they make up their mind. They don't have to stay for a long time, but we need them now. . . . We don't like to recruit this way 'cause it ain't good for the long run, but this is necessary now because in order for us to expand our business in this area we got to get control, and in order to do that we got to have members who live in the neighborhood. We can't be building no structure to defend ourselves against the Wings [the rival gang in the area], or set up some communications in the area, or set up a connection with the community. We can't do shit unless we got a base and we ain't going to get any base without people. It's that simple.

A third situation where a gang feels a need to use a coercive recruiting strategy involves gangs who are defending themselves against a hostile attempt to take over a portion of their territory. Under such conditions, the gang defending its interests will need to bolster its ranks in order to fend off the threat. This will require that the embattled gang recruit rapidly. Often, a gang that normally uses the fraternity type of recruitment will be forced to abandon it for the more coercive type. The actions of these gangs can be compared to those of nation-states when they invoke universal conscription (certainly a form of coercion) during times when they are threatened and then abrogate it when they believe they have recruited a sufficient number to neutralize the threat, or, more usually, when a threat no longer exists. The comments of M.R. and Rider represent those who are recruited using coercion. M.R. was a nineteen-year-old ex-gang member from Los Angeles:[17]

I really didn't want to be in any gang, but one day there was this big blowout [fight] a few blocks from here. A couple of O Streeters who were from another barrio came and shot up a number of the Dukes [local gang's name]. Then it was said that the O Streeters wanted to take over the area as theirs, so a group of the Dukes went around asking people to join for a while till everything got secure. They asked me, but I still didn't want to get involved because I really didn't want to get killed over something that I had no interest in. But they said they wanted me and if I didn't join and help they were going to mess me up. Then the next day a couple of them pushed me around pretty bad, and they did it much harder the following day. So I thought about it and then decided I'd join. Then after some gun fights things got secure again and they told me thanks and I left.

Rider was a sixteen-year-old member of an Irish gang from New York:

Here one day I read in the paper there was fighting going on between a couple of gangs. I knew that one of the gangs was from a black section of the city. Then some of the Greenies [local Irish gang] came up to me and told me how some of the niggers from this gang were trying to start some drugs in the neighborhood. I didn't want the niggers coming in, but I had other business to tend to first. You know what I mean? So I said I thought they could handle it themselves, but then about three or four Greenies said that if I didn't go with them that I was going to be ground meat and so would members of my family. Well, I know they meant business because my sister said they followed her home from school and my brother said they threw stones at him on his way home. So, they asked me again and I said OK. . . . Then after we beat the niggers' asses, I quit. . . . Well, the truth is that I wanted to stay, but after the nigger business was over, they didn't want me. They just said that I was too crazy and wouldn't work out in the group.

This last interview highlights the gang's movement back to their prior form of recruitment after the threat was over. Rider wanted to stay in the gang but was asked to leave. Many of the members of the gang felt Rider was too crazy, too prone to vicious and outlandish acts, simply too unpredictable to trust. The gang admired his fighting ability, but he was the kind of person who caused too much trouble for the gang. As T.R., an eighteen-year-old leader of the gang, said:

There's lots of things we liked about Rider. He sure could help us in any fight we'd get in, but he's just too crazy. You just couldn't tell what he'd do. If we kept him, he'd have the police on us all the time. He just had to go.

There is also a fourth situation in which coercion is used in recruiting. Sometimes a gang that has dominated a particular area has declined to such an extent that it can no longer control all its original area. In such situations, certain members of this gang often decide to start a new one. When this occurs, the newly constituted gang often uses coercive techniques to recruit members and establish authority over its defined territory. Take the comments of Rob and Loan Man, both of whom were leaders of two newly constituted gangs. Rob was a sixteen-year-old leader of the gang, said:

> There was the Rippers [old gang's name], but so many of their members went to jail that there really wasn't enough leadership people around. So a number of people decided to start a new gang. So then we went around the area to check who wanted to be in the gang. We only checked out those we really wanted. It was like pro football scouts, we were interested in all those that could help us now. Our biggest worry was getting members, so when some of the dudes said they didn't want to join, we had to put some heavy physical pressure on them; because if you don't get members, you don't have anything that you can build into a gang. . . . Later after we got established we didn't need to pressure people to get them to join.

Loan Man was a twenty-five-year-old member of a gang in New York:

> I got this idea to start a new gang because I thought the leaders we had were all fucked up. You know, they had shit for brains. They were ruining everything we built up and I wasn't going to go down with them and lose everything. So I talked to some others who didn't like what was going on and we decided to start a new club [gang] in the neighborhood we lived in. So we quit. . . . Well, we got new members from the community, one way or the other. . . . You know we had to use a little persuasive muscle to build our membership and let the community know we were able to take control and hold it, but after we did get control, then we only took brothers who wanted us [they used the fraternity type of recruiting].

In sum, the coercive method of recruitment is used most by gangs that find their existence threatened by competitor gangs. During such periods, the gang considers that its own needs must override the choice of the individual and coercion is used to induce individuals to join their group temporarily. . . .

WHO DOES NOT JOIN A GANG?

Who does not become involved with a gang, and why? There are two answers. First, individuals who see no personal advantages in participating in a gang do not become involved. These individuals can be separated into two distinct groups. The first is those who possess all the characteristics associated with defiant individualism, but have decided that participation in a gang is not to their advantage at the present. Most of these individuals are involved in a variety of economic ventures (usually illegal) that they hope will make them rich and do not perceive any advantage to becoming involved with the gang's activities, but the vast majority of them will become involved with a gang at some time in the future. The comment of Cover, a seventeen-year-old who lives in New York, illustrates this point:

Right now, man, there ain't no reason for me to join the Black Widows. I got some good business going and I'm making some decent money. If the police don't mess me up, I can get some good cash flow going. So right now there ain't any incentive to join, you know what I mean? Now I ain't saying that I won't join sometime, 'cause they gots some good business going themselves. But right now I want to keep with what I'm doing and see where it goes. [He joined the gang one year later.]

There are also those who not only see no advantage to becoming involved in a gang, but also see significant disadvantages: the risks of being killed or imprisoned associated with gang life are too great in their view. They want to get out of the area they live in, but they have developed a strategy for doing so that does not involve the gang. Some of these people will seek socioeconomic mobility through sports, placing their hopes for a better life in their ability to become professional athletes. Ironically, however, there is probably less likelihood of their achieving such mobility through sports than through becoming involved in some illegal type of business venture.

Others from low-income neighborhoods who seek socioeconomic mobility but do not want to join a gang are those prepared to take the risk that investing their time and money in some type of formal training (formal education or the trades) will produce mobility for them. The comments of Phil, the eighteen-year-old son of a window washer in Los Angeles, represent this group:

No, I don't want to join any gang. I know you can make money by being in, but frankly I don't want to take the risks of being killed or something. I mean some of the dudes in the gang make a whole lot of money, but they take some big risks too. I just don't want to do that. I want to get out of this neighborhood, so I'll just take my chances trying to get out by studying and trying to go to college. I know there are risks with that too. I mean even if you go to college don't mean you going to make a fortune. My cousin went to college and he started a business and it failed, so I know there is risks that I won't make doing it my way, but at least they don't include getting shot or going to prison.

The individuals in both these groups (those seeking mobility through sports and those who seek it through training) possess some of the characteristics associated with defiant individualism, but not the full defiant individualist character structure. Why some people from low-income areas have only a few of the characteristics of defiant individualism and others have them all has to do with a number of contingent factors related to exactly how each individual has experienced his or her social environment.

The second answer to the question of who does not become involved in gangs and why has to do with the fact that gangs do not want everyone who seeks membership. They will reject people if they are not good fighters, cannot be trusted, or are unpredictable (cannot be controlled), as well as if the gang itself already has too many members (in which case additional members would create difficulties in terms of social control and/or a burden in providing the services expected by the membership). Take the comment of Michael, the eighteen-year-old son of a street sweeper in New York:

> Sure, I wanted to join the Spears and I hung out with them, but they never invited me to formally join. You have to get a formal invitation to join, and they didn't give me one. They just told me that they had too many members right now, maybe sometime later.

WHAT HAPPENS TO GANG MEMBERS?

What is the trajectory of the individual who is involved in gangs? Thrasher and most of the subsequent studies on gangs believed that individuals who were in gangs simply matured out of them as they got older.[18] However, evidence from the present study suggests that the real story is more complicated. I found there are seven possible outcomes, some of which are not necessarily exclusive of each other. First, some people stay in the gang. As of this writing some individuals in their late thirties were still members of gangs. What will happen to them is open to question.

Second, some members will drop out of their gangs and pursue various illegal economic activities on their own.

Third, a number of gang members will move on to another type of organization or association. Many of the Irish will join Irish social clubs. Others will move to various branches of organized crime.

Fourth, there will be individuals who move from gangs and become involved in smaller groups like "crews" where they can receive a larger take of the money they have stolen than if they were in a gang.

Fifth, some will be imprisoned for a considerable part of their lives. While this will negate their involvement in the gangs of the streets, they will remain involved in the prison gangs.

Sixth, there will be those whose fate will be death as a consequence of a drug overdose, a violent confrontation, or the risks of lower-class life.[19]

Seventh, a large number will take the jobs and live the lives they were trying to avoid. While this may appear to be what Thrasher and others have previously reported, it is hardly accurate to think of it as "maturing out of the gang."

What do these future paths mean for the gang? Gangs are composed of individuals with defiant individualist characters who make decisions on the basis of what is good for them. On an individual level, this means that gang members will often come and go throughout long periods of their lives. Take the comment of Clip, a thirty-six-year-old member of a Los Angeles gang:

> I've been in gangs since I was fifteen. I joined and then quit and joined again. I did different things, I been married twice, but I come back to the gang 'cause there is always a chance if you get some business going, you

can make some big money and live in leisure. That's been my goal and always will be.

On the aggregate level this means that coming and going is merely an integral part of the organizational environment.

CONCLUSION

Who joins a gang and why have been central concerns of many studies having to do with gangs. One set of studies concentrates on delinquency, asking why individuals are inclined to engage in illegal acts. Another incorporates gangs into a larger analysis of community. Neither approach directly addresses the gang itself. If one begins not from delinquency or community but from the defiant individualist character of gang recruits, one sees that defiant individuals make rational decisions as to what is best for them. Although previous studies have argued that all individuals have the same reason for becoming involved with gangs, prospective members in fact have a variety of reasons for doing so. However, this chapter also shows that the varying motives of defiant individuals do not suffice to explain who joins a gang and why. Gang involvement is also determined by the needs and desires of the organization. The answer to the question of who joins a gang and why depends on the complex interplay between the individual's decision concerning what is best for him and the organization's decision as to what is best for it.

NOTES

1. See Frederic Thrasher, *The Gang* (Chicago: University of Chicago Press, 1928); Gerald D. Suttles, *The Social Order of the Slum* (Chicago: University of Chicago Press, 1968); John Hagedorn, *People and Folks* (Chicago: Lakeview Press, 1988).

2. Of course, some of the studies cited here overlap these categories, and I have therefore placed them according to the major emphasis of the study. See Richard A. Cloward and Lloyd B. Ohlin, *Delinquency and Opportunity* (New York: Free Press, 1960); Hagedorn, *People and Folks*; Joan Moore, *Homeboys* (Philadelphia: Temple University Press, 1978); James F. Short, Jr., and Fred L. Strodtbeck, *Group Process and Gang Delinquency* (Chicago: University Press, 1965).

3. Here again it is important to restate that many of these studies overlap the categories I have created, but I have attempted to identify them by what seems to be their emphasis. See Herbert A. Block and Arthur Niederhoffer, *The Gang* (New York: Philosophical Library, 1958); Lewis Yablonsky, *The Violent Gang* (New York: Macmillan, 1966).

4. See the qualifying statement in nn. 2 and 3 above. See Ruth Horowitz, *Honor and the American Dream* (New Brunswick, N.J.: Rutgers University

Press, 1983); Albert Cohen, *Delinquent Boys* (Glencoe, Ill.: Free Press, 1955); Walter B. Miller, "Lower Class Culture as a Generating Milieu of Gang Delinquency," *Journal of Social Issues* 14 (1958): 5–19; James Diago Vigil, *Barrio Gangs* (Austin: University of Texas Press, 1988).

5. See Jack Katz, *The Seduction of Crime: Moral and Sensual Attractions in Doing Evil* (New York: Basic Books, 1988), p. 9.

6. Although the present study is not a quantitative study, the finding reported here and the ones to follow are based on observation of, and conversations and formal interviews with, hundreds of gang members.

7. For a discussion of the political machine's role in providing psychological and financial support for poor immigrant groups, see Robert K. Merton, *Social Theory and Social Structure* (New York: Free Press, 1968), pp. 126–36. Also see William L. Riordan, *Plunkitt of Tammany Hall* (New York: Dutton, 1963).

8. There are numerous examples throughout the society of social clubs using the lodge or clubhouse as one of the incentives for gaining members. There are athletic clubs for the wealthy (like the University Club and the Downtown Athletic Club in New York), social clubs in ethnic neighborhoods, the Elks and Moose clubs, the clubs of various veterans' associations, and tennis, yacht, and racket ball clubs.

9. See Anne Campbell, *Girls in the Gang* (New York: Basil Blackwell, 1987).

10. For the use of drugs as recreational, see Vigil, *Barrio Gangs*; and Jeff Fagan, "Social Organization of Drug Use and Drug Dealing among Urban Gangs," *Criminology* 27 (1986): 633–70, who reports varying degrees of drug use among various types of gangs. For studies that report the monitoring and/or prohibition of certain drugs by gangs, see Vigil, *Barrio Gangs*, on the prohibition of heroin use in Chicano gangs; and Thomas Mieczkowski, "Getting Up and Throwing Down: Heroin Street Life in Detroit," *Criminology* 24 (November 1986): 645–66.

11. See Thrasher, *The Gang*, pp. 84–96. He also discusses the gang as a source of recreation.

12. See Moore, *Homeboys*, ch. 2; Horowitz, *Honor and the American Dream*, ch. 8; Vigil, *Barrio Gangs*; Hagedorn, *People and Folks*.

13. For a discussion of these types of jobs, see Michael J. Piore, *Notes for a Theory of Labor Market Stratification*, Working Paper no. 95 (Cambridge, Mass.: Massachusetts Institute of Technology, 1972).

14. See Horowitz, *Honor and American Dream*; and Ruth Horowitz and Gary Schwartz, "Honor, Normative Ambiguity and Gang Violence," *American Sociological Review* 39 (April 1974): 238–51. There are many other studies that could have been cited here. These two are given merely as examples.

15. The testing of potential gang members as to their fighting ability was also observed by Vigil. See his *Barrio Gangs*, pp. 54–55.

16. This quotation was recorded longhand, not tape-recorded.

17. I first met M.R. when he was in one of the gangs that I was hanging around with. He subsequently left the gang, and I stayed in touch with him by talking to him when our paths crossed on the street. This quotation is from a long conversation that I had with him during one of our occasional encounters.

18. See Thrasher, *The Gang*, pp. 66–67. A great number of studies take a similar position. I shall mention but a few. See Suttles, *Social Order of the Slum*, and William F. Whyte, *Street Corner Society* (Chicago: University of Chicago Press, 1943). The work of Ruth Horowitz represents a modified exception to the other findings. While she does imply that gang members grow old and abandon their street lives, she also reports that involvement in gangs may last well into the thirties for individuals. See her *Honor and the American Dream*, pp. 177–97.

19. Some of these people have died from illnesses, food poisoning, and various accidents.

35

Marks of Mischief

Becoming and Being Tattooed

CLINTON R. SANDERS

A person's physical appearance is a central element affecting his or her self-definition, identity, and interaction with others (Cooley, 1964: 97–104, 175–178, 183; Stone, 1970; Zurcher, 1977: 44–45, 175–178). People use appearance to place each other into categories which aid in the anticipation and interpretation of behavior and to make decisions about how best to coordinate social activities (Goffman, 1959: 24–25; McCall and Simmons, 1982: 214–216; Ruesch and Kees, 1972: 40–41, 57–65).

How closely one meets the cultural criteria for beauty is an appearance factor of key social and personal import. Being defined as attractive has considerable impact on our social relationships. We think about attractive people more often, define them as being more healthy, express greater appreciation

From: "Marks of Mischief: Becoming and Being Tattooed," Clinton R. Sanders, *Journal of Contemporary Ethnography*, Vol. 16, No. 4, 1988. Reprinted by permission of Sage Publications, Inc.

for their work, and find them to be more appealing interactants (Jones et al., 1984: 53–56). Attractive people are more adept at establishing relationships (Brislin and Lewis, 1968) and enjoy more extensive and pleasant sexual interactions than do those who are not as physically appealing (Hatfield and Sprecher, 1986). Their chances of economic success are greater (Feldman, 1975), and they are consistently defined by others as being of high moral character (Needleman and Weiner, 1977).

Enjoying more frequent positive interactions, attractive people have correspondingly more positive self-definitions. In general, they express more feelings of general happiness (Berscheid et al., 1973), have higher levels of self esteem and are less likely than the relatively unattractive to expect that they will suffer from mental illness in the future (Napoleon et al., 1980).

Clothing, cosmetics, and hair styling are mechanisms for altering appearance that have in common the relative ease with which one can change one's social "vocabulary" if the message communicated becomes outdated, undesirably stigmatizing or otherwise worthy of reconsideration. In general, the cross-cultural literature on adornment and body alteration indicates that non-permanent decorative forms (principally costume and body paint) are most commonly associated with transitional statuses or specific and limited social situations. The major forms of permanent alteration—body sculpture, infibulation (piercing), cicatrization (scarification) and tattooing—are, on the other hand, connected to permanent statuses (e.g., gender, maturity), life-long social connections (e.g., clan or tribal membership) or conceptions of beauty that show considerable continuity from generation to generation (see Polhemus, 1978: 149–173).

Those who choose to permanently modify their bodies in ways that violate prevailing appearance norms—or who reject culturally prescribed alterations—risk being defined as socially or morally inferior. Public display of voluntarily acquired, symbolic physical deviance effectively communicates a wealth of information that shapes the social situation in which interaction takes place (Goffman, 1963a; Lofland, 1973: 79–87).

This article focuses on tattooing as a form of permanent body alteration in contemporary society. Choosing to mark one's body in this way changes the tattooee's experience of his or her physical self and has significant potential for altering social interaction. Because of the historical course of tattooing in the West the tattoo is conventionally defined as an indication of the bearer's alienation from mainstream norms and social networks. It is *voluntary stigma* that symbolically isolates the bearer from "normals." Since tattooees are deemed to be responsible for their "deviant" physical condition, the mark is especially discrediting (Jones et al., 1984: 56–65).

Like most stigmatizing conditions, however, tattooing also has an affiliative effect; it identifies the bearer as a member of a select group. When publically displayed the tattoo may act as a source of mutual accessibility (Goffman, 1963b: 131–139). Fellow tattooees commonly recognize and acknowledge their shared experience, decorative tastes and relationship to conventional society. Tattooing also has affiliative impact in that it is routinely employed to demonstrate one's indelible connection to primary associates (e.g., name tattoos) or groups whose members share specialized interests and activities (e.g., motorcycling, use of illegal drugs, or involvement with a specific youth gang). . . .

METHOD

I first became interested in tattooing in San Francisco in 1979. Having a bit of time on my hands, I decided to explore the more obscure museums listed in the telephone directory. Climbing the dingy stairway leading to Lyle Tuttle's Tattoo Art Museum, I found myself in a new and fascinating world of cultural production. After looking at the sizeable collection of tattoo memorabilia, I entered the tattoo studio adjacent to the museum and, like many first-time visitors to tattoo establishments, impulsively decided to join the ranks of the tattooed. After choosing a small scarab design from the wall "flash," I submitted to the unexpectedly painful tattoo experience. Although the resident tattooist was not very forthcoming in response to the questions I forced out between clinched teeth, I did realize that this was a phenomenon which combined my interests in both social deviance and art worlds and offered a research experience which would provide a much-needed escape from the polite confines of academia.

Returning to the east coast I visited a small "street shop" in a "transitional neighborhood" located a few minutes from my office. The owner was flattered that a "professor" would want to hang out and listen to him talk about himself, and I soon became a regular participant in the shop, observing the work, talking to the participants, and—despite my original vow to never again undergo the pain of indelible body alteration—eventually receiving considerable tattoo "work" from a variety of renowned tattoo artists with whom I came into contact during the subsequent seven years.

The following discussion is based primarily on data collected during participant observation in four tattoo "studios" located in or near major urban centers in the east. Three were traditional shops specializing in the formulaic images favored by military personnel, bikers, laborers, and occasional groups of college students and secretaries. One establishment was a "custom" studio, in which a tattooist with extensive professional experience in a variety of artistic media created original and unique works of art for a more select, monied, and aesthetically sophisticated clientele.

For the most part, my role in the settings was that of one of a number of regular hangers-on who either lived in the neighborhood or were friends of the local artist. My participation in the establishment to which I originally gained access was considerably more extensive. In addition to (apparently) just standing around and chatting, I helped with the nontattooing business of the shop. I made change for the amusement games, provided information about cost and availability of designs, stretched the skin of customers who were receiving tattoos on body areas other than arms or legs, calmed the anxiety of first-time recipients, and made myself generally useful.

In addition to the field data, this discussion is based on a series of lengthy, semi-structured tape-recorded interviews conducted with tattoo recipients encountered during the course of the research. I collected interviews with 16 people (10 men and 6 women) who were representative of the sex, age, and social status categories I encountered in the field settings. Their average age was 24 (from 17 to 39); as a group they carried 35 tattoos (9 had one, 3 had two, and 4 had three or more).

A somewhat more structured body of data was drawn from 163 four-page questionnaires completed by tattooees contacted in three separate settings. 56 were filled out by tattoo "enthusiasts" attending the 1984 convention of the National Tattoo Association in Philadelphia, 44 were returned by clients in the "artistic" studio, and 63 questionnaire respondents completed the instrument following their tattoo experience in the street shop in which I began to collect field data. Sixty-eight percent of the questionnaire respondents were men and 32 percent were women. They ranged in age from 17 to 71, with an average of 30 years. Sixty-two percent of the respondents had received some education past high school and 5% had graduate degrees. Skilled craftwork, machine operation, and general laboring were the most common occupations pursued by the men; service and clerical work was most heavily represented among the women. Twelve percent of the men and 6% of the women were involved in professional or technical occupations.

THE PROCESS OF BECOMING
A TATTOOED PERSON

Initial Motives

Becoming tattooed is a highly social act. The decision to acquire a tattoo (and, as we will see in a later section, the image that is chosen), like most major consumer products is motivated by how the recipient defines him or her self. The tattoo becomes an item in the tattooee's personal "identity kit" (Facetti and Fletcher, 1971; Goffman, 1961: 20–21), and in turn it is used by those with whom the individual interacts to place him or her into a particular, interaction-shaping social category (see Csikszentmihalyi and Rochberg-Halton, 1981; Solomon, 1983).

When asked to describe how they decided to get a tattoo, the vast majority of respondents made reference to another person or group. Family members, friends, business associates, and other people with whom they regularly interacted were described as being tattooed. Statements such as "Everyone I knew was really into tattoos. It was a peer decision. Everyone had one, so I wanted one" and "My father got one when he was in the war and I always wanted one, too" were typical. Entrance into the actual tattooing "event," however, has all of the characteristics of an impulse purchase. It typically is based on very little information or previous experience (58% of the questionnaire respondents reported *never* having been in a studio prior to the time they received their first tattoo). While tattooees commonly reported having "thought about getting (a tattoo) for a long time," they usually drifted into the actual experience when they "didn't have anything better to do," had sufficient money to devote to a nonessential purchase, and were, most importantly, in the general vicinity of a tattoo establishment. The following accounts were fairly typical.

> We were up in Maine and a bunch of us were just talking about getting tattoos—me and my friends and my cousins. One time my cousin came back from the service with one and I liked it. . . . The only place I knew about was _____ _____ 's down in Providence. We were going right by there on our way back home, so we stopped and all got them.
>
> My friends were goin' down there to get some work, you know. That was the only place I knew about, anyway. My friends said there was a tattoo parlor down by the beach. Let's go! I checked it out and seen something I liked. I had some money on me so I said, "I'll get this little thing and check it out and see how it sticks." I thought if I got a tatty it might fade, you know. You never know what's goin' to happen. I don't want anything on my body that is goin' to look fucked up.

The act of getting the tattoo itself is usually, as seen in these quotes, a social event experienced with close associates. Sixty-nine percent of the interviewees, (11 of 16) and 64% of the questionnaire sample reported having received their first tattoo in the company of family members or friends. These close associates act as "purchase pals" (Bell, 1967). They provide social support for the decision, help to pass anxiety-filled waiting time, offer opinions regarding the design and body location, and commiserate with or humorously ridicule the recipient during the tattoo experience (see Becker and Clark, 1979; Sanders, 1985a).

The tattoo event frequently involves a ritual commemoration of a significant transition in the life of the recipient (compare Brain, 1979: 174–184; Ebin, 1979: 39–56; Van Gennep, 1960). The tattooee conceives of the mark as symbolizing change—especially achieving maturity and symbolically separating the self from individuals or groups (parents, husbands, wives, employers, etc.) who have been exercising control over the individual's personal choices. A tattoo artist related his understanding of his clients' motivations in this way:

> I do see that many people get tattooed to find out again . . . to say, "Who was I before I got into this lost position?" It's almost like a tattoo pulls you back to a certain kind of reality about who you are as an individual. Either that or it transfers you to the next step in your life—the next plateau. A woman will come in and say, "Well, I just went through a really ugly

divorce. My husband had control of my fucking body and now I have it again. I want a tattoo. I want a tattoo that says that I have the courage to get this, that I have the courage to take on the rest of my life. I'm going to do what I want to do and do what I have to do to survive as a person." That's a motivation that comes through the door a lot.

One interviewee expressed her initial reason for acquiring her first tattoo in almost exactly the same terms:

(My friend and I) both talked semiseriously about getting (a tattoo). I mentioned it to my husband and he was adamantly opposed—only certain seedy types get tattoos. He didn't want someone else touching my body intimately, which is what a tattoo would involve . . . even if it was just my arm. He was against it, which made me even more for it. . . . I finally really decided some time last year when my marriage was coming apart. It started to be a symbol of taking my body back. I was thinking that about the time I got divorced would be a good time to do it.

Locating a Tattooist

Like the initial decision to get a tattoo, the tattooist one decides to patronize commonly is chosen through information provided by members of the individual's personal networks. The shop in which they received their first tattoo was located by 58% of the questionnaire respondents through a recommendation provided by a friend or family member. Since in most areas establishments that dispense tattoos are not especially numerous, many first-time tattooees choose a studio on a very practical basis—it is the only one they know about or it is the studio which is closest to where they live (20% of the questionnaire sample chose the shop on the basis of location, 28% because it was the only one they knew about).

The central importance of personal recommendation as the source of tattoo clients is well-known to tattooists. All tattooists have business cards that they hand out quite freely (one maintained that he had dispensed over 50,000 cards in the past two years). Listing one's services in the telephone directory is the other major means employed to draw customers, since it provides locating information for those who, for a variety of reasons, do not have interpersonal sources.

Most first-time tattooees enter the tattoo setting with little information about the process or even the relative skill of the artist. Rarely do recipients spend as much time and effort acquiring information about a process that is going to indelibly mark their bodies as they would were they preparing to purchase a TV set or other far less significant consumer item.

Consequently, tattooees usually enter the tattoo setting ill-informed and experiencing a considerable degree of anxiety. Their fears center around the anticipated pain of the process and the permanence of the tattoo. Here, for example, is an interaction that took place while a young man received his first tattoo.

Recipient: Is this going to fade out much? There's this guy at work that has these tattoos all up and down his arms and he goes back to the guy that did them every couple of months and gets them recolored because they fade out. (general laughter)

Sanders: Does this guy work in a shop or out of his house?

R: He just does it on the side.

Tattooist: He doesn't know what the fuck he's doing.

R: This friend of mind told me that getting a tattoo really hurts. He said there would be guys in here hollering and bleeding all over the place.

S: Does he have any tattoos?

R: No, but he says he wants to get some. . . . Hey, this really doesn't hurt that much. It doesn't go in very deep, does it? It's like picking a splinter out of your skin. I was going to get either a unicorn or a Pegasus. I had my sister draw one up because I thought they just drew the picture on you or something. I didn't know they did it this way (with an acetate stencil). I guess this makes a lot more sense.

S: You ever been in a tattoo shop before this?

R: No, this is my first time. Another guy was going to come in with me, but he chickened out.

For the most part, tattooists are quite patient about answering the questions clients ask with numbing regularity (pain, price, and permanence). This helps to put the recipient more at ease, smooths the service delivery interaction, and increases the chances that a satisfied customer—who will recommend the shop to his/her friends and perhaps return again for additional work—will leave the establishment (for extended discussions of in-shop interaction see Becker and Clark, 1979; Govenar, 1977; St. Clair and Govenar, 1981; and Sanders, 1983 and 1985b).

CHOOSING A DESIGN AND BODY LOCATION

Tattooees commonly stated their basic motivations for becoming tattooed in very general terms. Wearing a tattoo connected the person to significant others who were similarly marked, made one unique by separating him or her from those who were too convention-bound to so alter their bodies, symbolized freedom or self-control, and satisfied an aesthetic desire to decorate the physical self.[1]

The image one chooses, on the other hand, is usually selected for a specific reason. Typically, design choice is related to the person's connection to other people, his or her definition of self or, especially in the case of women, the desire to enhance and beautify the body.

One of the most common responses which tattoo clients gave to my routine question, "How did you go about deciding on this particular tattoo?" was to make reference to a personal association with whom they had a close emotional relationship. Some chose a particular tattoo because it was like that worn by a close friend or a member of their family. Others chose a design which incorporated the name of their boy/girl friend, spouse or child or a design associated with that person (e.g., zodiac signs):

> I had this homemade cross and skull here and I needed a coverup. [The tattooist] couldn't just do anything, so I thought to myself, "My daughter was born in May, and that's the Bull." I'm leaving the rest of this arm clean because it is just for my daughter. If I ever get married, I'll put something here [on the other arm]. I'll get a rose or something for my wife.
>
> This tattoo is a symbol of friendship. Me and my best friend—I've known him since I could walk—came in together and we both got bluebirds to have a symbol that when we do part we will remember each other by it.

The ongoing popularity of "vow tattoos," such as the traditional heart with "MOM" or flowers with a ribbon on which the loved one's name is written, attests to the importance of tattooing as a way of symbolically expressing love and commitment (see Anonymous, 1982).

Similarly, tattoos are used to demonstrate connection and commitment to a group. For example, military personnel pick tattoos which relate to their particular service, motorcycle gang members choose club insignia, and members of sports teams enter a shop en masse and all receive the same design.

Tattoos are also employed as symbolic representations of how one conceives of the self or interests and activities which are key features of self definition. Tattooees commonly choose their birth sign or have their name or nickname inscribed on their bodies. Others choose more abstract symbols of the self.

> I put a lot of thought into this tattoo. I'm an English lit major, and I thought that the medieval castle had a lot of significance. I'm an idealist, and I thought that was well expressed by a castle with clouds. Plus, I'm blond and I wanted something blue.
>
> [Quote from field notes] Two guys in their twenties come in and look at the flash. After looking around for a while one of the guys comes over to me and asks if we have any bees. I tell him to look through the book [of small designs] because I have seen some bees in there. I ask, "Why do you want a bee? I don't think I have ever seen anyone come in here for one." He replies, "I'm allergic to bees. If I get stung by one again I'm going to die. So I thought I'd come in here and have a big, mean looking bee put on. I want one that has this long stinger and these long teeth and is coming in to land. With that, any bee would think twice about messing with me."

Tattooees commonly represent the self by choosing designs which symbolize important personal involvements, hobbies, occupational activities, and so forth. In most street shops, the winged insignia of Harley-Davidson motorcycles and variants on that theme are the most frequently requested images. During one particularly busy week in the major shop in which I was observing, a rabbit breeder acquired a rabbit tattoo, a young man requested a cartoon frog because the Little League team he coached was named the "Frogs," a fireman received a fire fighter's cross insignia surrounded by flame, and an optician chose a flaming eye.

No matter what the associational or self-definitional meaning of the chosen tattoo, the recipient is commonly aware of the decorative-aesthetic function of the design. When I asked tattooees to explain how they went about choosing a particular design, they routinely made reference to aesthetic criteria—they "liked the colors" or they "thought it was pretty."

> [I didn't get this tattoo] because of being bad or cool or anything like that. It's like a picture. You see a picture you like and you put it in your room or your house or something like that. It's just a piece of work that you like. I like the art work they do here. I like the color [on my tattoo]. It really brings it out—the orange and the green. I like that—the colors.

On their part, tattooists tend to recognize the aesthetic importance of their work as seen by their clients. One tattooist, for example, observed:

> If you ask most people why they got (a particular tattoo) they aren't going to have any deep Freudian answers for you. The most obvious reason that someone gets a tattoo is because they like it for some reason and just want it. I mean, why do people wear rings on their fingers or any sort of non-functional decorative stuff—put on makeup or dye their hair? People have the motivation to decorate themselves and be different and unique. . . . Tattooing is really the most intimate art form. You carry it on your body. The people that come in here are really mostly just "working bumpkins." They just want to have some art they can understand. This stuff in museums is bullshit. Nobody ever really sees it. It doesn't get to "the people" like tattoo art.[2]

A number of factors determine a tattooee's decision about where on the body the tattoo will be located. The vast majority of male tattooees choose to have their work placed on the arm. In his study of the tattoos carried by 2,000 members of the Royal Navy, Scutt found that 98% had received their tattoo(s) on the arm (Scutt and Gotch, 1974: 96). In my own research, 55% of the questionnaire respondents received their first tattoo on the arm or hand (71% of the males and 19% of the females). The 16 interviewees had, all together, 35 tattoos, 27 of which were carried by the 10 males. Eighty-one percent (22) of the men's tattoos were on their arms (of the remainder 2 were on hips, one was on the back, one on the face and one on the recipient's chest). The 6 women interviewees possessed 8 tattoos—3 on the back or the shoulder area, 3 on the breast, 1 on an arm and 1 on the lower back. Thirty-five percent of female questionnaire respondents received their first tattoo on the breast, 13% on the back or shoulder and 10% on the hip.

Clearly, there is a definite convention affecting the decision to place the tattoo on a particular part of the body—men, for the most part, choose the arm while women choose the breast, hip, lower abdomen or back/shoulder. To some degree the tendency for male tattooees to have the tattoo placed on the arm is determined by technical features of the tattoo process. Tattooing is a two-handed operation. The tattooist must stretch the skin with one hand while inscribing the design with the other. This operation is most easily accomplished when the tattoo is being applied to an extremity. Tattooing the torso is more difficult and, commonly, tattooists have an assistant who stretches the client's skin when work is being done on that area of the body. Technical difficulty, in turn, affects price. Most tattooists charge 10 to 25% more for tattoos placed on body parts other than the arm or leg. The additional cost factor probably has some effect on the client's choice of body location.

Pain is another factor shaping the tattooee's decision. The tattoo machine contains needle groups which superficially pierce the skin at high speed, leaving small amounts of pigment in the tiny punctures. Obviously, this process will cause more or less pain depending on the sensitivity of the area being tattooed. In general, tattooing arms or legs is less painful than marking body areas with a higher concentration of nerve endings or parts of the body where the bones are not cushioned with muscle tissue.[3]

The different symbolic functions of the tattoo for males versus females appears to be a major issue affecting the sex-based conventions regarding choice of body site. Women tend to regard the tattoo (commonly a small, delicate design) as a permanent body decoration primarily intended for personal pleasure and the enjoyment of those with whom they are most intimate. The chosen tattoos are, therefore, placed on parts of the body most commonly seen by those with whom women have primary relationships. Since tattoos on women are especially stigmatizing, placement on private parts of the body allows women to retain unsullied identities when in contact with casual associates or strangers (see Goffman, 1963a: 53–55, 73–91). Here, for example, is a portion of a brief conversation with a young woman who carried an unconventional design (a snake coiled around a large rose) on what is, for women, an unconventional body location (her right bicep).

Sanders: How did you decide on that particular design?

Woman: I wanted something really different and I'd never seen a tattoo like this on a woman before. I really like it, but sometimes I look at it and wish I didn't have it.

S: That's interesting. When do you wish you didn't have it?

> W: When I'm getting real dressed up in a sleeveless dress and I want
>   to look . . . uh, prissy and feminine. People look at a tattoo and
>   think you're real bad . . . a loose person. But I'm not.

Another interviewee described the decision-making process she had gone
through in choosing to acquire a small rose design on her shoulder, emphasizing aesthetic issues and stigma control.

> The only other place I knew of that women got tattoos was on the breast.
> I didn't want it on the front of my chest because I figured if I was at work
> and had an open blouse or a scoop neck, then half would show and half
> wouldn't. I wanted to be able to control when I wanted it to show and
> when I didn't. If I go for a job interview I don't want a tattoo on my
> breast. I didn't want it, like, on my thigh or on the lower part of my
> stomach. I didn't like how they look there. I just thought it would look
> pretty on my shoulder. . . . The main reason is that I can cover it up if I
> want to.

Men, on the other hand, typically are less inclined than women to define
the tattoo primarily as a decorative and intimate addition to the body. Instead,
the male tattoo is an identity symbol—a more public display of interests, associations, separation from the normative constraints of conventional society
and, most generally, masculinity. The designs chosen by men are usually larger
than those favored by women and, rather than employing the gentle imagery
of nature and mythology (flowers, birds, butterflies, unicorns and so forth),
they frequently symbolize more violent impulses. Snakes, bloody daggers,
skulls, dragons, grim reapers, black panthers and birds of prey are dominant
images in the conventional repertoire of tattoo designs chosen by men.
Placement of the image on the arm allows both casual public display and,
should the male tattooee anticipate a critical judgment from someone whose
negative reaction could have untoward consequences (mostly commonly, an
employer), the tattooed arm can be easily hidden with clothing. One male
interviewee spoke about the public meaning of tattoos and expressed his
understanding of the difference between male and female tattoos as follows:

> You fit into a style. People recognize you by your hair style or by your
> tattoo. People look at you in public and say, "Hey, they got a tattoo.
> They must be a particular kind of person," or, "He's got his hair cropped
> short (so) he must be a different kind of person." The person with a tat-
> too is telling people that he is free enough to do what he wants to do. He
> says, "I don't care who you think I am. I'm doing what I want to do."
> (The tattoo) symbolizes freedom. It says something about your personali-
> ty. If a girl has a skull on her arm—it's not feminine at all—that would
> symbolize vengeance. If a woman gets a woman's tattoo, that's normal. If
> she gets a man's tattoo symbolizing vengeance or whatever, I feel that is
> too far over the boards. A woman should act like a woman and keep her

tattoos feminine. Those vengeance designs say, "Look out." People see danger in them.

THE INTRAPERSONAL AND INTERPERSONAL EXPERIENCE OF WEARING A TATTOO

Impact on Self-Definition

As indicated in the foregoing presentation of the initial motives which prompt the decision to acquire a tattoo, tattooees consistently conceive of the tattoo as having impact on their definition of self and demonstrating to others information about their unique interests and social connections. Interviewees commonly expressed liking their tattoo(s) because it (they) made him or her "different" or "special" (see Goffman, 1963a: 56–62):

> Having a tattoo changes how you see yourself. It is a way of choosing to change your body. I enjoy that. I enjoy having a tattoo because it makes me different from other people. There is no one in the whole world who has a right arm that looks anything like mine. I've always valued being different from other people. Tattooing is a way of expressing that difference. It is a way of saying, "I am unique."

In describing his understanding of his client's motives, one tattoo artist employed the analogy of the customized car.

> Tattooing is really just a form of personal adornment. Why does someone get a new car and get all of the paint stripped off of it and paint it candy apple red? Why spend $10,000 on a car and then spend another $20,000 to make it look different from the car you bought? I associate it with ownership. Your body is one of the things you indisputably own. There is a tendency to adorn things that you own to make them especially yours.

> Interviewees also spoke of the pleasure they got from the tattoo as related to having gone through the mysterious and moderately painful process of being tattooed. The tattoo demonstrated courage to the self ("For some people it means that they lived through it and weren't afraid"). One woman, when asked whether she intended to acquire other tattoos in the future, spoke of the excitement of the experience as the potential motivator of additional work.

> [Do you think you will have more work done after you add something to the one you have now?] Oh God! don't know why, but my initial reaction is, "I hope I don't, but I think I'm going to." I think getting a tattoo is so exciting and I've always been kind of addicted to excitement. It's fun. While it hurt and stuff it was a new experience and it wasn't that horrible for me. It was new and different.

In a poignant statement, another woman spoke similarly of the tattoo as memorializing significant aspects of her past experience.

[In the future] when I'm sitting around and bored with my life and I wonder if I was ever young once and did exciting things, I can look at the tattoo and remember.

Interactional Consequences

In general, tattooees' observations concerning the effect of having a tattoo and the process of being tattooed on their self-definitions were rather basic and off-hand. In contrast, all interviewees spoke at some length about their social experiences with others and how the tattoo affected their identities and interactions. Some stressed the affiliational consequences of being tattooed—the mark identified them as belonging to a special group.

> I got tattooed because I had an interest in it. My husband is a chef and our friends tend to be bikers, so it gets me accepted more into that community. They all think of me as "the college girl" and I'm really not. So this (tattoo) kind of brings the door open more. . . . The typical biker would tell you that you almost have to have tattoos to be part of the group.

Most took pleasure in the way the tattoo enhanced their identities by demonstrating their affiliation with a somewhat more diverse group—tattooed people.

> Having a tattoo is like belonging to a club. I love seeing tattoos on other people. I go up and talk with other people with tattoos. It gives me an excuse because I'm not just going up to talk with them. I can say, "I have one, too." I think maybe subconsciously I got (the tattoo) to be part of that special club.
>
> Having tattoos in some ways does affect me positively because people will stop me on the street and say, "Those are really nice tattoos," and show me theirs. We kind of . . . it is a way of having positive contact with strangers. We have something very much in common. We can talk about where we got them and the process of getting them and that sort of thing.

Given the symbolic meaning carried by tattoos in conventional social circles, all tattooees have the experience of being the focus of attention because of the mark they carry. The positive responses of others are, of course, the source of the most direct pleasure.

> People seem to notice you more when you walk around with technicolor arms. I don't think that everyone who gets tattooed is basically an exhibitionist, someone that walks down the street and says, "Hey, look at me!" you know. But it does draw attention to yourself. [How do people respond when they see your tattoos?] Well, yesterday we were sitting in a bar and the lady brings a beer over and she says, "That's gorgeous," and she's looking at the wizard and she's touching them and picking up my shirt. Everyone in the bar was looking and it didn't bother me a bit.

Not all casual encounters are as positive as this one. Revelation of the tattoo is also the source of negative attention when defined by others as a stigmatizing mark.

Sometimes at these parties the conversation will turn to tattoos and I'll mention that I have some. A lot of people don't believe it, but if I'm feeling loose enough I'll roll up my sleeve and show my work. What really aggravates me is that there will almost always be someone who reacts with a show of disgust. "How could you do that to yourself?" No wonder I usually feel more relaxed and at home with bikers and other tattooed people.

I think tattoos look sharp. I walk down the beach and people look at my tattoos. Usually they don't say anything. [When they do] I wish they would say it to my face . . . like, "Tattoos are ugly." But, when they say something behind my back . . . "Isn't that gross." Hey, keep your comments to yourself! If you don't like it, you don't like it. I went to the beach with my father and I said, "Hey, let's walk down the beach," and he said, "No, I don't feel like it." What are you, embarrassed to walk with me?

Given the negative responses that tattooees encounter with some frequency when casual associations or strangers become aware of their body decorations, most are selective about to whom they reveal their tattoos. This is particularly the case when the "other" is in a position to exercise control over the tattooee.

Usually I'm fairly careful about who I show my tattoos to. I don't show them to people at work unless they are really close friends of mine and I know I won't get any kind of hassle because of them. I routinely hide my tattoos. . . . I generally hide them from people who wouldn't understand or people who could potentially cause me trouble. I hide them from my boss and from a lot of the people I work with because there is no reason for them to know.

Tattooees commonly use the reactions of casual associates or relative strangers as a means of categorizing them. A positive reaction to the tattoo indicates social and cultural compatibility, while a negatively judgmental response is seen as signifying a narrow and convention-bound perspective.

I get more positive reactions than I do negative reactions. The negative reactions come from people who aren't like me—who have never done anything astray. It is the straight-laced, conservative person who really doesn't believe that this is acceptable in their set of norms.

It seems as though I can actually tell how I'm going to get along with people and vice-versa by the way they react to my tattoo. It's more or less expressive of the unconventional side of my character right up front. Most of the people who seem to like me really dig the tattoo too [quoted in Hill, 1972: 249].

While it is fairly easy to selectively reveal the tattoo in public settings when interacting with strangers or casual associates, hiding the fact that one is tattooed, thereby avoiding negative social response, is difficult when the "other" is a person with whom the tattooee is intimately associated. The majority of those interviewed recounted incidents in which parents, friends, and, especially for the women, lovers and spouses reacted badly when they initially became aware of the tattoo.

> [What did your husband say when he saw your tattoo?] He said he almost threw up. It grossed him out. I had asked him years ago, "What would you think if . . . " and he didn't like the idea. So, I decided not to tell him. It seemed a smart thing to do. He just looked rather grossed out by the whole thing; didn't like it. Now it is accepted, but I don't think he would go for another one.

Another woman interviewee recounted a similar post-tattoo experience with her boyfriend.

> I got a strange reaction from my boyfriend. We had a family outing to go to and there was going to be swimming and tennis and all this stuff and I was real excited about going. He said, "Are you going to go swimming?" I said, "Yeah." I was psyched, because I love to swim. He looked at me and said, "You know, your tattoo is going to show if you go swimming." Probably. He didn't want me to go swimming because he didn't want his parents to know that I had a tattoo. Lucky for him it was cloudy that day and nobody swam. I told him, "I'm sorry, but I know your parents can handle this kind of news." To boot, he's got a shamrock on his butt! So he has a tattoo—a real double standard there. He didn't say anything for a while after I first got it. It was subtle. He let me know he didn't like it but that because it was on me he could excuse it. He's got adjusted to it, though. He just let me know that he's never dated a girl who's got a tattoo before. He would prefer that I didn't have it, but there isn't much he can do about that now.

Given the negative social reaction often precipitated by tattoos, it would be reasonable to expect that tattooees who regretted their decision would have emphasized the unpleasant interactional consequences of the tattoo. Interviewees and questionnaire respondents rarely expressed any doubts about their decision to acquire a tattoo. Those that did indicate regret, however, usually did not focus on the stigmatizing effect of the tattoo. Instead, regretful tattooees most commonly were dissatisfied with the *technical quality* of the tattoo they purchased.[4] (See also Sanders, 1985a.) . . .

NOTES

1. Questionnaire respondents were given an open-ended question which asked them to speculate as to why people get tattooed. Of the 163

respondents, 135 provided some sort of reply to this item. Forty-four percent of those responding emphasized that becoming tattooed was motivated by a desire for self-expression (e.g., "vanity," "it's a personal preference," "a statement of who you are"), 21% emphasized tattooing as a mechanism for asserting uniqueness and individuality (e.g., "people like to be different," "personal originality," "it makes you special") and 28% made some form of aesthetic statement (e.g., "because it is beautiful," "a form of art that lasts forever," "body jewelry").

2. Of the 35 tattoos worn by the 16 interviewees, 14% (5) represented birds, 6% (2) represented mammals, 14% (5) represented mythical animals, 9% (3) represented insects, 3% (1) represented a human female, 17% (6) represented human males, 14% (5) were noncommercial symbols (hearts, crosses, military insignia, etc.), 14% (5) were floral, 3% (1) were names or vow tattoos, and 6% (2) were some other image. Questionnaire respondents were asked to indicate the design of their first tattoo. They were: 14% (23) bird, 11% (18) mammal, 12% (19) mythical animal, 10% (17) insect, 1% (2) human female, 6% (10) human male, 4% (6) commercial symbols, 8% (13) noncommercial symbols, 21% (34) floral/arborial, 4% (7) name/vow, and 9% (14) other.

3. The painfulness of the tattoo process is the most unpleasant element of the tattoo event. Only 33% (54) of the questionnaire respondents maintained that there was something about the tattoo experience that they disliked. One third of these (18) said that the pain was what they found most unpleasant. Fourteen of the 16 interviewees mentioned pain as a troublesome factor. Numerous observations of groups of young men discussing pain or stoically expressing little regard for the pain as they were receiving tattoos made it difficult not to see the tattoo event as having ritualized initiatory aspects. In some cases the tattoo process provides a situation in which the male tattooee can demonstrate his "manliness" to his peers. Here, for example, is a description of an incident in which five members of a local college football team acquired identical tattoos on their hips:

> I asked the guy nearest to me if they are all getting work done. "Yeah. He [indicates friend] was so hot for it he would have done it himself if we couldn't get it done today." [You all getting hip shots?] "Yeah, that's where all jocks get them. The coach would shit if he found out." The conversation among the jocks turns to the issue of pain. They laugh as the guy being worked on grimaces as W [artist] finishes the outline and wipes the piece down with alcohol. The client observes that this experience isn't bad compared to the time "I fucked up my hand in a game and had to have steel pins put in the knuckles. One of them got bent and the doctor had to cut it out. That was *bad*. I got the cold sweats." Some of the others join in by telling their "worst pain I ever experienced" stories. The guy being worked on is some-

thing of a bleeder and the others kid him about this. As W begins shading one of them shouts, "Come on, really grind it in there."

The cross-cultural literature on body alteration indicates that the pain of the process is an important factor. Ebin (1979: 88–89), for example, in discussing tattooing in the Marquesas Islands, states:

> The tattoo was not only an artistic achievement: it also demonstrated that its recipient could bear pain. On one island, the word to describe a person who was completely covered with tattoos is *ne'one'o,* based on a word meaning either "to cry for a long time" or "horrific." One observer in the Marquesas noted that whenever people discussed the tattoo design, they emphasized the pain with which it was acquired.

See also Becker and Clark, 1979: 10, 19; Brain, 1979: 183–184; Ross and McKay, 1979: 44–49, 67–69; St. Clair and Govenar, 1981: 100–135.

4. Other than simply accepting the regretted mark, there are a few avenues of resolution open to dissatisfied tattooees. At the most extreme, the tattooee may try to obliterate the offending mark with acid or attempt to cut it off. A somewhat more reasoned (and considerably less painful) approach entails seeking the aid of a dermatologist or plastic surgeon who will medically remove the tattoo. However, the most common alternative chosen by regretful tattooees is to have the technically inferior piece redone or covered with another tattoo created by a more skilled practitioner. Tattooists estimate that 40 to 50% of their work entails reworking or applying cover-ups to poor-quality tattoos. See Goldstein et al., 1979 and Hardy, 1983.

REFERENCES

Anonymous. (1982) "The name game," Tattootime 1: 50–54.

Becker, N., and R. Clark (1979) "Born to raise hell: an ethnography of tattoo parlors." Presented at the meetings of the Southwestern Sociological Association, March.

Bell, G. (1967) "Self-confidence, persuasability and cognitive dissonance among automobile buyers," pp. 442–468 in D. Cox (ed.) Risk-Taking and Information Handling in Consumer Behavior. Boston: Harvard University Graduate School of Business Administration.

Berscheid, E., et al. (1973) "Body image, physical appearance and self-esteem." Presented at the annual meetings of the American Sociological Association.

Brain, D. (1979) The Decorated Body. New York: Harper & Row.

Brislin, R., and S. Lewis (1968) "Dating and physical attractiveness: a replication." Psych. Reports 22: 976–984.

Cooley, C. H. (1964 [1902]) Human Nature and the Social Order. New York: Schocken.

Csikszentmihalyi, M., and E. Rochberg-Halton (1981) The Meaning of Things. Cambridge: Cambridge Univ. Press.

Ebin, V. (1979) The Body Decorated. London: Thames & Hudson.

Facetti, G., and A. Fletcher (1971) Identity Kits: A Pictorial Survey of Visual Signals. New York: Van Nostrand Reinhold.

Feldman, S. (1975) "The presentation of shortness in everyday life," pp. 437–442 in S. Feldman and G. Thielbar (eds.) Life Styles. Boston: Little, Brown.

Goffman, E. (1959) Presentation of Self in Everyday Life. Garden City, NY: Doubleday.

———. (1961) Asylums. Garden City, NY: Doubleday.

———. (1963a) Stigma. Englewood Cliffs, NJ: Prentice-Hall.

———. (1963b) Behavior in Public Places. New York: Free Press.

Goldstein, N., et al. (1979) "Techniques of removal of tattoos." J. of Dermatological Surgery and Oncology 5: 901–910.

Govenar, A. (1977) "The acquisition of tattooing competence: an introduction." Folklore Annual of the University Folklore Association 7 and 8: 43–63.

Hardy, D. (1983) "Inventive cover work." Tattootime 2: 12–17.

Hatfield, E., and S. Sprecher (1986) Mirror, Mirror . . . Albany: State Univ. of New York Press.

Hill, A. (1972) "Tattoo renaissance," pp. 245–249 in G. Lewis (ed.) Side-Saddle on the Golden Calf. Pacific Palisades, CA: Goodyear.

Jones, E., et al. (1984) Social Stigma: The Psychology of Marked Relationships. New York: Freeman.

Lofland, L. (1973) A World of Strangers. New York: Basic Books.

McCall, G., and J. Simmons (1982) Social Psychology. New York: Free Press.

Napoleon, T., et al. (1980) "A replication and extension of 'physical attractiveness and mental illness.'" J. of Abnormal Psychology 89: 250–253.

Needleman, B., and N. Weiner (1977) "Appearance and moral status in the arts." Presented at the annual meetings of the Popular Culture Association.

Polhemus, T. (ed.) (1978) The Body Reader. New York: Pantheon.

Ross, R., and H. McKay (1979) Self-Mutilation. Lexington, MA: Lexington Books.

Ruesch, J., and W. Kees (1972) Nonverbal Communication. Berkeley: Univ. of California Press.

St. Clair, L., and A. Govenar (1981) Stoney Knows How: Life as a Tattoo Artist. Lexington: Univ. Press of Kentucky.

Sanders, C. (1983) "Drill and fill: client choice, client typologies and interactional control in commercial tattoo settings." Presented at the Art of the Body Symposium, UCLA.

————. (1985a) "Tattoo consumption: risk and regret in the purchase of a socially marginal service," pp. 17–22 in E. Hirshman and M. Holbrook (eds.) Advances in Consumer Research, Vol. XII. New York: Association for Consumer Research.

————. (1985b) "Selling deviant pictures: the tattooist's career and occupational experience." Presented at the Conference on Social Theory, Politics and the Arts, Adelphi University, October.

Scutt, R., and R. Gotch (1974) Art, Sex and Symbol. New York: Barnes.

Solomon, M. (1983) "The role of products as social stimuli: a symbolic interactionist perspective." J. of Consumer Research 10 (December): 319–329.

Stone, G. (1970) "Appearance and the self," pp. 394–414 in G. Stone and H. Faberman (eds.) Social Psychology through Symbolic Interaction. Waltham, MA: Xerox.

Van Gennep, A. (1960) The Rites of Passage. Chicago: Univ. of Chicago Press.

Zurcher, L. (1977) The Mutable Self. Newbury Park, CA: Sage.

36

Managing Deviance: Hustling, Homophobia, and the Bodybuilding Subculture

ALAN M. KLEIN

INTRODUCTION

Some of American society's most exclusively male institutions, such as the military and organized sports, where masculinity is fashioned and most exaggerated, have rarely been studied in depth, and seldom critically.[1] The following case study examines the social and psychological dimensions of masculinity within the sport subculture of southern California bodybuilding.

From: "Managing Deviance: Hustling, Homophobia, and the Bodybuilding Subculture," Alan M. Klein, *Deviant Behavior*, Vol. 10, 1989. Reprinted by permission of Taylor and Francis, Inc. All rights reserved.

There, one can find a variety of behaviors and conventions that exaggerate, yet reflect, the larger society's notions of masculinity. One of the most intriguing complexes of behavior is the condemned but prevalent practice of "hustling." Hustling is the selling of sex to gays, and is the behavioral conflux for a variety of male traits: hypermasculinity, homophobia, and narcissism. I argue that while functioning to hold disparate segments of the bodybuilding community together, hustling (a) is partly a temporary response to an economic crisis in the competitive bodybuilder's pursuit of success, (b) fulfills some bodybuilders' need for admiration, (c) can support the comicbook notion of masculinity that is so prevalent in bodybuilding, and (d) is highly conflicting for those involved as they must maintain a self-perception of heterosexuality while engaging in homosexual practices. Borrowing from Reiss (1971) and Matza (1969), I show how hustling is carried out and how its practitioners juggle self-identity.

Riding on the coattails of the health movement, cultural fears of aging, and an increased cultural receptivity to mass spectacle, bodybuilding has experienced unprecedented growth in the past decade (Klein, n.d). This popularity is in part due to the societal resurgence of an atavistic notion of masculinity (e.g., films such as *Rambo, Terminator;* New Wave fashion and hair styles) that has articulated what bodybuilders have always accepted as a standard for men. Comicbook masculinity depicts men one-dimensionally as stoic, brave to a fault, always in control, aggressive, and competitive. These qualities become synonymous with a well-built man. No form of sport or popular culture seeks to replicate the trappings of this notion of masculinity more than bodybuilding.

The function of sports in establishing male identity as a stereotype has been thoughtfully presented by Sabo and Runfola (1980) and Sabo (1985). This study, however, probes more deeply into a male subculture which simultaneously values the conventions and trappings of macho while enabling its members to engage in seemingly opposing homosexual acts.[2] In particular, for hustling bodybuilders struggling to juggle these opposing notions, the psychological consequences can lead to a serious crisis.

THE SETTING

Field Methods

Based on anthropological fieldwork carried out between 1979 and 1985 at one of southern California's foremost gyms (here given the fictitious name of Olympic Gym), this study and its findings are part of a larger ethnographic work on the subculture of bodybuilding. In six years of intermittent study, 55 formal interviews, and approximately 120 informal interviews (field interviews) were carried out with men and women from all strata at the gym. Of the core community of 150 (with approximately 90 to 100 men), I was able to secure formal interviews with 6 hustlers and informal (designated as field interviews herein) interviews with another 6. Quotes from hustlers are distinguished from other members of the core community by a capital "H" after citing the type of interview. During field stays, observations were conducted daily, and some portion of each day was spent in active participation (for one six-month period, the author trained with two others who were preparing for an upcoming Mr. America contest; at another time the author served as a judge of a contest). Observations and interviews were conducted outside of the gym as well as in a wide variety of external contexts.

Some use was made of questionnaires in the early phase of the study, but high turnover of members made it more desirable to use in-depth interviews with long-term committed members. These interviews consisted of both closed and open-ended questions covering a wide range of areas: biographical, training, and interpersonal relationships (both within and outside of the gym). Additionally, open-ended questions were asked on a number of topics that interested the author at the time (e.g., hustling, steroid use, and bodybuilding politics).

The Social Setting of Olympic Gym

A few blocks from Muscle Beach is Olympic Gym, which is located among some of the world's elite gyms and is home to many world-class bodybuilders. The proximity of these facilities to each other, as well as the pull of Hollywood's media industry, exaggerates many of the flamboyant qualities of bodybuilding subculture.

Despite a membership of over 1500, the "real" bodybuilding community at Olympic Gym consists of a core of only about 150 devotees who follow the lifestyle of bodybuilding more or less full-time. This means working out seriously enough to develop a physique of a bodybuilder, and having the gym become the center of one's life. One's social relations, economic opportunities, and psychological balance revolve around the gym, much as the church or street corner functions for the religiously observant (Aschenbrenner, 1975) or urban social group (Kaiser, 1979).

The core community, however, is constantly in flux. Few bodybuilders wind up staying at Olympic Gym longer than five years. The women, not properly part of this study, now comprise about 30% of the population and are substantially different from the men in their social relations, goal orientations, and backgrounds (Klein, 1985a). At Olympic Gym most range in age between 19 and 25. Racial composition is increasingly mixed, and race relations are generally good.

For all the change in Olympic Gym, there is a definite social structure, one that reflects its position in the bodybuilding hierarchy (Klein, 1985b). Six strata can be distinguished and ranked in terms of status. The respective numbers of each of these groupings increases as one moves down the hierarchy, forming a pyramid. The largest grouping has the least status, the smallest has the most. From top down the groups are (a) owners and managers, (b) professional bodybuilders, (c) amateur bodybuilders, (d) gym rats, (e) members-at-large, and (f) pilgrims and on-lookers (Klein, 1985b).

Southern California bodybuilders have, over decades, evolved a subculture replete with its own shared terminology, behavior, and values. As with all subcultures there is a cultural and social separation necessary to forge a collective identity—separation that often comes from being considered marginal or deviant by the larger society. Southern California bodybuilding has had a difficult relationship with the wider society. These days, however, under the guise of superior fitness and cultural popularity, bodybuilders relate to the outside world through a veneer of arrogance; an outward show of superiority that sprang from deep-seated inferiority. This insecurity is rooted in popular suspicions held about bodybuilders.

BODYBUILDING AND ITS
PSYCHOLOGICAL UNDERPINNINGS

Many men gravitate to bodybuilding because of low self-esteem. The physique, it could be argued, is a mask or wall between low self-esteem and a potentially threatening outside world. If done within the supportive environment of the subculture of bodybuilding, the process of building oneself up can boost self-confidence. This is often projected as arrogance, so characteristic of their comments about outsiders (e.g., referring to them as pencilnecks), or their public posturing.

It is noteworthy that some of the traits that the public most disdains are imperative to the bodybuilder's self-esteem. Narcissism, for instance, as in the bodybuilder who constantly gazes at himself in the mirror, has ironically been shown to have a positive function: bolstering a flagging self-esteem (Klein, 1987). The hyper-masculine behavior of many bodybuilders that works to keep the subculture at arms-length from the mainstream also allows for development of the subculture.

The insecurity regarding traits such as shortness or physical impediments (e.g., hearing impairment) are seen by bodybuilders as "afflictions" which continue to pull people into bodybuilding decades after the first Charles Atlas ads appeared in comicbooks and men's magazines. Hence, unlike most sports that use positive impulses to recruit its devotees, bodybuilding has always recruited on the basis of what sport psychologist Butts (1975:56) termed the "neurotic element." Typical of this were comments as, "I don't know, I guess I wanted some size. I was, you know, real skinny." (Interview 5/12/83). Another bodybuilder said:

> Yeah, I guess there's an element of insecurity in me that will always make me unsatisfied with my build. I think it's true of most of the guys around here. You don't think that he (pointing) walks like that (exaggerating his musculature) because he's secure do you? (Interview, 11/5/79)

Because bodybuilding has the overt function of shoring up cracks in the egos of the people who come to it, the needs of the bodybuilders for acknowledgement and admiration is much greater than for the public-at-large.

The subculture of bodybuilding institutionalizes many of the concerns of bodybuilders, objectifying their neuroses so that they may partially overcome them. This is done by promoting a shared world-view in which accruing size is equated with psychological enhancement. Yet for some the need for admiration and/or acknowledgement continues to go unmet.

Hypermasculinity and Bodybuilding Subculture

Feelings of weakness and insecurity are often masked by a veneer of power. The institutions of bodybuilding not only fetishize the look of power but also foster identification with and reliance upon figures of power. The author (1985a, 1987) has pointed this out both at the level of bodybuilding politics as well as within the institutionalized narcissism of bodybuilding. Both hypermasculinity and homophobia are in part reactions against feelings of powerlessness. Psychologist T. Adorno (1950:405, 428), who interpreted the "authoritarian personality," associated it with, among other things, overly muscled bodies, boastfulness, swaggering independence, and worship of power seen as cloaking feelings of weakness.

The background as well as behavior of many bodybuilders meshes with Adorno's characterization. Pleck (1982:34) goes so far as to claim bodybuilding as, "perhaps the archetypal expression of male identity insecurity." Administering the Cattel 16PF psychological test to a sample of men and women at Olympic Gym in 1982, Sprague (n.d.) too found a high degree of insecurity.

HOMOPHOBIA

One longtime professional at Olympic Gym noted the homophobia present as:

> . . . natural, whenever you have a nicely developed body exposed to the public, you'll have people flocking around to see it. As far as men being attracted to me and that whole notion, I could care less. For so many this (bodybuilding) is a macho trip; and for the traditional male in America there's still a lot of homophobia. I know it abounds at Olympic Gym. (Field interview, 11/24/79)

The fear of homosexuality functions to socialize males.[3] Lehne points out that homophobia curbs the range of responses in men so as to dichotomize between sexes, but he goes on to note athletes as the only male exception to many homophobic prohibitions:

> Only athletes and women are allowed to touch and hug each other in our culture. Athletes are only allowed this because presumably their masculinity is beyond doubt. (Lehne, 1976:84)

The athletic exception turns on the same fundamental contradiction between homoeroticism and a rigid heterosexual identity, both of which are informed by mysogyny. This is poignantly revealed in the autobiography of Dave Kopay:

David Kopay's story raises the question of not how could he emerge from his super masculine society as a homosexual, but how could any man come through it as purely heterosexual after spending so much time idealizing and worshipping the male body, while denigrating and ridiculing the female. (Kopay and Young, 1977:117)

Kopay, a 10-year veteran of the National Football League emerged from the closet and onto the front pages of many major daily newspapers when he suddenly admitted being gay. His termination from the NFL was as extreme as the excessive and vindictive new coverage. American male anxieties had been hit at the core. Pro-football players especially, it was thought, don't become gay because they are the gatekeepers of masculinity.

Other examples that juxtapose exaggerated masculinity and homoeroticism can be found in the widespread locker-room banter that often centers on homosexuality. Clichés on the subject abound and were common at Olympic Gym as well. However, because some of bodybuilding's practices such as hustling, wearing very brief posing trunks, removal of body hair, unusual posing, etc., generate suspicion, the joking about homosexuality is often tense. Often what begins as a joke evokes very serious and menacing responses. Ironically, then, rather than reducing anxiety, homophobia in a bodybuilding context works to generate anxiety around homosexuality. The practice of hustling is the activity that most brings together these disparate elements into a particularly thorny set of sociological and psychological issues for those who engage in it.

HUSTLING AND BODYBUILDING

Definition

The selling of implicit or explicit sex by a bodybuilder to someone gay is called hustling, and it appears to be widespread among southern California's most competitive bodybuilders. Both hustlers and the overwhelming majority of bodybuilders in southern California interviewed do not consider hustlers gay, however. Because they seek to retain heterosexuality as their avowed sexual preference, hustlers must find a way to justify the homosexual practices with heterosexual identities. The few who are gay, are seen as gay rather than as hustlers, even though they all engage in similar exchanges. Hustling is seen as something one does out of economic necessity; it is an economic strategy in a world of few options.

Actually, hustling can cover a range of behaviors from popping out of a cake nude at gay parties, to nude photography or pornography, to explicit sex with another man. "You do what your conscience lets you do" was the way one man put it. All hustling is paid for, however, and all hustling is with gay men.[4]

Incidence of Hustling

Hustling is widely condemned yet, judging by interviews and observation, quite common. Estimates from a wide variety of Olympic Gym's core community claim that anywhere from 50% to 80% have or do now hustle:

Respondent: Here [pointing at the gym], there's a lot of hustling going on.

Author: What percentage [of bodybuilders] do you estimate are hustling?

Respondent: I don't know, but I'd say somewhere around half. S. over there still does it. So does M. and K., and A. actually gets off on it. (Field interview, 8/1/84)

Even if we were to half the estimate of the incidence of hustling, it would still be significant enough to be considered widespread. People freely admit it goes on, but almost never the person interviewed. This makes accurate assessment impossible. The stigma attached to hustling is so great that only six hustlers granted me formal interviews on the subject. Six others I knew hustled and admitted it but limited their comments. Still others were observed engaging in transactions (e.g., setting up dates and negotiating with potential clients). Despite the formal disapproval of and discomfort associated with hustling, it seems to be increasing, prompting one well-placed member in the core community to comment:

When I first came here I wasn't aware of half of the situation, even though I had a few people hit on me. Before, people had to prove to me that they were gay, now you gotta prove to me that you're straight. Yes, if anything hustling's increased. (Field interview, 12/13/81)

A set of 12 confirmed hustlers out of more than 90 men is small, but the fact that it took so long to get them to participate in this study underscores the sensitivity of the matter and a reluctance to discuss the subject.[5]

Hustling as an Economic Strategy

Both the money involved and extensiveness of the network make hustling the bulwark of bodybuilding's underground economy. As such hustling functions to enable competitors, particularly amateurs, to subsidize their training and lifestyle until such time as they can succeed in turning professional or drop out of competition. In the 12 cases of hustlers for which I was able to document a shift in status from amateur to professional, all but one quit hustling. In short, hustling is an important economic strategy in an environment where access to resources is very limited.

Were one to work for a living, training would have to fit into the few hours the gym is open before or after work. Given the modest educational background of most men in my sample, finding employment that allows work flexibility and pays enough for training is not likely.[6] Certain jobs, however, do run in the gym community and get passed from one to another. Bouncers in bars, bodyguards, and bill collectors, all are jobs that make use of the large size of the bodybuilder, yet allow for his need for flexible hours.

Compared to hustling, though, these jobs pay little. The market mentality that sees hustling as a means-ends relationship is evident in this hustler's typical comment:

It's not that I'm trying to make it [hustling] okay for me. This is a constant conflict in myself, because I don't have to be one. But I trained for Mr.

America for eight hours a day—eight hours of some sort of training. I couldn't do that working twelve hours a day in some shipyard. I simply couldn't do it. (Field interview, H, 1/30/84)

The economic connection between gays and bodybuilders is clear. Segments of the gay community have been bankrolling aspiring bodybuilders in southern California for decades. Gays can be found on the margins of the subculture, as well as in positions of importance (e.g., entrepreneurs, contest promoters, gym owners, judges, competitors, and gym members).

Hustling as a Psychological Strategy

The need for admiration that many bodybuilders have may be only partially satisfied within the institutional and cultural confines of the subculture. Some crave additional acknowledgement and, for them, a natural bridge exists to segments of the gay community. Whether they exploit it or not is another matter, but this potential source of psychological gratification gains impetus both from personal insecurity and the difficulty of succeeding in the southern California bodybuilding scene. Competition for titles is keen, getting into the magazines is very difficult, and the caliber of bodybuilders is world class. Informants often languished in obscurity in Los Angeles after having been highly successful elsewhere. However, in gay circles one can receive admiration or even become a minor celebrity. The sexual, economic, and psychological bond between gays and hustlers is complex, and at times a non-hustling element can dominate the relationship:

S. and I would see these guys in the gym and we'd say, "Okay, the guy looks down. Let's take him out." We took C. out one day. Took him to Griffith Park. We didn't know it, but it was his birthday and later he told us that it was one of the nicest times of his life. We didn't want a thing to do with him sexually. (Field interview, H, 2/22/86)

Drifting into Hustling

Matza's (1964, 1969) and Davis's (1971) notion of drifting into deviant identity was also evident in some of the people I interviewed. Drifting allows one to account for the time and psychological processes needed to alter thinking and take on a new identity.

The high risk in coming to Los Angeles to compete is coupled with the youthfulness and naivete of many of the men taking the trek. Their economic vulnerability quickly becomes apparent to any veteran bodybuilder, as well as the gays on the fringes of the community. Many veterans (gay and straight) offer these new arrivals tips on survival; some offering places to stay and jobs to get started. This is commonly done in the spirit of camaraderie. Some, however, are looking to "hit on" (make sexual advances toward) these youthful questers. They too may appear to be offering something out of friendship. It may be as innocent-appearing as one of the many ads on the bulletin board at the gym: "Bodybuilder Needed for Photographer. Good money! Call _____." Old timers assume these photos will be nude, and probably done by a gay photographer. They may warn the new arrival, but not always. Serious photographers worry that their ads will also be misconstrued and feel compelled to underscore the fact that they are "straight." Other advances may be more informal:

> I knew a gay guy back in New York. One gay guy, and I liked him. I respected him. When I came out here I had guys hittin' on me, and I didn't even know it. I had a guy hand me a card that said he was a photographer. He said, "If you wanna make some bucks, give me a call." I said, "Geez thanks. That's great! Here's a guy doin' me a favor, wants to take photos of me." I didn't have any [photos] in my portfolio. That's great. Boy did I learn. It was my ass he wanted. (Field interview, 10/5/81)

The material needs of the young bodybuilder predispose him to accept casual offers if they are forthcoming from acceptable quarters. When these offers are accepted and they come from men seeking sex, the pressures to reciprocate sexually can exert a powerful influence. Once sexual conventions have been violated by giving in to the pressure to hustle, the young man, much like a female prostitute in a similar situation, slips into the deviant identity (Davis, 1971).

Once involved, hustlers move into a network of gays who seek out bodybuilders. Hustlers begin to place and answer ads (some explicitly for sex, others for "escort service"), in the larger gay newspapers. Most, however, stay within the very personal network of the subculture or the larger network of gays from Palm Springs to San Francisco. Typically this means a "repeat business" of calling to make arrangements or waiting to be called. Some hustlers establish a large clientele that they guard against others, and very well known hustlers can command as much as $500 a visit. On a few occasions hustlers have parlayed hustling into lucrative gay-porno film careers. New arrivals to the scene are lured in more easily when they are informed of these successes. Expecting this kind of success, some novices will prematurely ask for outrageous fees:

> A friend of mine who had never "done anything" before got into the idea of it. I told him, "Don't get into that sick shit," but he got some rich guy originally from Oklahoma City to dig him. He demanded $500 and a plane ticket. Like if you get a lot of cake, it makes it [hustling] alright. Well, there are rates for that sort of thing, and the guy told my buddy to screw off. Within two days he was going for $50. (Field interview, H, 2/25/86)

THE CRISIS OF HUSTLING IN BODYBUILDING

For bodybuilders who hustle, their greatest contradiction lies in juxtaposing hustling and heterosexual identity. Reiss's (1971) study of street delinquents as hustlers and the norms they generate to separate themselves from homosexuals is particularly applicable here. Compared to the delinquents, the norms bodybuilders use to separate themselves do not work as well.

It is ironic that so many strands of the American male psyche are brought together in hustling: wanting to be seen as virile and sexually desirable; male bonding; homophobic currents; competition; and aggression. One hustler stated it this way:

> It's kinda sad. We put ourselves in a bad social position. I know people who hire us for posing, but there's more expected than that. It puts bodybuilding in a shitty position—to be laughed at. Who's gonna help bodybuilders? A bunch of homosexuals, that's who. We're everything the U.S. is supposed to stand for—strength, determination, everything to be admired. But it's not the girls that like us, it's the fags! (Field interview, H, 10/4/81)

The awareness of this situation by hustlers themselves must be weighed against the needs that are met by hustling. The ease with which money is made is one compelling factor; but there is, as I will show, also a compelling psychological reward fostered by the hustler's personal needs. There is a great deal of anxiety around hustling. The difficulty of juggling one's self-concept and rationalizing how others feel about you makes it difficult to handle. Many deny doing it. One top professional claimed it is just this avoidance behavior that makes hustlers so easy to spot. He noted that hustlers often won't look him in the eye as a result of a public stand he took on hustling. Sometimes, he pointed out, the hustler is so troubled that he develops nervous twitches.

In some instances, the anxiety of being a hustler who is acting straight reaches crisis proportions. During my stay at Olympic Gym, there were three reported suicide attempts as well as a greater number of bodybuilders who, in response to their conflicts, repudiated the subculture altogether. These men and women would often very suddenly become Born Again Christians. Finding God, however, is not the final solution. So long as one remains a competitive bodybuilder in southern California there is a strong pull exerted to re-engage in the repudiated behavior.

Understandably, promoters, organizers, and owners of publishing houses for bodybuilding products seek to conceal or downplay the institution of hustling. They see it as a threat to their vested interests. These people often cut a bodybuilder off from exposure in magazines and, thereby, wipe out the mail order businesses so vital to the up and coming bodybuilder. At Olympic Gym, hustling is officially repudiated, though there is a tacit understanding that it is imperative for survival.

Psychological Coping and Hustling

Creating a psychological framework that permits hustling while rejecting the possibility that they are gay, bodybuilders make primary use of compartmentalization as a defense. Other mechanisms, discussed below, are seen as derivative.

Hustlers compartmentalize by separating their hustling from their straight life, prompting one young hustler to claim, "Hey, it's tougher for gay guys to hustle cuz they gotta be into it. But me, I can get it on with anybody. It's like I'm two different people." (Field interview, H, 11/30/80) Since the gym is an intimate universe, however, the separation so important to identity management is always in jeopardy, waiting only for the first angry outburst by another bodybuilder.

Competitive bodybuilders who hustle have difficulty handling both heterosexual and hustling relations. The time and emotional commitment hustling demands would create a host of problems for a serious heterosexual relationship. From the perspective of the women involved, simultaneous hetero-and-hustling relations of their partners might prove problematic since the women not only have to accept infidelity from their partners but also the homosexual nature of it. From the men's perspective, mixing the two kinds of relations can prove too complicated and, as this hustler phrased it, too morally problematic as well.

> On any given time I can go out with a woman. But it's not very satisfying, like a regular kind of relationship. Women demand time, and I'm too involved with bodybuilding . . . I miss her [pointing to pictures of an ex-mate that are all around his apartment]. I lived with that girl for a year and a half. But it's not that good. Several [women] know what I'm doing. Some can handle it, but some can't and that's another reason. I couldn't lie, that's why I'm not living with anybody. (Field interview, H, 3/19/82)

One novice hustler confessed to the woman he had been involved with about his "California" activities. After apologizing, he got her to come to Los Angeles from the Midwest. She seemed to understand, yet whenever they argued she would seize on his hustling past. Driving past his apartment, she would scream, "Charlie! You're a goddamn faggot!" Other hustlers make the transition back to their heterosexual relations more easily.

Sexual activity among competitive bodybuilders is an area of life that must be carefully parceled out. The rigors of dieting, training, and more importantly, of excessive steroid use severely curb the capacity and will for sex. Trying to juggle homosexual and heterosexual contacts exacerbates this situation. One hustler (who was also gay) who was close to two other hustlers and their female mates pointed out some previously unknown wrinkles in this problem:

> I used to hear these guys going on about their girls. The girls did this and that, and how great they were sexually. But I knew their girls real well, and they'd talk about how these guys would only go down on them in order to get them off [performing oral sex], ya know? But they, the guys, couldn't get it up. They couldn't get hard-ons no matter what. The girls were always goin' on about being horny. (Field interview, H, 2/22/86)

Compartmentalizing makes use of the ideology of heterosexuality. In this way the hustler may cling to his self-description of being heterosexual in the absence of any heterosexuality. It was not uncommon for men who hadn't had a heterosexual relationship for two or three years to continue to refer to themselves as straight, by invoking their past relations (pointing to pictures of ex-mates, or excessive reference to them as if they were momentarily coming back).

The need to distance oneself from loved ones and/or family may make use of psychological and/or social distance. As Reiss (1971) pointed out, the need for the hustler to deny the possibility of being emotionally dependent or involved in any way is critical. Kirkham's (1971) study of homosexuality in prisons also shows that for most inmates who maintained their heterosexual identity despite homosexual activity, two conditions must be met. First, they must be clear that they engage in the act only because they lack the opportunity for heterosexual contact. Second, they must be emotionally distant from the act. To make sure of this they often punctuate their acts with violence or macho toughness.

Social distance can be generated through the creation of intricate rules of behavior. Other gay-hustler studies point to a perception among hustlers that, despite engaging in homosexual behavior, they are not gay (Humphreys, 1970; Reiss, 1971). To promote this perception, intricate rules must be followed by both parties. As with Reiss's "peer-queer" relations, hustlers at the gym restrict their practices to oral sex, with the hustlers being fellated. However, there are more accounts in the gym of hustlers resorting to a wider series of behaviors than reported in Reiss's work. Some nude dancing at parties, as well as sex acts beyond the norm worked out between hustlers and homosexuals, are reported. Generally, it is the hustler who lays down the rules, but if the gay male is particularly assertive or very powerful in bodybuilding circles, this may be altered. The result is a good deal of jockeying for control in the relationship.

The Nexus of the Hustler-Gay Relationship

The nature of this relationship, despite its symbiotic qualities, is negative. This is sufficient to create a distance, which in this context is adaptive. Among their own, each side denigrates the other. Hustlers prefer to see themselves as exploiting gays for quick money, hence they seek to get as much as possible while giving little. For gays it is a sense of being able to buy these men, and so control them, that lets gays feel superior. Each side also feels stigmatized by the relationship. For gay men the stigma comes from having to buy sex when they ought to be desirable enough to have it offered to them. They, in turn, project their self-loathing onto bodybuilders whom they see as brutish and vulgar. Because it is the bodybuilder who often tries to establish the ground rules, there is a tendency for gays to see them as overly aggressive; while bodybuilders, having to deal with their homophobia, view gays as the source of their corruption. This of course exacerbates bodybuilders' homophobia.

As viewed by two gay bodybuilders, who are intimate with the scene in southern California, the predatory characterization of gays is erroneous, and is sometimes the fabrication of the hustler who needs to protect himself from doubts about his sexual preference or activities. Here, the hustler may mistakenly impute thoughts and behavior to gays:

> Truth is that there are a lot of gays around bodybuilding who are kind, giving people. We didn't want a thing to do with most of the young hustlers. But they'd hang around us. It got so bad that we'd hear them coming up the stairs and go, "Oh no, don't answer the door." They'd even paw us, literally, and try to do other things that they thought we'd like, you know, just to get our attention. (Field interview, H, 2/22/86)

Homophobic relations are sometimes violent (Dundes, 1985:354). At Olympic Gym homophobia also, at times, spills over into abuse.

> I remember Stan G. He'd grab gays who came into the gym to watch bodybuilders. He'd grab them and say, "Okay, fag. I want you out." Well, I saw him in the Village in New York, and he said he didn't really feel that way, but he felt like he was expected to do it. (Field interview, H, 3/25/83)

> Don S. was this Marine who's now a cop. He would ask me why he never saw me at the bars. I'd tell him that I wasn't into that anymore. He'd encourage me to come in. But at other times he'd go around beating the shit out of gays and calling them "queers" and all. (Field interview, H, 2/22/86)

Hustling and homophobia become an instrumental complex. Engaging in homosexual behavior works to perpetuate homophobia, and homophobia as an escape valve thrives on this form of homosexual prostitution.

> I'll tell you, being involved in it [hustling] reaffirmed my whole thing with straightness. I remember in San Francisco, I was involved in all this, and you start seeing these people as leeches and vicious. That's okay for some people, but that's not the way I wanna go. (Field interview, H, 12/29/81)

In this revealing statement we see that hustling can enhance heterosexual identification by amplifying and giving immediate, concrete focus to homophobia. As long as the hustler remains emotionally removed from the homosexual relationship by keeping it at the level of exchange, he distinguishes himself from the gay male who would do it for lust or love. He dislikes the men who have seduced him into homosexual acts, and this resentment convinces him, despite all evidence to the contrary, that he is not like them (i.e., straight). In addition, the needs he has for esteem, and for being physically appreciated—a need met primarily through men—can be realized as he affirms his heterosexuality. Hustling, then, is in the novel position of both resolving and creating crises in self-esteem and self-definition. Small wonder that despite condemnation from every side, hustling proves so tenacious. Said one man, "You don't think about it while you're doin' it [hustling]. It's after you stop

that it gets really heavy. You don't know how hard it is to stop hustling."
(Field interview, 5/2/82)

CONCLUSION

Bodybuilding is central to certain male anxieties which are accentuated by our
society's restrictive notions about masculinity. While fostering behavior and
values that underscore virility and macho posturing, the bodybuilding subculture simultaneously creates new problems stemming from some of those same
sources. This examination of hustling was shown to be a conflux for contradictory characteristics, in part an outgrowth of juggling disparate roles. But the
study of bodybuilding subculture can also inform the larger society's handling
of masculinity. One important result, though it is beyond the strict scope of
this study, is the restricted capacity of men to live a more meaningful and
complete emotional life caused, in part, by this restrictive mindset. Since male
traits like homophobia, hypermasculinity, and gender narcissism all exist in a
dialectical relationship with female traits, what we see in our examination of
men is an underevaluation of women, a disdain for the effeminate, and a loss
of all that women stand to offer by not being men.

NOTES

1. Exceptions do occur. Rustad's (1982) work on women in the military is
   an excellent case in point, as are various chapters in Fussel's (1975) work.
   In sports one can cite a number of critical case studies (e.g., Brower,
   1976; Yablonsky and Brower, 1979), though fewer case studies on men
   in sport (e.g., Messner, 1985).

2. Hypermasculinity and homophobia, traits that are closely associated with
   rigid and authoritarian personalities (Smith, 1971; Sherril, 1974), are also
   related to the hustling behavior of men at Olympic Gym. In this article
   their role in psychological maintenance is somewhat abbreviated.

3. Elliott Gorn's (1986) insightful study of bare knuckle fighting in 19th
   century America, and the author's work (n.d.) also describe a form of
   gender narcissism that functions also to foster a disdain for women, or
   what is perceived of as effeminate. The implications for male socialization
   are obvious.

4. Only one case of a man hustling a woman was recorded during my field-
   stays, and this was also by a man who hustled other men. Female bodybuilders do not have to hustle because they tend to have higher status jobs
   and more education. There is also no history of female hustling in the
   subculture, making it more difficult to start it up.

5. Ironically, the outbreak of AIDS has made it easier to discuss the subject
   of hustling with them. On a related note AIDS has curbed hustling somewhat. However, the more enterprising hustler-bodybuilders have begun

making videos of themselves posing suggestively either nude or semi-nude. These are sold in place of sex acts by some.

6. Male bodybuilders tend to come from blue-collar backgrounds. Their educational levels, relative to that of female bodybuilders is significantly lower. In one sample taken after my first year of fieldwork (1979–80) 16% of the man (n = 40) graduated from college, as compared to 40% (n = 38) of the women. The sample size is smaller because it was taken early on in the study, but it shows real discrepancies between the sexes regarding education achieved and work history. Work histories consistently showed men in menial jobs, while women had professionally oriented careers.

REFERENCES

Adorno, Theodore. 1950. The Authoritarian Personality. New York: Wiley.

Aschenbrenner, J. 1975. Lifelines: Black Families in Chicago. New York: Holt, Rinehart, and Winston.

Brower, Jonathan. 1976. "Little League Baseball and Little Leaguism: A Critique of Sport." Paper presented to the annual meeting of the Pacific Sociological Conference, San Francisco.

Butts, Susan Dorcas. 1975. The Psychology of Sport. New York: Van Nostrand Reinhold.

Davis, Nanette. 1971. "The Prostitute: Developing a Deviant Identity." Pp. 135–157 in Studies in the Sociology of Sex, edited by James Henslin. New York: Appleton-Century-Crofts.

Dundes, Alan. 1985. "The American Game of 'Smear the Queer' and the Homosexual Component of Male Competitive Sport and Warfare." Journal of Psychoanalytic Anthropology 8:115–131.

Fusell, Paul. 1975. The Great War and Modern Memory. London: Oxford University Press.

Gorn, Elliot. 1986. The Manly Art: Bare Knuckle Fighting in 19th Century America. Ithaca: Cornell University Press.

Humphreys, Laud. 1970. Tearoom Trade: Impersonal Sex in Public Places. Chicago: Aldine Publishing.

Kaiser, Lincoln. 1979. The Vice Lords: Warriors of the Streets. New York: Holt, Rinehart, and Winston.

Kirkham, George. 1971. "Homosexuality in Prison." Pp. 204–230 in Studies in the Sociology of Sex, edited by James Henslin. New York: Appleton-Century-Crofts.

Klein, Alan M. 1985a. "Pumping Iron." Society 22:68–76.

———. 1985b. "Muscle Manor: The Use of Sport Metaphor and History in Sport Sociology." Journal of Sport and Social Issues 9:4–17.

———. 1986. "Pumping Irony: Crisis and Contradiction in Body-Building Subculture." Sport Sociology Journal 3:3–23.

———. 1987. "Fear and Self-Loathing in Venice: Narcissism, Facism, and Bodybuilding." Journal of Psychoanalytic Anthropology 10:117–137.

———. Unpublished. No Pain, No Gain: The Ethnography of Bodybuilding.

Kopay, David, and Paul Young. 1977. The David Kopay Story. New York: Ann Arbor Press.

Lehne, Gregory. 1976. "Homophobia among Men." Pp. 57–70 in the Forty-Nine Percent Majority: The Male Sex Role, edited by David and R. Brannon. Reading, MA: Addison-Wesley.

Matza, David. 1964. Delinquency and Drift. New York: Wiley.

———. 1969. Becoming Deviant. Englewood Cliffs, NJ: Prentice-Hall.

Messner, Michael. 1985. Masculinity and Sports: An Exploration of Changing Meaning of Male Identity in the Lifecourse of the Athlete. Doctoral Dissertation, University of California at Berkeley.

Pleck, Joseph. 1982. The Myth of Masculinity. Cambridge, MA: MIT Press.

Reiss, David. 1971. "The Social Integration of Peers and Queers." Pp. 395–406 in Deviance: The Interactionist Perspective, edited by Earl Rubington and Martin Weinberg. New York: Macmillan.

Rustad, Michael. 1982. Women in Khaki: The American Enlisted Women. New York: Praeger.

Sabo, Donald. 1985. "Sport, Patriarchy, and Male Identity: New Questions about Men and Sport." Arena Review 9:1–30.

Sabo, Donald, and Richard Runfola (eds.). Jock: Sports and Male Identity. Englewood Cliffs, NJ: Prentice-Hall.

Sherril, J. 1974. "Homophobia and Social Psychology." Journal of Homosexuality 1:9–21.

Smith, Kenneth. 1971. "Homophobia: A Tentative Personality Profile." Psychological Reports 29:1091–1094,

Sprague, Homer. Unpublished. "Psychological Testing on Bodybuilders."

Yablonsky, G., and Jonathan Brower. 1979. The Little League Game. New York: Times Press.

37

The Intangible Rewards from Crime:
The Case of Domestic
Marijuana Cultivation

RALPH A. WEISHEIT

S tudies of the drug business in the United States have had an evolving
focus, generally paralleling shifts in the explanation of crime more generally. . . . This study examines the rewards from the drug business
which are neither monetary nor psychopharmacological. The findings are
based on interviews with 31 commercial marijuana growers, 30 law enforcement officials familiar with domestic cultivation, and about a dozen others
with connections to marijuana cultivation.

THE SCALE OF DOMESTIC
MARIJUANA GROWING

In 1986, the federal government estimated there were between 90,000 and
150,000 commercial marijuana growers in the United States, and over 1 million people who grew for personal use (cited in Gettman 1987). The National
Organization for the Reform of Marijuana Laws (NORML) calculated the
number of growers by combining crop production estimates with a projection
of the amount that typical growers produce in a year (5 pounds for personal
use growers and 13 pounds for commercial growers). Using this approach,
NORML estimated there were approximately 250,000 commercial growers
and over 2 million personal use growers (Gettman 1987).

By the late 1980s it was estimated that between 25% and 50% of the marijuana consumed in the United States was produced here (Slaughter 1988;
Gettman 1987), and that the cash value of the domestic marijuana crop was as
high as $60 billion a year, easily making it the largest cash crop in the United
States (Weisheit 1990a). The domestic marijuana industry may also be producing marijuana for export. In 1983 and 1984, for example, when the U.S.
industry was just being recognized, it was estimated that 10% of Canada's
marijuana supply came from the United States (Stamler, Fahlman, and Vigeant
1985).

From: "The Intangible Rewards from Crime: The Case of Domestic Marijuana
Cultivation," Ralph A. Weisheit, *Crime and Delinquency,* Vol. 37, No. 4, 1991. Reprinted
by permission of Sage Publications, Inc.

Despite the substantial contribution of the domestic marijuana industry to the U.S. marijuana market, and perhaps to the markets of other countries, little is known about domestic growers. Several factors contribute to this gap. First is the relative youth of the domestic marijuana industry. Having first been recognized by the federal government in 1982, it was not until the late 1980s that authorities identified even loosely structured multistate growing operations. Second, during much of the mid- to late 1980s, public attention was drawn to cocaine and crack cocaine. Other drugs, including heroin, marijuana, and LSD were largely ignored by the press and by politicians. Third, among illegal drugs marijuana is least likely to be defined by the public as deadly, addictive, or tied to violent crimes. Even among law enforcement officers, marijuana arrests afford less status than arrests for heroin or cocaine. Fourth, the domestic industry has often operated in rural settings. The national press, special police drug units, and researchers are all more often centered in urban areas and have focused on drugs in the urban environment. To urban observers of the drug scene, the domestic marijuana industry was largely invisible.

THE STUDY

Between November 1988 and August 1990 data were assembled on all cases in which someone had been arrested for growing marijuana in Illinois between 1980 and the spring of 1990. Only those cases involving 20 or more plants were included, so that the primary focus was on commercial growers. Of course, some growers with fewer than 20 plants may be growing for profit, and some with more than that may seek no financial gain from their activities. However, 20 plants seemed a reasonable cutoff.

Newspapers throughout the state were examined and letters of inquiry were sent to prosecutors and local sheriffs. Through this process, 74 cases were identified. Of these, 23 were excluded for a variety of reasons (case was pending, could not locate growers, no record of case in circuit court, and so on). Of the remaining 51 cases, growers were contacted by letter and then, where possible, by telephone. Of these 51 cases, 31 agreed to be interviewed. Interviews lasted from 45 minutes to several hours and usually took place in the subject's home.

In addition, 20 Illinois law enforcement officials familiar with marijuana cases were interviewed about patterns of growing in Illinois and 10 officials from surrounding states were interviewed so that patterns in Illinois could be placed in a larger context. A more complete description of the methodology and sample can be found in Weisheit (1990a).

THE GROWING OPERATIONS

Growers in this study had been in the business for an average of 5 years, with one grower having been involved for 18 years. The largest operation had over 6,000 plants, and the median size of operation was 75 plants. Nearly all of them cultivated sinsemilla, a technique of cultivation in which only female plants are grown and only the flowering buds are harvested (the product is considerably more potent than that from only leaves). Most expected to harvest from 1/2 to 3/4 of a pound of marijuana from each mature plant. Though growers were reluctant (and usually unable) to provide detail about their income from marijuana growing, most expected to receive between $700 and $1,500 a pound for their crops. During periods when the domestic marijuana was scarce, they could receive as much as $2,000 a pound. A conservative estimate of $750 for each mature plant would mean the median operation in this study could expect a gross income of over $56,000 a year from their operation. Further, an operation of this size could be run with little or no hired help and, if grown outdoors, at an almost negligible operating cost. (The elaborate irrigation systems reported in some California operations are seldom necessary in the Midwest).

Six of the 31 growers grew all or part of their crops indoors, which made detection more difficult and which allowed them to harvest crops year round. Because of the expense of equipment and utilities, indoor operations are usually much smaller than outdoor operations, and the decision to focus on growers with 20 or more plants probably reduced the representation of indoor growers in this sample. This was not a concern, however, because the focus was on larger commercial growing operations.

TYPES OF GROWERS

Understanding the rewards from commercial marijuana growing requires an understanding of the types of people involved in growing and the factors which motivate them to enter the business. Elsewhere I have identified three types of marijuana growers: the hustler, the pragmatist, and the communal grower (Weisheit 1990b, 1990c). A brief description of these groups will facilitate understanding the discussion of rewards.

Hustlers

These marijuana growers are entrepreneurs by instinct. They may have used marijuana and may have engaged in some dealing for the challenge of it, though neither activity is a necessary prerequisite for their involvement in growing. Some use farmland they already own, others purchase or rent farmland for the explicit purpose of growing marijuana, and still others enlist the aid of land owners who are having financial problems. For some, the risks associated with growing are themselves part of the appeal of growing. Although the money they make may serve as an indicator of success, they are generally less motivated by the money itself than by the challenge of being a successful entrepreneur. They could just have easily set up business in any one of a dozen legitimate or illegitimate enterprises and once the challenge wears off, they are likely to move out of marijuana growing.

A 1985 Texas case illustrates this type of grower. A 63-year-old farmer was arrested with six hundred pounds of processed marijuana, two hundred pounds of seeds, and 22,284 marijuana plants growing in a barn equipped with grow lights. He was also among the best-known businessmen in his small Texas town and had a reputation for hard work. In addition to owning almost five thousand acres of land, he also owned a local feed store, a chemical company, and a company which sold satellite dishes, hog pellets, and sheep manure (Applebome 1985).

Numerically, hustlers probably represent a very small proportion of growers, though their grandiose schemes push them to large growing operations. Thus, they have a disproportionately large impact on the marijuana market (Weisheit 1990a).

The current study includes at least one such entrepreneur (though he may not define himself this way). At the time of his arrest he had the largest marijuana crop ever seized in Illinois and received the largest fine ever imposed on an arrested grower. He seemed proud of these facts, holding them up as evidence of the scale of his operation. At the time of his arrest he told the sheriff that he (the sheriff) should be grateful because closing such a large operation would likely guarantee his re-election. In prison he studied horticulture. Following his release he wrote a book on marijuana growing and served as a paid consultant to those interested in "organic gardening." He agreed to be interviewed for the study only after renegotiating the standard fee paid interview subjects.

A second case fits this category, although no interview was conducted. The individual had contracted with several others to grow very large marijuana plots on their farms. He provided the seeds, technical expertise, and had planned to expand at least one of the operations to as many as 30,000 plants per year. At the time of the study he was a fugitive from the police (the farmers having been arrested) and his whereabouts were unknown.

Pragmatists

These growers are driven to marijuana production by economic necessity. Importantly, they approach marijuana with no moral or philosophical righteousness. They often see what they are doing as both legally and morally wrong, but feel they have few options. Unlike hustlers, they are in the marijuana business to help themselves through tough economic times, not to become wealthy.

G04:   I didn't plan on getting rich or anything, I just wanted to hang on to what I had. I had no illusions of buying a yacht or anything like that. I was just trying to hang on to what I had. . . . I believe most of the people who are trying to make money off of marijuana are doing it because they're broke. And, they are usually like me, I was faced with foreclosure, even though I was not a farmer. A lot of farmers around here, you know, lose 10 to 15 thousand a year growing corn and make it up growing reefer.

G29: I was laid off at the coal mine, and hell, had house payments and every-
       thing else coming in. And, I just couldn't make it farming and stuff, and
       I couldn't find a job. So I thought I'd try that [marijuana growing] first.

Warner's (1986) account of marijuana growers described a grower and his
wife as staunch Republicans who had never even tried marijuana. He had
"punched out" his son for smoking marijuana 6 or 7 years earlier but now
used his son's help in growing marijuana. He described his feelings about marijuana and himself as a grower in these terms:

> I don't even consider [smoking] it. I'm not going to fry my brains. . . . I'm
> just involved in producing a saleable product. I'm just trying to make
> money. . . . I'm antimarijuana. . . . I feel that marijuana is destroying
> young people by changing their attitudes about working. I feel so strongly
> about that I'm almost ashamed of growing it. (Warner 1986, p. 229)

Pragmatists are a particularly interesting group, for they demonstrate that
growing requires no commitment to a drug lifestyle or even a "liberal" or tolerant attitude toward drugs in general. Potter, Gaines, and Holbrook (1990)
note that in Kentucky, marijuana growing is most common in regions which
are both impoverished and characterized by conservatism. They have noted
one large-scale grower who did not smoke cigarettes or drink alcohol and was
a leader in a conservative snake-handling church (G. Potter and L. Gains, personal communication, April 4, 1990).

Although more common than hustlers, pragmatists are, in most regions,
probably not the most common type of grower. The size of their operations
varies greatly, depending primarily on their economic need. Chapple (1984)
described a woman who fits this category and grew only seven plants per year.
She carefully calculated how much money she would need to meet her basic
living expenses, how much she would receive per mature plant, and then
grew precisely enough to meet that need.

Communal Growers

These persons probably represent the single largest category of grower, though
their operations are often very small. They cultivate marijuana as part of a
larger lifestyle, of which marijuana use is often an important part. Some seek
to retreat from society by living in remote areas, and others are active mainstream members of society for whom marijuana growing is a personal statement of rebellion or independence. For many, growing makes a social statement. They have little interest in the business side of marijuana growing, or in
accumulating money from growing marijuana, although the money may help
them through short-term financial problems. A few are holdovers from the
late 1960s and early 1970s, but more commonly they are indistinguishable
from "ordinary" citizens in their outward appearance and involvement in
conventional activities. In the current study, several were involved in volunteer work with such groups as the Red Cross and the Special Olympics.

Although some of these individuals gradually drift into large-scale production, most have more modest goals. As Warner (1986) observed:

> Many of them are motivated as much by ideas of self-sufficiency (the satisfaction of growing their own) or thrift (not having to buy pot) as by making money selling their extra ounces or pounds. What money they make seems to go to a few purchases they could not afford otherwise, and to good times. (p. 200)

As their comments indicate, a number of the growers interviewed for this study could be classified as communal growers:

G07:   I probably grew about 10–15 pounds per year and about three of those pounds were for me personally. Another 4–5 I would give away, and I'd save five pounds every year and around Christmas time I sold it to the same guy every year, and I used that money to buy presents, something we couldn't otherwise afford. . . . I sold it well under commercial prices. Because you see, that's one of the main reasons I got into it in the first place, because commercial prices were ridiculous, and plus I don't believe in supporting organized crime that's into all this other shit, too. There's a philosophical argument about smoking marijuana that could go on for days, but personally I think it's a person's—if somebody wants to, it's no big deal, if you are an adult. I don't think kids should, of course, but I think if you are an adult and you want to, it's no different than drinking a beer, or whisky, or wine, and whatever else. And, it's none of the damn state's business.

Q: You weren't growing mainly for profit then?

G21:   No, no. Not really, Most of what I sold, I sold because people nagged me for it.

Q: Mostly friends?

G21:   Yeah, My friends nagged me for it so much. I'd turn them on to some and then they'd want some, you know. I usually gave in and would sell them some. I gave away a lot. If I was in it for the money I would have sold all that I could, but I gave away a lot.

The idea of simply giving away marijuana is common among communal growers, but would be unthinkable for hustlers, unless it was to attract future business. Several communal growers also saw their activities as having political and economic implications. They felt that domestic marijuana growing helped the overall balance of trade with other countries, benefited the local economy, and kept money from going to organized crime—which they were certain would happen if they bought marijuana through an importer.

G21:   See, my reasons behind growing were monetary, but just for the sake of not having to spend money that I worked for on pot. But there were a lot of other reasons involved. I liked to keep my money in the locality

where I live. I considered going across the river to shop, but I didn't. I
shop in my own country. Another reason was that I didn't like supporting the black market. Most of the pot or drugs you pay for goes up
to Chicago and from Chicago who knows where it goes. You know, a
lot of it goes to foreign countries. I just felt I was helping the current
[economic] problem rather than hurting society, because I happened to
keep the demand [for foreign marijuana] smaller. And, if I support
myself, I didn't have to rely on foreign sources. So in turn, that actually
helped out.

These economic and political issues were secondary considerations, and
may have been nothing more than rationalizations to justify their illegal activities to themselves. Nevertheless, these arguments cannot be dismissed as irrelevant to the process if the growers truly believed them, and it appeared they did.

Approaching production as they do, communal growers view other growers as kindred spirits rather than threatening competitors. When communal
growers get together they often enjoy sharing experiences and technical information about growing, and "war stories" about their brushes with the law and
with marijuana thieves. Similarly, this group is less likely to engage in the violence sometimes associated with marijuana growing, such as setting booby
traps and guarding crops with automatic weapons or dogs (Lawren 1985;
Warner 1986; Raphael 1985).

THE REWARDS OF GROWING

Growing marijuana can be a lucrative business, depending on the variety
grown, whether the product is sinsemilla or commercial grade, and current
market conditions. The value of the crop to the grower tends, everything else
being equal, to be directly related to the amount of effort and technical expertise the grower applies to the process. At the bottom of the scale are those
who only harvest wild growing marijuana, also known as ditchweed. This
practice takes little effort and can yield as little as $50 per pound. At the other
extreme are those who cultivate potent strains and grow the plants as sinsemilla, which can yield as much as $2,000 per pound.[1] Most growers in this study
focused on sinsemilla and expected to receive an average of $1,000 per mature
plant. Although demanding and time consuming, it is possible to develop
operations which gross over $1,000,000 per year. By their sheer size, which
itself increases the risk of detection and usually require hiring help, which itself
increases the risk of arrest and theft. These large operations are generally
undertaken only by the most ambitious (such as a hustler) or the most desperate for money (such as the pragmatist).

Although money is important, and perhaps the most important reason for
commercial growing, it is not the only motivating factor. This is particularly
true for communal growers who are less likely to be driven by internal (hustlers) or external (pragmatists) forces to use growing to make large sums of

money. For many growers in this study, the expected cash return from growing was modest, and the intangible rewards of growing rivaled cash benefits as motivating factors.

INTANGIBLE REWARDS

Aside from cash benefits, there were at least three ways in which the process of growing was itself rewarding and satisfying. Growers tended to emphasize either the spiritual, social, or intrinsic rewards from growing.

Spiritual Rewards

Several spoke of the spiritual rewards from growing, using religious or almost religious descriptions of their feelings. The process of growing, as distinct from the money or the effects of using marijuana, was something they spoke of with great passion.

Q: Aside from the money, was there anything else about growing marijuana that you found satisfying or enjoyable?

G03: This plant of all the Lord's plants is the most intriguing plant that I know of. Yes, I would grow this plant as some people grow roses. I think roses are dumb even though I grow them and like them. Bananas are a lovable plant and so is papaya, but marijuana is in a category all by itself. I would class it as one of the more sophisticated plants. Yes I took great pride in producing a high quality product for my fellow man to enjoy. I have brought much happiness and well being into this world. To me this is one of the miracles of this world that the Lord has given us such a worthy plant for us to enjoy.

G02: There is something about growing a marijuana plant, unless you do it you don't understand it, because it is a beautiful experience. I'm talking from a spiritual sense in that you nurture, you work, you learn, and then you—I tell you, I used to go out late at night, and lay down under my plants and watch the moon pass through those beautiful buds and just smoke one and lay there and watch the moon pass through them, just amazed. You know there are moments in that period that I just, I loved, I just loved it.

It is ironic, perhaps, that in several ancient civilizations marijuana was treated as a mystical plant provided by the gods (Chopra 1969; Abel 1980). Similarly, the contemporary hemp movement seems to grant such mystical qualities to the plant, seeing it as a solution to a variety of problems, from global warming to faltering rural economies. Perhaps the most explicit statement of this position is Jack Herer's book *The Emperor Wears No Clothes* (1990) which is subtitled: "The Authoritative Historical Record of the Cannabis Plant, Hemp Prohibition, and How Marijuana Can Still Save the World."

Social Rewards

More common was the emphasis on social rewards from growing. Many enjoyed impressing their friends with the quality of their crop and there was often friendly competition among growers. What was important was not the fact of their growing, but the quality of the product they grew and the status from having others recognize a job well done.

G21: Once I started growing I gave away a lot to close friends that was just as good as the commercial stuff that they were paying for on the street. I just gave that stuff away.

Q: Did you have a lot of pride in that?

G21: Yeah, I did. So many people couldn't believe the potency of it and if you had any sense of pride and accomplishment you felt it. . . . Me and my girlfriend did a really good job manicuring the leaves and we took pride in that. A lot of the other growers didn't do a very good job at it. But everybody, when they looked at the appearance of it, they could tell that we had done a good job and they complimented us on that, too. So, yeah, we had intense pride. Also we always tried to outdo the other growers and we usually succeeded there. It was obvious people knew about it from what they said so we had a sense of pride about it.

Q: And, you felt a lot of pride in the quality of plants you were growing?

G10: Yeah. In the plants I got caught with in the house. I had budded those out in mid-summer in the house and we had been smoking it all summer. And I was letting other people smoke, friends of mine. They said, "Oh, this is the best pot we've ever smoked." You know, I had this great feeling of accomplishment. When you get done, come out back and look at my vegetable garden, and you'll see. I've got radishes and onions, and all that, ready all ready. Yeah, well, you see, that year I didn't grow any garden, cause I was obsessed with growing pot.

I go out for the awards. I want to be the best. The same thing happened to me with the pot. I got obsessed with growing it. I wanted to outdo all those guys. I wanted to have the best pot, you know. When you grow a garden you want to have the biggest tomatoes or whatever. As a hobbyist, you know how hobbyists get involved, so sometimes they were obsessed with their hobbies too. Well, that was my hobby that year, and I thought at the same time I could have enough stashed to last me a year or two. And, I even had in mind that I could make some money, cause things were financially tough because I went through a bankruptcy.

Q: You felt a lot of pride in the quality of the plants you were growing?

G09: Oh, yeah. You take a big pride, especially when people get it, you got better than anything else, than somebody else around, I don't know how to put this tactfully. I mean, you're not a woman, you can't satisfy someone in that way, but no matter if they're a whore on the streets, some people are going to be, cause they need that relief, that satisfaction. You give them a form of relief and they're kind of grateful to you for it, even if they are paying a ridiculous price for it.

Q: So, aside from the money, you were helping people out?

G09: Oh, yeah, in a way.

Q: Did you use hydroponics or anything?

G09: Oh, yeah, I used hydroponics. When I first got into it, I did a lot of experiments. I learned how it works. Yeah, it was great, it was fun, it was a blast. And of course, I wasn't worried about getting busted. Then all of your friends get into it, you know. It's something everybody starts getting into. Something to sit around and talk about. Some people get into cars, some people get into plants.

The social rewards from the drug business have been noted by others (Mouledoux 1972; Goode 1970). For example, in his study of psychedelic drug dealers, Langer (1977) observed that:

The desire to obtain profit was played down and those dealers suspected of being interested in drugs for monetary reasons were subject to severe sanctions and ostracized by customers. Generally, the "value" of drugs was as much a social concern as an economic one. (p. 378)

Similarly, Adler's (1985) study of upper-level cocaine dealers recognized the importance of social relations and informal networks which both reinforce and facilitate involvement in the drug business.

Intrinsic Rewards

Finally, a number of growers reported that growing provided them with intrinsic rewards. For these growers the process provided the kind of self-satisfaction that many people find in hobbies with which they become deeply enmeshed:

Q: Aside from the money, was there anything about growing that you found enjoyable or satisfying?

G22: I was growing a high quality product. It was kind of rewarding to see all your work turn into something that was going to be nice.

Q: You obviously got more out of it than just saving money.

G21: Well, I enjoyed doing it a lot. I just took great interest in it. I've always had one main hobby. A long time ago as a teenager it was building model airplanes, small engines like solid engines. Then I went to motorcycling for a while, dirt bikes, then firearms. When I got my fill I had reloading equipment, small arms, handguns. I just had to give all that up. I had $2,500 of reloading equipment alone. My girlfriend used to help me with that. In the meantime we did grow but on a small scale. I had a good job. I made about $15.00 an hour. It didn't bother me to buy pot. But after I saw that film [on growing] I found myself more involved in my pot growing hobby. It kind of got out of hand. This judge, you know, he wouldn't understand my view on it. Basically it was just an extensive hobby with rewards that I could enjoy. Not only could I save money but I could get high too and I enjoyed doing that. In the long run I'm paying for it.

Q: Aside from saving money on buying and all that, is there anything else about growing marijuana that you found satisfying or enjoyable?

G08: Oh, yeah, I liked to grow it, it's just like growing a nice patch of sweet corn and getting kind of proud of it. Maybe you have a friend come over who enjoys smoking a little marijuana, and you say "Hey I grew this," and he says "Oh wow, that's nice shit." You know that's just like taking a nice bunch of tomatoes to your grandma's. So, I guess I more or less took pride in a good job, everybody does that, whether it be building a bird house, or growing something. Well I like just being a nature person, you know I enjoy it, whether it be pot or some other plant. I just seemed to always be growing things and I'm kind of a green thumb, you know, a horticulturist I guess. My grandmother had flowers of every variety. I used to help her, so maybe that's where I got the interest. I've got all kinds of house plants, it's just another plant to me.

Q: Was there anything really enjoyable about growing marijuana?

G07: Oh, sure. Just the same thing that anybody finds enjoyable about having a nice garden, you want to take pride. I always have a good garden, you know, nice tomato plants and pepper plants, and like that. The same that anybody takes in growing anything that they do a good job on.

Q: Aside from the money, was there anything you found really enjoyable or satisfying about growing marijuana?

G09: Probably the most therapeutic thing I've ever done, just planting it. A lot of people get into planting for the first time in their life and say this is something you've done, you can see for yourself. Of course, then another aspect of marijuana is what it does to the inside of your mind, just what it does to you. The more you get into it, you learn you can control the different varieties of highs or whatever, and it seems like the more you get into it, the more you learn, the more exciting it is. There's just an initial therapeutic thing being out there, and watching your plants grow. That self-satisfaction does something, even on its own, cause I even got into pot when I was a kid. I mean I didn't smoke it. I didn't know anything about drugs, but it had always grown around there everywhere, it was so thick, and I was kinda interested in it. I tried it a couple of times, the wild stuff, it gave me a headache, and I never could figure out why anyone would pay money for the stuff. But, it had always gotten my curiosity up.

G16:   Well, I enjoy doing it. Yeah. It's just like I grow a big garden. We have a garden out here with sweet corn, all kinds of stuff. I like growing things, always have. We farmed all our lives.

. . . [One] grower even used his operation as an opportunity for spending "quality time" with his young son:

Q: You did all the harvesting yourself?

G18:   Me and my little boy. He'd be down there with his Tonka trucks. I'd say "bring it over here." I'd clear off some buds and load them up in his Tonka truck and he'd take it over there. We had a lot of fun. He felt bad because they [the police] took my plants too. He knew it was pot. I did explain to him what it was, and he was mad because the officers took my garden.

Intrinsic rewards are not simply the product of putting a seed into the ground and watching it grow, but also of the fact that growers can take a simple "hobby" and make it as complex as they wish. The nature of the marijuana plant makes it particularly suited for this (Weisheit 1990a).

G21:   Well, I think it was 1985 when I ordered my first batch of seed [through the mail]. And from '85 up until the spring of '89 when I got my last batch, I placed four orders. All together I had 16 different varieties. See, they kept coming up with new varieties. I was just curious and interested in the different varieties they had. They had quite a bit to choose from. You can obtain different kinds of high, whether its Indica plant, which gives you like a narcotic high, kind of a sleepy, heavy high. Or a Sativa plant that give you a more uppity high, kind of a tricky high they call it. Some of it you can almost trip on. And in between, there were hybrids. But there was more than just the reason of trying to obtain a certain type of high. The Indica was the fastest growing, the easiest to grow, and the heaviest yielding. But the Sativa grew tall and sparse. It didn't yield that much, and it took a long time to grow, but it had the best high. So then, you tried to find something in between. And that's what I was doing. I just kept going around trying to find which would be the best to grow, and I was going to narrow it down to two or three varieties. That was one of the reasons I had so much at the time. I started with about six different varieties for one; you have to narrow it down to your best plants. Then you just grow them for awhile and make sure it's what you want. Then you can just keep one plant and make that your mother plant, grow it in a separate building and just keep cloning from that, from then on out. That mother plant will last three to five years. Then that way you can grow nothing but females when you clone off the female mother plant. They're nothing but females and they're all pretty much identical. They grow just like the mother plant you take it off of. That's what I was trying to achieve, but I had so many different varieties, I hadn't quite narrowed it down. I tried explaining this to the judge. He couldn't comprehend why I had so much. I was trying to tell him this was the process you had to go through.

The complexity of these operations often went well beyond what was needed to simply make money. Several growers used technical how-to books for marijuana growing, but quickly found them inadequate and moved on to general books and magazines about horticulture and plant botany. One grower shared his technical skills with others to help them deal with plant diseases, insects, and general problems with growing marijuana. He did not charge for his services, although it was common for growers to show their appreciation by sharing samples of their marijuana. . . .

CONCLUSIONS

This study has shown how a crime which on the surface appears to be a purely economic enterprise, can also provide a strong level of personal enrichment. The business of drugs has been examined from a number of perspectives. Early studies viewed dealers as pathetic addicts who were little more than passive reactors to their environment. More recent research has shifted the focus to dealers as hard working entrepreneurs who share much of the drive and motivation of the legitimate businessman. This study has drawn attention to yet another aspect of the issue, the drug business as a means of personal enrichment and satisfaction. Rather than some manifestation of pathology, these nonmonetary rewards are directly comparable to those which derive from legitimate hobbies. Like the enthusiastic sports fan or the dedicated gardener, these marijuana growers are addicted (if that is the right term) to the experience of growing marijuana. The "high" comes from a process rather than a chemical substance, in a manner consistent with Peele's (1985) conception of addiction.

Katz (1988) has also observed that crime can be fulfilling for the offender. Like growers in this study, Katz found that for some property crimes "the experience is not simply utilitarian and practical; it is eminently magical" (1988, p. 53). Unlike growers in this study, Katz's property offenders found crime thrilling *because* it was illegal—an act of defiance and defilement of the victim. That so many marijuana growers were also heavily involved in growing legal plants suggests that the process of growing was itself rewarding, quite apart from the fact that it was illegal.

These findings have implications for policies aimed at ending the drug business. First, treatment cannot focus only, or even primarily, on the pharmacological effects of drugs. Because the addictive potential of marijuana is comparatively low, the fascination these growers have with the process of growing cannot be explained simply in terms of chemical dependency. Second, policies aimed at taking the profits out of the drug business may well drive entrepreneurs and pragmatists from domestic cultivation, but the impact of such policies will be far less significant for communal growers. Such policies could deter them from selling, or from using their knowledge to expand their operation in the wake of a financial crisis. These policies are unlikely, however, to have much impact on casual or small-scale growing, which probably includes the vast majority of growers in the United States. It is likely that many, perhaps most, communal growers will not leave marijuana growing because they are forced from it by the law, or because they undergo "treatment," but because they find an alternative "hobby" that consumes their attention. For these growers, cultivating marijuana is much more than simply a business or vocation, it is an avocation as well.

NOTE

1. Throughout the study there were periodic reports of growers receiving as much as $3,000 per pound for their crop. These rumors proved impossible to substantiate, and may have simply been part of the folklore surrounding growing, much as the mythical "china white" heroin was once part of the folklore of heroin users. Although $3,000 per pound is possible during shortages and at the retail level, it seems unlikely that growers were receiving this during the period covered by the study. Marijuana prices have risen dramatically over time, however, and the days of $3,000 per pound domestic marijuana may not be far away.

REFERENCES

Abel, Ernest L. 1980. *Marihuana: The First Twelve Thousand Years.* New York: Plenum Press.

Adler, Patricia A. 1985. *Wheeling and Dealing: An Ethnography of an Upper-Level Drug Dealing and Smuggling Community.* New York: Columbia University Press.

Applebome, Peter. 1985. "What Was Farmer Brown Doing with All That Pot?" *Texas Monthly* 13:10–11.

Chapple, Steve. 1984. *Outlaws in Babylon: Shocking True Adventures on America's Marijuana Frontier.* London: Angus & Robertson.

Chopra, Gurbakhsh S. 1969. "Man and Marijuana." *The International Journal of the Addictions* 4:215–47.

Gettman, Jon B. 1987. *Marijuana in America—1986.* Washington, DC: National Organization for the Reform of Marijuana Laws (NORML).

Goode, Erich. 1970. *The Marijuana Smokers.* New York: Basic Books.

Herer, Jack. 1990. *The Emperor Wears No Clothes.* Van Nuys, CA: HEMP Publishing.

Katz, Jack. 1988. *Seductions of Crime.* New York: Basic Books.

Langer, John. 1977. "Drug Entrepreneurs and Dealing Culture." *Social Problems* 24:377–86.

Lawren, Bill. 1985. "Killer Weed." *Omni* 7:16, 106.

Mouledoux, Joseph. 1972. "Ideological Aspects of Drug Dealership." Pp. 110–22 in *Society's Shadow: Studies in the Sociology of Countercultures,* edited by Kenneth Westhues. Toronto: McGraw-Hill.

Peele, Stanton. 1985. *The Meaning of Addiction.* Lexington, MA: Lexington Books.

Potter, Gary, Larry Gaines, and Beth Holbrook. 1990. "Blowing Smoke: An Evaluation of Marijuana Eradication in Kentucky." *American Journal of Police* 9:97–116.

Raphael, Ray. 1985. *Cash Crop: An American Dream.* Mendocino, CA: Ridge Times Press.

Slaughter, James B. 1988. "Marijuana Prohibition in the United States: History and Analysis of a Failed Policy." *Colombia Journal of Law and Social Problems* 51:417–74.

Stamler, R. T., R. C. Fahlman, and H. Vigeant. 1985. "Illicit Traffic and Abuse of Cannabis in Canada." *Bulletin on Narcotics* 37:37–49.

Warner, Roger. 1986. *Invisible Hand: The Marijuana Business.* New York: Beech Tree Books.

Weisheit, Ralph. 1990a. *Cash Crop: A Study of Illicit Marijuana Growers.* (Draft of report for grant #88-IJ-CX-0016). Washington, DC: National Institute of Justice.

———. 1990b. "Domestic Marijuana Growers: Mainstreaming Deviance." *Deviant Behavior* 11:107–29.

———. 1990c. "Marijuana as a Cash Crop: Drugs and Crime on the Farm." Pp. 17–35 in *Criminal Investigation: Essays and Cases,* edited by James N. Gilbert. Columbus, OH: Merrill.

38

Shifts and Oscillations in Deviant Careers: The Case of Upper-Level Drug Dealers and Smugglers

PATRICIA A. ADLER

AND PETER ADLER

The upper echelons of the marijuana and cocaine trade constitute a world which has never before been researched and analyzed by sociologists. Importing and distributing tons of marijuana and kilos of cocaine at a time, successful operators can earn upwards of a half million dollars per year. Their traffic in these so-called "soft"[1] drugs constitutes a potentially lucrative occupation, yet few participants manage to accumulate any substantial sums of money, and most people envision their involvement in drug trafficking as only temporary. In this study we focus on the career paths followed by members of one upper-level drug dealing and smuggling community. We discuss the various modes of entry into trafficking at these upper levels, contrasting these with entry into middle- and low-level trafficking. We then describe the pattern of shifts and oscillations these dealers and smugglers experience. Once they reach the top rungs of their occupation, they begin periodically quitting and re-entering the field, often changing their degree and type of involvement upon their return. Their careers, therefore, offer insights into the problems involved in leaving deviance.

Previous research on soft drug trafficking has only addressed the low and middle levels of this occupation, portraying people who purchase no more than 100 kilos of marijuana or single ounces of cocaine at a time (Anonymous, 1969; Atkyns and Hanneman, 1974; Blum et al., 1972; Carey, 1968; Goode, 1970; Langer, 1977; Lieb and Olson, 1976; Mouledoux, 1972; Waldorf et al., 1977). Of these, only Lieb and Olson (1976) have examined dealing and/or smuggling as an occupation, investigating participants' career developments. But their work, like several of the others, focuses on a population of student dealers who may have been too young to strive for and attain the upper levels of drug trafficking. Our study fills this gap at the top by describing and analyzing an elite community of upper-level dealers and smugglers and their careers.

We begin by describing where our research took place, the people and activities we studied, and the methods we used. Second, we outline the process of becoming a drug trafficker, from initial recruitment through learning the trade. Third, we look at the different types of upward mobility displayed by dealers and smugglers. Fourth, we examine the career shifts and oscillations which veteran dealers and smugglers display, outlining the multiple, conflicting forces which lure them both into and out of drug trafficking. We conclude by suggesting a variety of paths which dealers and smugglers pursue out of drug trafficking and discuss the problems inherent in leaving this deviant world.

SETTING AND METHOD

We based our study in "Southwest County," one section of a large metropolitan area in southwestern California near the Mexican border. Southwest County consisted of a handful of beach towns dotting the Pacific Ocean, a location offering a strategic advantage for wholesale drug trafficking.

Southwest County smugglers obtained their marijuana in Mexico by the ton and their cocaine in Colombia, Bolivia, and Peru, purchasing between 10 and 40 kilos at a time. These drugs were imported into the United States along a variety of land, sea, and air routes by organized smuggling crews. Southwest County dealers then purchased these products and either "middled" them directly to another buyer for a small but immediate profit of approximately $2 to $5 per kilo of marijuana and $5,000 per kilo of cocaine, or engaged in "straight dealing." As opposed to middling, straight dealing usually entailed adulterating the cocaine with such "cuts" as manitol, procaine, or inositol, and then dividing the marijuana and cocaine into smaller quantities to sell them to the next-lower level of dealers. Although dealers frequently varied the amounts they bought and sold, a hierarchy of transacting levels could be roughly discerned. "Wholesale" marijuana dealers bought directly from the smugglers, purchasing anywhere from 300 to 1,000 "bricks" (averaging a kilo in weight) at a time and selling in lots of 100 to 300 bricks. "Multi-kilo" dealers, while not the smugglers' first connections, also engaged in upper-level trafficking, buying between 100 to 300 bricks and selling them in 25 to 100 brick quantities. These were then purchased by middle-level dealers who filtered the marijuana through low-level and "ounce" dealers before it reached the ultimate consumer. Each time the marijuana changed hands its price increase was dependent on a number of factors: purchase cost; the distance it was transported (including such transportation costs as packaging, transportation equipment, and payments to employees); the amount of risk assumed; the quality of the marijuana; and the prevailing prices in each local drug market. Prices in the cocaine trade were much more predictable. After purchasing kilos of cocaine in South America for $10,000 each, smugglers sold them to Southwest County "pound" dealers in quantities of one to 10 kilos for $60,000 per kilo. These pound dealers usually cut the cocaine and sold pounds

($30,000) and half-pounds ($15,000) to "ounce" dealers, who in turn cut it again and sold ounces for $2,000 each to middle-level cocaine dealers known as "cut-ounce" dealers. In this fashion the drug was middled, dealt, divided and cut—sometimes as many as five or six times—until it was finally purchased by consumers as grams or half-grams.

Unlike low-level operators, the upper-level dealers and smugglers we studied pursued drug trafficking as a full-time occupation. If they were involved in other businesses, these were usually maintained to provide them with a legitimate front for security purposes. The profits to be made at the upper levels depended on an individual's style of operation, reliability, security, and the amount of product he or she consumed. About half of the 65 smugglers and dealers we observed were successful, some earning up to three-quarters of a million dollars per year.[2] The other half continually struggled in the business, either breaking even or losing money.

Although dealers' and smugglers' business activities varied, they clustered together for business and social relations, forming a moderately well-integrated community whose members pursued a "fast" lifestyle, which emphasized intensive partying, casual sex, extensive travel, abundant drug consumption, and lavish spending on consumer goods. The exact size of Southwest County's upper-level dealing and smuggling community was impossible to estimate due to the secrecy of its members. At these levels, the drug world was quite homogeneous. Participants were predominantly white, came from middle-class backgrounds, and had little previous criminal involvement. While the dealers' and smugglers' social world contained both men and women, most of the serious business was conducted by the men, ranging in age from 25 to 40 years old.

We gained entry to Southwest County's upper-level drug community largely by accident. We had become friendly with a group of our neighbors who turned out be heavily involved in smuggling marijuana. Opportunistically (Riemer, 1977), we seized the chance to gather data on this unexplored activity. Using key informants who helped us gain the trust of other members of the community, we drew upon snowball sampling techniques (Biernacki and Waldorf, 1981) and a combination of overt and covert roles to widen our network of contacts. We supplemented intensive participant-observation, between 1974 and 1980,[3] with unstructured, taped interviews. Throughout, we employed extensive measures to cross-check the reliability of our data, whenever possible (Douglas, 1976). In all, we were able to closely observe 65 dealers and smugglers as well as numerous other drug world members, including dealers' "old ladies" (girlfriends or wives), friends, and family members. . . .

SHIFTS AND OSCILLATIONS

. . . Despite the gratifications which dealers and smugglers originally derived from the easy money, material comfort, freedom, prestige, and power associated with their careers, 90 percent of those we observed decided, at some point, to quit the business. This stemmed, in part, from their initial perceptions of the career as temporary ("Hell, nobody wants to be a drug dealer all their life"). Adding to these early intentions was a process of rapid aging in the career: dealers and smugglers became increasingly aware of the restrictions and sacrifices their occupations required and tired of living the fugitive life. They thought about, talked about, and in many cases took steps toward getting out of the drug business. But as with entering, disengaging from drug trafficking was rarely an abrupt act (Lieb and Olson, 1976:364). Instead, it more often resembled a series of transitions, or oscillations,[4] out of and back into the business. For once out of the drug world, dealers and smugglers were rarely successful in making it in the legitimate world because they failed to cut down on their extravagant lifestyle and drug consumption. Many abandoned their efforts to reform and returned to deviance, sometimes picking up where they left off and other times shifting to a new mode of operating. For example, some shifted from dealing cocaine to dealing marijuana, some dropped to a lower level of dealing, and others shifted their role within the same group of traffickers. This series of phase-outs and re-entries, combined with career shifts, endured for years, dominating the pattern of their remaining involvement with the business. But it also represented the method by which many eventually broke away from drug trafficking, for each phase-out had the potential to be an individual's final departure.

Aging in the Career

Once recruited and established in the drug world, dealers and smugglers entered into a middle phase of aging in the career. This phase was characterized by a progressive loss of enchantment with their occupation. While novice dealers and smugglers found that participation in the drug world brought them thrills and status, the novelty gradually faded. Initial feelings of exhilaration and awe began to dull as individuals became increasingly jaded. This was the result of both an extended exposure to the mundane, everyday business aspects of drug trafficking and to an exorbitant consumption of drugs (especially cocaine). One smuggler described how he eventually came to feel:

> It was fun, those three or four years. I never worried about money or anything. But after awhile it got real boring. There was no feeling or emotion or anything about it. I wasn't even hardly relating to my old lady anymore. Everything was just one big rush.

This frenzy of overstimulation and resulting exhaustion hastened the process of "burnout" which nearly all individuals experienced. As dealers and smugglers aged in the career they became more sensitized to the extreme risks they faced. Cases of friends and associates who were arrested, imprisoned, or killed began to mount. Many individuals became convinced that continued drug trafficking would inevitably lead to arrest ("It's only a matter of time before you get caught"). While dealers and smugglers generally repressed their awareness of danger, treating it as a taken-for-granted part of their daily existence, periodic crises shattered their casual attitudes, evoking strong feelings of fear. They temporarily intensified security precautions and retreated into near-isolation until they felt the "heat" was off.

As a result of these accumulating "scares," dealers and smugglers increasingly integrated feelings of "paranoia"[5] into their everyday lives. One dealer talked about his feelings of paranoia:

> You're always on the line. You don't lead a normal life. You're always looking over your shoulder, wondering who's at the door, having to hide everything. You learn to look behind you so well you could probably bend over and look up your ass. That's paranoia. It's a really scary, hard feeling. That's what makes you get out.

Drug world members also grew progressively weary of their exclusion from the legitimate world and the deceptions they had to manage to sustain that separation. Initially, this separation was surrounded by an alluring mystique. But as they aged in the career, this mystique became replaced by the reality of everyday boundary maintenance and the feeling of being an "expatriated citizen within one's own country." One smuggler who was contemplating quitting described the effects of this separation:

> I'm so sick of looking over my shoulder, having to sit in my house and worry about one of my non-drug world friends stopping in when I'm doing business. Do you know how awful that is? It's like leading a double life. It's ridiculous. That's what makes it not worth it. It'll be a lot less money [to quit], but a lot less pressure.

Thus, while the drug world was somewhat restricted, it was not an encapsulated community, and dealers' and smugglers' continuous involvement with the straight world made the temptation to adhere to normative standards and "go straight" omnipresent. With the occupation's novelty worn off and the "fast life" taken-for-granted, most dealers and smugglers felt that the occupation no longer resembled their early impressions of it. Once they reached the upper levels of the occupation, their experience began to change. Eventually, the rewards of trafficking no longer seemed to justify the strain and risk involved. It was at this point that the straight world's formerly dull ambiance became transformed (at least in theory) into a potential haven.

Phasing Out

Three factors inhibited dealers and smugglers from leaving the drug world. Primary among these factors were the hedonistic and materialistic satisfactions the drug world provided. Once accustomed to earning vast quantities of money quickly and easily, individuals found it exceedingly difficult to return to the income scale of the straight world. They also were reluctant to abandon the pleasure of the "fast life" and its accompanying drugs, casual sex, and power. Second, dealers and smugglers identified with, and developed a commitment to, the occupation of drug trafficking (Adler and Adler, 1982). Their self-images were tied to that role and could not be easily disengaged. The years invested in their careers (learning the trade, forming connections, building reputations) strengthened their involvement with both the occupation and the drug community. And since their relationships were social as well as business, friendship ties bound individuals to dealing. As one dealer in the midst of struggling to phase-out explained:

> The biggest threat to me is to get caught up sitting around the house with friends that are into dealing. I'm trying to stay away from them, change my habits.

Third, dealers and smugglers hesitated to voluntarily quit the field because of the difficulty involved in finding another way to earn a living. Their years spent in illicit activity made it unlikely for any legitimate organizations to hire them. This narrowed their occupational choices considerably, leaving self-employment as one of the few remaining avenues open.

Dealers and smugglers who tried to leave the drug world generally fell into one of four patterns.[6] The first and most frequent pattern was to postpone quitting until after they could execute one last "big deal." While the intention was sincere, individuals who chose this route rarely succeeded; the "big deal" too often remained elusive. One marijuana smuggler offered a variation of this theme:

> My plan is to make a quarter of a million dollars in four months during the prime smuggling season and get the hell out of the business.

A second pattern we observed was individuals who planned to change immediately, but never did. They announced they were quitting, yet their outward actions never varied. One dealer described his involvement with this syndrome:

> When I wake up I'll say, "Hey, I'm going to quit this cycle and just run my other business." But when you're dealing you constantly have people dropping by ounces and asking, "Can you move this?" What's your first response? Always, "Sure, for a toot."

In the third pattern of phasing-out, individuals actually suspended their dealing and smuggling activities, but did not replace them with an alternative source of income. Such withdrawals were usually spontaneous and prompted by exhaustion, the influence of a person from outside the drug world, or problems with the police or other associates. These kinds of phase-outs usually lasted only until the individual's money ran out, as one dealer explained:

> I got into legal trouble with the FBI a while back and I was forced to quit dealing. Everybody just cut me off completely, and I saw the danger in continuing, myself. But my high-class tastes never dwindled. Before I knew it I was in hock over $30,000. Even though I was hot, I was forced to get back into dealing to relieve some of my debts.

In the fourth pattern of phasing out, dealers and smugglers tried to move into another line of work. Alternative occupations included: (1) those they had previously pursued; (2) front businesses maintained on the side while dealing or smuggling; and (3) new occupations altogether. While some people accomplished this transition successfully, there were problems inherent in all three alternatives.

(1) Most people who tried resuming their former occupations found that these had changed too much while they were away. In addition, they themselves had changed: they enjoyed the self-directed freedom and spontaneity associated with dealing and smuggling, and were unwilling to relinquish it.

(2) Those who turned to their legitimate front business often found that these businesses were unable to support them. Designed to launder rather than earn money, most of these ventures were retail outlets with a heavy cash flow (restaurants, movie theaters, automobile dealerships, small stores) that had become accustomed to operating under a continuous subsidy from illegal funds. Once their drug funding was cut off they could not survive for long.

(3) Many dealers and smugglers utilized the skills and connections they had developed in the drug business to create a new occupation. They exchanged their illegal commodity for a legal one and went into import/export, manufacturing, wholesaling, or retailing other merchandise. For some, the decision to prepare a legitimate career for their future retirement from the drug world followed an unsuccessful attempt to phase-out into a "front" business. One husband-and-wife dealing team explained how these legitimate side businesses differed from front businesses:

> We always had a little legitimate "scam" [scheme] going, like mail-order shirts, wallets, jewelry, and the kids were always involved in that. We made a little bit of money on them. Their main purpose was for a cover. But [this business] was different; right from the start this was going to be a legal thing to push us out of the drug business.

About 10 percent of the dealers and smugglers we observed began tapering off their drug world involvement gradually, transferring their time and money into a selected legitimate endeavor. They did not try to quit drug trafficking altogether until they felt confident that their legitimate business could support them. Like spontaneous phase-outs, many of these planned withdrawals into legitimate endeavors failed to generate enough money to keep individuals from being lured into the drug world.

In addition to voluntary phase-outs caused by burnout, about 40 percent of the Southwest County dealers and smugglers we observed experienced a "bustout" at some point in their careers.[7] Forced withdrawals from dealing or smuggling were usually sudden and motivated by external factors, either financial, legal, or reputational. Financial bustouts generally occurred when dealers or smugglers were either "burned" or "ripped-off" by others, leaving them in too much debt to rebuild their base of operation. Legal bustouts followed arrest and possibly incarceration: arrested individuals were so "hot" that few of their former associates would deal with them. Reputational bustouts occurred when individuals "burned" or "ripped-off" others (regardless of whether they intended to do so) and were banned from business by their former circle of associates. One smuggler gave his opinion on the pervasive nature of forced phase-outs:

> Some people are smart enough to get out of it because they realize, physically, they have to. Others realize, monetarily, that they want to get out of this world before this world gets them. Those are the lucky ones. Then there are the ones who have to get out because they're hot or someone else close to them is so hot that they'd better get out. But in the end when you get out of it, nobody gets out of it out of free choice; you do it because you have to.

Death, of course, was the ultimate bustout. Some pilots met this fate because of the dangerous routes they navigated (hugging mountains, treetops, other aircrafts) and the sometimes ill-maintained and overloaded planes they flew. However, despite much talk of violence, few Southwest County drug traffickers died at the hands of fellow dealers.

Re-Entry

Phasing-out of the drug world was more often than not temporary. For many dealers and smugglers, it represented but another stage of their drug careers (although this may not have been their original intention), to be followed by a period of reinvolvement. Depending on the individual's perspective, re-entry into the drug world could be viewed as either a comeback (from a forced withdrawal) or a relapse (from a voluntary withdrawal).

Most people forced out of drug trafficking were anxious to return. The decision to phase-out was never theirs, and the desire to get back into dealing or smuggling was based on many of the same reasons which drew them into the field originally. Coming back from financial, legal, and reputational bustouts was possible but difficult and was not always successfully accomplished. They had to re-establish contacts, rebuild their organization and fronting arrangements, and raise the operating capital to resume dealing. More difficult was the problem of overcoming the circumstances surrounding their departure. Once smugglers and dealers resumed operating, they often found their former colleagues suspicious of them. One frustrated dealer described the effects of his prison experience:

> When I first got out of the joint [jail], none of my old friends would have anything to do with me. Finally, one guy who had been my partner told me it was because everyone was suspicious of my getting out early and thought I made a deal [with police to inform on his colleagues].

Dealers and smugglers who returned from bustouts were thus informally subjected to a trial period in which they had to re-establish their trustworthiness and reliability before they could once again move in the drug world with ease.

Re-entry from voluntary withdrawal involved a more difficult decision-making process, but was easier to implement. The factors enticing individuals to re-enter the drug world were not the same as those which motivated their original entry. As we noted above, experienced dealers and smugglers often privately weighed their reasons for wanting to quit and wanting to stay in. Once they left, their images of and hopes for the straight world failed to materialize. They could not make the shift to the norms, values, and lifestyle of the straight society and could not earn a living within it. Thus, dealers and smugglers decided to re-enter the drug business for basic reasons: the material perquisites, the hedonistic gratifications, the social ties, and the fact that they had nowhere else to go.

Once this decision was made, the actual process of re-entry was relatively easy. One dealer described how the door back into dealing remained open for those who left voluntarily:

> I still see my dealer friends, I can still buy grams from them when I want to. It's the respect they have for me because I stepped out of it without being busted or burning someone. I'm coming out with a good reputation, and even though the scene is a whirlwind—people moving up, moving down, in, out—if I didn't see anybody for a year I could call them up and get right back in that day.

People who relapsed thus had little problem obtaining fronts, re-establishing their reputations, or readjusting to the scene.

Career Shifts

Dealers and smugglers who re-entered the drug world, whether from a voluntary or forced phase-out, did not always return to the same level of transacting or commodity which characterized their previous style of operation. Many individuals underwent a "career shift" (Luckenbill and Best, 1981) and became involved in some new segment of the drug world. These shifts were sometimes lateral, as when a member of a smuggling crew took on a new specialization, switching from piloting to operating a stash house, for example. One dealer described how he utilized friendship networks upon his re-entry to shift from cocaine to marijuana trafficking:

> Before, when I was dealing cocaine, I was too caught up in using the drug and people around me were starting to go under from getting into "base" [another form of cocaine]. That's why I got out. But now I think I've got myself together and even though I'm dealing again I'm staying away from coke. I've switched over to dealing grass. It's a whole different circle of people. I got into it through a close friend I used to know before, but I never did business with him because he did grass and I did coke.

Vertical shifts moved operators to different levels. For example, one former smuggler returned and began dealing; another top-level marijuana dealer came back to find that the smugglers he knew had disappeared and he was forced to buy in smaller quantities from other dealers.

Another type of shift relocated drug traffickers in different styles of operation. One dealer described how, after being arrested, he tightened his security measures:

> I just had to cut back after I went through those changes. Hell, I'm not getting any younger and the idea of going to prison bothers me a lot more than it did 10 years ago. The risks are no longer worth it when I can have a comfortable income with less risk. So I only sell to four people now. I don't care if they buy a pound or a gram.

A former smuggler who sold his operation and lost all his money during phase-out returned as a consultant to the industry, selling his expertise to those with new money and fresh manpower:

> What I've been doing lately is setting up deals for people. I've got foolproof plans for smuggling cocaine up here from Colombia; I tell them how to modify their airplanes to add on extra fuel tanks and to fit in more week, coke, or whatever they bring up. Then I set them up with refueling points all up and down Central America, tell them how to bring it up here, what points to come in at, and what kind of receiving unit to use. Then they do it all and I get 10 percent of what they make.

Re-entry did not always involve a shift to a new niche, however. Some dealers and smugglers returned to the same circle of associates, trafficking activity, and commodity they worked with prior to their departure. Thus, drug dealers' careers often peaked early and then displayed a variety of shifts, from lateral mobility, to decline, to holding fairly steady.

A final alternative involved neither completely leaving nor remaining within the deviant world. Many individuals straddled the deviant and respectable worlds forever by continuing to dabble in drug trafficking. As a result of their experiences in the drug world they developed a deviant self-identity and a deviant *modus operandi*. They might not have wanted to bear the social and legal burden of full-time deviant work but neither were they willing to assume the perceived confines and limitations of the straight world. They therefore moved into the entrepreneurial realm, where their daily activities involved some kind of hustling or "wheeling and dealing" in an assortment of legitimate, quasi-legitimate, and deviant ventures, and where they could be their own boss. This enabled them to retain certain elements of the deviant lifestyle, and to socialize on the fringes of the drug community. For these individuals, drug dealing shifted from a primary occupation to a sideline, though they never abandoned it altogether.

LEAVING DRUG TRAFFICKING

This career pattern of oscillation into and out of active drug trafficking makes it difficult to speak of leaving drug trafficking in the sense of final retirement. Clearly, some people succeeded in voluntarily retiring. Of these, a few managed to prepare a post-deviant career for themselves by transferring their drug money into a legitimate enterprise. A larger group was forced out of dealing and either didn't or couldn't return; the bustouts were sufficiently damaging that they never attempted re-entry, or they abandoned efforts after a series of unsuccessful attempts. But there was no way of structurally determining in advance whether an exit from the business would be temporary or permanent. The vacillations in dealers' intentions were compounded by the complexity of operating successfully in the drug world. For many, then, no phase-out could ever be definitely assessed as permanent. As long as individuals had skills, knowledge, and connections to deal they retained the potential to re-enter the occupation at any time. Leaving drug trafficking may thus be a relative phenomenon, characterized by a trailing-off process where spurts of involvement appear with decreasing frequency and intensity.

SUMMARY

Drug dealing and smuggling careers are temporary and fraught with multiple attempts at retirement. Veteran drug traffickers quit their occupation because of the ambivalent feelings they develop toward their deviant life. As they age in the career their experience changes, shifting from a work life that is exhilarating and free to one that becomes increasingly dangerous and confining. But just as their deviant careers are temporary, so too are their retirements. Potential recruits are lured into the drug business by materialism, hedonism, glamor, and excitement. Established dealers are lured away from the deviant life and back into the mainstream by the attractions of security and social ease. Retired dealers and smugglers are lured back in by their expertise, and by their ability to make money quickly and easily. People who have been exposed to the upper levels of drug trafficking therefore find it extremely difficult to quit their deviant occupation permanently. This stems, in part, from their difficulty in moving from the illegitimate to the legitimate business sector. Even more significant is the affinity they form for their deviant values and lifestyle. Thus few, if any, of our subjects were successful in leaving deviance entirely. What dealers and smugglers intend, at the time, to be a permanent withdrawal from drug trafficking can be seen in retrospect as a pervasive occupational pattern of mid-career shifts and oscillations. More research is needed into the complex process of how people get out of deviance and enter the world of legitimate work.

NOTES

1. The term "soft" drugs generally refers to marijuana, cocaine and such psychedelics as LSD and mescaline (Carey, 1968). In this paper we do not

address trafficking in psychedelics because, since they are manufactured in the United States, they are neither imported nor distributed by the group we studied.

2. This is an idealized figure representing the profit a dealer or smuggler could potentially earn and does not include deductions for such miscellaneous and hard-to-calculate costs as: time or money spent in arranging deals (some of which never materialize); lost, stolen, or unrepaid money or drugs; and the personal drug consumption of a drug trafficker and his or her entourage. Of these, the single largest expense is the last one, accounting for the bulk of most Southwest County dealers' and smugglers' earnings.

3. We continued to conduct follow-up interviews with key informants through 1983.

4. While other studies of drug dealing have also noted that participants did not maintain an uninterrupted stream of career involvement (Blum *et al.*, 1972; Carey, 1968; Lieb and Olson, 1976; Waldorf *et al.*, 1977), none have isolated or described the oscillating nature of this pattern.

5. In the dealers' vernacular, this term is not used in the clinical sense of an individual psychopathology rooted in early childhood traumas. Instead, it resembles Lemert's (1962) more sociological definition which focuses on such behavioral dynamics as suspicion, hostility, aggressiveness, and even delusion. Not only Lemert, but also Waldorf *et al.* (1977) and Wedow (1979) assert that feelings of paranoia can have a sound basis in reality, and are therefore readily comprehended and even empathized with others.

6. At this point, a limitation to our data must be noted. Many of the dealers and smugglers we observed simply "disappeared" from the scene and were never heard from again. We therefore have no way of knowing if they phased out (voluntarily or involuntarily), shifted to another scene, or were killed in some remote place. We cannot, therefore, estimate the numbers of people who left the Southwest County drug scene via each of the routes discussed here.

7. It is impossible to determine the exact percentage of people falling into the different phase-out categories: due to oscillation, people could experience several types and thus appear in multiple categories.

REFERENCES

Adler, Patricia A., and Peter Adler. 1982. "Criminal commitment among drug dealers." Deviant Behavior 3:117–135.

Anonymous. 1969. "On selling marijuana." Pp. 92–102 in Erich Goode (ed.), Marijuana. New York: Atherton.

Atkyns, Robert L., and Gerhard J. Hanneman. 1974. "Illicit drug distribution and dealer communication behavior." Journal of Health and Social Behavior 15(March):36–43.

Biernacki, Patrick, and Dan Waldorf. 1981. "Snowball sampling." Sociological Methods and Research 10(2):141–163.

Blum, Richard H., and Associates. 1972. The Dream Sellers. San Francisco: Jossey-Bass.

Carey, James T. 1968. The College Drug Scene. Englewood Cliffs, NJ: Prentice-Hall.

Douglas, Jack D. 1976. Investigative Social Research. Beverly Hills, CA: Sage.

Goode, Erich. 1970. The Marijuana Smokers. New York: Basic.

Langer, John. 1977. "Drug entrepreneurs and dealing culture." Social Problems 24(3):377–385.

Lemert, Edwin. 1962. "Paranoia and the dynamics of exclusion." Sociometry 25(March):2–20.

Lieb, John, and Sheldon Olson. 1976. "Prestige, paranoia, and profit: On becoming a dealer of illicit drugs in a university community." Journal of Drug Issues 6(Fall):356–369.

Luckenbill, David F., and Joel Best. 1981. "Careers in deviance and respectability: The analogy's limitations." Social Problems 29(2):197–206.

Mouledoux, James. 1972. "Ideological aspects of drug dealership." Pp. 110–122 in Ken Westhues (ed.), Society's Shadow: Studies in the Sociology of Countercultures. Toronto: McGraw-Hill, Ryerson.

Riemer, Jeffrey W. 1977. "Varieties of opportunistic research." Urban Life 5(4):467–477.

Waldorf, Dan, Sheigla Murphy, Craig Reinarman, and Bridget Joyce. 1977. Doing Coke: An Ethnography of Cocaine Users and Sellers. Washington, DC: Drug Abuse Council.

Wedow, Suzanne. 1979. "Feeling paranoid: The organization of an ideology." Urban Life 8(1):72–93.

39

The Professional Ex-: An Alternative
for Exiting the Deviant Career

J. DAVID BROWN

T his study explores the careers of professional ex-s, persons who have exited their deviant careers by replacing them with occupations in professional counseling. During their transformation professional ex-s utilize vestiges of their deviant identity to legitimate their past deviance and generate new careers as counselors.

Recent surveys document that approximately 72% of the professional counselors working in the over 10,000 U.S. substance abuse treatment centers are former substance abusers (NAADAC 1986; Sobell and Sobell 1987). This attests to the significance of the professional ex- phenomenon. Though not all ex-deviants become professional ex-s, such data clearly suggest that the majority of substance abuse counselors are professional ex-s.[1]

Since the inception of the notion of deviant career by Goffman (1961) and Becker (1963), research has identified, differentiated, and explicated the characteristics of specific deviant career stages (e.g., Adler and Adler 1983; Luckenbill and Best 1981; Meisenhelder 1977; Miller 1986; Shover 1983). The literature devoted to exiting deviance primarily addresses the process whereby individuals abandon their deviant behaviors, ideologies, and identities and replace them with more conventional lifestyles and identities (Irwin 1970; Lofland 1969; Meisenhelder 1977; Shover 1983). While some studies emphasize the role of authorities or associations of ex-deviants in this change (e.g., Livingston 1974; Lofland 1969; Volkman and Cressey 1963), others suggest that exiting deviance is a natural process contingent upon age-related, structural, and social psychological variables (Frazier 1976; Inciardi 1975; Irwin 1970; Meisenhelder 1977; Petersilia 1980; Shover 1983).

Although exiting deviance has been variously conceptualized, to date no one has considered that it might include adoption of a legitimate career premised upon an identity that embraces one's deviant history. Professional ex-s exemplify this mode of exiting deviance.

Ebaugh's (1988) model of role exit provides an initial framework for examining this alternative mode of exiting the deviant career. Her model suggests that former roles are never abandoned but, instead, carry over into new

From: "The Professional Ex-: An Alternative for Exiting the Deviant Career," J. David Brown, *Sociological Quarterly,* Vol. 32, No. 2, 1991. Reprinted by permission of JAI Press Inc.

roles. I elaborate her position and contend that one's deviant identity is not an obstacle that must be abandoned prior to exiting or adopting a more conventional lifestyle. To the contrary, one's lingering deviant identity facilitates rather than inhibits the exiting process.

How I gathered data pertinent to exiting, my relationship to these data, and how my personal experiences with exiting deviance organize this article, follow. I then present a four stage model that outlines the basic contours of the professional ex- phenomenon. Finally I suggest how the professional ex- phenomenon represents an alternative interpretation of exiting deviance that generalizes to other forms of deviance.

METHODS

Data for this research consists of introspective and qualitative material.

Introspective Data

My introspections distill 20 years of experience with substance abuse/alcoholism, social control agents/agencies, and professional counselor training. I spent 13 years becoming a deviant drinker and entered substance abuse treatment in 1979. For 5 years (1981–1986), I was a primary therapist and family interventionist for a local private residential treatment facility.

"Systematic sociological introspection" (Ellis 1987, 1990), "auto-ethnography" (Hayano 1979), and "opportunistic research" (Reimer 1977) accessed the introspective data. Each group status—abuser, patient, therapist—indicates the "complete membership role" (Adler and Adler 1987) that combines unique circumstances with personal expertise to enhance research. The four stage model of exiting described later is, in part, informed by reexamination of the written artifacts of my therapeutic/recovery experiences (e.g., alco-biography, moral inventory, daily inventory journal) and professional counselor training (e.g., term paper, internship journal).

Qualitative Data

Qualitative data were collected over a six month period of intensive interviews with 35 counselor ex-s employed in a variety of community, state, and private institutions that treat individuals with drug, alcohol, and/or eating disorder problems.[2]

These professional ex-s worked in diverse occupations prior to becoming substance abuse counselors. A partial list includes employment as accountants, managers, salespersons, nurses, educators, and business owners. Although they claimed to enter the counseling profession within two years of discharge from therapy, their decision to become counselors usually came within one year. On the average they had been counselors for four and one half years. Except one professional ex- who previously counseled learning disabled children, all claimed they had not seriously considered a counseling career before entering therapy.

THE EXIT PROCESS

Ebaugh (1988) contends that the experience of being an "ex" of one kind or another is common to most people in modern society. Emphasizing the sociological and psychological continuity of the ex-phenomenon she states, "[I]t implies that interaction is based not only on current role definitions but, more important, past identities that somehow linger on and define how people see and present themselves in their present identities" (p. xiii). Ebaugh defines the role exit as the "process of disengagement from a role that is central to one's self-identity and the reestablishment of an identity in a new role that takes into account one's ex-role" (p. 1).

Becoming a professional ex- is the outcome of a four stage process through which ex-s capitalize on the experience and vestiges of their deviant career in order to establish a new identity and role in a respectable organization. This process comprises emulation of one's therapist, the call to a counseling career, status-set realignment, and credentialization.

Stage One: Emulation of One's Therapist

The emotional and symbolic identification of these ex-s with their therapists during treatment, combined with the deep personal meanings they imputed to these relationships, was a compelling factor in their decisions to become counselors. Denzin (1987, pp. 61–62) identifies the therapeutic relationship's significance thus: "Through a process of identification and surrender (which may be altruistic), the alcoholic may merge her ego and her self in the experiences and the identity of the counselor. The group leader . . . is the group ego ideal, for he or she is a successful recovering alcoholic. . . . An emotional bond is thus formed with the group counselor. . . ."

Professional ex-s not only developed this emotional bond but additionally aspired to have the emotions and meanings once projected toward their therapists ascribed to them. An eating disorders counselor discussed her relationship with her therapist and her desire to be viewed in a similar way with these words:

> My counselor taught me the ability to care about myself and other people. Before I met her I was literally insane. She was the one who showed me that I wasn't crazy. Now, I want to be the person who says, "No, you're not crazy!" I am the one, now, who is helping them to get free from the ignorance that has shrouded eating disorders.

Counselors enacted a powerfully charismatic role in professional ex-s' therapeutic transformation. Their "laying on of verbal hands" provided initial comfort and relief from the ravaging symptoms of disease. They came to represent what ex-s must do both spiritually and professionally for themselves. Substance abuse therapy symbolized the "sacred" quest for divine grace rather than the mere pursuit of mundane, worldly, or "profane" outcomes like abstinence or modification of substance use/abuse behaviors; counselors embodied the sacred outcome.

Professional ex-s claimed that their therapists were the most significant change agent in their transformation. "I am here today because there was one very influential counselor in my life who helped me to get sober. I owe it all to God and to him," one alcoholism counselor expressed. A heroin addiction counselor stated, "The best thing that ever happened in my life was meeting Sally [her counselor]. She literally saved my life. If it wasn't for her I'd still probably be out there shootin' up or else be in prison or, dead."

Subjects' recognition and identification of a leader's charismatic authority, as Weber (1968) notes, is decisive in validating that charisma and developing absolute trust and devotion. The special virtues and powers professional ex-s perceived in their counselors subsequently shaped their loyalty and devotion to the career.

Within the therapeutic relationship, professional ex-s perform a priestly function through which a cultural tradition passes from one generation to the next. While knowledge and wisdom pass downward (from professional ex- to patient), careers build upward (from patient to professional ex-). As the bearers of the cultural legacy of therapy, professional ex-s teach patients definitions of the situation they learned as patients. Indeed, part of the professional ex- mystique resides in once having been a patient (Bissell 1982). In this regard,

> My counselor established her legitimacy with me the moment she disclosed the fact that she, too, was an alcoholic. She wasn't just telling me what to do, she was living her own advice. By the example she set, I felt hopeful that I could recover. As I reflect upon those experiences I cannot think of one patient ever asking me about where I received my professional training. At the same time, I cannot begin to count the numerous times that my patients have asked me if I was "recovering."

Similar to religious converts' salvation through a profoundly redemptive religious experience, professional ex-s' deep career commitment derives from a transforming therapeutic resocialization. As the previous examples suggest, salvation not only relates to a changed universe of discourse; it is also identified "with one's personal therapist."[3]

At this stage, professional ex-s trust in and devote themselves to their counselors' proselytizations as a promissory note for the future. The promise is redemption and salvation from the ever-present potential for self-destruction or relapse that looms in their mental horizon. An eating disorders counselor shared her insights in this way:

> I wouldn't have gotten so involved in eating disorders counseling if I had felt certain that my eating disorder was taken care of. I see myself in constant recovery. If I was so self assured that I would never have the problem again there would probably be less of an emphasis on being involved in the field but I have found that helping others, as I was once helped, really helps me.

The substance abuse treatment center transforms from a mere "clinic" occupied by secularly credentialed professionals into a moral community of single believers. As Durkheim (1915) suggests, however, beliefs require rites and practices in order to sustain adherents' mental and emotional states.

Stage Two: The Call to a Counseling Career

At this juncture, professional ex-s begin to turn the moral corner on their deviance. Behaviors previously declared morally reprehensible are increasingly understood within a new universe of discourse as symptoms of a much larger disease complex. This recognition represents one preliminary step toward grace. In order to emulate their therapist, however, professional ex-s realize they must dedicate themselves to an identity and lifestyle that ensure their own symptoms' permanent remission. One alcoholism counselor illustrated this point by stating:

> I can't have my life, my health, my family, my job, my friends, or any-thing, unless I take daily necessary steps to ensure my continued recovery. My program of recovery has to come first. Before I can go out there and help my patients I need to always make sure that my own house is in order.

As this suggests, a new world-view premised upon accepting the contingencies of one's illness while maintaining a constant vigilance over potentially recurring symptoms replaces deviant moral and social meanings. Professional ex-s' recognition of the need for constant vigilance is internalized as their moral mission from which their spiritual duty (a counseling career) follows as a natural next step.

Although professional ex-s no longer engage in substance abuse behaviors, they do not totally abandon deviant beliefs or identity. "Lest we become complacent and forget from whence we came," as one alcoholism counselor indicated the significance of remembering and embracing the past.

Professional ex-s' identification with their deviant past undergirds their professional, experiential, and moral differentiation from other professional colleagues. A heroin addiction counselor recounted how he still identified himself as an addict and deviant:

> My perspective and my affinity to my clients, particularly the harder core criminals, is far better than the professor and other doctors that I deal with here in my job. We're different and we really don't see things the same way at all. Our acceptance and understanding of these people's diseases, if you will, is much different. They haven't experienced it. They don't know these people at all. It takes more than knowing about something to be effective. I've been there and, in many respects, I will always be there.

In this way, other counselors' medical, psychiatric, or therapeutic skills are construed as part of the ordinary mundane world. As the quotation indicates, professional ex-s intentionally use their experiential past and therapeutic transformations to legitimate their entrance into and authority in counseling careers.

Professional ex-s embrace their deviant history and identity as an invaluable, therapeutic resource and feel compelled to continually reaffirm its validity in an institutional environment. Certainly, participating in "12 Step Programs"[4] without becoming counselors could help others but professional ex-s' call requires greater immersion than they provide. An alcoholism counselor reflected upon this need thus:

> For me, it was no longer sufficient to only participate "anonymously" in A.A. I wanted to surround myself with other spiritual and professional pilgrims devoted to receiving and imparting wisdom.

Towards patients, professional ex-s project a saintly aura and exemplify an "ideal recovery." Internalization of self-images previously ascribed to their therapist and now reaffirmed through an emotional and moral commitment to the counseling profession facilitate this ideation. Invariably, professional ex-s' counseling careers are in institutions professing treatment ideologies identical to what they were taught as patients. Becoming a professional ex- symbolizes a value elevated to a directing goal, whose pursuit predisposes them to interpret all ensuing experience in terms of relevance to it.

Stage Three: Status-Set Realignment

Professional ex-s' deep personal identification with their therapist provides an ego ideal to be emulated with regard to both recovery and career. They immerse themselves in what literally constitutes a "professional recovery career" that provides an institutional location to reciprocate their counselors' gift, immerse themselves in a new universe of discourse, and effectively lead novitiates to salvation. "I wouldn't be here today if it wasn't for all of the help I received in therapy. This is my way of paying some of those people back by helping those still in need," one alcoholism counselor related this.

Professional ex-s' identities assume a "master status" (Hughes 1945) that differs in one fundamental respect from others' experiencing therapeutic resocialization. Specifically, their transformed identities not only become the "most salient" in their "role identity hierarchy" (Stryker and Serpe 1982), but affect all other roles in their "status-sets" (Merton 1938). One alcoholism counselor reflected upon it this way:

> Maintaining a continued program of recovery is the most important thing in my life. Everything else is secondary. I've stopped socializing with my old friends who drink and have developed new recovering friends. I interact differently with my family. I used to work a lot of overtime but I told my old boss that overtime jeopardized my program. I finally began to realize that the job just didn't have anything to do with what I was really about. I felt alienated. Although I had been thinking about becoming a counselor ever since I went through treatment, I finally decided to pursue it.

Role alignment is facilitated by an alternative identity that redefines obligations associated with other, less significant, role identities. In the previous example, the strains of expectations associated with a former occupation fostered a role alignment consistent with a new self-image. This phenomenon closely resembles what Snow and Machalek (1983, p. 276) refer to as "embracement of a master role" that "is not merely a mask that is taken off or put on according to the situation. . . . Rather, it is central to nearly all situations. . . ." An eating disorders counselor stated the need to align her career with her self-image, "I hid in my former profession, interacting little with people. As a counselor, I am personally maturing and taking responsibility rather than letting a company take care of me. I have a sense of purpose in this job that I never had before."

Financial renumeration is not a major consideration in the decision to become a professional ex-. The pure type of call, Weber (1968, p. 52) notes, "disdains and repudiates economic exploitation of the gifts of grace as a source of income. . . ." Most professional ex-s earned more money in their previous jobs. For instance, one heroin addiction counselor stated:

> When I first got out of treatment, my wife and I started an accounting business. In our first year we cleared nearly sixty thousand dollars. The money was great and the business showed promise but something was missing. I missed being around other addicts and I knew I wanted to do more with my life along the lines of helping out people like me.

An additional factor contributing to professional ex-s' abandonment of their previous occupation is their recognition that a counseling career could resolve lingering self-doubts about their ability to remain abstinent. In this respect becoming a professional ex- allows "staying current" with their own recovery needs while continually reaffirming the severity of their illness. An eating disorders counselor explained:

> I'm constantly in the process of repeating insights that I've had to my patients. I hear myself saying, to them, what I need to believe for myself. Being a therapist helps me to keep current with my own recovery. I feel that I am much less vulnerable to my disease in this environment. It's a way that I can keep myself honest. Always being around others with similar issues prevents me from ignoring my own addiction clues.

This example illustrates professional ex-s' use of their profession to secure self-compliance during times of self-doubt. While parroting the virtues of the program facilitates recognition that they, too, suffer from a disease, the professional ex- role, unlike their previous occupations, enables them to continue therapy indirectly.

Finally, the status the broader community ascribes to the professional ex- role encourages professional ex-s' abandonment of previous roles. Association with an institutional environment and an occupational role gives the professional ex- a new sense of place in the surrounding community, within which form new self-concepts and self-esteem, both in the immediate situation and in a broader temporal framework.

The internal validation of professional ex-s' new identity resides in their ability to successfully anticipate the behaviors and actions of relevant alters. Additionally, they secure validation by other members of the professional ex-community in a manner atypical for other recovering individuals. Affirmation by this reference community symbolizes validation by one's personal therapist and the therapeutic institution, as a heroin addiction counselor succinctly stated:

> Becoming a counselor was a way to demonstrate my loyalty and devotion to helping others and myself. My successes in recovery, including being a counselor, would be seen by patients and those who helped me get sober. It was a return to treatment, for sure, but the major difference was that this time I returned victorious rather than defeated.

External validation, on the other hand, comes when others outside the therapeutic community accord legitimacy to the professional ex- role. In this regard, a heroin addiction counselor said:

> I remember talking to this guy while I was standing in line for a movie. He asked me what I did for a living and I told him that I was a drug abuse counselor. He started asking me all these questions about the drug problem and what I thought the answers were. When we finally got up to the door of the theater he patted me on the back and said, "You're doing a wonderful job. Keep up the good work. I really admire you for what you're trying to do." It really felt good to have a stranger praise me.

Professional ex-s' counseling role informs the performance of all other roles, compelling them to abandon previous work they increasingly view as mundane and polluting. The next section demonstrates how this master role organizes the meanings associated with their professional counselor training.

Stage Four: Credentialization

One characteristic typically distinguishing the professions from other occupations is specialized knowledge acquired at institutions of higher learning (Larson 1977; Parsons 1959; Ritzer and Walczak 1986, 1988). Although mastering esoteric knowledge and professional responsibilities in a therapeutic relationship serve as gatekeepers for entering the counseling profession, the moral and emotional essence of being a professional ex- involves much more.

Professional ex-s see themselves as their patients' champions. "Knowing what it's like" and the subsequent education and skills acquired in training legitimate claims to the "entitlements of their stigma" (Gusfield 1982), including professional status. Their monopoly of an abstruse body of knowledge and skill is realized through their emotionally lived history of shame and guilt as well as the hope and redemption secured through therapeutic transformation. Professional ex-s associate higher learning with their experiential history of deviance and the emotional context of therapy. Higher learning symbolizes rediscovery of a moral sense of worth and sacredness rather than credential acquisition. This distinction was clarified by an alcoholism counselor:

Anymore, you need to have a degree before anybody will hire you. I entered counseling with a bachelors but I eventually received my MSW about two years ago. I think the greatest benefit in having the formal training is that I have been able to more effectively utilize my personal alcoholism experiences with my patients. I feel that I have a gift to offer my patients which doesn't come from the classroom. It comes from being an alcoholic myself.

These entitlements allow professional ex-s to capitalize on their deviant identity in two ways: the existential and phenomenological dimensions of their lived experience of "having made their way from the darkness into the light" provide their experiential and professional *legitimacy* among patients, the community, and other professionals, as well as occupational *income*. "Where else could I go and put bulimic and alcoholic on my resume and get hired?" one counselor put it.

Professional ex-s generally eschew meta-perspective interpretations of the system in which they work. They desire a counseling method congruent with their fundamental universe of discourse and seek, primarily, to perpetuate this system (Peele 1989; Room 1972, 1976). The words of one educator at a local counselor training institute are germane:

> These people [professional ex-s] . . . are very fragile when they get here. Usually, they have only been in recovery for about a year. Anyone who challenges what they learned in therapy, or in their program of recovery [i.e., A.A., Narcotics Anonymous, Overeaters Anonymous] . . . is viewed as a threat. Although we try to change some of that while they're here with us, I still see my role here as one of an extended therapist rather than an educator.

Information challenging their beliefs about how they, and their patients, should enact the rites associated with recovery is condemned (Davies 1963; Pattison 1987; Roizen 1977). They view intellectual challenges to the disease concept as attacks on their personal program of recovery. In a Durkheimian sense, such challenges "profane" that which they hold "sacred."

Within the walls of these monasteries professional ex-s emulate their predecessors as one generation of healers passes on to the next an age old message of salvation. Although each new generation presents the path to enlightenment in somewhat different, contemporary terms, it is already well lit for those "becoming a professional ex-."

DISCUSSION

Focusing on their lived experiences and accounts, this study sketches the central contours of professional ex-s' distinctive exit process. More generally, it also endeavors to contribute to the existing literature on deviant careers.

An identity that embraces their deviant history and identity undergirds the professional ex-s' careers. This exiting mode is the outcome of a four stage process enabling professional ex-s to capitalize on their deviant history. They do not "put it all behind them" in exchange for conventional lifestyles, values, beliefs, and identities. Rather, they use vestiges of their deviant biography as an explicit occupational strategy.

My research augments Ebaugh's (1988) outline of principles underlying role exit in three ways. First, her discussion suggests that people are unaware of these guiding principles. While this holds for many, professional ex-s' intentional rather than unintentional embracement of their deviant identity is the step by which they adopt a new role in the counseling profession. Second, Ebaugh states that significant others' negative reactions inhibit or interrupt exit. Among professional ex-s, however, such reactions are a crucial precursor to their exit mode. Finally, Ebaugh sees role exit as a voluntary, individually initiated process, enhanced by "seeking alternatives" through which to explore other roles. Professional ex-s, by contrast, are compelled into therapy. They do not look for this particular role. Rather, their alternatives are pre-scribed through their resocialization into a new identity.

Organizations in American society increasingly utilize professional ex-s in their social control efforts. For example, the state of Colorado uses prisoners to counsel delinquent youth. A preliminary, two year, follow-up study sug-gests that these prisoner-counselors show only 13% recidivism (Shiller 1988) and a substantial number want to return to college or enter careers as guid-ance counselors, probation officers, youth educators, or law enforcement consultants. Similarly, a local effort directed toward curbing gang violence, the Open Door Youth Gang Program, was developed by a professional ex- and uses former gang members as counselors, educators, and community rela-tions personnel.

Further examination of the modes through which charismatic, albeit licensed and certified, groups generate professional ex- statuses is warranted. Although the examples just described differ from the professional ex-s exam-ined earlier in this research in terms of therapeutic or "medicalized" resocial-ization, their similarities are even more striking. Central to them all is that a redemptive community provides a reference group whose moral and social standards are internalized. Professional ex- statuses are generated as individuals intentionally integrate and embrace rather than abandon their deviant biogra-phies as a specific occupational strategy.

NOTES

1. Most individuals in substance abuse therapy do not become professional ex-s. Rather, they traverse a variety of paths not articulated here includ-ing (1) dropping out of treatment, (2) completing treatment but returning to substance use and/or abuse, and (3) remaining abstinent after treatment but feeling no compulsion to enter the counseling profession. Future research will explore the differences among persons by mode of exit.

Here, however, analysis and description focus exclusively on individuals committed to the professional ex- role.

2. I conducted most interviews at the subject's work environment, face-to-face. One interview was with a focus group of 10 professional ex-s (Morgan 1988). Two interviews were in my office, one at my home, and one at a subject's home. I interviewed each individual one time for approximately one hour. Interviews were semi-structured, with open-end questions designed to elicit responses related to feelings, thoughts, perceptions, reflections, and meanings concerning subjects' past deviance, factors facilitating their exit from deviance, and their counseling career.

3. I contend that significantly more professional ex-s pursue their careers due to therapeutic resocialization than to achieving sobriety/recovery exclusively through the 12 Step Program (e.g., A.A.). It is too early, however, to preclude that some may enter substance abuse counseling careers lacking any personal therapy. My experiences and my interviews with other professional ex-s suggest that very few professional ex-s enter the profession directly through their contacts with the 12 Step Program. The program's moral precepts—that "sobriety is a gift from God" that must be "given freely to others in order to assure that one may keep the gift"—would appear to discourage rather than encourage substance abuse counseling careers. Financial renumeration for assisting fellow substance abusers directly violates these precepts. Further, professional ex-s are commonly disparaged in A.A. circles as "two hatters" (cf. Denzin 1987). They are, therefore, not a positive reference group for individuals recovering exclusively through the 12 Step Program. Sober 12 Step members are more inclined to emulate their "sponsors" than pursue careers with no experiential referents or direct relevance to their recovery. Further data collection and analysis will examine these differences. Extant data, however, strongly indicate that therapeutic resocialization and a professional role model provide the crucial link between deviant and substance abuse counseling careers.

4. "12 Step Program" refers to a variety of self-help groups (e.g., A.A., Narcotics Anonymous, Overeaters Anonymous) patterning their recovery model upon the original 12 Steps and 12 Traditions of A.A.

REFERENCES

Adler, Patricia, and Peter Adler. 1983. "Shifts and Oscillations in Deviant Careers: The Case of Upper-Level Drug Dealers and Smugglers." *Social Problems* 31: 195–207.

———. 1987. *Membership Roles in Field Research.* Newbury Park, CA: Sage.

Becker, Howard. 1963. *Outsiders: Studies in the Sociology of Deviance.* New York: Free Press.

Best, Joel, and David F. Luckenbill. 1982. *Organizing Deviance*. Englewood Cliffs, NJ: Prentice-Hall.

Bissell, LeClair. 1982. "Recovered Alcoholism Counselors." Pp. 810–817 in *Encyclopedic Handbook of Alcoholism,* edited by E. Mansell Pattison and Edward Kaufman. New York: Gardner.

Davies, D. L. 1963. "Normal Drinking in Recovered Alcoholic Addicts" (comments by various correspondents). *Quarterly Journal of Studies on Alcohol* 24: 109–121, 321–332.

Denzin, Norman. 1987. *The Recovering Alcoholic*. Beverly Hills: Sage.

Durkheim, Emile. 1915. *The Elementary Forms of the Religious Life*. New York: Free Press.

Ebaugh, Helen Rose Fuchs. 1988. *Becoming an Ex: The Process of Role Exit*. Chicago: University of Chicago Press.

Ellis, Carolyn. 1987. "Systematic Sociological Introspection and the Study of Emotions." Paper presented to the annual meetings of the American Sociological Association, Chicago.

————. 1990. "Sociological Introspection and Emotional Experience." *Symbolic Interaction* 13(2): Forthcoming.

Frazier, Charles. 1976. *Theoretical Approaches to Deviance*. Columbus: Charles Merrill.

Glassner, Barry, Margret Ksander, Bruce Berg, and Bruce D. Johnson. 1983. "A Note on the Deterrent Effect of Juvenile vs. Adult Jurisdiction." *Social Problems* 31: 219–221.

Goffman, Erving. 1961. *Asylums*. Garden City, NY: Anchor.

Gusfield, Joseph. 1982. "Deviance in the Welfare State: The Alcoholism Profession and the Entitlements of Stigma." *Research in Social Problems and Public Policy* 2: 1–20.

Hayano, David. 1979. "Auto-Ethnography: Paradigms, Problems and Prospects." *Human Organization* 38: 99–104.

Hughes, Everett. 1945. "Dilemmas and Contradictions of Status." *American Journal of Sociology* L: 353–359.

Inciardi, James. 1975. *Careers in Crime*. Chicago: Rand McNally.

Irwin, John. 1970. *The Felon*. Englewood Cliffs: Prentice-Hall.

Larson, Magali. 1977. *The Rise of Professionalism*. Berkeley: University of California Press.

Livingston, Jay. 1974. *Compulsive Gamblers*. New York: Harper and Row.

Lofland, John. 1969. *Deviance and Identity*. Englewood Cliffs: Prentice-Hall.

Luckenbill, David F., and Joel Best. 1981. "Careers in Deviance and Respectability: The Analogy's Limitations." *Social Problems* 29: 197–206.

Meisenhelder, Thomas. 1977. "An Exploratory Study of Exiting from Criminal Careers." *Criminology* 15: 319–334.

Merton, Robert. 1938. *Social Theory and Social Structure*. Glencoe: Free Press.

Miller, Gale. 1986. "Conflict in Deviant Occupations." Pp. 373–401 in *Working: Conflict and Change,* 3rd ed., edited by George Ritzer and David Walczak. Englewood Cliffs: Prentice-Hall.

Morgan, David L. 1988. *Focus Groups as Qualitative Research*. Beverly Hills: Sage.

NAADAC. 1986. *Development of Model Professional Standards for Counselor Credentialing*. National Association of Alcoholism and Drug Abuse Counselors. Dubuque: Kendall-Hunt.

Parsons, Talcott. 1959. "Some Problems Confronting Sociology as a Profession." *American Sociological Review* 24: 547–559.

Pattison, E. Mansell. 1987. "Whither Goals in the Treatment of Alcoholism." *Drugs and Society* 2/3: 153–171.

Peele, Stanton. 1989. *The Diseasing of America: Addiction Treatment Out of Control*. Toronto: Lexington.

Petersilia, Joan. 1980. "Criminal Career Research: A Review of Recent Evidence." Pp. 321–379 in *Crime and Justice: An Annual Review of Research,* vol. 2, edited by Norval Morris and Michael Tonry. Chicago: University of Chicago Press.

Reimer, Jeffrey. 1977. "Varieties of Opportunistic Research." *Urban Life* 5: 467–477.

Ritzer, George, and David Walczak. 1986. *Working: Conflict and Change*. 3rd ed. Englewood Cliffs: Prentice-Hall.

———. 1988. "Rationalization and the Deprofessionalization of Physicians." *Social Forces* 67: 1–22.

Roizen, Ron. 1977. "Comment on the Rand Report." *Quarterly Journal of Studies on Alcohol* 38: 170–178.

Room, Robin. 1972. "Drinking and Disease: Comment on the Alcohologist's Addiction." *Quarterly Journal of Studies on Alcohol* 33: 1049–1059.

———. 1976. "Drunkenness and the Law: Comment on the Uniform Alcoholism Intoxication Treatment Act." *Quarterly Journal of Studies on Alcohol* 37: 113–144.

Shiller, Gene. 1988. "A Preliminary Report on SHAPE-UP." Paper presented to the Colorado District Attorneys Council, Denver.

Shover, Neil. 1983. "The Later Stages of Ordinary Property Offenders' Careers." *Social Problems* 31: 208–218.

Snow, David, and Richard Machalek. 1983. "The Convert as a Social Type." Pp. 259–289 in *Sociological Theory 1983,* edited by Randall Collins. San Francisco: Jossey-Bass.

Sobell, Mark B., and Linda C. Sobell. 1987. "Conceptual Issues Regarding Goals in the Treatment of Alcohol Problems." *Drugs and Alcohol* 2/3: 1–37.

Stryker, Sheldon, and Richard Serpe. 1982. "Commitment, Identity Salience, and Role Behavior: Theory and Research Example." Pp. 199–218 in *Personality, Roles, and Social Behavior,* edited by William Ickes and Eric S. Knowles. New York: Springer-Verlag.

Volkman, Rita, and Donald Cressey. 1963. "Differential Association and the Rehabilitation of Drug Addicts." *American Journal of Sociology* 69: 129–142.

Weber, Max. 1968. *On Charisma and Institution Building.* Edited by S. N. Eisenstadt. Chicago: University of Chicago Press.

References

Becker, Howard S. 1963. *Outsiders: Studies in the Sociology of Deviance:* New York: Free Press.

Cloward, Richard, and Lloyd Ohlin. 1960. *Delinquency and Opportunity.* Glencoe, IL: Free Press.

Cohen, Albert. 1955. *Delinquent Boys.* Glencoe, IL: Free Press.

Davis, Fred. 1961. "Deviance Disavowal: The Management of Strained Interaction by the Visibly Handicapped." *Social Problems* 9: 120–32.

Goffman, Erving. 1961. *Asylums.* New York: Doubleday.

———. 1963. *Stigma.* Englewood Cliffs, NJ: Prentice-Hall.

Henry, Jules. 1964. *Jungle People.* New York: Vintage.

Hughes, Everett. 1945. "Dilemmas and Contradictions of Status." *American Journal of Sociology* (March): 353–59.

Kitsuse, John. 1962. "Societal Reactions to Deviant Behavior: Problems of Theory and Method." *Social Problems* 9: 247–56.

———. 1980. "Coming Out all Over: Deviants and the Politics of Social Problems." *Social Problems* 28: 1–13.

Lemert, Edwin. 1951. *Social Pathology.* New York: McGraw-Hill.

———. 1967. *Human Deviance, Social Problems and Social Control.* Englewood Cliffs, NJ: Prentice-Hall.

Lyman, Stanford M. 1970. *The Asian in the West.* Reno/Las Vegas, NV: Western Studies Center, Desert Research Institute.

Matza, David. 1964. *Delinquency and Drift.* New York: Wiley.

Merton, Robert. 1938. "Social Structure and Anomie." *American Sociological Review* 3(October): 672–82.

Miller, Walter. 1958. "Lower Class Culture as a Generating Milieu of Gang Delinquency." *Journal of Social Issues* 14(3): 5–19.

Schur, Edwin. 1979. *Interpreting Deviance*. New York: Harper and Row.

Scott, Marvin, and Stanford Lyman. 1968. "Accounts." *American Sociological Review* 33(1): 46–62.

Sellin, Thorsten. 1938. "Culture Conflict and Crime." A Report of the Subcommittee on Delinquency of the Committee on Personality and Culture, *Social Science Research Council Bulletin* 41. New York.

Sumner, William. 1906. *Folkways*. New York: Vintage.

Sutherland, Edwin. 1934. *Principles of Criminology*. Philadelphia: J. B. Lippincott.

Sykes, Gresham, and David Matza. 1957. "Techniques of Neutralization: A Theory of Delinquency." *American Sociological Review* 22: 664–70.

Tannenbaum, Frank. 1938. *Crime and the Community*. Boston: Ginn.

Turner, Ralph H. 1972. "Deviance Avowal as Neutralization of Commitment." *Social Problems* 19(Winter): 308–21.

Weatherford, Jack. 1986. *Porn Row*. New York: Arbor House.